Geochemistry of Clay-Pore Fluid Interactions

Mineralogical Society Series

Series editor
Professor Geoffrey D. Price

The aim of the series is to provide up-to-date reviews through the selected but specialized contributions of leading experts. A particularly attractive feature of the series is the tutorial element, making it suitable for third year undergraduates and researchers of all levels not solely within the Earth Sciences. Each volume is purpose-designed, highly illustrated and serves as an excellent reference tool.

TITLES AVAILABLE

1 **Deformation Processes in Minerals, Ceramics and Rocks**
Edited by D. J. Barber and P. G. Meredith

2 **High-temperature Metamorphism and Crustal Anatexis**
Edited by J. R. Ashworth and M. Brown

3 **The Stability of Minerals**
Edited by Geoffrey D. Price and Nancy L. Ross

4 **Geochemistry of Clay-Pore Fluid Interactions**
Edited by D. A. C. Manning, P. L. Hall and C. R. Hughes

Geochemistry of Clay-Pore Fluid Interactions

Edited by D. A. C. Manning BSc, PhD
Lecturer in Geology, Department of Geology, University of Manchester

P. L. Hall BSc, PhD
Schlumberger Cambridge Research, Cambridge

and

C. R. Hughes BSc, PhD
Experimental Officer, Department of Geology, University of Manchester

Published in association with
The Mineralogical Society of Great Britain and Ireland

CHAPMAN & HALL

London · Glasgow · New York · Tokyo · Melbourne · Madras

Published by Chapman & Hall, 2–6 Boundary Row, London SE1 8HN

Chapman & Hall, 2–6 Boundary Row, London SE1 8HN, UK

Blackie Academic & Professional, Wester Cleddens Road, Bishopbriggs, Glasgow G64 2NZ, UK

Chapman & Hall, Inc. 29 West 35th Street, New York NY10001, USA

Chapman & Hall Japan, Thomson Publishing Japan, Hirakawacho Nemoto Building, 6F, 1-7-11 Hirakawa-cho, Chiyoda-ku, Tokyo 102, Japan

Chapman & Hall Australia, Thomas Nelson Australia, 102 Dodds Street, South Melbourne, Victoria 3205, Australia

Chapman & Hall India, R. Seshadri, 32 Second Main Road, CIT East, Madras 600 035, India

---

First edition 1993

Typeset in 10/12 Times by Interprint Ltd, Malta
Printed in Great Britain at the University Press, Cambridge.

ISBN 0 412 48980 5

A catalogue record for this book is available from the Brisith Library

Library of Congress Cataloging-in-Publication data

Geochemistry of clay-pore fluid interactions/edited by D. A. C. Manning,
P. L. Hall, and C. R. Hughes. – 1st ed.
p. cm. – (Mineralogical Society series; 4)
'Published in association with the Mineralogical Society of Great Britain and Ireland.'
Proceedings of a meeting held in Sept. 1990.
Includes bibliographical references and index.
ISBN 0-412-48980-5
1. Soil mechanics – Congresses. 2. Clays – Congresses.
3. Geochemistry – Congresses. 4. Fluid dynamics – Congresses.
I. Manning, D. A. C. II. Hall, P. L. III. Hughes, C. R. IV. Series.
TA710.G435 1993
624.1′5136–dc20 92-39558
CIP

Contents

Preface

This book contains the proceedings of a very successful meeting on the Geochemistry of Clay-Pore Fluid Interactions organized jointly by the Mineralogical Society (Clay Minerals and Geochemistry Groups) and the Geological Society, which was held at Burlington House in September 1990. The meeting attracted scientists and engineers from a wide range of backgrounds, particularly in the petroleum and radioactive waste disposal industries, as well as academics. Almost all of the speakers at the meeting have contributed to the written proceedings; in line with the interdisciplinary spirit of the conference, authors were invited to write for an audience from outside their specialist fields, and so many of the papers presented here contain valuable elements of review. The editors are particularly grateful for the help they have received from a number of referees: S. Bottrell, C. D. Curtis, J. I. Drever, W. M. Edmunds, W.-L. Huang, I. Hutcheon, H. Irwin, M. D. Lewan, J. H. S. MacQuaker, D. A. Polya, P. E. Rosenberg, D. K. Ross, D. Savage, M. Tranter and S. D. Weedman. The meeting was also well supported by industrial sponsors; we would like to thank BP Research, Enterprise Oil and GeoScience Ltd in this respect.

D. A. C. Manning
Manchester
March 1992

Contributors

A.C. Aplin	Newcastle Research Group in Fossil Fuels and Environmental Geochemistry, Drummond Building, The University, Newcastle upon Tyne NE1 7RU, UK.
M. P. Atkins	BP Research Centre, Sunbury-on-Thames, Middlesex TW16 7LN, UK.
A. H. Bath	Fluid Processes Group, British Geological Survey, Keyworth, Nottingham NG12 5GG, UK.
P. J. Bourke	Radwaste Disposal Division, AEA Decommissioning and Radwaste, Harwell Laboratory, Oxon OX11 0RA, UK.
J. C. Dore	Physics Laboratory, University of Kent, Canterbury CT2 7NR, UK.
D. C. Entwisle	Engineering Geology and Geophysics Group, British Geological Survey, Keyworth, Nottingham NG12 5GG, UK.
A. Fontanive	ENEA, Centro Energia Casaccia, Casella Postale n. 2400, 00100 Rome, Italy.
G. Furrer	Institute of Terrestrial Ecology, ETH Zürich, CH-8092 Zürich, Switzerland.
A. Gautschi	National Cooperative for the Disposal of Radioactive Waste (NAGRA), Hardstrasse 73, CH-5430 Wettingen, Switzerland.
R. Gragnari	ENEA, Centro Richerche Energia Casaccia, Casella Postale n. 2400, 00100 Rome, Italy.
P. L. Hall	Schlumberger Cambridge Research, P.O. Box 153, Cambridge CB3 0HG, UK.
R. G. G. Holmes	Sellafield Technical Department, British Nuclear Fuels plc, Sellafield, Seascale, Cumbria CA20 1PG, UK.
J. M. Huggett	Department of Geology, Imperial College of Science, Technology and Medicine, London SW7 2BP, UK.
C. R. Hughes	Department of Geology, University of Manchester, Manchester M13 9PL, UK.
N. L. Jefferies	Radwaste Disposal Division, AEA Decommissioning and Radwaste, Harwell Laboratory, Oxon OX11 0RA, UK.

D. A. Lever	Radwaste Disposal Division, AEA Decommissioning and Radwaste. Harwell Laboratory, Oxon OX11 0RA, UK.
T. R. Lineham	Radwaste Disposal Division, AEA Decommissioning and Radwaste, Harwell Laboratory, Oxon OX11 0RA, UK.
D. A. C. Manning	Department of Geology, University of Manchester, Manchester M13 9PL, UK.
J. D. C. McConnell	Department of Earth Sciences, University of Oxford, Parks Road, Oxford OX1 3PR, UK.
C. Mignuzzi	ENEA, Centro Ricerche Energia Casaccia, Casella Postale n. 2400, 00100 Rome, Italy.
A. N. North	Physics Laboratory, University of Kent, Canterbury CT2 7NR, UK.
T. J. Primmer	BP Research Centre, Sunbury-on-Thames, Middlesex TW16 7LN, UK.
E. I. C. Rae	Department of Geology, University of Manchester, Manchester M13 9PL, UK.
S. Reeder	Analytical Chemistry Group, British Geological Survey, Keyworth, Nottingham NG12 5GG, UK.
K. Refson	Department of Earth Sciences, University of Oxford, Parks Road, Oxford OX1 3PR, UK.
C. Ross	Fluid Processes Group, British Geological Survey, Keyworth, Nottingham NG12 5GG, UK.
P. W. Schindler	Department of Inorganic Chemistry, University of Bern, Freiestrasse 3, CH-3000 Bern, Switzerland.
A. Scholtis	National Cooperative for the Disposal of Radioactive Waste (NAGRA), Hardstrasse 73, CH-5430 Wettingen, Switzerland.
I. C. Scotchman	Enterprise Oil plc, Grand Buildings, Trafalgar Square, London WC2N 5EJ, UK.
B. K. Sharma	BP Research Centre, Sunbury-on-Thames, Middlesex TW16 7LN, UK.
H. F. Shaw	Department of Geology, Imperial College of Science, Technology and Medicine, London SW7 2BP, UK.
N. Skipper	Department of Earth Sciences, University of Oxford, Parks Road, Oxford OX1 3PR, UK.
J. S. Small	Department of Geology, University of Manchester, Manchester M13 9PL, UK.
G. Spat	ENEA, Centro Ricerche Energia Casaccia, Casella Postale n. 2400, 00100 Rome, Italy.
D. P. Trivedi	Technical and Safety Department, British Nuclear Fuels plc, Risley, Warrington WA3 6AS, UK.

E. A. Warren	BP Research Centre, Sunbury-on-Thames, Middlesex TW16 7LN, UK.
M. Zysset	Department of Inorganic Chemistry, University of Bern, Freiestrasse 3, CH-3000 Bern, Switzerland.

Introduction

P. L. Hall, D. A. C. Manning and C. R. Hughes

The interactions of aqueous pore fluids with clay minerals in upper crustal rocks and surface environments are of major importance in a number of scientific and engineering fields. These interactions include mineral dissolution, precipitation of neo-formed clays and natural cements, mineral transformation reactions, ion exchange reactions, molecular filtration, surface diffusion and physical sorption/desorption. In highly compacted systems, semi-permeable membrane phenomena may also occur, including anion exclusion, osmosis and reverse osmosis. Individually or collectively, such reactions may strongly influence the microfabric, porosity and permeability of clay-bearing sediments and argillaceous rocks, and hence affect the transport of ions and dissolved species into and through such materials.

Clay-fluid interactions occur on both geological and human timescales. They therefore influence not only the natural products of weathering and diagenesis, but also alteration processes associated with:

1. the storage of toxic and radioactive wastes;
2. the disposal of non-hazardous wastes;
3. petroleum exploration and production;
4. the exploitation of geothermal power.

Clay-pore fluid interactions are, therefore, a subject of importance to mineralogy, hydrogeology, soil science, inorganic and organic fluid geochemistry, civil and petroleum engineering, and petroleum geology. They also play a part in manufacturing industry where clays are used as catalysts.

The characterization of clay-pore fluid interaction processes inevitably requires detailed information regarding a variety of factors, including:

1. The mineralogy and geochemistry of the host rock or sediment (clay types and contents, cation exchange capacity, organic content, etc.).
2. The physical characteristics of the host rocks, i.e. pore sizes and size distributions, porosity and permeability, specific surface area, anisotropy, etc.
3. Detailed pore fluid geochemistry, i.e. analyses of ion content and dissolved inorganic and organic species, pH, Eh, redox potential, etc.
4. An understanding of the fluid and ionic transport processes which occur (diffusive, convective, osmotic, etc.).

Geochemistry of Clay-Pore Fluid Interactions. Edited by D. A. C. Manning, P. L. Hall and C. R. Hughes. Published in 1993 by Chapman & Hall, London. ISBN 0 412 48980 5

Detailed knowledge of the compositions of pore fluids and clay mineral assemblages is often difficult to obtain. Pore fluids are usually complex mixtures of many different chemical species, while most clay-bearing assemblages are very fine grained, poorly ordered and intimately admixed with various oxides, oxyhydroxides, sulphides, carbonates and organic material.

There are three approaches to understanding clay-pore interactions, involving case studies on natural or engineered systems, experimental studies or theoretical studies involving numerical modelling. By characterizing the current mineralogical and chemical compositions of the rocks and fluids, it is possible to infer the sequence and timing of the alteration processes which may have occurred, whether in diagenetic or engineered environments. This approach is represented by the 'traditional' descriptive petrological approach to the study of diagenetic processes (e.g. Potter *et al.* 1980; Chamley, 1989), which now involves some very sophisticated techniques and extends current analytical technology to its limits. The unambiguous characterization of clay minerals frequently requires a number of different analytical techniques. There are several texts which deal with clay compositions and their determination; Wilson (1987) and Deer *et al.* (1962) provide a useful introduction. More detailed works include those by Grim (1968), Van Olphen (1977), Brindley and Brown (1980), Newman (1987), Moore and Reynolds (1989), the Royal Society volume edited by Fowden *et al.* (1984) and the Clay Minerals Society workshop lecture notes (Pevear and Mumpton, 1989; MacKinnon and Mumpton, 1990; Stucki *et al.* 1990). Weaver (1989) has published a comprehensive reference work on the sedimentology and diagenesis of clay minerals in a variety of natural environments. Practical aspects of the compaction of clay-bearing sediments, in the context of engineering applications, are covered in the books by Rieke and Chilingarian (1974) and Magara (1978).

The experimental and theoretical approaches involve the forward prediction of diagenetic reactions and their influence on permeability and fluid transport from a knowledge of present rock and fluid compositions, representing an even more difficult challenge. To achieve this goal requires not only a complete understanding of the thermodynamics of all the possible reactions (in particular the stability fields of the possible neo-formed phases) but also the relative kinetics of these reactions and their net influence on pH, redox potential, ionic speciation and other factors. While the thermodynamic stability fields (in terms of temperature, pressure and solution composition) are known for many minerals, there are relatively few experimental determinations of the stability fields of the clay minerals, especially those which are non-stoichiometric and of variable chemical composition. In the absence of experimental data, a number of schemes have been developed to estimate the free energy of formation of clays by either summing the values for their individual oxide components (Tardy and Garrels, 1974) or by assuming ideal solid solution behaviour between stoichiometric end-members (Tardy and Fritz, 1981; Aagaard and Helgeson, 1983). Numerical models of clay structures can be used to provide hypothetical, independent estimates of certain structural or bulk properties. There is still much work to be done in these fields.

The present volume consists of 17 invited and contributed review and research papers which were presented at the joint meeting of the Geochemistry and Clay Minerals Groups of the Mineralogical Society of Great Britain and Ireland and the Petroleum Group of the Geological Society of London, which was held at Burlington House, Piccadilly, London, in September 1990. The papers are drawn from many areas of research, sharing a common interest in clay-pore fluid reactions. They extend from theoretical studies through experimental studies to investigations of natural materials and engineered clays. From these papers, readers will obtain a snapshot view of the methods and capabilities of current research into clay-pore fluid reactions, and be able to use the book to extend their own work. Certain papers include a large review component, providing introductions to the current state of research in their fields. They are accompanied by research papers providing examples of current individual research projects; some interesting links between different fields become apparent. All papers have been subjected to a rigorous review procedure, following the procedures of academic journals.

Experimental and theoretical characterization of mudrocks and clays

Our understanding of the physical and chemical properties of clays and rocks containing clays depends heavily on knowledge of their structure and physical properties. In this section, Dore and North (University of Kent) review the potential of modern physical techniques for mudrock characterization, using low-angle scattering of both X-rays and neutrons to determine the pore structure and anisotropy of porous solids and particle aggregates. This chapter particularly highlights the way in which the concepts of fractal geometry and scale invariance (Mandelbrot, 1982) are capable of enhancing the characterization of random and chaotic structures.

The following two papers extend away from the experimental characterization of bulk properties by presenting the initial steps in their prediction using molecular simulation techniques. Skipper, Refson and McConnell (Oxford University) present some exciting advances in describing both the structure and dynamics of interlayer water in Na- and Mg-exchanged smectites using Monte Carlo computer simulation. These give structures with interlayer spacings which are in good agreement with those found by X-ray diffraction techniques (Brindley and Brown, 1980) and show clearly the different preferred locations of hydrated monovalent and divalent cations near the smectite surface. Refson, Skipper and McConnell (Oxford University) also report the results of molecular dynamics simulations which yield water self-diffusion coefficients and thermal activation energies in good agreement with previous measurements by quasi-elastic neutron scattering (Tuck *et al.*, 1984). Both types of simulation essentially confirm, and lend further insights into, the picture of the smectite–water system which has emerged from a number of spectroscopic studies (e.g. Sposito, 1984).

Natural processes of diagenesis and weathering

The solid raw materials of sedimentation are the products of erosion; in a review, Aplin from Newcastle examines the geographical patterns of clay mineral distribution within rivers and oceans, and the formation of authigenic clays within marine and lacustrine sediments. He emphasizes the importance of unstable amorphous or poorly crystalline aluminosilicate, oxide or oxyhydroxide phases as precursors to clay mineral growth. The contributed papers follow on from this, with petrographic and geochemical studies of Jurassic mudrocks from UK outcrops and the North Sea.

Huggett and Shaw (Imperial College) describe a complex sequence of diagenetic events from Jurassic mudrocks from the North Sea, based on the mineralogical analysis (mainly using SEM and XRD techniques) of their samples, which reflect variation in the proportions of authigenic and detrital clays. They illustrate clearly some of the petrographic complexities of mudrocks.

Scotchman (Enterprise Oil) combines petrographic results with stable isotope data for carbon and oxygen in his work on concretions from the Kimmeridgian, which demonstrates a transition from carbonate cementation to silica cementation, just prior to oil generation. This study of concretions within mudstones yields information about cementation which may relate to the larger scale phenomenon of quartz and clay cementation within sandstones. Scotchman's paper emphasizes the possibility that the enhanced silica solubilities which are required to account for the observed cements may arise as a consequence of the generation of organic acids within the decarboxylation zone of diagenesis.

Experimental studies of diagenesis and weathering

Experimental studies are used to provide the basic data from which thermodynamic quantities are deduced; they can also be used to explore possible diagenetic phenomena, opening additional areas of study. Primmer and co-workers (BP Research, Sunbury-on-Thames) report the results of an experimental study to determine the stability of smectite, illite and kaolinite in the synthetic system Na_2O–K_2O–Al_2O_3–SiO_2–H_2O at 50 MPa and 150 and 200°C. Their results have interesting implications for the thermodynamic modelling of neo-formed clays in terms of ideal solid solution models.

Clay morphology is addressed in a similar experimental study reported by Small and Manning (Manchester University), who succeeded in synthesizing fibrous and lath-shaped illites under controlled conditions at 50 MPa and temperatures of up to 250°C. The fibrous morphology arises when crystallization commences outside the stability field for illite, from fluids which are initially far from equilibrium with illite, and is enhanced by the initial presence of oxalate in the starting fluids. These results emphasize the possibility that organic and inorganic processes during diagenesis are synergetic (Surdam and Crossey, 1985; Surdam *et al.*, 1989), and may begin to answer some of the observed features of the smectite-to-illite transformation (Keller *et al.*, 1986; Inoue *et al.*, 1987).

In contrast to the relatively simple compositions reported for the preceding studies, Rae, Manning and Hughes (Manchester University) present the results of an exploratory study of petroleum source rock diagenesis, using the Kimmeridgian Blackstone Band and Permian Marl Slate (Kupferschiefer) as starting materials. Results for both organic and inorganic products are given; the behaviour of calcium was effectively controlled by carbonate mineral solubilities, and the fluids derived were saturated with respect to amorphous silica. The results demonstrate the possible importance of organic-rich mudrocks as sources of inorganic cement-forming solutes as well as petroleum fluids.

Weathering processes are specifically addressed by Furrer, Zysset and Schindler (Bern), in their report on the kinetics of dissolution of montmorillonite under surface conditions of P and T. This paper provides a model example of a laboratory study of mineral dissolution kinetics. It concludes that the dissolution of montmorillonite takes place predominantly at particle edges, and is chiefly controlled by pH. Inhibition of dissolution is observed when Al surface complexes are adsorbed at particle edges at pH values exceeding 3, or when particles are aggregated.

Applications in civil and petroleum engineering

To varying extents, clay-pore fluid interactions affect all engineering operations which take place within wet environments in the presence of clays. Usually these effects can be ignored for practical purposes, but for petroleum production and projects which involve very demanding specifications (such as hazardous and radioactive waste disposal) it is essential that they are taken into account. Within sedimentary basins, the presence of mudrock units contributes to the development of overpressuring (with potentially disastrous and expensive consequences for drilling), by acting as seals and sources of pressurizing fluids (gas or water). The phenomenon of overpressuring has many implications for natural processes within sedimentary basins, as it controls the build-up (and release) of fluid pressure, affecting fluid flow and solute transport during cementation, fault development, hydrocarbon migration and ore mineralization of Mississippi Valley type. In this book, Hall (Schlumberger Cambridge Research) provides an updated review of the mechanisms of overpressure generation, and attempts to propose a broad classification of the overpressured environments.

Returning to near-surface environments, Bath (British Geological Survey) presents a useful review of the role of clay-pore fluid interactions in the engineering design of waste disposal sites, including those used for hazardous and radioactive waste. This topic is rarely covered in texts aimed at geochemists, and Bath provides an excellent supply of references for further reading.

Specific studies are addressed by other authors. Bourke and colleagues (AEA Industrial Technology, Harwell) examine the mechanisms of flow through compacted samples of London Clay, using neutral (HDO and HTO) and anionic (I^-) tracers. Diffusion experiments showed a marked reduction in diffusion coefficients compared with bulk solution, but the anionic and neutral tracers showed different behaviour reflecting anion exclusion and illustrating osmosis.

Trivedi and Holmes (British Nuclear Fuels Ltd, Risley and Sellafield) report the results of long-term monitoring of clay pore-water chemistry in an 'engineered clay' (a sodium bentonite–carbonate mix) beneath a concrete vault used for storage of low-level radioactive waste. In this case, groundwater chemistry is dominated by carbonate mineral equilibria, with high fluid sodium contents as a consequence of dissolution of artificially added sodium carbonate, or ion exchange. Two complementary papers (by Entwisle and Reeder of the British Geological Survey and Fontanive, Gragnari, Mignuzzi and Spat of ENEA, Rome) describe equipment and procedures for obtaining fluid samples from clays and mudrocks by high-pressure squeezing techniques. Fontanive's group report copious analytical data for Italian argillaceous formations of Plio-Pleistocene age, providing abundant source material for modellers of fluid chemistry. Finally Gautschi, Ross and Scholtis (NAGRA, Wettingen, Switzerland) report a study of pore water and groundwater compositional relationships in shales and limestones from northern Switzerland, using Entwisle and Reeder's procedures to obtain their samples. They demonstrate mixing of connate and meteoric waters, and discuss the relatively low mobility of waters hosted by the Jurassic Opalinus Clay Formation.

This book should be of interest to scientists in a variety of research and development environments. For students it represents an opportunity to obtain a broad impression of work with clays in many different situations, and illustrates the potential for transfer of ideas and experience between different applications. The material presented is directly relevant to current issues in the fields of oil and gas exploration and production, the nuclear power industry and those environmental agencies involved with the regulation of waste disposal and environmental monitoring. An understanding of rock–fluid and clay–fluid interactions becomes increasingly crucial when placed into the context of growing evidence of the progressive and complex environmental stresses which human economic and social activities have already inflicted on the Earth's fragile and interlinked ecosystems. It is our sincere wish that the dialogue between the various communities represented by the contributors to this book will increase in the future.

References

Aagaard, P. and Helgeson, H. C. (1983) Activity/composition relationships among silicates and aqueous solutions: II. Chemical and thermodynamic consequences of ideal mixing of atoms on homological sites in montmorillonites, illites and mixed-layered clays. *Clays and Clay Minerals*, **31**, 207–17.

Brindley, G. W. and Brown, G. (1980) *Crystal Structures of Clay Minerals and their X-ray Identification*, Mineralogical Society, London.

Chamley, H. (1989) *Clay Sedimentology*, Springer Verlag, Berlin.

Deer, W. A., Howie, R. A. and Zussman, J. (1962) *Rock-forming Minerals 3. Sheet Silicates*, Longman, London.

Fowden, L., Barrer, R. M. and Tinker, P. B. (1984) *Clay Minerals: their Structure, Behaviour and Use*, Royal Society, London.

Grim, R. E. (1968) *Clay Mineralogy*, 2nd edn, McGraw Hill, New York.

Inoue, A., Kohyama, N., Kitigawa, R. and Watanabe, T. (1987) Chemical and morphological evidence for the conversion of smectite to illite. *Clays and Clay Minerals*, **35**, 111–20.

Keller, W. D., Reynolds, R. C. and Inoue, A. (1986) Morphology of clay minerals in the smectite-to-illite conversion series by scanning electron microscopy. *Clays and Clay Minerals*, **34**, 187–97.

MacKinnon, I. D. R. and Mumpton, F. A. (eds) (1990) *CMS Workshop Lectures*, vol. 2. *Electron-optical Methods in Clay Science*, Clay Minerals Society, Evergreen, Colorado.

Magara, K. (1978) *Developments in Petroleum Science*, vol. 9, *Compaction and Fluid Migration*, Elsevier, Amsterdam.

Mandelbrot, B. B. (1982) *The Fractal Geometry of Nature*, Freeman, New York.

Moore, D. M. and Reynolds, R. C. (1989) *X-ray Diffraction and the Identification and Analysis of Clay Minerals*, Oxford University Press, New York.

Newman, A. C. (ed.) (1987) *Chemistry of Clays and Clay Minerals*, Monograph No. 6, Mineralogical Society, Longman, London.

Pevear, D. R. and Mumpton, F. A. (eds) (1989) *CMS Workshop Lectures*, vol. 1, *Quantitative Mineral Analysis of Clays*, Clay Minerals Society, Evergreen, Colorado.

Potter, P. E., Maynard, J. B. and Pryor, W. A. (1980) *Sedimentology of Shale*, Springer Verlag, Berlin.

Rieke, H. H. III and Chilingarian, G. V. (1974) *Developments in Sedimentology* vol. 16, *Compaction of Argillaceous Sediments*, Elsevier, Amsterdam.

Sposito, G. (1984) *The Surface Chemistry of Soils*, Oxford University Press, New York.

Stucki, J. W., Bish, D. L. and Mumpton F. A. (eds) (1990) *CMS Workshop Lectures*, vol. 3, *Thermal Analysis in Clay Science*, Clay Minerals Society, Evergreen, Colorado.

Surdam, R. C. and Crossey, L. J. (1985) Organic–inorganic interactions during progressive burial: key to porosity and permeability enhancement and preservation. *Philosophical Transactions of the Royal Society of London*, **A315**, 135–56.

Surdam, R. C., Crossey, L. J., Hagen, E. S. and Heasler, E. P. (1989) Organic–inorganic interactions and sandstone diagenesis. *American Association of Petroleum Geologists Bulletin*, **73**, 1–23.

Tardy, Y. and Fritz, B. (1981) An ideal solid solution model for calculating the solubility of clay minerals. *Clay Minerals*, **16**, 361–73.

Tardy, Y. and Garrels, R. M. (1974) A method of estimating the Gibbs free energy of formation of layer silicates. *Geochimica et Cosmochimica Acta*, **38**, 1101–16.

Tuck, J. J., Hall, P. L., Hayes, M. H. B., Ross, D. K. and Poinsignon, C. (1984) Quasi-elastic neutron scattering studies of the dynamics of intercalated molecules in charge deficient layer silicates. I. Temperature dependence of the scattering from water in Ca(2+) montmorillonite. *Journal of the Chemical Society Faraday Transactions I*, **80**, 309–24.

Van Olphen, H. (1977) *An Introduction to Clay Colloid Chemistry*, Wiley-Interscience, New York.

Weaver, C. E. (1989) *Clays, Muds and Shales*, Elsevier, Amsterdam.

Wilson, M. J. (ed.) (1987) *A Handbook of Determinative Methods in Clay Mineralogy*, Blackie, London.

PART ONE

Experimental and Theoretical Characterization of Mudrocks and Clays

CHAPTER ONE

Characterization of porous solids and particle aggregates by small-angle scattering techniques

J. C. Dore and A. N. North

1.1 Introduction

Most natural geochemical systems, such as argillaceous sediments and sedimentary rocks, are complex and inhomogeneous materials which exhibit variations of density or chemical composition on a relatively microscopic scale. These materials also contain pores of widely varying sizes, from hundreds of microns down to nanometres. The nature of the physical and chemical heterogeneities of such materials might be expected to depend strongly on the processes and environmental settings involved in their evolution, including erosion, particle transport, deposition, burial and compaction, and subsequent chemical modification (diagenesis and metamorphism). Also, their detailed microstructural properties play a significant role in the continuing processes such as fluid flow and mass transport through these materials.

The microstructural features of such materials have conventionally been characterized by determining void volumes, specific surface areas and pore size distributions by adsorption and desorption of liquids and gases (Gregg and Sing, 1982). The microstructural features of heterogeneous materials can also be investigated by small-angle scattering techniques, using radiation of suitable wavelengths. Inhomogeneities on length scales between 20 Å and 2 μm have been investigated in many materials using small-angle scattering of X-rays (SAXS, USAXS) or of thermal neutrons (SANS). The application of these techniques to the characterization of random or self-similar systems has been enhanced in recent years by the development of the theoretical concepts of self-similarity and fractal geometry (Mandelbrot, 1982; Falconer, 1990).

The objectives of the present paper are to give a brief survey of the theoretical principles and experimental techniques of SAXS, USAXS and SANS, and to illustrate their application by a number of suitable examples. Relatively little work has yet been done with geological materials using these techniques, but it is anticipated that these methods will often have a number

Geochemistry of Clay-Pore Fluid Interactions. Edited by D. A. C. Manning, P. L. Hall and C. R. Hughes. Published in 1993 by Chapman and Hall, London. ISBN 0 412 48980 5

of advantages over more established techniques. More detailed descriptions of X-ray and neutron scattering techniques can be found elsewhere (Kostorz, 1979; Squires, 1978; and Glatter and Kratky, 1982).

1.2 Formalism for small-angle scattering

A scattering experiment is based on the use of an incident beam of monochromatic radiation of wavelength λ which interacts with the sample and is elastically scattered through an angle θ to the detector as shown schematically in Figure 1.1. The scattering vector $\boldsymbol{Q}$ is defined by the vector triangle $\boldsymbol{Q}=\boldsymbol{Q}_s-\boldsymbol{Q}_0$ and has a magnitude given by:

$$Q=|k-k'|=\frac{4\pi}{\lambda}\sin\frac{\theta}{2}. \tag{1}$$

The detector may be moved over a range of angles in order to define the SAS intensity distribution, $I(Q)$ for a suitable Q-range. This function contains information on the spatial distribution of the scattering centres in the material. In most modern equipment the single detector is replaced by a position-sensitive area (or linear) detector so that the full $I(Q)$ function is measured simultaneously.

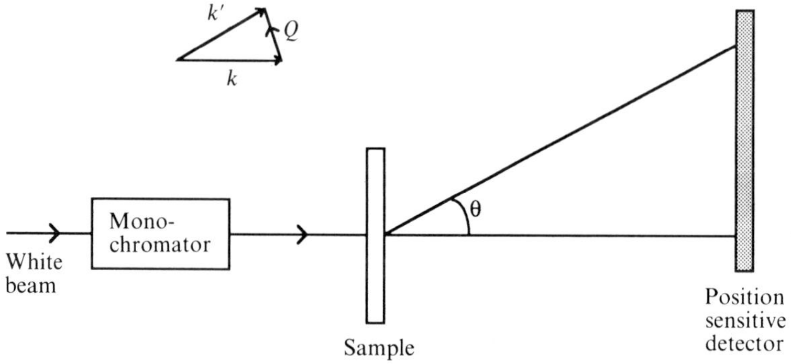

Figure 1.1 Schematic layout for a small-angle scattering experiment; the scattering vector, $\boldsymbol{Q}$, is defined by the vector triangle shown as an inset.

The intensity of the scattered radiation is dependent on the 'contrast' between different regions of the sample material. The mean coherent scattering amplitude per unit volume, known as the coherent scattering length density, ρ, can be evaluated from a knowledge of the composition of the material. In the case of X-rays this is given by the relative electron density for the different regions and in the case of neutrons by the coherent scattering amplitudes (b-values) of the individual nuclei. Tables of the b-values for thermal neutron scattering are available (Sears, 1984) and are listed for natural elements and individual isotopes. For studies involving liquids it is possible to vary the contrast between different regions of the sample by changing the isotopic composition; these principles will be considered in more detail in Section 1.4.

The amplitude of the scattered wave may be evaluated for a general scattering distribution $\rho(\boldsymbol{r})$ by summing the contributions from the individual elements in the scattering volume as shown schematically in Figure 1.2. The relative phase $\Delta\phi$ of the scattered wave is given by the expression $\exp(i\boldsymbol{Q}\cdot\boldsymbol{r})$, where $\boldsymbol{r}$ is the vector between the element and arbitrary origin, so that the scattering amplitude becomes

$$a(\boldsymbol{Q})=\int \rho(\boldsymbol{r})\,\mathrm{e}^{i\boldsymbol{Q}\cdot\boldsymbol{r}}\,\mathrm{d}^3\boldsymbol{r} \tag{2}$$

where the integral is taken over the volume of the sample.

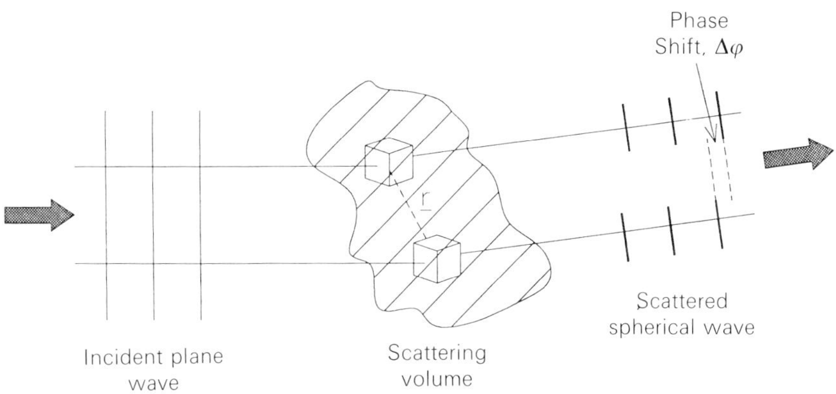

Figure 1.2 The interference of coherently-scattered waves from a distributed scattering medium.

This expression indicates that $a(\boldsymbol{Q})$ is a three-dimensional Fourier transform of $\rho(\boldsymbol{r})$ and the observed intensity may be written as

$$I(\boldsymbol{Q})=K\,|\,a(\boldsymbol{Q})\,|^2, \tag{3}$$

where the constant K is dependent on sample volume and detector geometry. For most materials there is no preferred orientation in the sample and the scattering is, therefore, radially symmetric. The volume integral then simplifies to a radial integration and the intensity becomes

$$I(Q)=K\left|\,4\pi\int r^2\rho(r)\,\frac{\sin Qr}{Qr}\,\mathrm{d}r\,\right|^2. \tag{4}$$

This expression can be evaluated for simple geometries and the scattering from a homogeneous spherical particle of radius, R, and scattering density, ρ_{P}, in a medium, ρ_M, is

$$I_S(Q) = P_S(Q) = V^2(\rho_P - \rho_M)^2\left(\frac{\sin QR - QR\cos QR}{Q^3R^3}\right)^2, \quad (5)$$

where $V(=4\pi R^3/3)$ is the volume of the sphere and $P(Q)$ is a particle form factor. It is notable that the intensity at $Q=0$ is proportional to the (volume)^2 and also dependent on the difference or contrast, $\Delta\rho$ $(=\rho_P - \rho_M)$, between the two materials. If the system consists of a densely packed distribution of spheres such as a colloidal suspension, there is an additional effect due to interference between the particles. This is represented as a static structure factor, $S(Q)$ which was originally developed in the theoretical formalism for description of atomic liquids (Egelstaff, 1967; Hansen and MacDonald, 1976). The structure of disordered assemblies of atoms, such as argon, requires the introduction of the pair correlation function $g(r)$ which describes the relative probability of finding two atoms a distance r apart. The number of atoms in a shell of radius r and thickness dr is then defined as

$$\mathrm{n}(r) = 4\pi \mathrm{r}^2 \rho_N g(r)\,\mathrm{d}r,$$

where ρ_N is the mean number density; the relations are illustrated in Figure 1.3(a). The $S(Q)$ function is again obtained by a Fourier transformation and may be written in the form

$$S(Q) = 1 + \frac{2\pi^2\rho_N}{Q}\int r[g(r) - 1]\sin Q\mathrm{r}\,\mathrm{d}Q \quad (6)$$

for an assembly with no preferred orientation. Typical shapes of the $g(r)$ and $S(Q)$ functions are shown in Figure 1.3(b).

For particulate systems the structure of the individual entities defines the form factor $P(Q)$ and the spatial arrangement of the particle centres defines the structure factor $S(Q)$. In a dilute gas-like system $S(Q)$ is unity, but as the particles are brought into close proximity there is a broad peak in the function which occurs at a Q-value of $2\pi/d$, where d is the mean separation of the particles. Under these circumstances the measured SAS intensity becomes

$$I(Q) = S(Q)P(Q), \quad (7)$$

which is the basic equation for either an ordered or disordered aggregate of identical particles. As the packing density is increased, the $S(Q)$ function develops a sharper peak of increasing magnitude which becomes shifted to larger Q-values. More general expressions can also be written for distributions of particles with varying shapes or sizes and can be found in the literature

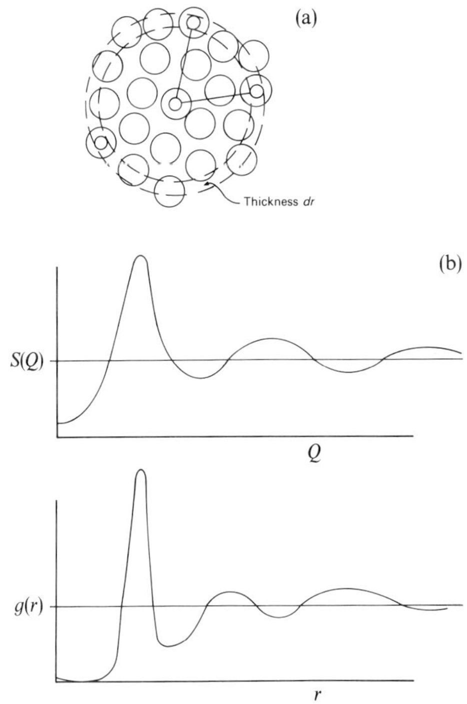

Figure 1.3 (a) The shell construction defining the coordination number, $n(r)$ of a disordered assembly of identical particles; (b) the pair correlation function and structure factor.

(e.g. Glatter and Kratky, 1982). It can also be seen that the functions are dependent on the product QR so that large particles give an $I(Q)$ function which is restricted to low Q-values and vice versa. An experimental measurement must, therefore, be made in the appropriate Q-range to probe the dimensions of interest.

Porous materials may be regarded as a distribution of solid material with voids. The $\rho(r)$ distribution is, therefore, dependent on the shape and size of the pore distribution function which will normally consist of a complicated network of cavities and connecting channels. The characterization of such materials seems initially to be very difficult to represent mathematically but recent theoretical advances have led to remarkable new insights into this type of structure through the concept of fractal geometry. There are many new books (Stanley and Ostrowsky, 1985; 1991; Feder, 1988; Falconer, 1990) based on these ideas, following the initial publication of Mandelbrot's *The Fractal Geometry of Nature* in 1982. The basic mathematics underlying the description of fractal objects is based on the ideas of scaling and self-similar structures. These concepts may be illustrated by a few simple examples. Figures 1.4(a)–(e) show a series of repeated constructions in which an equilateral triangle is constructed on a line joining two points. At each stage the length of the line

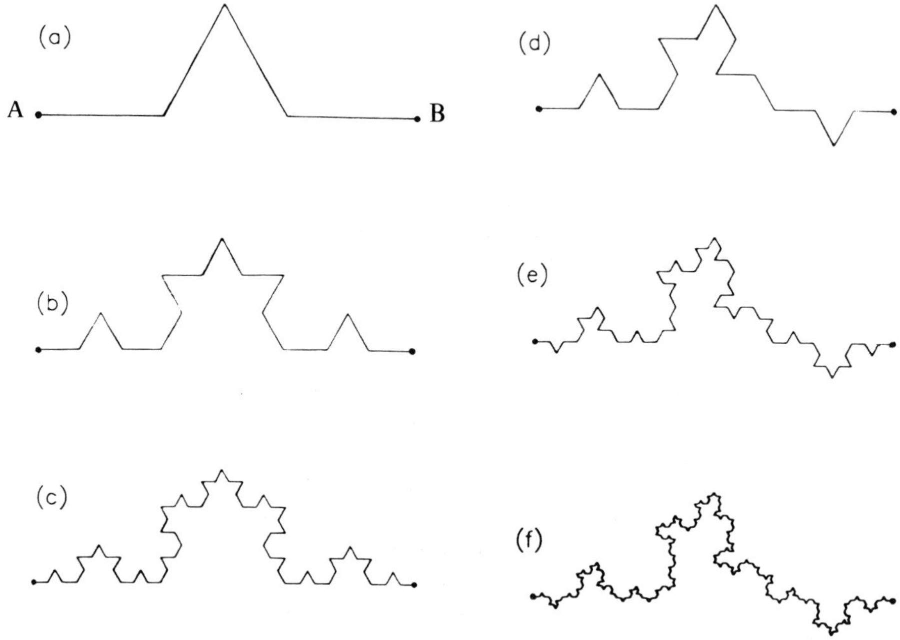

Figure 1.4 Fractal geometry: (a)–(c) stages in the construction of a von Koch 'snowflake' curve; (d), (e) the introduction of disorder into the construction.

joining A to B increases by a factor (4/3). We may now imagine that the process is repeated indefinitely so that the length of the line diverges towards infinity. This curve, which is often termed the von Koch snowflake, is an example of scaling. The length of the curve is dependent on the length of the divider or ruler used to measure successive points. Furthermore if any portion of the curve is magnified to see its shape in more detail it looks identical to the original curve.

In this case the constructions have been carried out in an ordered manner but there is no need to do this and a more random process is shown in Figures 1.4(d)–(f). The basic properties governing the length measurement are not influenced by this change, so the line which now appears to have a more irregular form is still self-similar. Other constructions are possible but the length of the line can always be expressed in the form

$$L=\left(\frac{\Sigma l_n}{r_0}\right)^D,$$

where r_0 is the fundamental length or gauge of the measuring unit and D is termed the fractal dimensionality. The snowflake curve is just one example of a one-dimensional case but the principles can also be applied to area as shown by the Sierpinski gasket construction of Figure 1.5. Each iteration removes one

Figure 1.5 The Sierpinski gasket, showing fractal properties for area construction.

of four triangles so that the area density obeys a scaling law. A more familiar case of varying area occurs in the measurement of a rough surface where the effective area is dependent on surface texture. A smooth surface will have a fractal dimensionality of 2 but this will increase towards 3 for increased roughness. Homogeneous objects in three dimensions will obviously have a dimensionality of three as the mass scales linearly with $(\text{length})^3$. However, a porous material may have a lower mass fractal dimensionality D_M dependent on its detailed characteristics; if D_M is less than 2, it will be possible to see through the gaps in the aggregated material!

The small-angle scattering from fractal structures can be readily evaluated (Allen and Schofield, 1986; Teixeira, 1988, 1990) and has a power-law dependence. Real porous materials will have fractal properties over a limited range of values (if at all) so there are breaks in the linearity of a log I vs. log Q plot and the general form will be as shown in Figure 1.6. The initial region has a slope of $-D_M$ for a mass fractal and ends at a Q-value corresponding to the dimension, l, of the individual building blocks of the aggregate, where $a \cong 2\pi/Q_1$. The second region corresponds to the texture of the surface with a slope of $(6-D_s)$ extending to a Q-value characteristic of the correlation length, $\xi \cong 2\pi/Q_2$. This line then changes to the asymptotic value given by Porod's law (Allen and Schofield, 1986) for a smooth interface and has a slope of -4. These

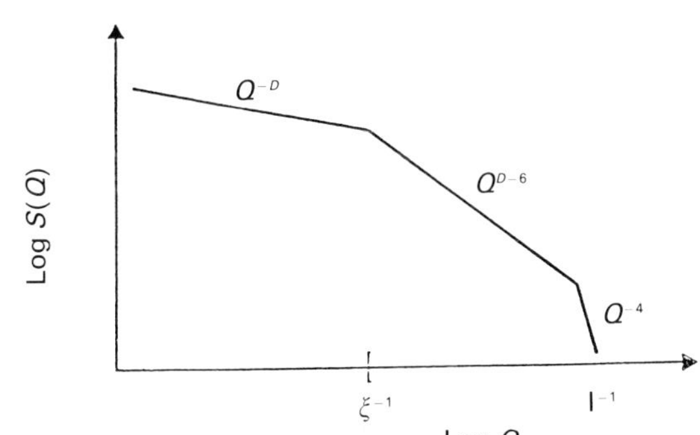

Figure 1.6 The general form of SAS intensity for a typical porous material of fractal character in an aggregated particle assembly (Mildner and Hall, 1986).

separate regions may not always be distinguished by an abrupt change in slope if the characteristic length scales (e.g. l and ξ) are similar or there is significant polydispersity in the basic structural unit. In recent years, there has been a tendency to see 'fractals' in everything and it should be emphasized that many porous materials do not have fractal properties.

Another aspect of the SAS intensity is the dependence on the contrast, $\Delta\rho$. If the pore system is filled with a liquid in which the ρ-value can be changed, there is a possibility of contrast-matching in which $\Delta\rho$ becomes zero. Under these circumstances the scattered intensity drops to zero, provided all the pores are filled and there are no residual inhomogeneities in the system. The contrast-matching technique is easily used for SANS studies by varying the H/D composition of the liquid and has now become a routine technique for a wide range of investigations (Kostorz, 1979). Its application to porous materials can be useful for determining the connectivity of the pore network since some voids may not be connected to the outer surface of the particle. In some cases, it is interesting to use a liquid in which ρ_L is zero. The introduction of the liquid does not, therefore, change the $\Delta\rho$-value, but the SANS intensity will be affected if there is any modification in the surface features of the material due to interaction with the liquid. These special techniques have only recently been introduced, primarily in neutron measurements, but offer interesting possibilities for studies of geological materials.

1.3 Experimental techniques

1.3.1 Neutrons

Small-angle neutron scattering instruments have been installed on all research reactors used for condensed-matter studies. Usually the source of the neutrons is a 'cold source' moderator in the reactor in which the energy distribution is shifted to low values ($\sim$1 meV) so that the operative wavelength range is 5–15 Å. The incident 'white' beam, often directed by neutron guides, then passes through a velocity selector which produces a monochromatic beam with a typical spread of 5 or 10%. After passing through the sample the neutrons are detected in a position-sensitive area detector as shown in the schematic diagram, (Figure 1.7). Electronic coding and data processing by dedicated computer systems allows the SANS intensity pattern to be displayed during the experiment although more detailed treatment is required to extract the final $I(Q)$ data. The sample dimensions are typically for an irradiated area of 10 mm × 10 mm and a thickness of 1 mm; powdered samples are contained in a silica cell which gives negligible scattering except at low Q-values ($< 5 \times 10^{-3}$ Å^{-1}). Sufficient statistical precision can usually be obtained in less than or around 20 min for simple unfilled porous solids. Longer times are needed if the contrast-variation technique is used, depending on the Q-range to be covered.

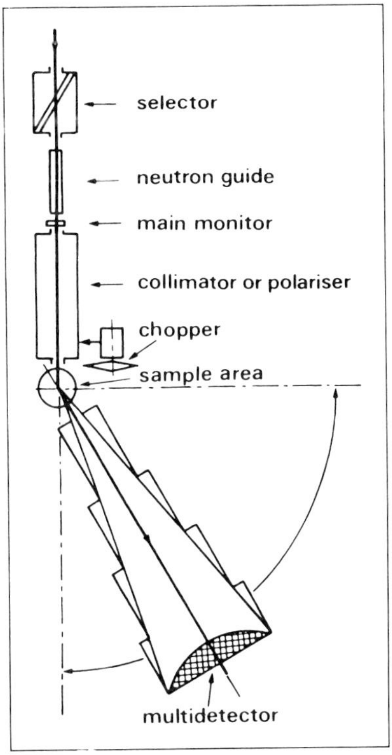

Figure 1.7 The experimental layout for a SANS experiment using a reactor source.

1.3.2 X-rays

Small-angle X-ray scattering was initially based on laboratory X-ray generators using the Kratky system (Kratky and Leopold, 1970) but the advent of synchrotron radiation sources has led to considerable improvements over recent years. Synchrotron radiation is produced by orbiting electrons of high energy in a circular electron accelerator. Owing to relativistic effects the emitted radiation is tangential to the curved orbit and has a characteristic intensity distribution extending from the X-ray region into the infrared. The Synchrotron Radiation Source (SRS) at the Daresbury Laboratory in Cheshire, England, shown in Figure 1.8, has beam-lines which each have up to three instruments that can be used simultaneously. The SAXS facilities are based on the use of a focussing mirror and crystal monochromator to produce a beam of only 0.5 mm (high) $\times$ 3 mm (wide) at the sample position; a fixed wavelength of $\sim$1.5 Å is used for most experiments. Since the beam is polarized in the horizontal plane, the SAXS scattering must be observed in the vertical plane and a linear position-sensitive detector is normally used although a quadrant area detector is also available for studying anisotropic scattering. A schematic diagram of the experimental layout is given in Figure 1.9. Owing to the much

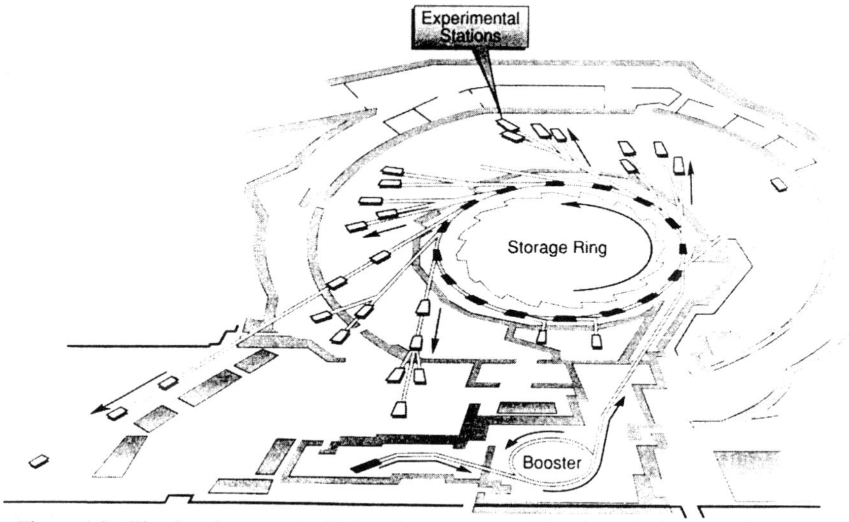

Figure 1.8 The Synchrotron Radiation Source at the Daresbury Laboratory (reproduced by courtesy of the Daresbury Laboratory).

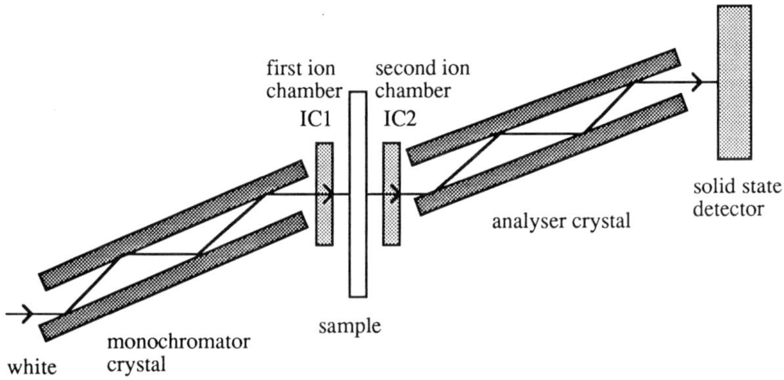

Figure 1.9 The experimental layout for a USAXS experiment using synchrotron radiation.

smaller beam size and higher incident flux the sample dimensions can be smaller than for the neutron case so much less material is needed and the counting times are shorter. Powdered samples are held between thin mica sheets. The absorption coefficient for X-rays is higher than for neutrons so that a transmission monitor (ion chamber) is used to measure the attenuation of the beam through the sample. Measurements on unfilled porous materials may be as short as 5 min and it is sometimes necessary to attenuate the main beam to avoid saturation of the detector electronics.

Small-angle scattering instruments in routine usage cover a Q-range of 2×10^{-3}–0.3 Å^{-1} corresponding to spatial dimensions of 20–3000 Å. The lower limit to the Q-range is set by the collimation defining the beam geometry and the size of the beam stop. For measurements at lower Q values a different technique may be used in which the effective collimation is obtained by the use of crystal optics. The principles have been known for many years and were first presented as the Bonse–Hart camera (Bonse and Hart 1965; 1966) in which multiple reflections from wafers of a monolithic single crystal such as silicon or germanium could be used to define a very highly collimated monochromatic beam. A second crystal was used to analyse the scattered beam and intensity measurements could then be made at very small angles. An instrument based on these principles of ultra-low small-angle X-ray scattering (USAXS) has been constructed at the Daresbury Laboratory. The Q-range extends down to $\sim 5 \times 10^{-5}$ Å^{-1}. The range of spatial dimensions that can be studied is, therefore, raised to $\sim 2\,\mu$m and is comparable to that available from both static and dynamic light-scattering methods. However, the X-ray technique is universal in the sense that it can be used for materials that are not transparent in the visible region, which is usually the case for geological materials; the uses of the SRS for various investigations in condensed-matter physics have been given by North *et al.* (1988). Some examples are given in the following sections. USAXS is a relatively new technique which has been used extensively in a biological context (Nave *et al.*, 1986; Bordas *et al.*, 1987) but has not yet been used very widely in other research areas (North *et al.*, 1990). The corresponding development of a comparable USANS instrument is complicated by the broader geo- metry of the neutron beams but work is progressing and several prototype instruments have been constructed (Zeyen and Allen, 1991, private communication).

1.4 Examples

The following sections illustrate some of the principles presented in the previous paragraphs. There is a relatively small amount of data on geological materials but this is used whenever possible. Limitations in the present studies are examined and projected into future opportunities in the final section.

1.4.1 Particle size in dilute suspensions

Most conventional SANS and SAXS measurements concern solid particles or liquid droplets with dimensions of 20–500 Å and the study of larger sizes requires measurements at much smaller Q-values. The importance of the Q-range is well illustrated by a series of SAXS and USAXS experiments on a dilute suspension of spherical Haloflex particles with a diameter of 210 nm (North and Dore, 1991, unpublished work). The results are shown in Figures 1.10(a)–(d) and cover an intensity range of four decades with over 20 orders clearly observed in the full diffraction pattern. It should be noted that USAXS technique is necessary to observe the first three orders as the Q-values are outside the region accessible by SAXS. The particles themselves are also highly

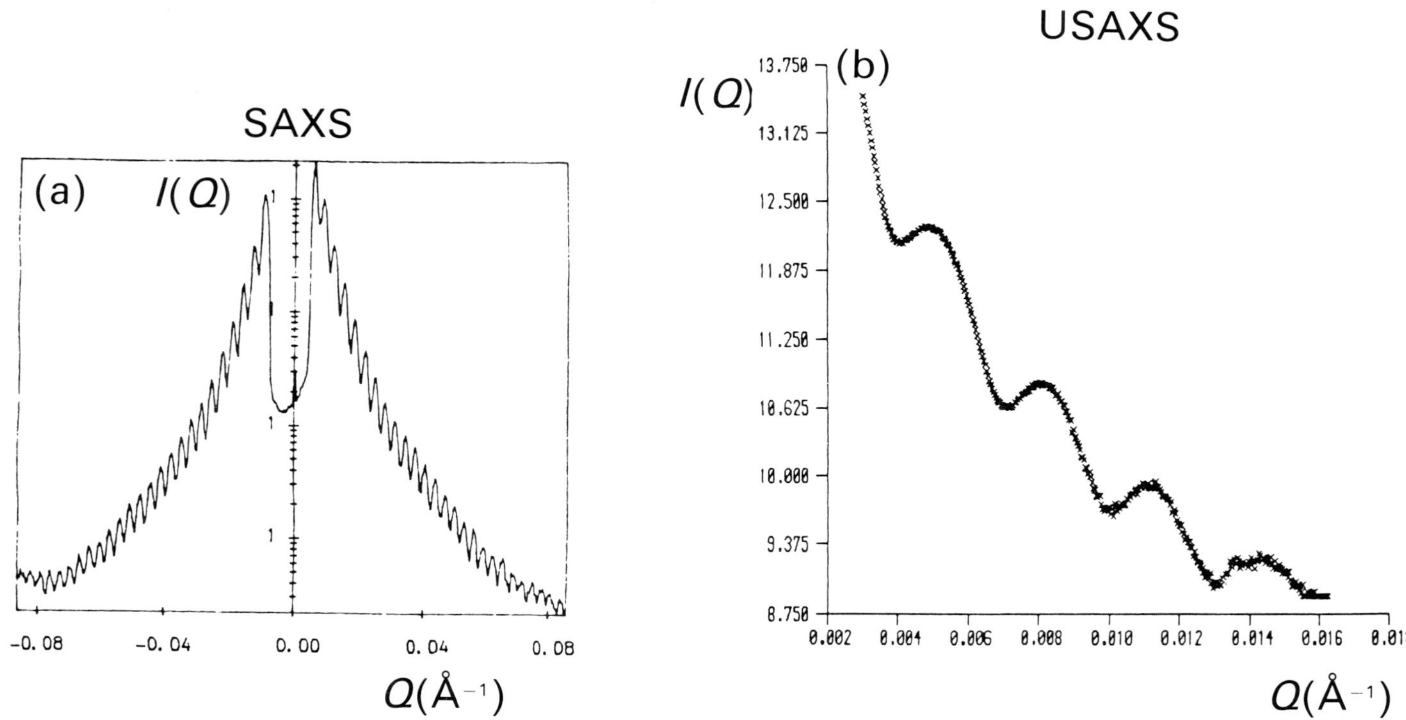
SAXS
(a)
I(Q)
-0.08
-0.04
0.00
0.04
0.08
Q(Å⁻¹)
USAXS
(b)
I(Q)
13.750
13.125
12.500
11.875
11.250
10.625
10.000
9.375
8.750
0.002
0.004
0.006
0.008
0.010
0.012
0.014
0.016
0.018
Q(Å⁻¹)

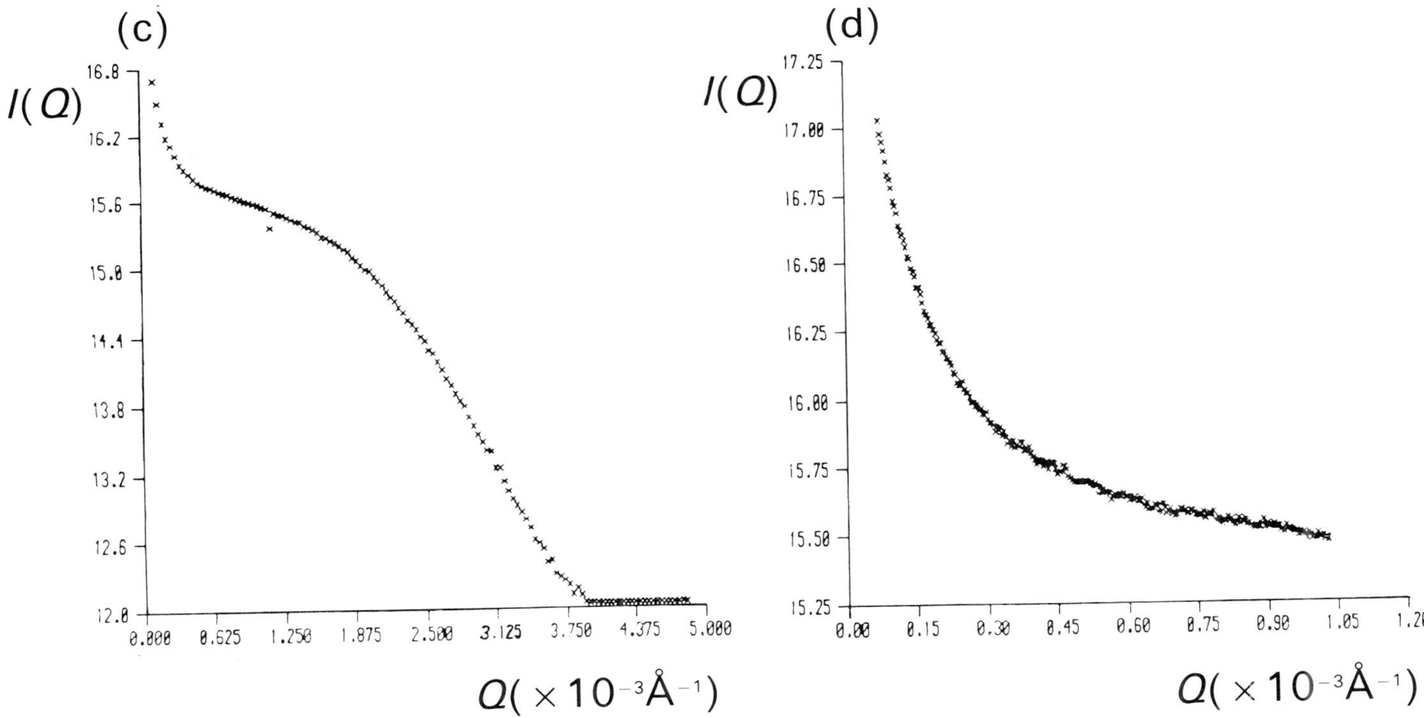

Figure 1.10 SAXS and USAXS measurements on a dilute suspension of polystyrene (Haloflex) spheres showing different Q-ranges.

monodisperse since a distribution of particle sizes would produce a different periodicity in the diffraction pattern (equation (5)) leading to a loss of definition at the larger Q-values. Another feature of the data is the large increase in intensity at low values so that even with the high angular resolution of the USAXS method it is necessary to apply convolution corrections for scattering geometry if precise quantitative information is required.

1.4.2 Porous materials of synthetic origin

If the concentration of a particulate system is increasingly raised there becomes a point when the individual units are aggregated to give a solidified mass. This is the basis of the sol–gel process used to produce porous silicas of high surface area. The scattering intensity $I(Q)$ for a typical silica gel is shown in Figure 1.11 and the extracted pore distribution function, $w(r)$, is given in Figure 1.11(inset). The appropriate Q-range again scales according to the size of the pores and various chemical processes can be used to control the characteristics. In recent years it has been possible to produce aggregates with very low densities, which

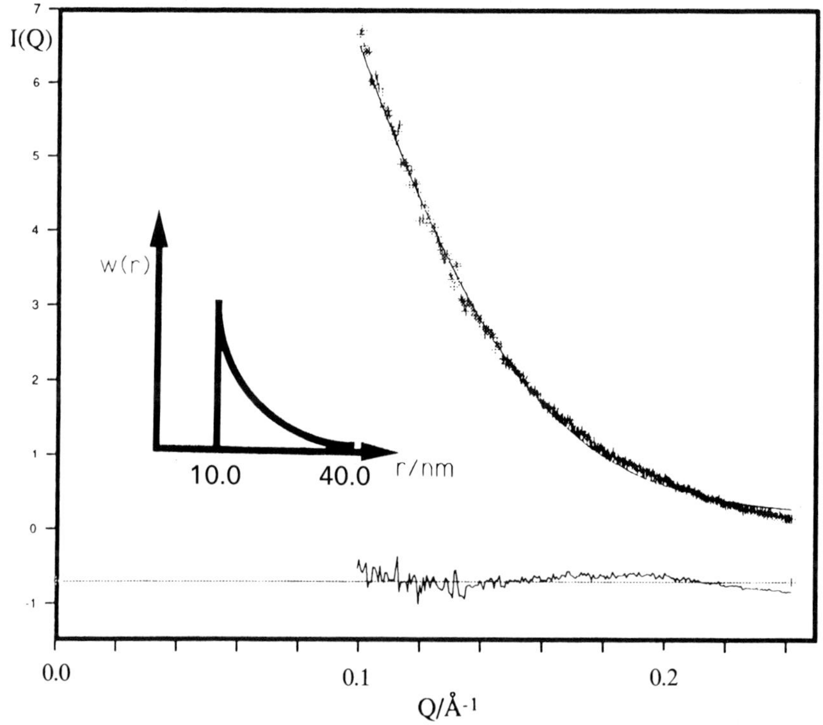

Figure 1.11 SANS intensity for a porous sol–gel silica and (inset) the fitted pore-size distribution function.

can be less than 10% of the bulk density of silica. These aerogels have a characteristic scattering intensity which is well represented by a fractal over a limited Q-range. The data of Figure 1.12 show this behaviour with a constant fractal dimension of 2.4 and varying position for the change in slope which defines the characteristic size of the units according to the density of the aerosol (Vacher *et al.*, 1987).

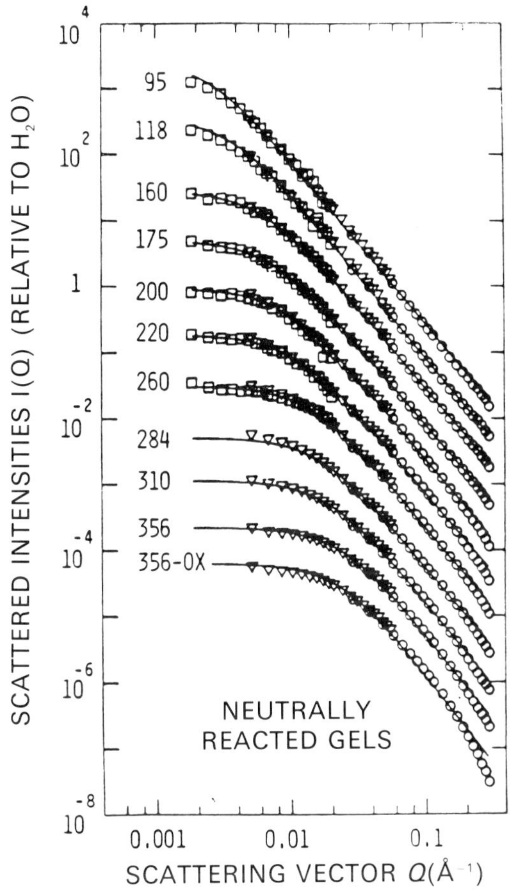

Figure 1.12 SANS intensity for silica aerogels of varying density (Vacher *et al.*, 1987).

Some porous silicas are made by a leaching technique following the removal of water-soluble material from a two-component glass that has undergone spinodal decomposition after a rapid quench from the melt*. The SANS intensities shown in Figure 1.13 are for the pure silica with empty pores and also when the pores are filled with water of varying isotopic (H_2O/D_2O) composition. The normalized

*The sample of leached glass was provided by P. Lange of Schott Glass (Mainz).

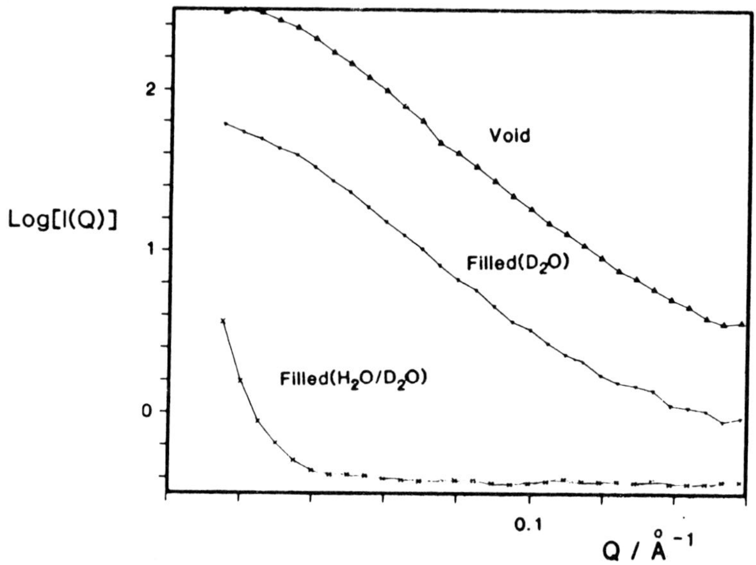

Figure 1.13 The variation of SANS intensity for porous silica containing water (H_2O/D_2O) of varying isotopic composition.

intensity is changed by the variable contrast ($\Delta\rho$) and the match-point is achieved for 65% D_2O. The analysis of this kind of data provides information on the accessibility of the pore volume and also possible changes in surface texture as a result of contact with the water (Benham *et al.*, 1989; Tuck *et al.*, 1990).

A further possibility arises in the case of orientated pore characteristics. A synthetic chemical fibre can be fabricated by a spinning technique and this produces a preferred axial alignment of the cylindrical pores*. The $I(Q)$ function is, therefore, anisotropic and can be investigated by measuring the scattering function in an area detector. Figure 1.14 gives a contour plot of the SANS distribution and detailed analysis of this pattern can yield information on the nature of the pore shape and degree of axial alignment.

For many of these measurements, the use of high-intensity beams provides data of good statistical precision on a timescale of minutes so it is sometimes possible to directly observe the gelation process itself by time-resolved techniques (North *et al.*, 1991, unpublished work). Most work has been conducted on silica systems but there is scope for wider studies.

1.4.3 Characterization of geological materials

The importance of fractal concepts for SAS studies was first recognized by Bale and Schmidt (1984) in X-ray measurements on lignite. The data,

*The alumina fibre sample was provided by M. H. Stacey (Advanced Ceramics, ICI).

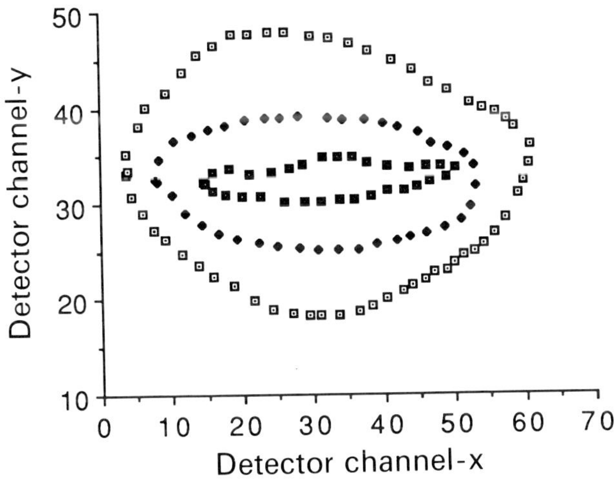

Figure 1.14 The intensity contours for SANS studies of alumina fibres showing anisotropic scattering.

shown on a log I vs. log Q plot in Figure 1.15, exhibit a linear relation over six decades in intensity and correspond to a surface fractal of $D_S = 2.62 \pm 0.03$.

Following this important work there were a number of developments around the mid-1980s in which two groups made substantial contributions. Hall, Mildner and collaborators presented results for a combined SANS and SAXS study of various sedimentary rocks including mudstone, shale, siltstone and slate (Hall *et al.*, 1986; Mildner and Hall, 1986; Mildner *et al.*, 1986). The porosity varied between 3 and 12% and measurements were made for thin slices cut parallel and perpendicular to the bedding plane. The initial analysis (Hall *et al.*, 1986) was based on the extraction of pore distribution functions and ellipsoidal parameters (for anisotropic scattering) as shown schematically in Figure 1.16(a). In a separate paper (Mildner *et al.*, 1986), the same data were re-analysed using a fractal formalism to give the fractal dimension, D_S and the correlation length, ξ, using a modified relation to allow for anisotropic effects. The results are given in Table 1.1 and indicate a systematic difference in the value of D_S for SAXS and SANS studies. This effect is attributed to the complex nature of the interface which contains water or hydrocarbon in the pore surfaces, resulting in a substantially different contrast profile. The variation in slope can be readily seen in Figure 1.16(b) and is clearly beyond the errors in the data points. The presence of hydrogen in the interface has an important effect for the neutron measurements due to the negative b-value so that the ρ-value of hydrogeneous material is much closer to that of a void (Table 1.1) and the surface appears to be more 'fractally rough' than for X-ray measurements. The authors, therefore, claim that the SANS measurements give a more

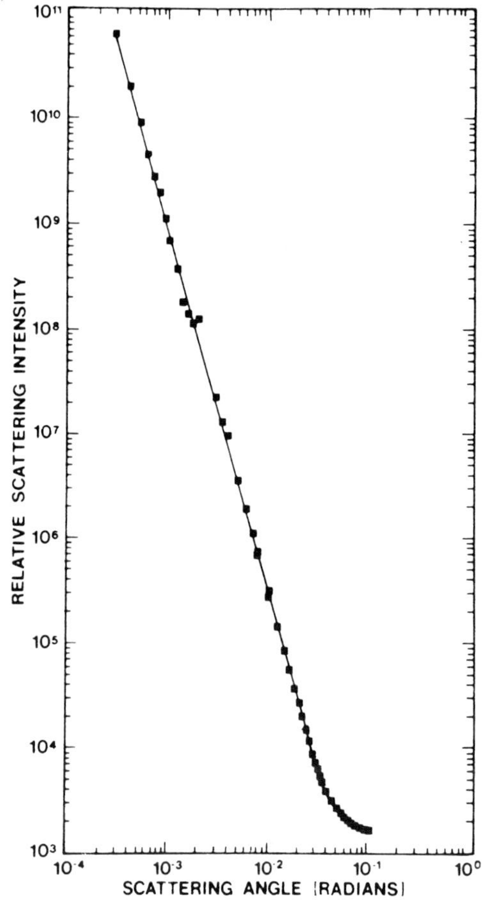

Figure 1.15 The fractal features of SAXS intensity for lignite observed by Bale and Schmidt (1984).

accurate representation of the matrix-pore interfaces (Mildner *et al.*, 1986). Likewise it is clear that the two measurements are in some respects complementary and a comparison gives a measure of the hydrogenous material in the interfacial region. In the case of the oil-bearing shaly rocks, this information is likely to have a direct bearing on the prospective yield at various stages of oil recovery.

Another SAS study of sedimentary rocks has been made by Wong, Lin and collaborators. The SANS measurements (Wong *et al.*, 1986) were initially made on several materials to resolve conflicting views obtained from studies using molecular adsorption (Avnir *et al.*, 1984) and scanning electron microscope (SEM) techniques (Katz and Thompson, 1985). The SANS data again provided convincing evidence for a structure based on fractal surface concepts (Wong and Lin, 1988) and the results are shown (in Figure 1.17(a)) for sandstones and

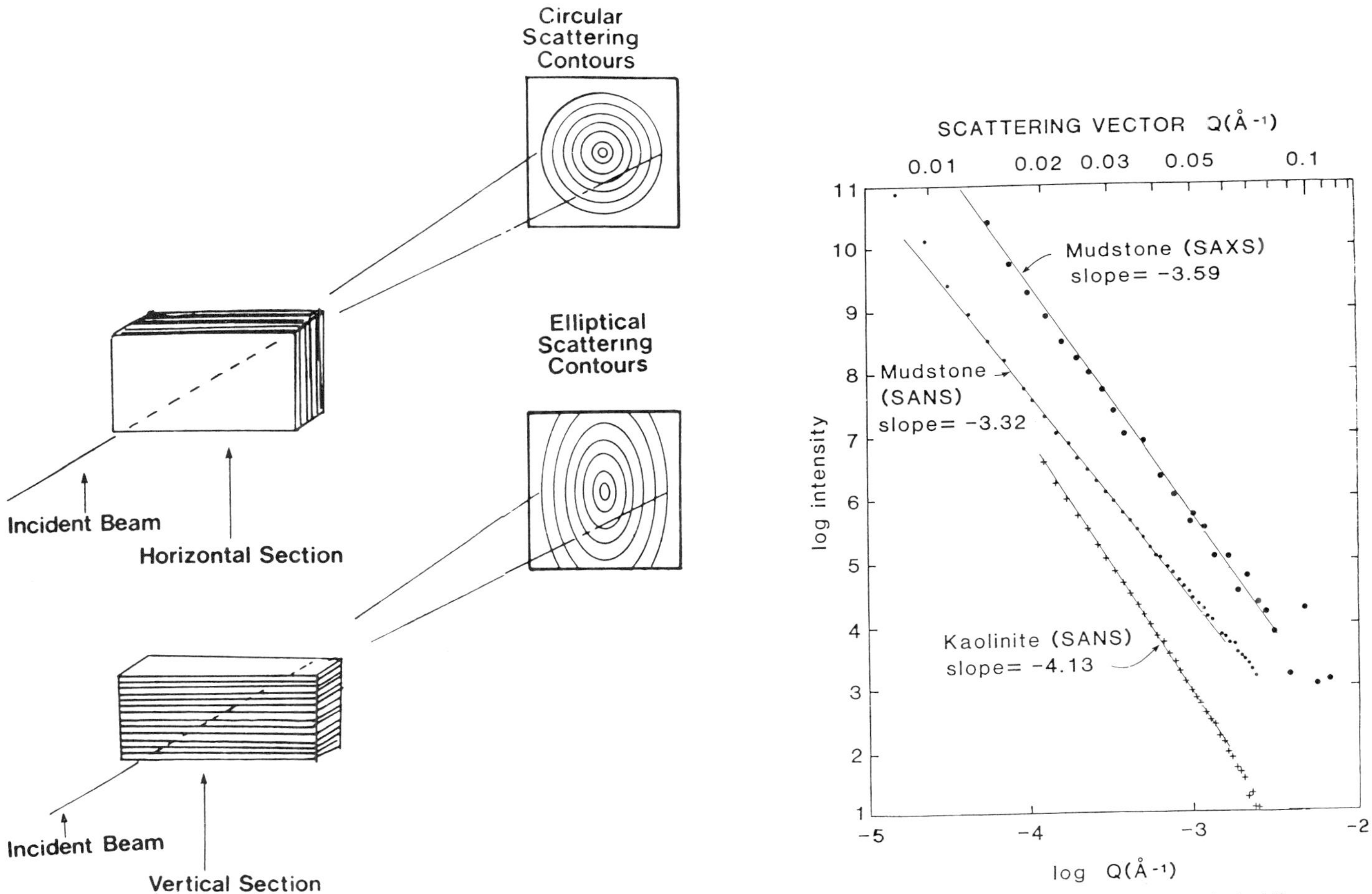

Figure 1.16 Structural studies of sedimentary rocks by Mildner, Hall and co-workers showing: (a) azimuthal variation perpendicular to the bedding plane, and (b) variation of slope for SANS and SAXS studies. (Mildner *et al.*, 1986; Hall *et al.*, 1986.)

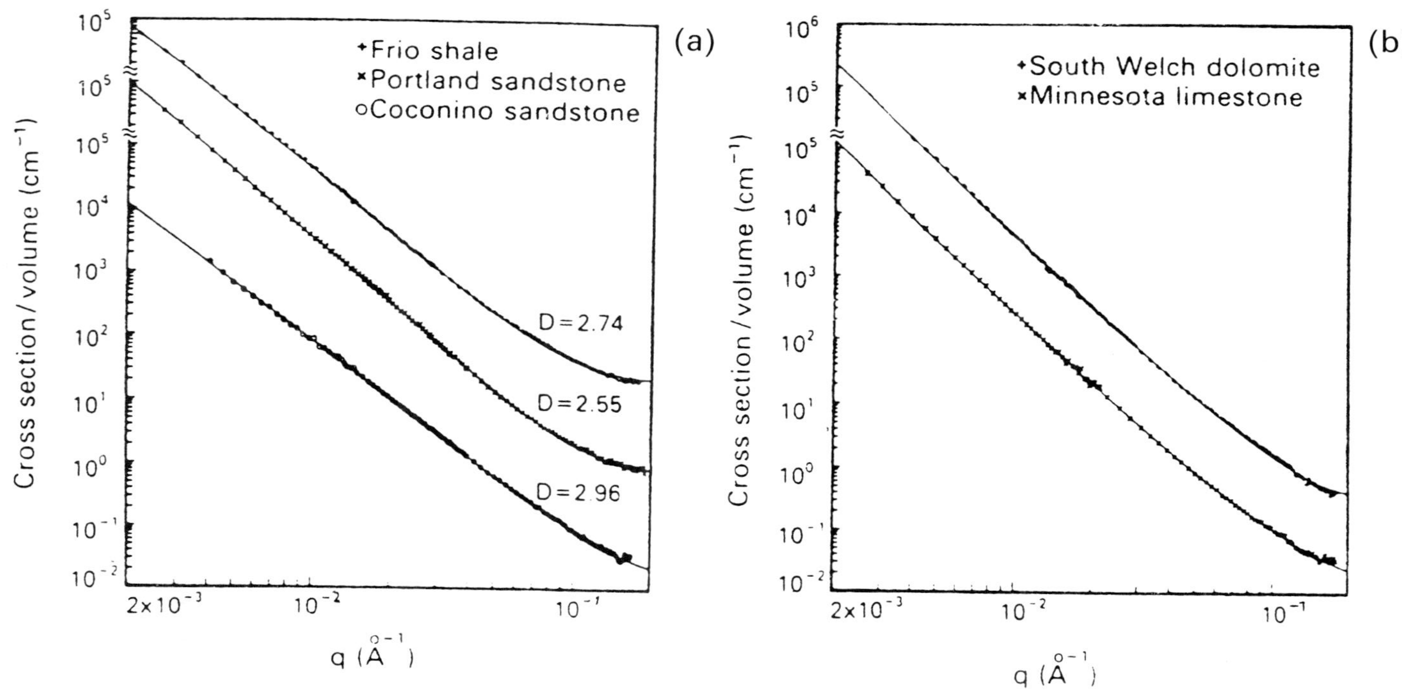

Figure 1.17 SAXS studies of geological materials by Wong *et al.* (1986): (a) sandstones and shales, (b) limestones and dolomites.

Table 1 Porod exponents from SAXS and SANS measurements on shaly rock samples (Mildner *et al.*, 1986).

Sample	Location and lithology	Porosity (ϕ)	Porod exponent	
			SAXS	SANS
1	Utah mudstone	0.082	3.59	3.32
2	Burbank shale	0.052	3.74	3.21
3	New Scotland shale	0.057	3.44	3.11
4	Texas oil shale	0.109	3.46	3.35
5	Weber siltstone	0.028	—	3.29
6	Bakken shale	0.043	3.36	3.17
7	Bakken shale	0.058	3.40	3.30

shales and (Figure 1.17(b)) for limestones and dolomites. The varying values of D_S can be explained in terms of the different components in the samples. The high value of 2.96 for Coconino sandstone is due to the highly textured irregular surface of illite clay. Electron micrograph pictures of illite covering a scale region of 1–50 μm reveal an intricate morphology as shown in Figure 1.18 for two natural samples (*cf.* Small and Manning in Chapter 8). Lower values of D_S were obtained for clay or kaolinite-bearing samples, containing only smooth surfaces; the authors suggest that it is primarily illite/smectite clays which are responsible for the fractal behaviour. The dolomite and limestone samples had a different behaviour which could be explained as a granular medium formed by the growth of small particles on larger ones. These carbonate-based rocks have a relatively simple growth mechanism dependent on dissolution and precipitation from the aqueous environment and appear to follow the characteristics of a roughening transition. The

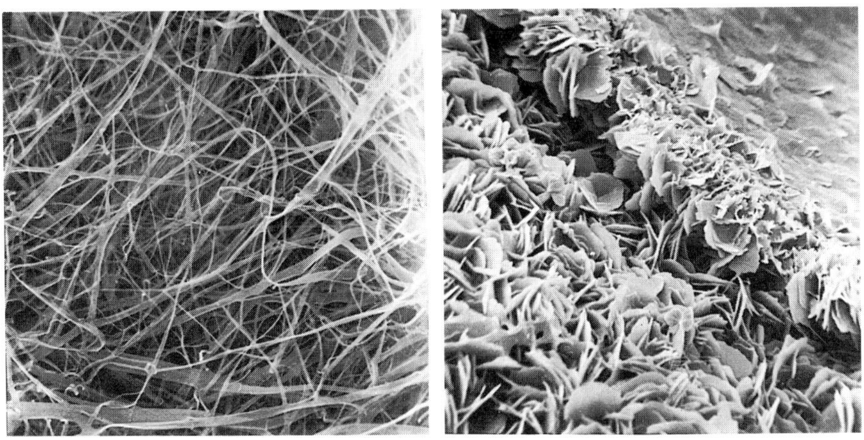

Figure 1.18 Scanning electron micrographs of (a) hairy illite and (b) platy illite from natural sandstones. (Photographs obtained by courtesy of Dr W. J. McHardy of the Macaulay Land Use Research Institute, Aberdeen, Scotland.)

characteristics of growth mechanisms and surface texture have received much attention in recent years (Stanley and Ostrowski, 1986).

The SAS data provide convincing evidence for the fractal nature of many geological materials for length scales in the 20–500 Å region. However, the question remains as to whether the linearity of the log–log plot extends into the range where electron microscopy can give a direct image of the surface structure. In order to probe this region it is necessary to use the USAXS technique for measurement at low Q-values. The results for a combined series of measurements on Bakken shale are shown in Figure 1.19 (Mildner *et al.*, 1986; North *et al.*, 1990). The slopes of the SANS and SAXS data are different, as previously observed, but the same slope is observed for both SAXS and USAXS data indicating that the fractal dimensionality is unchanged over a length scale extending from 20 Å to 2 μm (i.e. three decades). It is clear that, despite the obvious complexity of the interfaces when seen as images in electron micrographs, there is an underlying simplicity which is characterized by the fractal dimensionality, D_S. The problem is now to determine how the varying values of D_S can be related to the original growth of the interfaces and hence to link the experimental measurements to the geochemical processes responsible for the original formation. The present

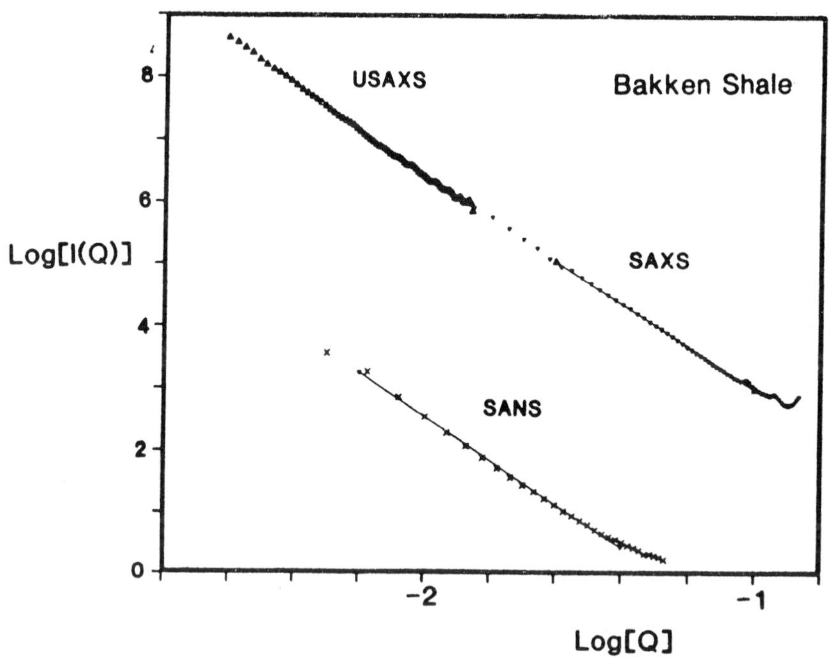

Figure 1.19 Extended Q-range for SAXS, USAXS and SANS studies of Bakken shale (North *et al.*, 1990).

data are too restricted to draw any definite conclusions, but further work should establish the general principles. The use of small-angle scattering methods could become a routine way of characterizing differences in geological specimens.

1.4.4 Mass transfer studies

The timescale for formation of geological materials is normally very long ($>10^3$ years), but other processes occur on a much shorter timescale, and these can be investigated by directly observing change in real-time experiments. For example, characterizing the flow of fluids through porous media is of fundamental importance to a wide range of geological and engineering processes.

The contrast-variation method of H/D isotopic substitution with SANS measurements (Section 1.2) is particularly useful in this context. Since the ρ_N value for water is systematically changed by variation of isotopic composition it is usually possible to obtain a contrast match with the substrate material. The effective contrast $\Delta\rho$ between water and substrate will, therefore, be zero at some H/D composition and this will be different for different materials; Figure 1.20 gives the calculated curves for several clay minerals (Allen *et al.*, 1988; Allen, 1991). The scattering intensity, which depends on $|\Delta\rho|^2$, therefore varies according to the isotopic composition of the pore water, although the shape of the pattern remains unaltered.

The principle of contrast variation has specific applications when there are also differences in contrast within the inhomogeneous medium and the technique is widely used for biological investigations. Similar methods could be used for inhomogeneous geological materials but this has not yet been attempted. The alternative use of contrast variation is through real-time observation of the scattering intensity in a permeation experiment (Allen *et al.*, 1988) where H_2O water is displaced by D_2O water. SANS experiments have been made on London Clay and calcium montmorillonite by Allen *et al.* (1989) in the context of an investigation for the disposal of radioactive waste. The normalized intensity plots, shown on a log–log plot as $Q^4I(Q)$ *vs.* Q for fully hydrogenated and fully deuterated samples are given in Figure 1.21. Time-dependent measurements were also taken in a diffusive (static) and convective mode (pressure-induced flow). The H/D exchange initially occurred quite rapidly (30% in 100 min), but slowed at later stages with the smaller voids taking longer to equilibrate than the larger ones. The data were used to check microstructural models of porosity and connectivity of pore volume arising from the accessibility to exchange with D_2O. The experiments were done with the specific purpose of testing the flow of groundwater through geological strata in order to assess retention of soluble radionuclides by sorption on to clay interfaces, but the methods used could be applied to

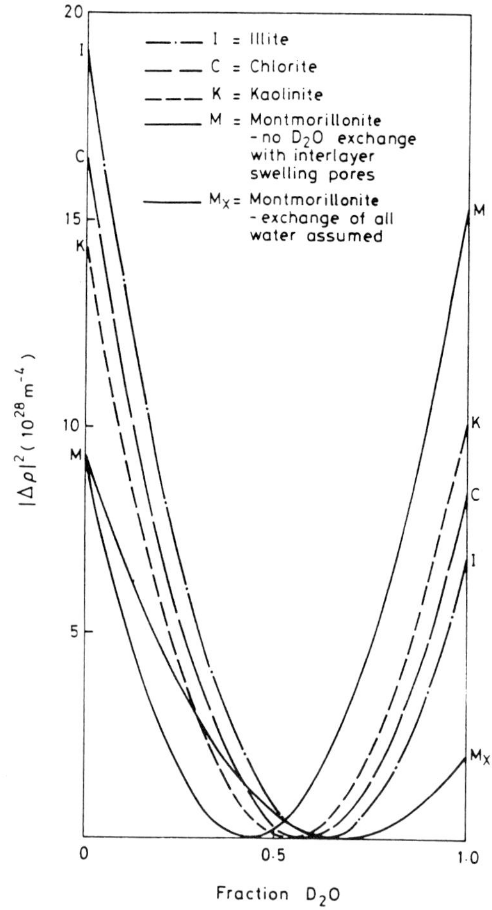

Figure 1.20 Variation of SANS contrast, $|\Delta\rho|^2$, for water of varying isotopic composition in selected clay minerals.

many other materials where flow and exchange features are important parameters.

1.5 Summary and future developments

The importance of small-angle scattering techniques for several topics of geological interest has been presented. Although the number of examples is small, it is apparent that the experimental data provide useful information to complement other methods of investigation. The available instrumentation also provides opportunities for further development. The use of contrast variation for liquids in natural porous rocks has not yet been carried through to completion in the sense that the fractal nature of the interface can be

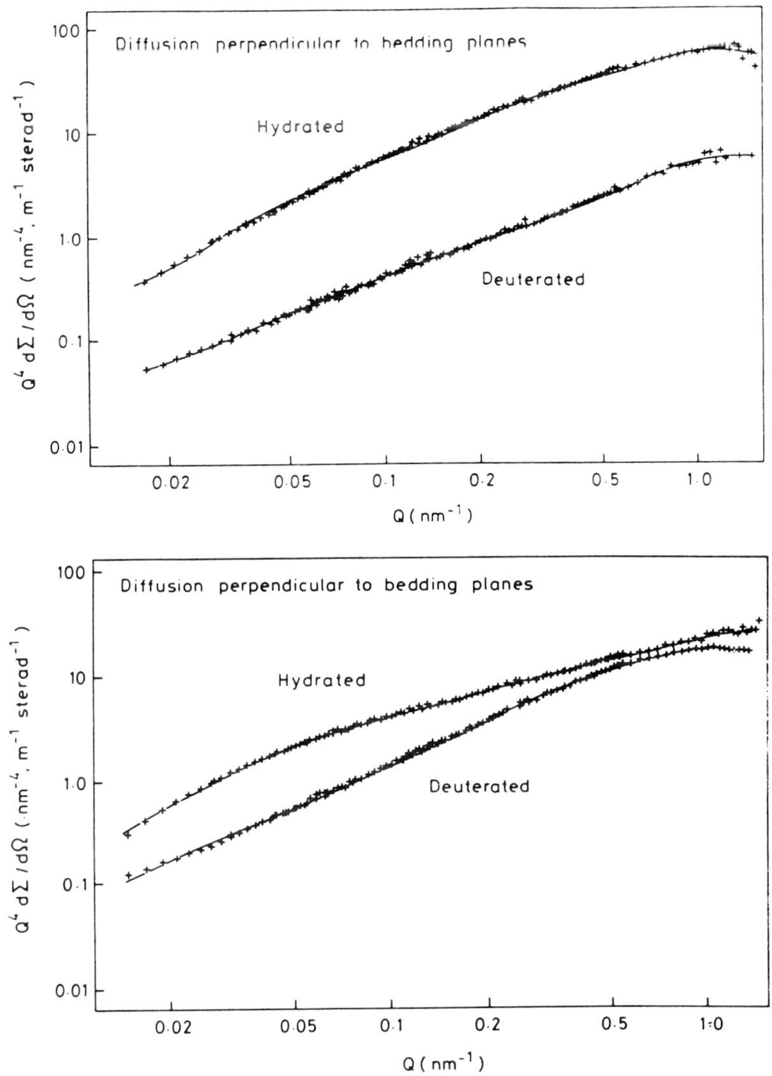

Figure 1.21 The normalized SANS intensity for H_2O and D_2O in: (a) London Clay, and (b) Oxford Clay.

further investigated in much the same way as for synthetic porous materials. The measurements of $I(Q)$ datasets for different contrast conditions are capable of revealing more than can be obtained from single measurements, and with suitable modelling it should be possible to produce a more accurate picture of the density profile within the surface region. This approach should give a quantitative basis to the penetration of liquids and also provide supplementary information on the presence of closed voids or inhomogeneities arising from unrelated effects. Similarly the anisotropic scattering from sedimentary rock sections cut perpendicular to the basal plane contains information on the microfabric and anisotropy of the rock, which may be influenced by compaction and diagenesis. At present, the information obtained from these experiments cannot be related directly to the process of formation and the fractal dimensionality merely serves as a parameter to distinguish different types of material. There is, therefore, a need to determine what additional features influence the texture of the deposited material and hence its D_S value. This kind of information can only be obtained from a study of correlations in a broad database for widely differing materials, or alternatively by the study of laboratory-produced materials under different environmental conditions (Chapter 8).

The USAXS method extends the study into a range which is not accessible to other methods. The limited measurements at present only serve to confirm that the fractal behaviour extends up to several microns but it is possible that further studies will show differing properties for this larger scale regime. Unfortunately X-ray techniques do not readily permit contrast-variation studies, but there are some prototype USANS instruments under development (Zeyen and Allen, 1991, private communication) so this may not be a limitation in future years. Another development, currently at the planning stage, is based on the use of the European Synchrotron Radiation Source, under construction at Grenoble (Riekel, 1989). With very high intensities in the short-wavelength region, it will be possible to obtain contrast variation for X-ray studies using anomalous scattering near an absorption edge to change the effective ρ-value as a function of wavelength.

The interest in fluid flow through porous material raises many questions of a theoretical and technical nature. The use of contrast variation under controlled conditions opens up a wide range of possibilities from invasion percolation and viscous fingering (Stanley and Ostrowski, 1986) to extraction efficiency in oil recovery. The Harwell group have demonstrated the effectiveness of the method for one specific application, but the study of systems involving two immiscible fluids raises many questions which can only be effectively studied by time-resolved SAS methods. It seems inevitable that many of these new techniques will become established as routine procedures over the next decade. The utilization of these powerful modern tools for geological investigations will depend on the readiness of groups of widely differing expertise to collaborate in extending the existing methods.

Acknowledgements

One of us (J.C.D.) would like to express thanks to David Manning for an invitation to attend this meeting and for the chance to discuss ways in which small-angle scattering may provide interesting new possibilities in the geological sciences. We (J.C.D. and A.N.N.) wish to thank Peter Hall for providing the sample of Bakken shale used in our experiments (Figure 1.19) and to both Peter Hall and Andrew Allen for providing source material used in this paper. We also hope that this review will help to focus attention on the availability of SAS techniques, both in this country and elsewhere. A.N.N. would like to thank SERC for financial support in the form of an Advanced Research Fellowship that made this work possible.

Note

Since this draft was prepared, there has been an interesting theoretical development associated with the analysis of SAXS measurements on a porous polymer resin. (T. Korb, private communication, C. Chachaty *et al.*, *Physics Review*, **B44**, 4778–93 (1991)). These materials are highly textured and involve specific cross-linking with metal ions. The SAXS regime exhibits both surface and mass fractal characteristics in the range investigated but in the reverse sequence from that shown in Figure 1.6. A recent study of kaolinite samples over a wide Q-range (S. Barnes, J. C. Dore, A. N. North and J. Adams, unpublished data, 1992) has shown that the fractal region (Figure 1.16) is restricted and the full curve has a complex distribution dependent on the sample origins.

References

Allen, A. J. (1991) Time-resolved phenomena in cements, clays and porous rocks. *Journal of Applied Crystallography*, **24**, 624–34.

Allen, A. J. and Schofield, P. (1986) Small angle scattering from fractal systems, in *Institute of Physics Conference Series*, No. 81, *Neutron Scattering Data Analysis*, (ed. M. W. Johnson), Institute of Physics, London, pp. 97–102.

Allen, A. J., Baston, A. H., Bourke, P. J. and Jefferies, N. L. (1988) Small angle neutron scattering studies of diffusion and permeation through pores in clays. *UK AEA Harwell Report*, No. AERE-R13261, UKAE, Harwell, Oxon.

Allen, A. J., Baston, A. H., Bourke, P. J. and Jefferies, N. L. (1989) The application of small-angle neutron scattering to the study of mass transfer in clays, in *Proceedings of the 6th International Symposium on Water–Rock Interactions*, Rotterdam, (ed. D. L. Miles), Balkema, Rotterdam. Also: *Harwell Report*, No. MPD/NBS 367.

Avnir, D., Farin, D. and Pfeiffer, P. (1984) Molecular fractal surfaces. *Nature*, **308**, 261–3.

Bale, H. D. and Schmidt, P. W. (1984) Small-angle X-ray scattering investigation of submicroscopic porosity with fractal properties. *Physical Review Letters*, **53**, 596–9.

Benham, M. J., Cook, J. C., Li, J-C., Ross, D. K., Hall, P. L. and Sarkissian, B. (1989) Small-angle neutron scattering study of adsorbed water in porous Vycor glass: supercooling phase transition and interfacial structure. *Physical Review*, **B39**, 633–6.

Bonse, U. and Hart, M. (1965) An X-ray interferometer. *Applied Physics Letters*, **6**, 155–6.

Bonse, U. and Hart, M. (1966) Small angle X-ray scattering by small particles of polystyrene and polyvinyltoluene. *Zeitschrift für Physik*, **189**, 151–62.

Bordas, J., Mant, G. R., Diakun, G. P. and Nave, C. (1987) X-ray diffraction evidence for the existence of 102.0 and 230.0 nm transverse periodicities in striated muscle. *Journal of Cell Biology*, **105**, 1311–18.

Chachaty, C., Korb, J. P., Van der Maurel, J. R. C., Bras, W. and Quinn, P. (1991) Fractal structure of a cross-linked polymer resin: a small-angle X-ray scattering, pulsed field-gradient and paramagnetic relaxation study. *Physical Review*, **B44**, 4778–93.

Egelstaff, P. A. (1967) *An Introduction to the Liquid State*, Academic Press, New York.

Falconer, K. (1990) *Fractal Geometry*, John Wiley & Sons, New York.

Glatter, O. and Kratky, O. (1982) *Small Angle X-ray Scattering*, Academic Press, New York.

Gregg, S. J. and Sing, K. W. S. (1982) *Adsorption, Surface Area and Porosity*, 2nd edn, Academic Press, Orlando.

Hall, P. L., Mildner, D. F. R. and Borst, R. L. (1986) Small-angle scattering studies of the pore spaces of shaly rocks. *Journal of Geophysical Research*, **B91**, 2183–92.

Hansen, J. P. and MacDonald, I. R. (1976) *Theory of Simple Fluids*, Academic Press, New York.

Katz, A. J. and Thompson, A. H. (1985) Fractal sandstone pores: implications for conductivity and pore formation. *Physical Review Letters*, **54**, 1325–8.

Kostorz, A. (1979) Small angle neutron scattering, in *Treatise on Materials Science and Technology*, (ed. A. Kostorz), vol. 15, *Neutron Scattering*, Academic Press, New York, pp. 227–89.

Kratky, O. and Leopold, H. (1970) A comparison between Bonse–Hart and block collimation systems. *Die Makromolekulare Chemie*, **133**, 181–195.

Mandelbrot, B. (1982) *The Fractal Geometry of Nature*, W. H. Freeman, San Francisco.

Mildner, D. F. R. and Hall, P. L. (1986) Small angle scattering from porous solids with fractal geometry. *Journal of Physics D*, **19**, 1535–45.

Mildner, D. F. R., Rezvani, R., Hall, P. L. and Borst, R. L. (1986) Small angle scattering of shaly rocks with fractal interfacial boundaries. *Applied Physics Letters*, **48**, 1314–16.

Nave, C., Diakun, G. P. and Bordas, J.. (1986) Ultra-small angle X-ray diffraction from muscle. *Nuclear Instruments and Methods*, **A246**, 609–12.

North, A. N., Dore, J. C., Heenan, R. K., Mackie, A. R., Howe, A. M., Robinson, B. H. and Nave, C. (1988) Small angle X-ray scattering studies of heterogeneous systems using synchrotron radiation techniques. *Nuclear Instruments and Methods*, **B34**, 188–202.

North, A. N., Dore, J. C., Mackie, A. R., Howe, A. M. and Harries, J. (1990) Ultra-small angle X-ray scattering studies of heterogeneous systems using synchrotron radiation techniques. *Nuclear Instruments and Methods*, **B47**, 283–90.

Riekel, C. (1989) Experimental possibilities in small-angle scattering at the European Synchrotron Radiation Facility. *Topics in Current Chemistry*, **151**, 205–29.

Sears, V. F. (1984) Thermal neutron scattering lengths and cross-sections for condensed matter research. *Chalk River Report*, No. AECL-8490.

Squires, G. L. (1978) *Introduction to the Theory of Thermal Neutron Scattering*, Cambridge University Press, Cambridge.

Stanley, H. E. and Ostrowsky, N. (1985) *Random Fluctuations and Pattern Growth*, Kluwer, Dordrecht.

Stanley, H. E. and Ostrowsky, N. (1986) *On Growth and Form*, Martinus Nijhoff, Amsterdam.

Stanley, H. E. and Ostrowsky, N. (1991) *Correlations and Connectivity*, Kluwer, Dordrecht.

Teixeira, J. T. (1988) Small-angle scattering from fractal systems. *Journal of Applied Crystallography*, **21**, 781–5.

Teixeira, J. T. (1990) Application of neutron scattering to the study of fractal systems. *New Journal of Chemistry*, **14**, 217–20.

Tuck, J. J., Ross, D. K., Li, J. C. and Sarkissian, B. (1990) SANS and QENS studies of Vycor containing H_2O/D_2O mixtures, in *Hydrogen-Bonded Liquids, NATO Advanced Studies Institute Series, Physical Sciences*, vol. 139, (ed. J. C. Dore and J. Teixeira), Kluwer, Dordrecht, pp. 159–70.

Vacher, R., Woignier, T., Pelous, J. and Courtens, E. (1987) Structure and self-similarity of silica aerogels. *Physical Review*, **B37**, 6500–03.

Wong, P.-Z., Howard, J. and Lin, J.-S. (1986) Surface roughening and the fractal nature of rocks. *Physical Review Letters*, **57**, 637–40.

Wong, P.-Z. and Lin, J.-S. (1988) Studying fractal geometry on submicron length scales by small-angle scattering. *Mathematical Geology*, **20**, 655–65.

CHAPTER TWO

Monte Carlo simulations of Mg- and Na-smectites

N. T. Skipper, K. Refson and J. D. C. McConnell

2.1 Introduction

Smectites are widespread among soils and sedimentary rocks, and have a number of important industrial uses (Grim, 1960). They have, therefore, been studied by a variety of experimental techniques (Sposito and Prost, 1982; Newman, 1987). In particular, X-ray diffraction has been used to determine the layer spacings and the layer structures (Brindley and Brown, 1980). However, it has not been possible to determine unambiguously the detailed interlayer structure of smectites, due to the small size of the clay platelets ($\approx 5\ \mu m^2$ in diameter) and the complexity of the systems (Hawkins and Egelstaff, 1980). Such microscopic information is essential to understand the bulk properties of these clays. We have undertaken, therefore, both a series of neutron diffraction experiments on vermiculite crystals, which are described elsewhere (Skipper *et al.*, 1990; 1991b), and a series of atomic level computer simulations concerning clay–water–cation systems, which we discuss here.

Monte Carlo and molecular dynamics computer simulations are used to calculate the properties of condensed matter from atomistic models (Allen and Tildesley, 1987). Unlike other theoretical methods the results produced are essentially exact. In the case of clay–water–cation systems, for example, they can be used to separate all components of the interlayer structure, and to calculate directly cation and water diffusion coefficients. Simulations can also be conducted at any temperature or pressure within the validity limits of the particular model potentials.

In this paper we describe Monte Carlo simulations of Mg- and Na-smectites. We use our recently derived SRM28 interaction potentials (Skipper *et al.*, 1991a) to describe the intermolecular interactions. The model systems contain a 21.12 Å by 18.28 Å clay sheet, $4Mg^{2+}$ or $8Na^{+}$ interlayer cations, and up to 64 water molecules (Figure 2.1). The calculations are conducted under conditions of constant uniaxial stress (p), temperature (T) and water content (n). The layer spacing is, therefore, allowed to reach equilibrium during the simulation.

Geochemistry of Clay-Pore Fluid Interactions. Edited by D. A. C. Manning, P. L. Hall and C. R. Hughes. Published in 1993 by Chapman & Hall, London. ISBN 0 412 48980 5

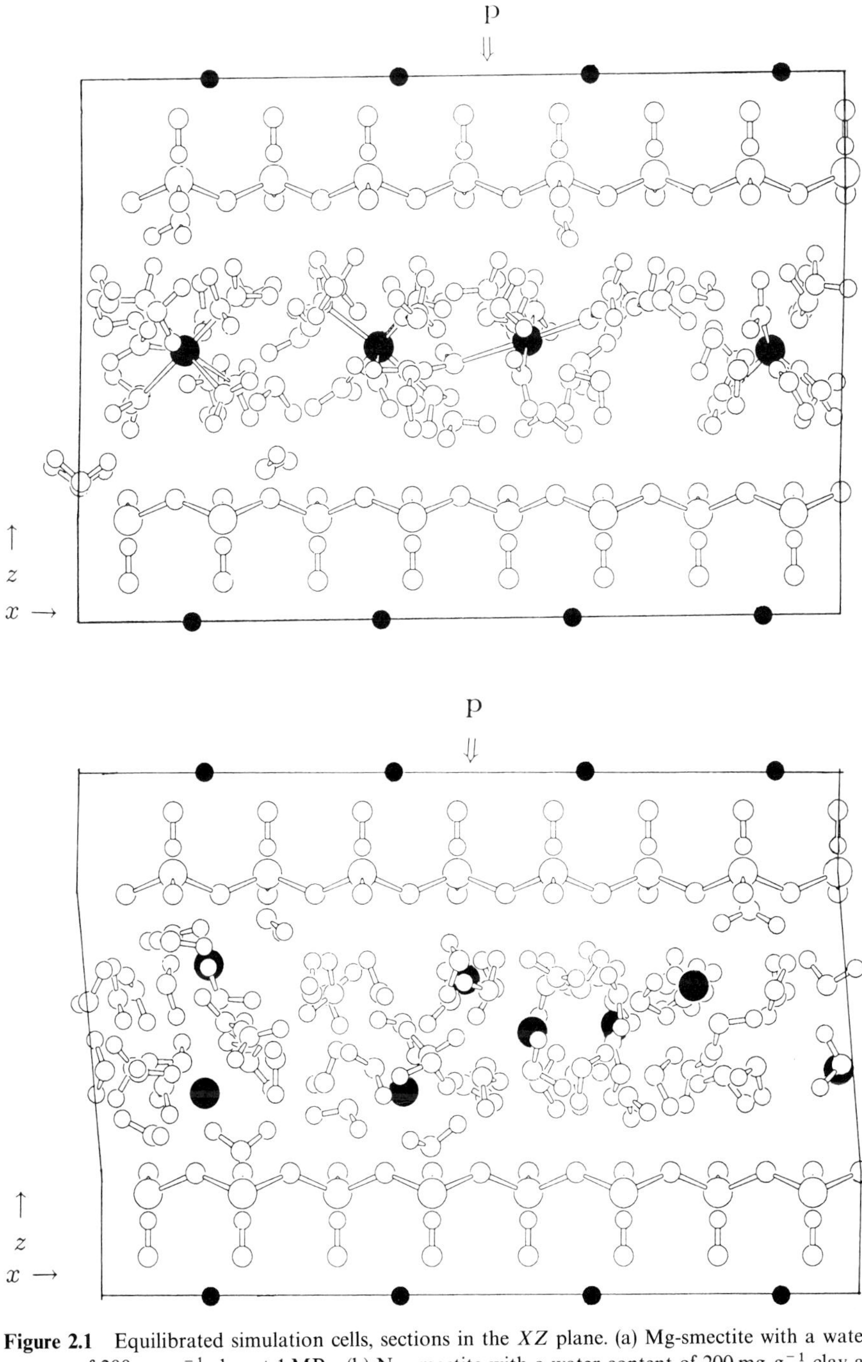

Figure 2.1 Equilibrated simulation cells, sections in the XZ plane. (a) Mg-smectite with a water content of 200 mg. g^{-1} clay at 1 MPa. (b) Na-smectite with a water content of 200 mg. g^{-1} clay at 1 MPa. The cations and layer charges sites are shown in bold.

We determine the layer spacing as a function of uniaxial pressure and water content, up to $p = 10\,000$ MPa and $n = 200$ mg. g^{-1} clay, at 300 K. Where appropriate we compare these calculated values with X-ray diffraction measurements. We then present results for the structure, density and energy of the interlayer water. Further, we discuss the structure around the interlayer cations, and in particular we demonstrate that Mg^{2+} and Na^{+} have different distributions between the model clay sheets.

2.2 Computer simulation method

Two methods of computer simulation are currently available for studying condensed matter: Monte Carlo and molecular dynamics (Allen and Tildesley, 1987). Both methods can be used to determine time-averaged thermodynamic quantities, such as the structure and potential energy, from an atomistic model. Molecular dynamics simulations, though more time consuming, can also give time-dependent quantities, such as diffusion coefficients (see Chapter 3). The simulations described in this paper use the Monte Carlo method of Metropolis *et al.* (1953). Implementation of this method for a clay–water–cation system is represented schematically in Figure 2.2, and is summarized below.

Each of the molecules in the model system interacts with the others through a set of effective pair potentials. These represent the interaction energy between

EFFECTIVE PAIR POTENTIALS

↓

≈100 molecules

↓

Periodic boundary conditions

↓

Infinite system

↓

COMPUTER SIMULATIONS

Monte Carlo

Molecular dynamics

↓

THERMODYNAMIC PROPERTIES

Average structures

Diffusion coefficients

Figure 2.2 Monte Carlo simulation method. A schematic diagram showing the main principles involved.

different sites on the molecules, $V_{ij}(r)$, as a function of their separation. To model a clay–water–cation system, six different sets of interaction potentials are required, namely water–water, water–cation, cation–cation, water–clay, cation–clay and clay–clay. In this work we use our recently derived SRM28 potentials, which have been discussed in detail elsewhere (Skipper *et al.*, 1991a). These potentials all have the functional form

$$V_{ij}(r) = -A_{ij}e^{(-B_{ij}r_{ij})} + C_{ij}e^{(-D_{ij}r_{ij})} + q_i q_j / r_{ij}. \quad (1)$$

The first term represents the attractive London or dispersion forces, the second the short-range repulsions and the third is the Coulomb force between charged sites. The parameters A_{ij}, B_{ij}, C_{ij} and q_{ij} describe the interactions between a particular pair of sites i and j.

Using effective pair potentials, such as the SRM28 model, it is possible to study systems of up to a few hundred molecules on a modern computer. If such a system were to be studied in isolation, edge effects would dominate. Periodic boundary conditions are therefore imposed, so that the unit cell is repeated to infinity in each direction. The molecules then interact with the others in the unit cell, and also with their periodic images. Long-range Coulombic interactions are calculated by the three-dimensional Ewald sum method (Allen and Tildesley, 1987). A real space cut off of 9 Å is used for the short-range Van der Waals and dispersion interactions (equation (1)). This requires periodic images outside the unit cell to be considered.

At the start of each simulation the molecules are placed in a disordered configuration. The energy of this initial state is calculated. A molecule is then chosen at random and moved a small* amount. If the energy of this new state is lower than the original one then the move is accepted. If the new energy is higher then it is accepted with a Boltzmann probability (Wood, 1968),

$$P = e^{(\Delta E/kT)}, \quad (2)$$

where

$$\Delta E = U_1 - U_2 - p\Delta V + kT\ln(V_1/V_2), \quad (3)$$

and T is the temperature, U_1 and U_2 are the initial and final potential energies, and V_1 and V_2 are the initial and final volumes, so

$$\Delta V = V_1 - V_2. \quad (4)$$

The clay systems described here typically reach equilibrium after about 250 000 moves (when the average total energy and layer spacing reach constant

*Small enough that about half the moves are accepted.

values). Ensemble-averaged quantities are then determined by continuing the simulation, typically for another 750 000 moves. Samples for thermodynamic averages are taken after every 1000 attempted updates.

Moves leading to a change in the layer spacing, z_L, are accompanied by a displacement, dz_i, of all the interlayer cations and water molecules,

$$dz_i = dz_L(z_i/z_L), \tag{5}$$

where z_i is the z-coordinate of the ith interlayer molecule. Changes in the simulation cell angles α and β are also permitted (the unit cell vectors $\boldsymbol{a}$, $\boldsymbol{b}$ and $\boldsymbol{c}$ are in the x, y and z directions, respectively, at the start of each simulation (Figure 2.1). α, β and γ are then angles between $\boldsymbol{b}$ and $\boldsymbol{c}$, $\boldsymbol{a}$ and $\boldsymbol{c}$ and $\boldsymbol{a}$ and $\boldsymbol{b}$, respectively). By allowing the simulation cell to change shape in this way adjacent clay layers are allowed to registrate.

In the present work the simulation cells contain up to 64 water molecules, $4Mg^{2+}$ or $8Na^{+}$ cations, and a section of clay sheet 21.12 Å by 18.28 Å and 6.54 Å thick (made up of the equivalent of 16 $Al_2Si_4O_{10}(OH)_2$ clay units). To give the clay layers a net negative charge, similar to those found in natural samples, a point charge of $-1.0\,e$ is placed systematically in one out of four of the octahedral cation sites (Figure 2.1). The layer charge density is therefore $-1.0\,e$ per $Al_4Si_8O_{20}(OH)_2$ unit. For comparison the charge densities for natural Chambers-, Otay- and Wyoming-montmorillonites are $-0.95\,e$, $-0.86\,e$ and $-0.75\,e$, respectively (Newman, 1987). We now present results for our simulations of these model smectites.

2.3 Results and analysis

We have conducted 24 separate simulations, which are summarized in Tables 2.1 and 2.2. These cover both Mg- and Na-smectites at uniaxial pressures between 0 and 10^{10} Pa and water contents between 0 and 200 mg. g^{-1} clay, at 300 K. For each system we calculate thermodynamic averages for the interlayer water (such as the density, ρ, and potential energy, u), relative to the dehydrated clays at the same pressure. Hence,

$$\rho = n_c m/\Delta V, \tag{6}$$

where n_c is the number of water molecules per unit cell, m is the mass of a single water molecule and ΔV is the change in unit cell volume relative to the dehydrated clay ($n_c = 0$), at the same pressure. Also,

$$u = \Delta U/n_c, \tag{7}$$

where ΔU is the change in potential energy relative to the dehydrated system. The errors in the quoted values of u and ρ are calculated from the RMS

Table 2.1 Comparison between simulated and experimental (Mooney *et al.*, 1952; Suquet *et al.*, 1975) layer spacings.

n	Mg-smectite		Na-smectite	
	$c_{experimental}$	$c_{simulated}$	$c_{experimental}$	$c_{simulated}$
0	11.2	11.1	9.8	9.8
100	14.4	14.0	12.4	11.9
200	14.6	14.7	15.2	14.2

n = water content (mg.g^{-1} clay).
c = layer spacing (Å ± 0.1 Å).

Table 2.2 Summary of simulation results.

p(MPa)	Cation	Mg			Na		
	n_c	0	32	64	0	32	64
	n	0	100	200	0	100	200
0	c	11.1	14.0	14.7	9.9	11.9	14.2
	ρ	0.0	0.85	1.38	0.0	1.23	1.15
	$-U$	4803	8218	9526	5141	7211	8327
	$-u$	0.0	107	74.0	0.0	64.8	49.7
1	c	11.1	14.0	14.7	9.8	11.9	14.2
	ρ	0.0	0.85	1.38	0.0	1.18	1.14
	$-U$	4807	8205	9526	5162	7206	8314
	$-u$	0.0	106	73.6	0.0	64.0	49.3
850	c	10.6	12.9	14.2	9.3	11.7	13.9
	ρ	0.0	1.08	1.38	0.0	1.03	1.08
	$-U$	4815	8218	9530	5154	7219	8335
	$-u$	0.0	106	73.6	0.0	64.4	49.7
10000	c	8.9	10.9	11.5	8.8	10.3	11.5
	ρ	0.0	1.24	1.91	0.0	1.65	1.84
	$-U$	3825	7482	7925	5033	6487	7587
	$-u$	0.0	114	64.0	0.0	45.1	39.7

n_c = number of water molecules per unit cell; n = water content (mg.g^{-1} clay); c = layer spacing (Å ± 0.1 Å); ρ = interlayer water density (g.cm^{-3} ± 0.1 g.cm^{-3}); U = total average potential energy (kJ.mol^{-1} ± 8 kJ.mol^{-1}); u = average potential energy per water molecule (kJ.mol^{-1}).

fluctuations (Allen and Tildesley, 1987), assuming that the samples taken after each 1000 attempted updates are independent.

In addition to the quantities shown in Table 2.2, the interlayer structure has been analysed. Water–water and cation–water radial distribution functions, $g_{\alpha\beta}(r)$, and the water and cation density profiles, $\rho_\alpha(z)$, are presented in Figures 2.3, 2.4 and 2.5. They are defined as:

$$\rho_\alpha(z) = z_L\, \mathrm{d}n_\alpha(z)/(N_\alpha\, \mathrm{d}z), \tag{8}$$

where $\mathrm{d}n_\alpha(z)$ is the number of α-type particles between the planes z and $z+\mathrm{d}z$; and

$$g_{\alpha\beta}(r) = \mathrm{d}n_{\alpha\beta}(r)/(4\pi\rho_\beta r^2\, \mathrm{d}r), \tag{9}$$

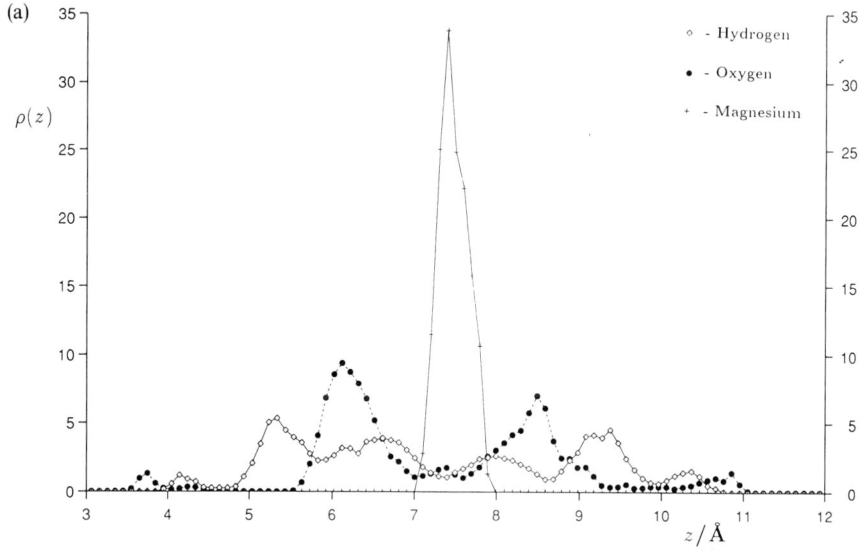

Figure 2.3 Interlayer water and cation density profiles, $\rho_O(z)$, $\rho_H(z)$, $\rho_{Mg}(z)$ and $\rho_{Na}(z)$. (a) Mg-smectite with a water content of 200 mg. g^{-1} clay at 1 MPa. (b) Na-smectite with a water content of 200 mg. g^{-1} clay at 1 MPa. (c) Mg- and Na-smectites with a water content of 200 mg. g^{-1} clay at 1 MPa. (d) Mg-smectite with a water content of 100 mg. g^{-1} clay at 1 MPa. (e) Na-smectite with a water content of 100 mg. g^{-1} clay at 1 MPa. (f) Mg-smectite with a water content of 200 mg. g^{-1} clay at 10 000 MPa. (g) Na-smectite with a water content of 200 mg. g^{-1} clay at 10 000 MPa. (h) Mg-smectite with a water content of 100 mg. g^{-1} clay at 10 000 MPa. (i) Na-smectite with a water content of 100 mg. g^{-1} clay at 10 000 MPa.

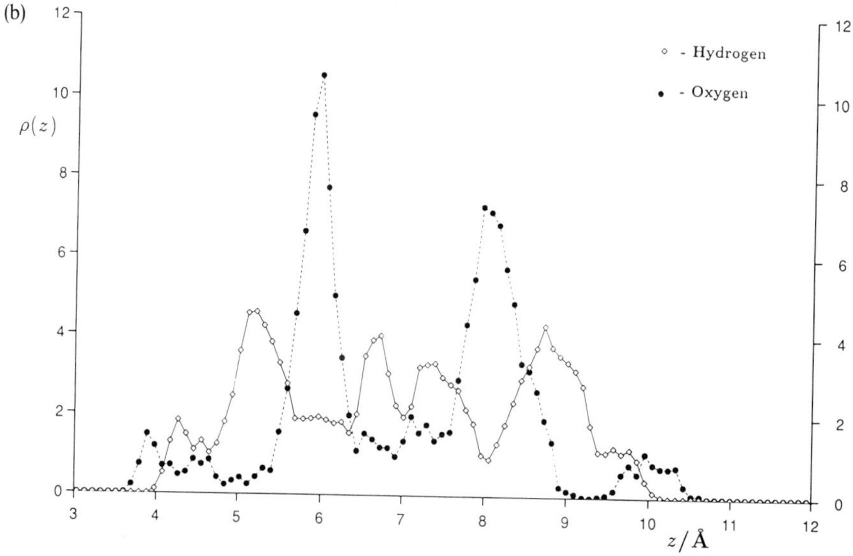

Figure 2.3 *Contd.*

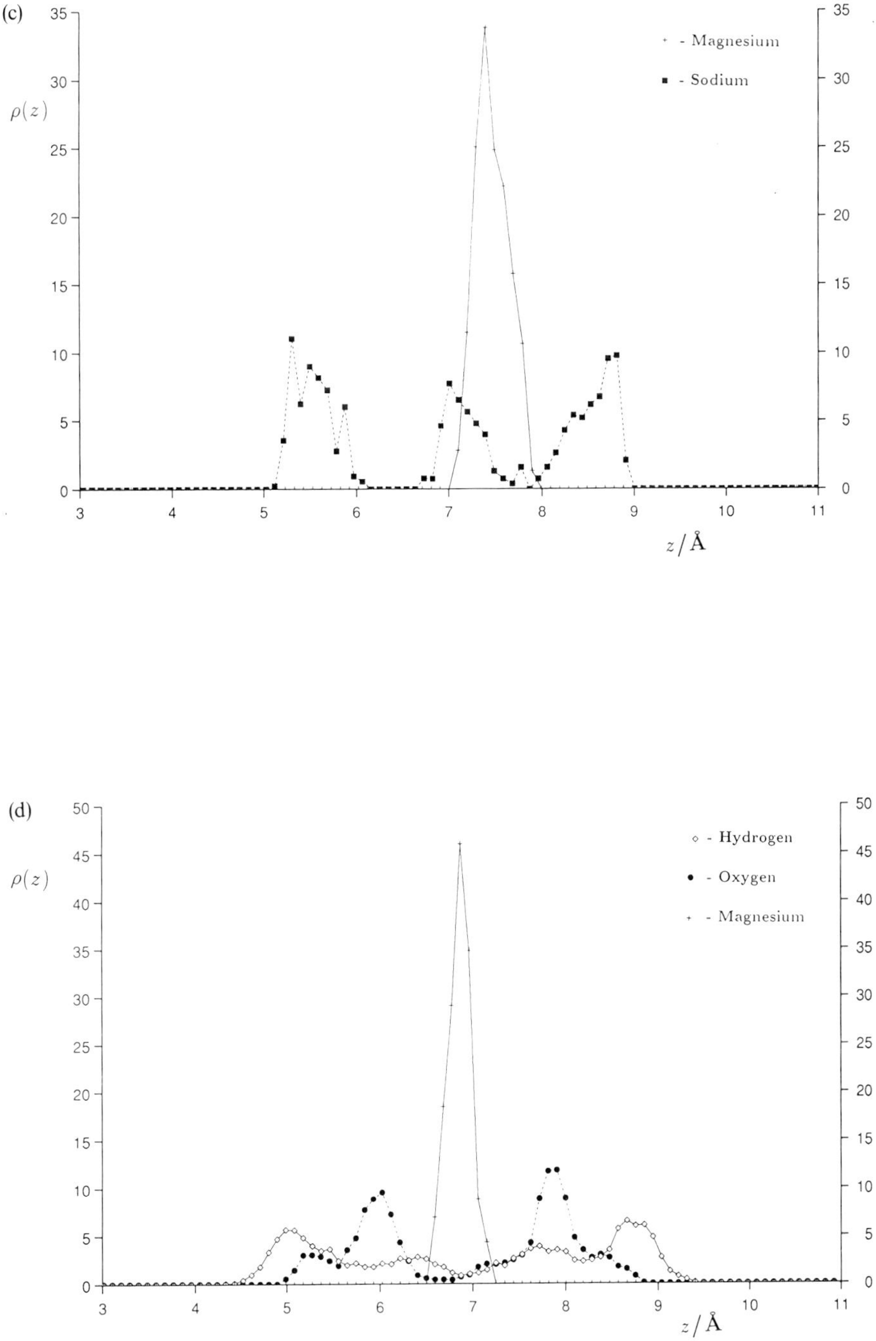

Figure 2.3 *Contd.*

(e)

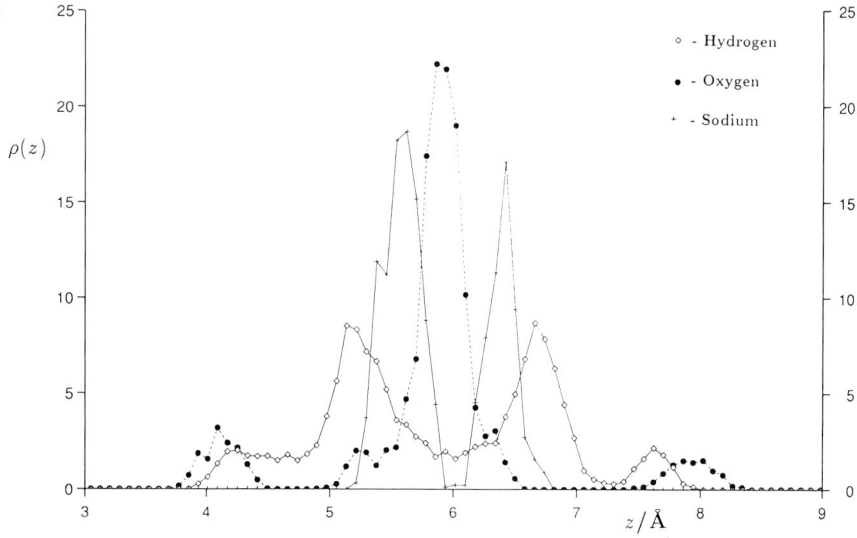

(f)

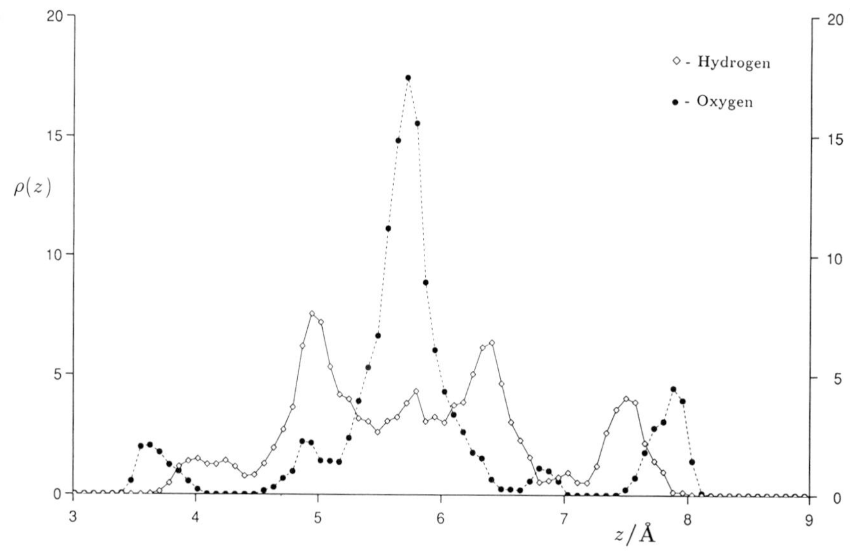

Figure 2.3 *Contd.*

(g)

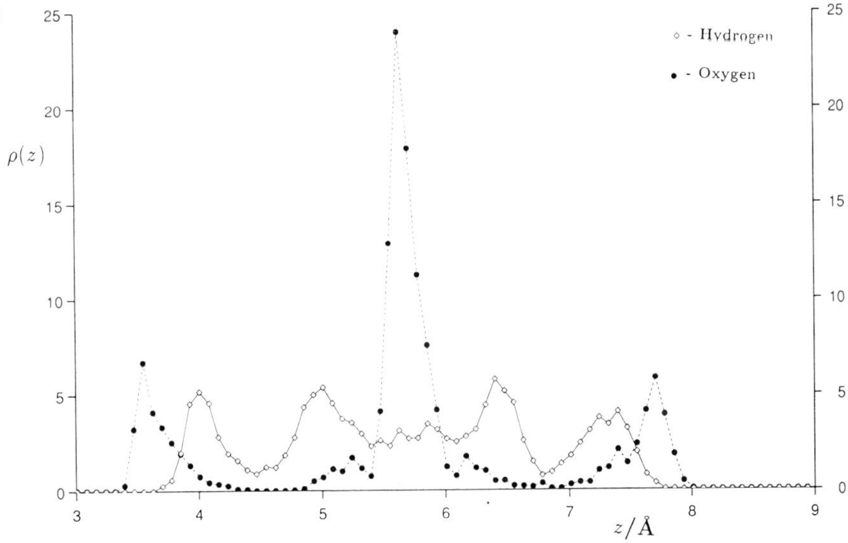

(h)

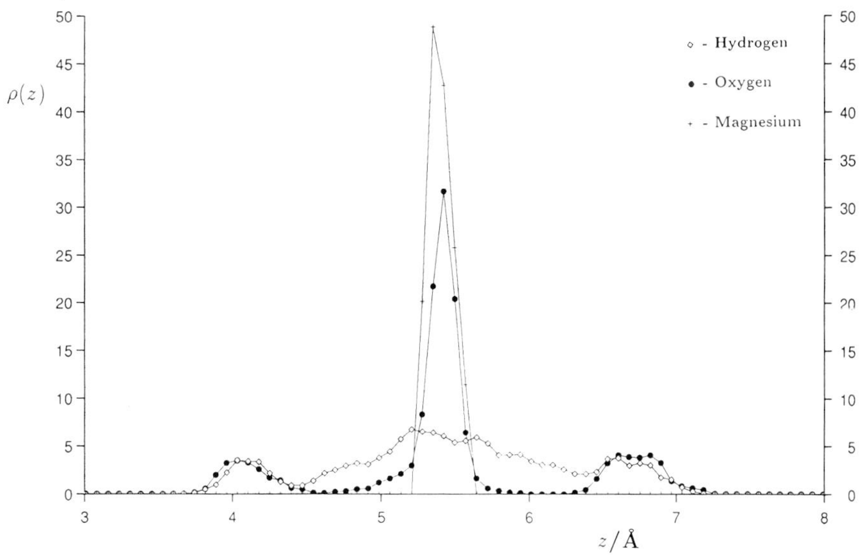

Figure 2.3 *Contd.*

(i)

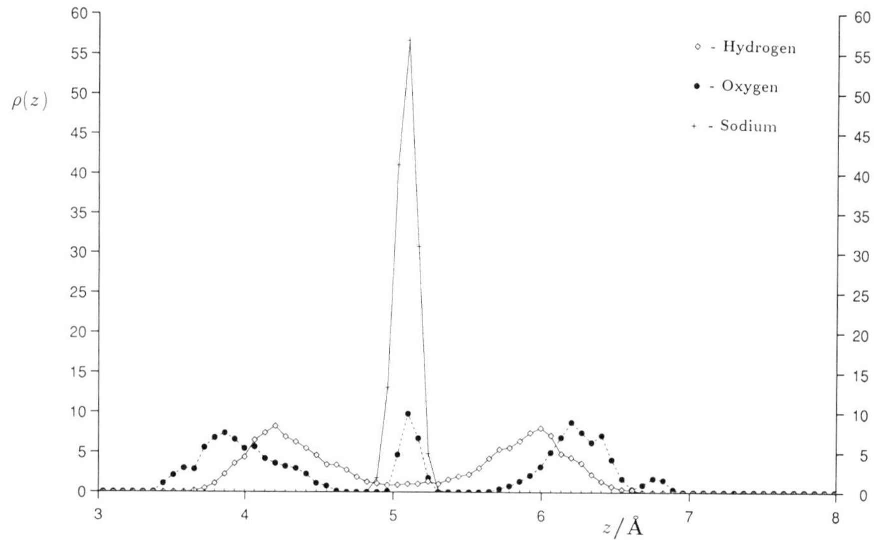

Figure 2.3 *Contd.*

where $dn_{\alpha\beta}(r)$ is the number of β-type particles in a spherical shell of radii r and $r+dr$ around each α-type particle; and

$$\rho_\beta = N_\beta / V, \tag{10}$$

where N_β is the total number of β-type particles and V is the volume. For analysis of the interlayer region α in $g_{\alpha\beta}(r)$ does not include layer atoms, whereas β does. Related to the radial distribution function is the running coordination number, $n_{\alpha\beta}(r)$. This is the average number of particles of type β around each particle of type α, up to a distance r. So,

$$n_{\alpha\beta}(r) = \int_0^r 4\pi\rho_\beta r^2 g_{\alpha\beta}(r)\, dr. \tag{11}$$

We have not conducted a point-by-point analysis of the errors in the structural functions $\rho_\alpha(z)$ and $g_{\alpha\beta}(r)$, as this would be extremely time consuming (Allen and Tildesley, 1987). However, as an example, the error in the height of the oxygen peak at 5.3 Å in Figure 2.3(a) is ± 0.1. Again this assumes that the samples taken after every 1000 moves are independent. This assumption may not be valid in the case of the cation density profiles, since the interlayer cations themselves are less mobile than the water molecules. We are currently conducting much longer simulations to assess the errors in the cation density profiles.

(a)

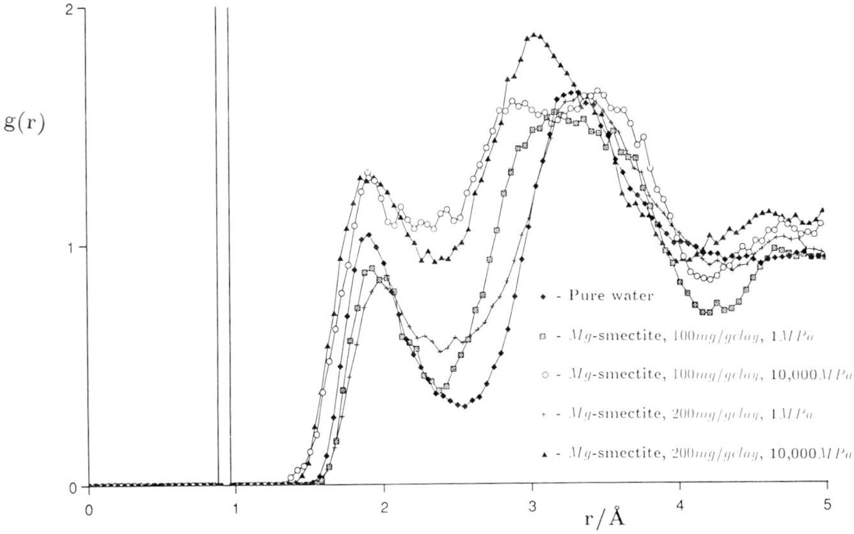

(b)

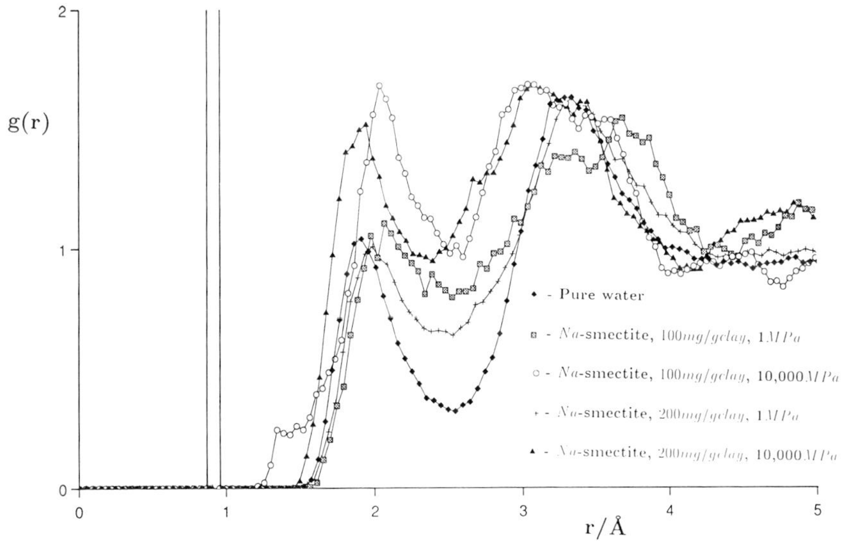

Figure 2.4 Interlayer water–water radial distribution functions, $g_{HO}(r)$. Oxygen atoms within the clay layer are included. (a) Pure water and Mg-smectites. (b) Pure water and Na-smectites.

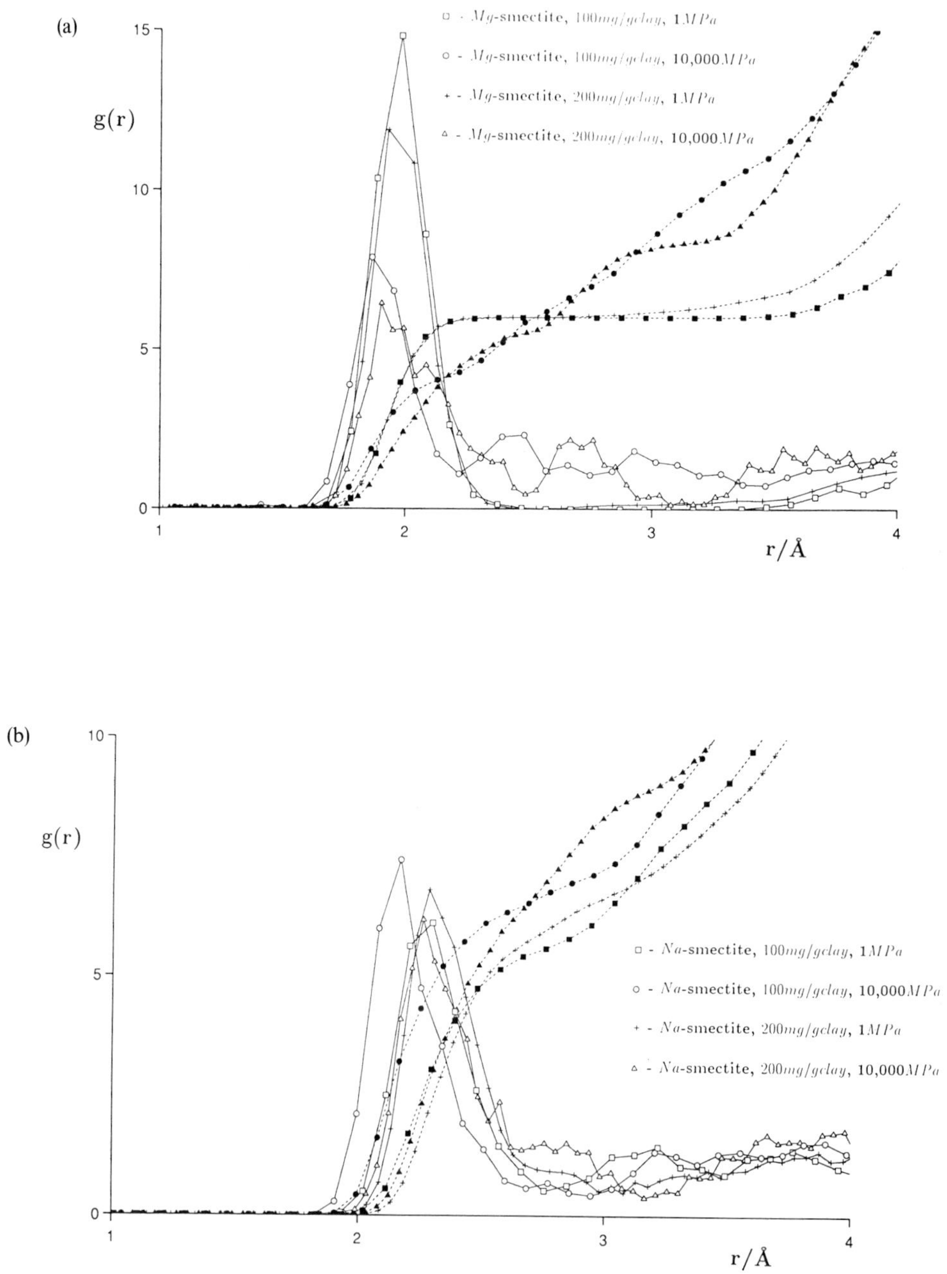

Figure 2.5 Interlayer cation–water radial distribution functions, $g_{CO}(r)$, and running coordination numbers, $n_{CO}(r)$. Oxygen atoms within the clay layers are included. (a) Mg-smectites. (b) Na-smectites. Empty symbols joined by solid lines are the distribution functions, solid symbols joined by the dashed lines are the corresponding cation–oxygen running coordination numbers.

2.4 Discussion

In our discussion of both the Mg-smectites and Na-smectites we first compare the layer specings calculated at 1 MPa with X-ray diffraction data. We then describe the microscopic structures developed, and use these to explain the bulk properties of the smectites.

2.4.1 Mg-Smectite

The calculated layer spacings for Mg-smectite at pressures of both 0 and 1 MPa are 11.1 Å, 14.0 Å and 14.7 Å, for water contents of 0 mg. g^{-1} clay, 100 mg. g^{-1} clay and 200 mg. g^{-1} clay, respectively. These spacings compare with experimental values of 11.2 Å, 14.4 Å and 14.6 Å, measured for Mg-saponite at relative humidities of 0%, 15% and 50%, respectively (Suquet *et al.*, 1975). These humidities correspond to water contents of about 0 mg. g^{-1} clay, 100 mg. g^{-1} clay and 200 mg. g^{-1} clay (Mooney *et al.*, 1952): the same as those used in the simulations (Table 2.1). We therefore conclude that as far as the layer spacing is concerned, the level of agreement between the simulations and experiment is good. Since the calculated layer spacings depend on all of the interaction potentials, we proceed to discuss the other results with some confidence in our model potentials.

The interlayer structures in the Mg-smectites are described by the density profiles, $\rho_O(z)$, $\rho_H(z)$ and $\rho_C(z)$, and the water and cation radial distribution functions, $g_{HO}(r)$ and $g_{CO}(r)$. These are plotted in Figures 2.3, 2.4 and 2.5. We first discuss the structures developed at 1 MPa, and then move on to those at 850 and 10 000 MPa.

The interlayer Mg^{2+} cations are located close to the central plane in all of the Mg-smectites studied (Figures 2.3(c)). At 1 MPa these cations are octahedrally coordinated to form $Mg^{2+}6(H_2O)$ complexes, with an Mg–O distance of 2.0 Å (Figure 2.5(a)). This is consistent with X-ray and neutron diffraction data for vermiculites (Mathieson and Walker, 1954; Skipper *et al.*, 1991b), and with simple ionic bonding models of vermiculite (Giese and Fripiat, 1979). The formation energy of the cation–water complexes is large ($\approx$1077 kJ. mol^{-1} (Curtis *et al.*, 1987)) and they are, therefore, able to hold the clay layers apart.

The water density profiles $\rho_O(z)$ and $\rho_H(z)$ (Figures 2.3(a)) show that at 1 MPa there is a high degree of hydrogen bonding between interlayer water molecules and the clay surfaces. For example, with a water content of 200 mg. g^{-1} clay, peaks in $\rho_O(z)$ occur at 6.1 Å and 8.5 Å (Figure 2.3(a)). These are due to water molecules whose oxygen atom lies about 2.8 Å from a clay surface. One hydrogen atom of each of these water molecules is then directed towards a clay layer, giving rise to peaks in $\rho_H(z)$ at 5.3 and 9.4 Å. We also find that a small number of water molecules lie within 0.5 Å of a clay surface, as has been reported in our recent neutron diffraction experiments on vermiculites (Skipper *et al.*, 1991b). These water molecules occupy sites directly above surface hydroxyl groups, within hexagonal rings of SiO_4 tetrahedra.

The water–water distribution functions, $g_{HO}(r)$ (Figure 2.4(a)), give further evidence of extensive hydrogen bonding among water molecules and to the clay layers. Both the distributions obtained from Mg-smectites at 1 MPa have a peak at about 1.9 Å. This peak, which is smaller than that obtained from pure water, is due to hydrogen bonded O–H pairs. The running H–O coordination number, equal to the number of oxygen atoms around each hydrogen atom (equation (11)), is about 0.8 in both cases, not including the intramolecular terms. This value shows that in Mg-smectites at 1 MPa about 80% of the hydrogen atoms can be involved in unstretched hydrogen bonds.

Our results for interlayer water structure can be compared with data from infrared and other spectroscopic techniques (Farmer and Russell, 1971; Sposito and Prost, 1982). It has been concluded from such experiments that those water molecules which are not directly bonded to an interlayer cation may hydrogen bond to a clay surface. The strength of these bonds is uncertain, but they are weaker in smectites (which have predominantly octahedral layer charge) than in vermiculites (which have predominantly tetrahedral layer charge). This may be the reason why the density profiles obtained from our simulations of smectites have broader peaks than those measured in neutron diffraction experiments on vermiculites (Skipper *et al.*, 1991b). We are currently carrying out simulations of vermiculites to allow direct comparison with these experimental data. We now discuss the effects of increasing the uniaxial pressure.

When the uniaxial pressure is increased to 850 MPa the layer spacing decreases to 12.9 Å, for a water content of 100 mg. g^{-1} clay and to 14.2 Å for a water content of 200 mg. g^{-1} clay (from 14.0 Å and 14.7 Å, respectively at 1 MPa, Table 2.2). We conclude that the interlayer $Mg^{2+}6(H_2O)$ complexes, which hold the layers apart, are beginning to distort at this pressure. This effect is most noticeable at the lower water content.

When the pressure is further increased to 10 000 MPa, the $Mg^{2+}6(H_2O)$ complexes collapse: the layer spacings at 10 000 MPa are 10.9 Å and 11.5 Å for water contents of 100 mg. g^{-1} clay and 200 mg. g^{-1} clay, respectively. The cation–water distribution functions then show that the cations are coordinated to about four water molecules (Figure 2.5(a)). These water molecules themselves lie close to the central plane, as can be seen from the water density profiles (Figures 2.3(f), (h)). We see also from these density profiles that at 10 000 MPa more of the water molecules occupy hexagonal ring sites: about 50% of the total when the water content is 100 mg. g^{-1} clay. These adsorbed water molecules lie almost flat in the plane of the clay sheets at the highest pressure.

Turning to the water–water distribution functions, $g_{HO}(r)$ (Figure 2.4(a)), we see that at 10 000 MPa the closest approach of water molecules is reduced to about 1.4 Å. However, there still remain peaks at about 1.9 Å, which correspond to hydrogen bonds among water molecules and to the clay surfaces. Indeed, the area under these peaks is larger at 10 000 MPa than at 1 MPa. This is due to closer water–clay, rather than water–water, approaches at the higher pressures.

The bulk properties of the clays are summarized in Table 2.2. We now discuss the trends in these quantities in terms of the models proposed for the interlayer structures. We deal first with the average potential energy, and then move to the water density.

The average potential energy of the water molecules in the 100 mg. g^{-1} clay system at 1 MPa is -106 kJ.mol^{-1}, which compares with a value of -30.8 kJ.mol^{-1} in bulk water (Wojcik, 1985). The strong binding of water molecules in the Mg-smectite occurs because 24 out of 32 water molecules are coordinated to an Mg^{2+} cation. The additional water molecules present in the 200 mg. g^{-1} clay system at 1 MPa form hydrogen bonds to other water molecules or clay surfaces, rather than binding directly to a cation. The average potential energy per water molecule is, therefore, increased relative to the 100 mg. g^{-1} clay case; to 74 kJ.mol^{-1}. Indeed, from Table 2.2 we find that the increment in potential energy for these extra water molecules amounts to only -39.7 kJ.mol^{-1}.

Increasing the pressure to 850 MPa has no measurable effect on the potential energy of the interlayer water. At first this may seem a surprising result, given that the layer spacing decreases to 12.9 Å for the 100 mg. g^{-1} system, and to 14.2 Å for the 200 mg. g^{-1} clay system. However, we find that even at 850 MPa the interlayer cations are still six-fold coordinated. Moreover, as the pressure is increasing, a larger number of water molecules occupy hexagonal ring sites. These sites, with a binding energy for water of at least -80 kJ. mol^{-1} (Skipper *et al.*, 1989), are energetically, though not entropically, very favourable.

At 10 000 MPa the layer spacing is 10.9 Å for a water content of 100 mg. g^{-1} clay, and 11.5 Å for a water content of 200 mg. g^{-1} clay. The corresponding potential energies are -114 kJ.mol^{-1} and -64 kJ.mol^{-1} per water molecule, respectively. In spite of the clay contracting as pressure is increased, the 100 mg. g^{-1} clay system, therefore, has a lower potential energy per water molecule at a high pressure.

There are two main reasons for this behaviour. First, the energies of the hydrated clays are calculated relative to the dry clays. At 10 000 MPa the dry clays are highly compacted, and therefore have a high potential energy. Second, at high pressures more water is forced to occupy hexagonal ring sites. As has been stated earlier, these are energetically, though not entropically, very favourable sites for water molecules. These two factors more than compensate for the breakdown of the six-fold cation coordination spheres, to four-fold. When the water content is increased to 200 mg. g^{-1} clay the potential energy per water molecules increases: there are few remaining hexagonal ring sites to be occupied, and the cations are already four-fold coordinated.

The interlayer water densities at 1 MPa (and 0 MPa) are 0.85 g. cm^{-3} and 1.38 g. cm^{-3} for the 100 mg. g^{-1} clay and the 200 mg. g^{-1} clay systems, respectively. The former density is lower than in the bulk liquid: the clay layers are expanded by the interlayer cation–water complexes, but there is little additional water to occupy any remaining space. This space is, however, occupied at a water content of 200 mg. g^{-1} clay. In this case the water density is therefore greater than in the bulk liquid. We now discuss the effect of increasing pressure.

As pressure is increased, first to 850 MPa and then to 10 000 MPa, the layer spacing decreases to 12.9 Å and 10.9 Å, respectively, for 100 mg. g^{-1} clay, and to 14.2 Å and 11.5 Å, respectively, for 200 mg. g^{-1} clay. The corresponding interlayer water densities at 850 MPa and 10 000 MPa are 1.08 g. cm^{-3} and 1.24 g. cm^{-3} for 100 mg. g^{-1} clay, and 1.38 g. cm^{-3} and 1.91 g. cm^{-3} for 200 mg. g^{-1} clay. This increase in density reflects the compression and then collapse of the octahedral cation–water complexes, and the occupation of hexagonal ring sites close to the clay layers. It does not, however, reflect a wholesale breakdown of hydrogen bonding in the interlayer region.

Finally, it is important to note that at 10 000 MPa the closest approach of interlayer water molecules is about 1.4 Å (Figure 2.4(a)). This implies that some amount of water dissociation would take place in a real system under the same conditions. It is not possible to quantify this effect with the present rigid molecule model.

2.4.2 Na-Smectite

The calculated layer spacings of Na-smectite at pressures of both 0 and 1 MPa are 9.8 Å, 11.9 Å and 14.2 Å, for water contents of 0 mg. g^{-1} clay and 200 mg. g^{-1} clay, respectively. These compare with basal plane spacings of 9.8 Å, 12.4 Å and 15.2 Å, obtained from X-ray diffraction experiments on saponite (Suquet *et al.*, 1975) at 0%, 30% and 65% relative humidity, respectively (Mooney *et al.*, 1952) (Table 2.1). Other smectites of different layer composition have basal plane spacings in the range 12.5–14.8 Å at 79% relative humidity (Brindley and Brown, 1980). Therefore, while our calculated spacings are lower than the experimental data for saponite, by about 5% at the higher humidities, they are within the range of spacings measured for Na-smectites in general. We therefore suggest that the differences between our calculated layer spacings and the experimental values for any particular Na-smectite are due more to the position and density of the layer charge than to inadequacy of the intermolecular potentials themselves. Bearing this in mind we proceed to discuss the other results of our simulations.

The most striking difference between the interlayer structures in Na-smectites and Mg-smectites is the position of the interlayer cations. Na^+ has a tendency to bind to the clay surfaces, whereas Mg^{2+} is always found close to the central plane. This is illustrated in Figures 2.3(c), (d) and (e), which show the cation density profiles, $\rho_C(z)$, at 1 MPa.

The different behaviour of Na^+ and Mg^{2+} is a result of the relative strengths of their hydration shells. The hydration enthalpies of Mg^{2+} and Na^+ are

$-1889\,\text{kJ.mol}^{-1}$ and $-389\,\text{kJ.mol}^{-1}$, respectively (Harrison, 1972). It is therefore much easier for Na^+ to partially dehydrate and bind to a clay surface. The site on the clay surface to which Na^+ binds is directly above an SiO_4 tetrahedron. In this site the Na^+ can still bond to between five and six oxygen atoms (see $g_{CO}(r)$ and $n_{CO}(r)$ in Figure 2.5(b)), as is the case in aqueous solution (Skipper and Neilson, 1989). More importantly, when the cations are bound to the clay sheets they cause less disruption to the water–water and water–clay hydrogen bonding. This is evident in the water density profiles and the water–water distribution functions.

Our sodium ion density profiles may be compared with results from infrared experiments. It has been proposed by Low (1981) that sodium ions lie in ditrigonal cavities on the clay surface, even when the clay itself is hydrated. We do not find this to be the case. Rather, on expansion of the clay the charge-balancing cations move into the interlayer region, where they are at least partially hydrated. This is consistent with the experiments of Sposito *et al.* (1983). However, X-ray diffraction (Slade *et al.*, 1985), NMR experiments (Hougardy *et al.*, 1976) and simple ionic bonding studies of two-layer hydrated sodium vermiculites (Giese and Fripiat, 1979) have been interpreted in terms of octahedrally coordinated interlayer sodium ions, located centrally between the clay layers. Unfortunately recent neutron diffraction studies have been unable to resolve the position of these interlayer sodium ions, due to their small neutron scattering amplitude (Skipper *et al.*, 1991b).

The water density profiles $\rho_O(z)$ and $\rho_H(z)$ at 1 MPa are presented in Figures 2.3(b) and (e). We see first that in the 100 mg. g^{-1} clay system there is only one layer of water molecules, as there are no strong cation–water pillars to hold the clay layers apart. However, as in the 200 mg. g^{-1} clay system, there is a high degree of hydrogen bonding among the water molecules and the clay surfaces. This is evident in the water–water distribution functions $g_{HO}(r)$ (Figure 2.4(b)). There are well-defined peaks in $g_{HO}(r)$ due to hydrogen bonding, at around 1.9 Å. The running coordination number of these peaks, up to the first minimum, is over 0.95 in both cases (equation 11). Over 95% of all interlayer hydrogen atoms are, therefore, involved in unstretched hydrogen bonds. This compares with about 80% in the Mg-smectites (see previous section).

The hydrogen-bonded networks of water molecules which we observe are consistent with conclusions drawn from infrared experiments (Farmer and Russell, 1971; Sposito and Prost, 1982). However, neutron diffraction experiments on Na-montmorillonite have been interpreted in terms of a much less-ordered water–clay structure, in which the hydrogen density profile is smoothed out by precession of the water molecules (Hawkins and Egelstaff, 1980). Unfortunately, the resolution of these neutron diffraction experiments is only 1 Å. It is therefore possible that structures such as those we obtain from our simulations would not be resolved in the experimental density profiles.

The effect of increasing pressure to 10 000 MPa is to produce highly ordered interlayer structures. The density profiles $\rho_O(z)$ and $\rho_H(z)$ calculated at

10 000 MPa are presented in Figures 2.3(g) and (i). In the 200 mg. g^{-1} clay system a single layer of oxygen atoms is found in the central plane, with hydogen atoms directed towards the clay layers. Other water molecules are located in the hexagonal ring sites. In the 100 mg. g^{-1} clay system the central layer of oxygen atoms has virtually disappeared, with most of these water molecules instead occupying the hexagonal ring sites. This dramatic change of structuring is also evident in the water–water and cation–water distributions, $g_{\mathrm{CO}}(r)$ and $g_{\mathrm{HO}}(r)$, shown in Figures 2.4(b) and 2.5(b).

The compaction of the 100 mg. g^{-1} clay Na-smectite system at 10 000 MPa is accompanied by a reduction in the shortest intermolecular distance in both $g_{\mathrm{HO}}(r)$ and $g_{\mathrm{CO}}(r)$ (to about 2.1 Å and 1.3 Å, respectively). This is due to closer approaches between the clay layers and the interlayer water molecules, and the clay layers and interlayer cations, respectively. We now discuss the effects of such structural changes on the bulk properties of the clays.

As with the Mg-smectites, we find that the potential energy of interlayer water in Na-smectites is constant up to 850 MPa. Since there is little variation in layer spacing up to 850 MPa this is not surprising. For the 100 mg. g^{-1} clay systems the energy is -64.8 kJ.mol^{-1}, while for the 200 mg. g^{-1} clay systems it is -49.7 kJ.mol^{-1}. These values are higher than those for Mg-smectites, reflecting the reduced hydration energy of Na^{+} compared with Mg^{2+}. At 10 000 MPa the energy has increased to -45.1 kJ.mol^{-1} for 100 mg. g^{-1} clay and to -39.7 kJ.mol^{-1} for 100 mg. g^{-1} clay. These values are due to closer packing of the water molecules.

The density of interlayer water molecules in the Na-smectites shows interesting variation as a function of pressure. At 1 MPa the water densities are 1.18 g.cm^{-3} and 1.14 g.cm^{-3} for 100 mg. g^{-1} clay and 200 mg. g^{-1} clay systems, respectively. In comparison with these values, experimental density determination on the system sodium vermiculite + water has indicated water packing densities about 2.5% higher in the vermiculite than in the bulk (Hougardy *et al.*, 1970).

On increasing pressure to 850 MPa the density decreases, to 1.08 g.cm^{-3} and 1.03 g.cm^{-3}, respectively. This decrease in density is due to the decrease of layer spacing in the dry clays, which are used as a reference system (equation (6)), rather than an increase in the layer spacing of the hydrated clays themselves. This can be seen by referring to the layer spacings quoted in Table 2.2. Only at 10 000 MPa does the interlayer water structure collapse, and the density, therefore, increases to 1.84 g.cm^{-3} and 1.65 g.cm^{-3}, for the 200 mg. g^{-1} clay and 100 mg. g^{-1} clay systems, respectively.

2.5 Conclusions

We have conducted 24 separate Monte Carlo computer simulations of Mg- and Na-smectites. They cover water contents in the range 0–200 mg. g^{-1} clay and uniaxial pressures in the range 0–10 000 MPa.

The layer spacings obtained from our simulations at 1 MPa are in satisfactory agreement with X-ray diffraction data. The calculated layer spacings of Mg-smectite are 11.1 Å, 14.0 Å and 14.7 Å for water contents of 0 mg. g^{-1} clay, 100 mg. g^{-1} clay and 200 mg. g^{-1} clay, respectively. These spacings compare with experimental values of 11.2 Å, 14.4 Å and 14.6 Å. The calculated layer spacings of Na-smectite are 9.8 Å, 11.9 Å and 14.2 Å for water contents of 0 mg. g^{-1} clay, 100 mg. g^{-1} clay and 200 mg. g^{-1} clay respectively. These compare with experimental values of the basal plane spacings of 9.8 Å, 12.4 Å and 15.2 Å.

The interlayer structures developed in Mg-smectites are dominated by cation–water interactions: Mg^{2+} hydrates strongly. In the expanded Mg-smectites at 1 MPa the interlayer cations lie close to the central plane and form $Mg^{2+}6(H_2O)$ complexes. These complexes hold the layers apart. This structure is in agreement with diffraction experiments on vermiculites. Other water molecules are either hydrogen bonded to a clay surface, or occupy hexagonal ring sites close to the clay sheets. At 10 000 MPa the octahedral $Mg^{2+}6(H_2O)$ complexes collapse. Each cation is then surrounded by about four water molecules in a planar geometry.

The density of the interlayer water in Mg-smectites increases from 0.85 g.cm^{-3} at 1 MPa and 100 mg. g^{-1} clay to 1.91 g.cm^{-3} at 10 000 MPa and 200 mg. g^{-1} clay. The corresponding average potential energies of the interlayer water are -107 kJ.mol^{-1} and 64 kJ.mol^{-1}.

In Na-smectites the interlayer Na^+ cations have a tendency to bind to a clay surface as well as to interlayer water molecules, even in the expanded clays. This behaviour is in sharp contrast to Mg^{2+} cations, which are strongly hydrated and lie close to the central plane. The fact that Na^+ ions tend to lie close to one of the clay surfaces may help to explain why Na-smectites disperse in water (Newman, 1987). The Na^+ ions bound to adjacent clay sheets would repel each other, thereby helping the layers to move apart.

The structures in Na-smectites are, therefore, dominated by hydrogen bonding of water molecules among themselves and to clay surfaces. The minimum density of interlayer water is 1.03 g.cm^{-3}, which occurs at 850 MPa and a water content of 100 mg. g^{-1} clay. The maximum is 1.84 g.cm^{-3}, at 10 000 MPa and 200 mg. g^{-1} clay. In these cases the corresponding average potential energies of the interlayer are -64.4 kJ.mol^{-1} and 41 kJ.mol^{-1}.

Having shown that Monte Carlo simulations can provide useful information about smectites, we intend to conduct a systematic study of clay–water–cation systems. In particular, we will investigate the effects on the interlayer region of increased temperature and layer charge.

References

Allen, M. P. and Tildesley, D. J. (1987) *Computer Simulation of Liquids*, Clarendon Press, Oxford.

Brindley, G. W. and Brown, G. (eds) (1980) *Crystal Structures of Clay Minerals and their X-ray Identification*, Mineralogical Society, London.

Curtis, L. A., Halley, J. W., Hautman, J. and Rahman, A. (1987) Nonadditivity of *ab initio* pair potentials for molecular dynamics of multivalent transition metals in water. *Journal of Chemical Physics*, **86**, 2319–27.

Farmer, V. C. and Russell, J. D. (1971) Interlayer complexes in layer silicates. The structure of water in lamellar ionic solutions. *Transitions of the Faraday Society*, **67**, 2737–49.

Giese, R. F. and Fripiat, J. J. (1979) Water molecule positions, orientations and motions in the dihydrates of Mg and Na vermiculites. *Journal of Colloid and Interface Science*, **71**, 441–50.

Grim, R. E. (1960) *Applied Clay Mineralogy*, McGraw Hill, New York.

Harrison, R. D. (ed.) (1972) *Nuffield Advanced Science Book of Data*, Longman, London.

Hawkins, R. K. and Egelstaff, P. A. (1980) Interfacial water structure in montmorillonite from neutron diffraction experiments. *Clays and Clay Minerals*, **28**, 19–28.

Hougardy, J., Stone, W. E. E. and Fripiat, J. J. (1976) NMR study of adsorbed water. I. Molecular orientation and protonic motions in two-layer hydrate of a montmorillonite. *Journal of Chemical Physics*, **64**, 3840–51.

Hougardy, J., Serratosa, J. M. Stone, W. E. E. and Van Olphen, H. (1970) Interlayer water in vermiculite: thermodynamic properties, packing density, nuclear pulse resonance, and infrared absorption. *Special Discussions of the Faraday Society*, **1**, 187–93.

Low, P. F. (1981) The swelling of clay: III. Dissociation of exchangeable cations. *Soil Science Society of America Journal*, **45**, 1074–8.

Mathieson, A. McL. and Walker, G. F. (1954) Crystal structure of magnesium vermiculite. *American Mineralogist*, **39**, 231–55.

Metropolis, N., Rosenbluth, A. W., Rosenbluth, M. N., Teller, A. H. and Teller, E. (1953) Equation of state calculations by fast computing machines. *Journal of Chemical Physics*, **21**, 1087–92.

Mooney, R. W., Keenan, A. G. and Wood, L. A. (1952) Absorption of water vapour by montmorillonite. II. Effect of exchangeable cations and lattice swelling as measured by X-ray diffraction. *Journal of the American Chemical Society* **74**, 1371–4.

Newman, A. C. D. (ed.) (1987) *Chemistry of Clays and Clay Minerals*, Mineralogical Society and Longman, London.

Skipper, N. T. and Neilson, G. W. (1989) X-ray and diffraction studies on concentrated aqueous solutions of sodium nitrate and silver nitrate. *Journal of Physics: Condensed Matter*, **1**, 4141–54.

Skipper, N. T., Refson, K. and McConnell, J. D. C. (1989) Computer calculation of water–clay interactions using atomic pair potentials. *Clay Minerals*, **24**, 411–25.

Skipper, N. T., Refson, K. and McConnell, J. D. C. (1991a) Computer simulation of water in 2:1 clays. *Journal of Chemical Physics*, **94**, 7434–45.

Skipper, N. T., Soper, A. K. and McConnell, J. D. C. (1991b) The structure of interlayer water in vermiculite. *Journal of Chemical Physics*, **94**, 5751–60.

Skipper, N. T., Soper, A. K., Refson, K. and McConnell, J. D. C. (1990) The structure of interlayer water in a hydrated 2:1 clay. *Chemical Physics Letters*, **166**, 411–25.

Slade, P. G., Stone, P. A. and Radoslovich, E. W. (1985) Interlayer structures of the two-layer hydrates of sodium- and calcium-vermiculites. *Clays and Clay Minerals*, **33**, 51–61.

Sposito, G. and Prost, R. (1982) Structure of water adsorbed on smectites. *Chemical Review*, **82**, 554–72.

Sposito, G., Prost, R. and Gaultier, J.-P. (1983) Infrared spectroscopic study of adsorbed water on reduced-charge Na/Li montmorillonites. *Clays and Clay Minerals*, **31**, 9–16.

Suquet, H., de la Calle, C. and Pezerat, H. (1975) Swelling and structural organisation of saponite. *Clays and Clay Minerals*, **23**, 1–9.

Wojcik, M. (1985) MD simulation of MCY water using LCAP. *Internal Report KGN-28*, IBM Corporation, Kingston, NY 12401, USA.

Wood, W. W. (1968) Monte Carlo calculations for hard discs in the isothermal-isobaric ensemble. *Journal of Chemical Physics*, **48**, 415–34.

CHAPTER THREE

Molecular dynamics simulation of water mobility in smectites

K. Refson, N. Skipper and J. D. C. McConnell

3.1 Introduction

The migration of water and cations in compacted smectite clays plays an important role in many industrial processes. In the shales which seal oil reservoirs the extremely low permeability of the smectite constituent provides a barrier to the movement of water and hydrocarbons. Smectites are also used to seal landfill sites for disposal of hazardous waste, where again low permeability is their key property. A related application is for radioactive waste disposal, where it is vital to control the migration of radionuclides over extremely long timescales.

The present work is concerned entirely with fluid in the interlayer region of smectites, where most of the water is to be found in a sufficiently homogeneous and well-compacted clay. In that case its properties will dominate the properties of the bulk material. In particular, in the absence of external 'pore' fluid the interlayer region provides the only route for water and cation migration. The two-fold aims of this research are to increase understanding of the molecular-scale phenomena which determine mobility and to predict the diffusion constants, the quantitative measure of mobility.

We have chosen to use the theoretical methods of atomistic computer simulation because of the severe difficulties associated with experimental determination of mobilities. Diffusion constants of the cations may be measured using electrical conductivity (Calvet and Mamy, 1971; Slade *et al.*, 1987) or radioactive tracer techniques (Lai and Mortland, 1968; Calvet and Chaussidon, 1969). Unfortunately these two methods give different results, leading Calvet and Mamy (1971) to conclude that the majority charge carriers are protons, rather than the counterions.

Nuclear Magnetic Resonance (NMR) spectroscopy gives information about dynamics on a timescale of 10^{-3} to 10^{-10}s. This technique has been used to study the rotational motion of water molecules (Fripiat, 1980) in hydrated clays, but does not provide a direct measurement of translational diffusion coefficients.

Geochemistry of Clay-Pore Fluid Interactions. Edited by D. A. C. Manning, P. L. Hall and C. R. Hughes. Published in 1993 by Chapman & Hall, London. ISBN 0 412 48980 5

The most fruitful method of measuring diffusion constants of the interlayer water has been quasi-elastic neutron scattering (Ross and Hall, 1980). Diffusion events in the system under study cause a broadening of the elastic peak in the frequency spectrum of the incoherent scattering, which may be measured using high-resolution neutron spectrometers. Analysis of these data requires designing a parameterized model of the diffusion processes and fitting the parameters to the measured peak widths. Since rotational as well as translational diffusion contributes to the broadening, it is not possible to separate the effects due to these two motions, except where they occur on different timescales (and hence energy scales). Therefore, it is difficult to obtain unambiguous values for the diffusion coefficients.

Nevertheless Tuck, Hall, Ross and others have performed a comprehensive series of experiments (Tuck *et al.*, 1984; 1985, and references therein) which give a good picture of diffusion processes in smectites. Their most recent experiments used the medium-resolution (20 μeV) spectrometer, IN5, and the high-resolution (1 μeV) IN10 instruments at the Institut Laue Langevin (ILL) in Grenoble. They found that there are two populations of water molecules; an immobile fraction which they identified with the hydration shell of the cations, and a mobile fraction of non-hydration-shell water. Because of their lower resolution, the IN5 results are sensitive to displacements of up to about 10 Å, whereas IN10 measures diffusion at rather longer distances. The 'effective' diffusion constants of unbound water in a Mg-montmorillonite at $p/p_0 = 0.76$ were $1.5 \times 10^{-5}\,cm^2.s^{-1}$ measured on IN5 and $0.42 \times 10^{-5}\,cm^2.s^{-1}$ on IN10 (Tuck *et al.*, 1985). This led the authors to suggest that water diffuses fairly rapidly in 'cages' of approximate dimension 7 Å and more slowly on longer length scales, probably because of hindrance by the cations and their hydration shells. However, those figures quoted are not the 'true' diffusion constants, but what the authors term 'effective' diffusion constants since the results were analysed on the assumption that rotational diffusion did not contribute to the observed broadening.

By contrast, computer simulation can calculate essentially exact theoretical results, which are systematically improvable. That is, there are no uncontrolled approximations; the limitations are due to length scales or time scales or to statistical accuracy. Molecular diffusion events occur on a scale typically of tens of angstroms and tens of picoseconds, which is just the regime accessible to molecular dynamics simulation. Because complete information on the configurations is available—all molecular positions and velocities—each contribution may be separately calculated. So rotational and translational diffusion may be distinguished and attributed to various populations such as hydration-shell or surface water.

Simulations based on atomistic models have been very successful in modelling the thermodynamic, structural and transport properties of condensed matter. A material of interest is regarded as an assembly of its constituent molecules, atoms or ions. The ingredients of this model are, therefore, a specification of the atomic, molecular or ionic species together with a description of the force law governing their interactions. So the model is not just representative of one or two specific properties (for example diffusion only) but rather of the system's behaviour as a whole. A model which gives good agreement with experiment over a sufficient range of experiments may, therefore, be expected to correctly compute other quantities less easily measured.

The intermolecular force law is usually regarded as a sum of more basic forces which act between pairs of atomic sites on two molecules. These forces are assumed to take a simple analytical form and to depend only on the site–site distance. They may, therefore, be written as derivatives of some *potential (energy) function* with a small number of adjustable parameters. Specific values of these parameters may be assigned either by fitting computed quantities to their experimental values (an *empirical* potential), or by calculation using first-principles quantum mechanics (an *ab initio* potential).

Our model of a smectite is necessarily highly heterogeneous and is one of the most complex which has been attempted to date. Ten different atomic species are required, compared with three or four for more homogeneous materials: the number of distinct interactions goes as the square of the number of species. To correctly determine so many parameters is a considerable task. Nevertheless we have succeeded in constructing a model which predicts swelling behaviour and layer spacings in good agreement with experiment (Chapter 2). We may, therefore, proceed to study the transport properties.

3.2 Simulation techniques and models

Molecular dynamics simulation (henceforth referred to as MD) is the method used in this study. Unlike the Monte Carlo technique of our previous paper (Chapter 2), MD simulates the correct dynamics of the model—a prerequisite to obtaining the transport properties. We give only a brief outline here and refer the interested reader to the book by Allen and Tildesley (1987, especially chapters 1 and 3). All of the simulations described herein were performed using the MD code 'Moldy', a more detailed description and copies of which may be obtained from the first author.

The goal of MD simulation is to solve the classical (Newton–Euler) equations of motion for each molecule in the model system given the interaction forces with other molecules*. Though an exact solution to this n-body problem is not possible, an approximate numerical solution may be obtained as follows. A complete record of the state of the model system at some instant is maintained in the computer program. (This record consists of the coordinates, velocities and accelerations of every molecule in the system and their rotational analogues.) The potential functions are evaluated using the current values of the site coordinates, and are used to calculate the total force (and torque) on

*For brevity we will use the word 'molecule' to refer to atoms and ions as well.

each molecule at that instant. The key step is to use all of this information to yield a good approximation to the state of the system some short time later. The interval is known as a *timestep* and is chosen so that the coordinates change by a only a small increment on each step. By repeating this procedure, the state of the system may be advanced one step at a time to give an approximation to the exact trajectory over, perhaps, ten or a hundred thousand steps.

The procedure for advancing from one state to the next is known as the integration algorithm and its properties are clearly vital to the accuracy of the simulation. In this work, we use a modified form of the Beeman algorithm (Beeman, 1976; Refson, 1985) which is known to give good stability for molecular systems. The interparticle forces were evaluated in a similar manner to that used in the Monte Carlo simulation (Skipper *et al.*, 1991) using the Ewald sum method with $\alpha = 0.35$ Å^{-1} and real and reciprocal space cutoffs of 9 Å and 2 Å^{-1}, respectively.

Thermodynamic quantities such as temperature and pressure may be calculated by averaging some 'instantaneous' T or P function over many timesteps (Allen and Tildesley, 1987, pages 46*ff.*). At first such averages are strongly determined by the initial state of the system, but after some time this influence decreases until the averages become representative of an equilibrium statistical ensemble. This *equilibration* process may be conveniently monitored by plotting the instantaneous value of the temperature. When this approaches a final value and its fluctuations decrease to a minimum the *production* period begins and results are recorded.

The measure of mobility is the self-diffusion constant. These constants are related to the mean-square displacements (abbreviated to msd*s*) undergone by individual molecules or ions as they migrate during the simulation. The continuum diffusion equation in three dimensions has the solution

$$\langle r^2 \rangle = \langle |r(t) - r(0)|^2 \rangle = 6Dt$$

(known as Fick's Law) where $r_i(t)$ is the position of molecule i and the angle brackets denote an average. Real molecular motion is of course more complex than the continuous diffusion of Fick's Law and may often proceed by periods of oscillation about some point, punctuated by rapid jumps or hops. But on a sufficiently long timescale the diffusion law will, in any case, approximate to continuous diffusion. Therefore, the best estimate of the diffusion constant will be given by the long-term behaviour of the derivative

$$D = \frac{1}{6} \lim_{t \to \infty} \frac{\mathrm{d}}{\mathrm{d}t} \langle r^2 \rangle. \tag{1}$$

So a plot of $\langle r^2 \rangle$ vs. t should asymptotically approach and fluctuate about a straight line of gradient $6D$ (which does not necessarily pass though the origin).

Of course water in the interlayer region of a clay is not free to diffuse in three dimensions. The clay layer is almost completely impermeable, and our idealized model of effectively infinite extent is an absolute barrier. Therefore, there can be no long-term diffusion perpendicular to the layers. The two-dimensional version of Fick's Law states that $\langle r^2 \rangle = 4Dt$ and, therefore, the gradient of the msd plot is $4D$. However in this work for purposes of comparison with experiment we calculate D by the above formula with the factor of 1/6, i.e. all diffusion constants are calculated as if they were three dimensional.

3.3 The model

The constituents of our model of a hydrated clay are water molecules, Na^+ or Mg^{2+} cations and the silicate clay sheet. Each water molecule and the clay sheet are treated as rigid bodies with no internal motion; water molecules may rotate and translate but the clay sheet is fixed in space. These molecular units are built from their constituent atoms — Si, O, H, Mg ions and 'pseudo-atomic' sites with electric charge only. All such sites interact with forces derived from potential functions specific to the types of the pair of sites involved. We have used the *ab initio* water–water potentials of Matsuoka *et al.* (1976), the sodium–water potentials of Bounds (1985) and the Mg^{2+}–water potential of Curtis *et al.* (1987). (The latter was actually developed as an Fe^{2+}–water potential and we use the fact that these two ions are isomorphic in solution (Neilson and Enderby, 1979).) Finally, we employ our recently derived SRM28 interaction potentials (Skipper *et al.*, 1991a) to describe the water–silicate and ion–silicate interactions.

The simulated system is constructed from the above species and consists of 64 water molecules, either 4 Mg^{2+} or 8 Na^+ cations and a single rectangular clay sheet. The latter consists of 8 formula units of $Al_4Si_8O_{20}(OH)_4$ smectite and has dimensions 21.12×18.28 Å. It fits exactly in an MD simulation cell of $21.12 \times 18.28 \times 14.68$ Å which is periodically repeated to form an effectively infinite sheet. These periodic boundary conditions eliminate edges whose effects would influence the calculated properties. Its geometry is idealized with perfectly hexagonal silicate rings. There is no Al/Si substitution in the tetrahedral layer and the octahedral layer contains electric charges only. The required layer charge of $-8\,e$ was achieved by placing a point charge of $-1\,e$ at one out of four aluminium sites in a regular fashion.

The water content of 64 molecules is equivalent to 200 mg. g^{-1} clay, which is experimentally achieved at a relative humidity p/p_0 of approximately 0.5 for magnesium cations and 0.67 for sodium (Mooney *et al.*, 1952). Because of the periodic boundary conditions the layer separation is identical to the c-axis of the simulation cell. This had the fixed value of 14.68 Å in all the runs, which were carried out under constant-volume conditions. A later version of the simulation program does allow the cell shape and size to fluctuate using the method of Parinello and Rahman (1981), to give a constant-pressure ensemble. Future work will reveal what difference, if any, this makes.

Our model was not intended to accurately describe any particular clay, but rather to be generally representative of a dioctahedral smectite. This enabled us to study the common properties of such materials without too much concern for the detailed differences attributable to composition. The natural clays which most closely approach this structure and charge density are Wyoming or Otay montmorillonites.

Runs were conducted at temperatures ranging from 280 to 350 K, for both Mg^{2+} and Na^{+} cations. A configuration from the Monte Carlo runs was used as the initial state for the first runs, and subsequent runs at different temperatures were started from the final state of the previous ones. Each run consisted of an equilibration period of 10,000 timesteps (of 0.25 fs) during which the temperature was adjusted to the desired value and the system allowed to settle down to thermodynamic equilibrium. This was followed by the production period during which configurational data were stored every 20 timesteps and was usually 60 000 timesteps long, giving a run length of 15 ps. The exceptions were the runs at 350 K when the asymptotic behaviour of the msd was not apparent after only 15 ps and longer runs were required. Data were only stored every 50 steps during these runs in order to limit the total volume of dump data.

3.4 Results and discussion

An examination of the trajectories followed by water molecules and cations during the simulation run gives a good qualitative picture of the kinds of behaviour present. Figure 3.1 shows the paths of selected water molecules during a run illustrating the three different environments encountered by water. Molecule number 1 forms part of the hydration shell of a Mg^{2+} ion. It remains close to the ion, to which it is tightly bound, for the entire duration of the run. It does not move significantly, ending up very close to its starting position. By contrast, molecule number 15 is not bound to any cation. It migrates more than 9 Å during the 15 ps of the run. It is such 'unbound' water molecules which are mainly responsible for water transport. The third environment for water may be seen from the trajectory of molecule number 7. Some time into the run this molecule becomes bound to a 'hole' site on the clay surface hydrogen bonded to a hydroxyl group in the centre of a hexagonal ring of SiO_4 tetrahedra. Its motion is highly restricted in the plane of the clay but has much more freedom in the perpendicular direction.

Turning our attention to the cations, and Figure 3.2, we see that at the lower temperatures both the Mg^{2+} and Na^{+}cations are fairly immobile. The traces displayed are typical; no cation migration is evident at this temperature. However at 350 K there *is* some diffusive motion of both cation species (see Figure 3.3). These examples are the extreme rather than typical cases, showing

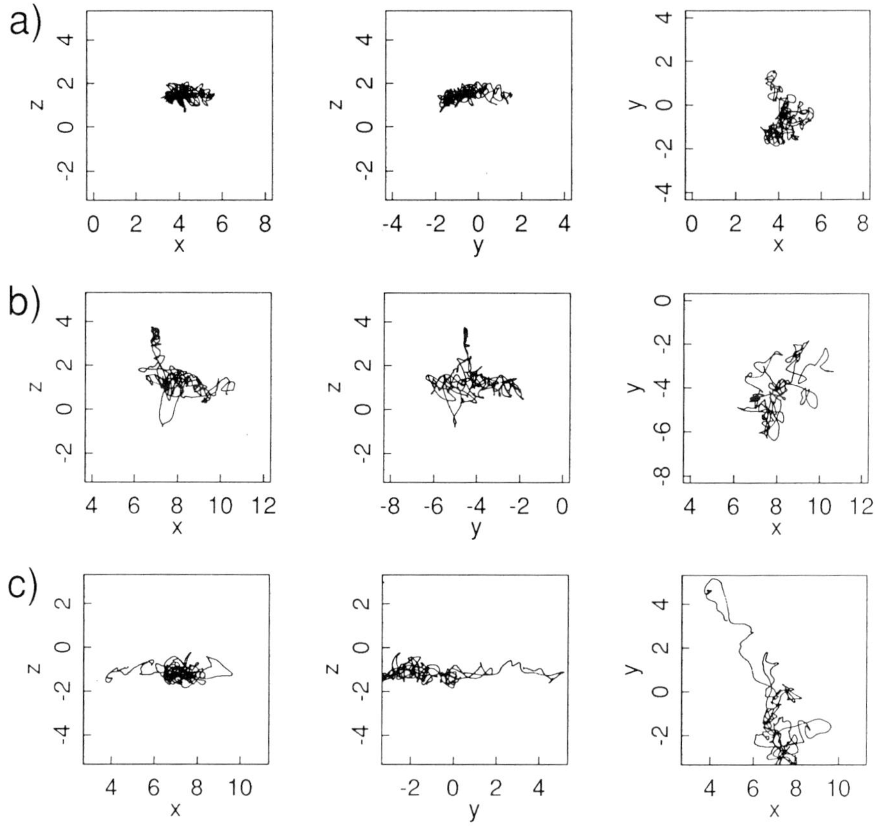

Figure 3.1 Trajectories of selected water molecules in Mg-montmorillonite at 315 K. (a) Molecule 1 is part of the hydration shell of a Mg^{2+} ion and has restricted motion. (b) Molecule 7 enters and leaves a surface 'hole' site. (c) Molecule 15 shows long-term diffusive motion. Projections are along all three axes and units are Å. The *z*-axis is perpendicular to the clay sheet and its zero is at the centre of the interlayer region.

the most mobile of the cations. In each case the total distance moved is of the same order as the periodicity in the structure of the clay surface, which indicates that at this temperature the cations have sufficient energy to surmount local potential barriers. However, it is not clear that this is an example of macroscopic diffusion, which takes place on a timescale longer than the 40 ps simulated time. Cation diffusion is at the limit of detectability on this timescale even at elevated temperatures.

Quantitative measures of the diffusion constants are obtained from the mean-square displacement vs. time plots of Figures 3.4 and 3.5. The 'best fit' line plotted on each graph was obtained from a least squares fit to approximately the last three-quarters of the data. (Ignoring the region about the origin gives a better indication of the long-term asymptotic behaviour.) Formula 1

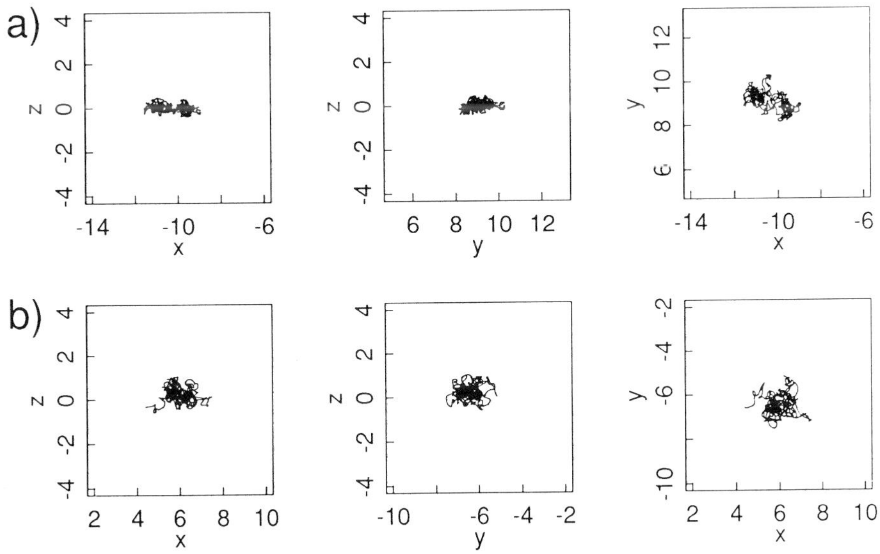

Figure 3.2 Selected cation trajectories in (a) Mg- and (b) Na-montmorillonite at 315 K. At this temperature there is no diffusive motion. Projections are along all three axes and units are Å. The *z*-axis is perpendicular to the clay sheet and its zero is at the centre of the interlayer region.

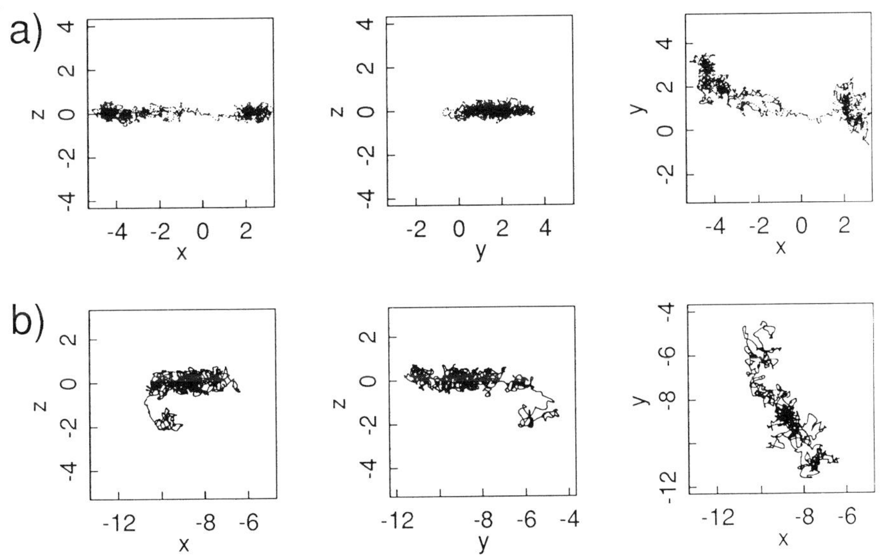

Figure 3.3 Selected cation trajectories in (a) Mg- and (b) Na-montmorillonite at 350 K. At this temperature some diffusion events ('hops') do occur, but only in a small fraction of the ions. Those with greatest diffusion are shown. Projections are along all three axes and units are Å. The *z*-axis is perpendicular to the clay sheet and its zero is at the centre of the interlayer region.

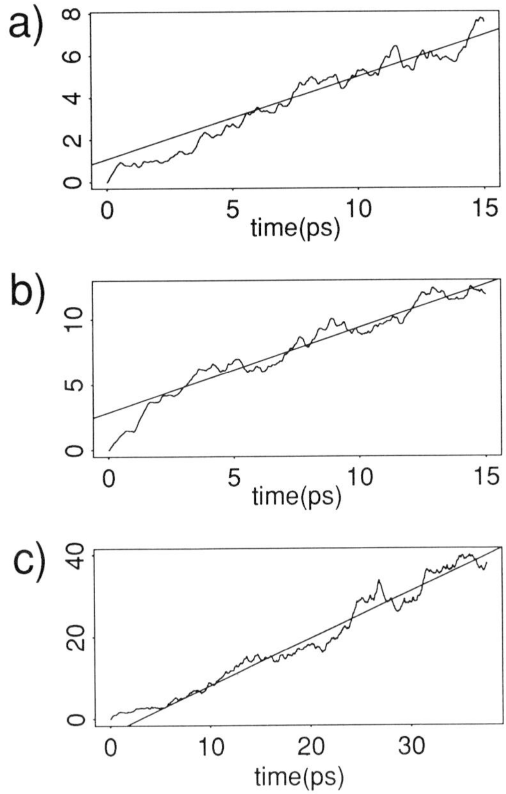

Figure 3.4 Mean-square displacement, $\langle r^2 \rangle$ of all water molecules in Mg-montmorillonite at (a) 282 K, (b) 316 K and (c) 350 K. The gradient increases with temperature. (Notice that the plots have different scales on both axes.)

was applied to calculate *D*. Uncertainties were estimated from the uncertainty in the gradient of the best fit lines to be 10% for the lower temperatures. The 350 K and 400 K results have a rather larger uncertainty because of infrequent cation jumps.

The results are set out in Table 3.1 and plotted in Figure 3.6(a) along with the experimental results of Tuck *et al.* (1984) on Ca-montmorillonite. This is the only study in the literature on the temperature dependence of water diffusion in clays. In an earlier work (Hall *et al.*, 1979) these authors found no difference in diffusion rate between Mg^{2+} and Ca^{2+} systems, so we expect these results to hold for Mg^{2+} as well. Those points are included in Table 3.1 and Figure 3.6(a). However, before we can compare the experimental and simulation values, we must take into account a difference in the definitions of *D*. The experimental values quoted in the original paper were for non-hydration-shell water only and have, therefore, been scaled by the measured free-water fraction to obtain the diffusion constant for the whole water population.

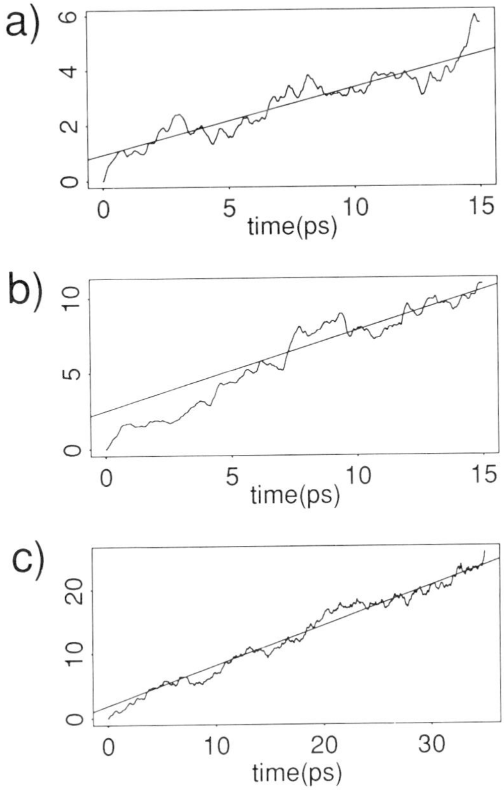

Figure 3.5 Mean-square displacement, $\langle r^2 \rangle$ of all water molecules in Na-montmorillonite at (a) 282 K, (b) 316 K and (c) 350 K. The gradient increases with temperature. (Notice that the plots have different scales on both axes.)

Table 3.1 and Figure 3.6(a) show a fairly good agreement between our calculated and these experimental values. It is notable that the simulation results agree with these data rather than with the lower value of 0.28×10^{-5} cm^2.s^{-1} (0.51 at 190 mg.g^{-1} multiplied by the quasi-elastic fraction of 0.55) obtained using the high-resolution IN10 instrument (Tuck *et al.*, 1985). This is easily explained by the range of displacements sampled by the simulated molecules. The maximum RMS displacement at 316 K is approximately 3.5 Å (Figure 3.4) though many individual water molecules migrated by up to 9 Å during the 15 ps of the run. In our simulation the cations were separated by around 9 Å so one would *not* expect to see the hindering effect postulated by Tuck *et al.* (1984) in simulation runs of this length.

We must, however, be cautious in comparing these experiments to our simulation results because they are taken at slightly different state points. First, our calculations were done with a slightly lower water content than the experiments, 200 mg.g^{-1} *vs.* 220 mg.g^{-1}. The effect of this may be judged from

Table 3.1 Computed and experimental diffusion constants of non-hydration-shell ('free') water in Mg- and Na-smectites. Experimental numbers have been multiplied by the measured quasi-elastic fraction to give 'all-water' coefficients. Uncertainties are approximately 10% except at 350 K and above when rare cation jumps may give larger but unquantifiable errors. Units are $10^{-5}\,cm^2.s^{-1}$.

T	D(Mg)	D(Na)	D(exp)
278	—	0.41	—
282	0.66	—	—
300	0.84	—	0.97
305	—	0.99	—
315	1.09	0.88	—
338	—	—	1.95
350	1.93	1.08	—
368	—	—	2.66
396	1.83	—	—

the IN10 results at differing vapour pressures (Tuck *et al.*, 1985) where decreasing the water content from 220 to 190 $mg.g^{-1}$ increased the 'long-range' effective diffusion constant from 0.42 to $0.51 \times 10^{-5} cm^2.s^{-1}$. We might, therefore, suppose that the experimental values at 200 $mg.g^{-1}$ would be around 10–15% higher than those reported. Secondly, the specific charge of the simulated system, $-1.0\ e$ per formula unit is slightly larger than that of the Wyoming montmorillonite used for the experiment, $-0.85\ e$. This means that the number of exchangeable cations, and hence the proportion of bound and immobile water is higher than in the experiments, which will lower the overall water diffusion rate by about 10%. It is fortuitous that this should approximately cancel the difference in water content.

In the Arrhenius plot of Figure 3.6(b), values of D_{H_2O} in Mg-smectite at 315 K and under lie on a straight line, indicating an activated process. The activation energy calculated from the gradient is 11.1 $kJ.mol^{-1}$, very close to Tuck *et al.*'s (1985) value of 11.5 $kJ.mol^{-1}$. However, the points at 350 and 400 K do not lie on this line. At these temperatures additional water diffusion processes, namely *cation* jumps, become important, as explained below. This has the double effect of changing the activation energy as well as making those calculated values statistically less reliable. Therefore, our quoted figure of $11.1 \pm 1\ kJ.mol^{-1}$ is only valid below about 320 K.

One feature of Figure 3.6(a) which warrants some discussion is the higher value of D_{H_2O} for Na-smectite at 305 K than at 315 K, contrary to the increasing trend with temperature. Though the difference is within the error bounds of the points, there is a more convincing explanation. The runs were performed in the order 316, 278, 350, and then 300 K: each run was started from the final state of the preceding one. The intention was to test whether the

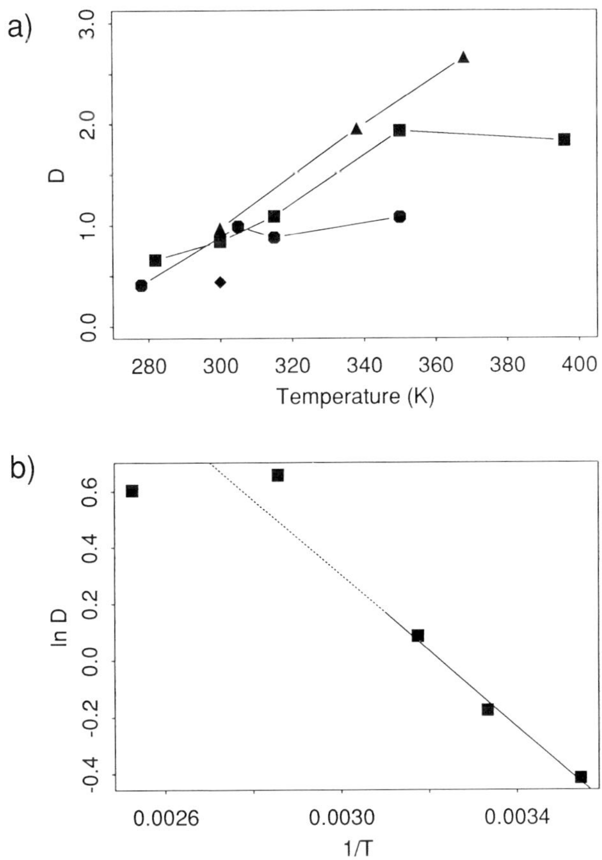

Figure 3.6 (a) Computed and experimental water diffusion constants. Calculated values for Mg-smectite are denoted by squares and Na-smectite by circles. Experimental results from QENS are denoted by triangles for the medium-resolution IN5 experiment (Tuck *et al.*, 1984) and diamonds for the high-resolution IN10 one (Tuck *et al.*, 1985). These numbers have been multiplied by the measured quasi-elastic fraction to give 'all-water' coefficients. (b) Arrhenius plot of Mg-smectite results. The line is fitted to the three lower temperature points only and gives an activation energy of 11.1 $kJ.mol^{-1}$.

system had truly achieved equilibrium by looking for any hysteresis. Clearly there *is* some small hysteresis which implies that at least one of those runs had not quite achieved equilibrium.

The nature of that departure from equilibrium is revealed by the sodium cation trajectories, of which examples are shown in Figure 3.2. The 315 K simulation began from a state with all eight sodiums in the centre of the interlayer region. Recall from our preceding paper (Chapter 2) that the sodiums are predominantly distributed in planes close to the surface of the clay, 2 Å from the centre of the interlayer region. During the 315 K run only two sodiums had migrated towards the clay sheet; the rest remained in the centre. But in the later, 300 K, run six sodiums were close to the clay sheet and only two remained in the centre, in accord with the Monte Carlo results. So we conclude that though the 315 K simulation did not reach equilibrium, the 300 K run did.

Since a sodium at the surface is partly hydrated by a surface oxygen, at least one extra water molecule will be released into the unbound water population. It may then participate in jump events and increase the diffusion constant. It is, therefore, expected that further runs would give slightly higher values for D_{H_2O} at 278 K and 315 K, possibly lying along the same curve as the magnesium smectite results.

Finally we turn to cation diffusion. Figure 3.7 shows the mean-square displacements of sodium and magnesium ions at 300 K and 350 K. In three cases the msd levels off after an initial period and merely fluctuates about some constant value. (Notice that the fluctuations are rather larger than for the water msds because there are fewer cations than water molecules.) There is no evidence of the steady increase with time which would signify diffusion. The exception is the Mg-smectite run at 350 K where there is a large jump at 24 ps, but this is not followed by an increasing trend either. An inspection of the trajectories shows that this increase is due to a single cation jump event, the one visible in Figure 3.3.

Tracer diffusion measurements (Nye, 1979) give cation diffusion constants of the order of $10^{-8}\,cm^2.s^{-1}$, and we should, therefore, expect no significant cation migration on the tens of picoseconds timescale of the simulation. The one jump event which did occur during the 350 K run was fortuitous: in an even hotter run at 400 K there were none at all. However, this event does have some influence on the computed water diffusion rate. As the cation moves it must drag its hydration shell of six water molecules along, so every cation jump is accompanied by six water molecule jumps. These contribute to the computed D_{H_2O}, and account for the higher value computed at 350 K than at 400 K.

3.5 Conclusions

We have shown that our model of a hydrated smectite exhibits the relevant transport properties of these systems. Three different environments for water molecules are found—'cation hydration shell', 'unbound' and 'hole' water. Only unbound water contributes to the macroscopic diffusion process.

Our calculated values of the interlayer water diffusion constants agree with the experimental results made at medium resolution. Those values were calculated using a 'pure translational' model to fit the data, and therefore may include the effects of rotational diffusion. That there is nevertheless such good agreement indicates that this effect is probably unimportant. This study has not addressed the longer term behaviour revealed by the high-resolution IN10 experiments, nor the rotational motions of the water and of complete hydration-shell rotations suggested by them. However, investigations of these phenomena are under way and will be reported at a later date. We hope that

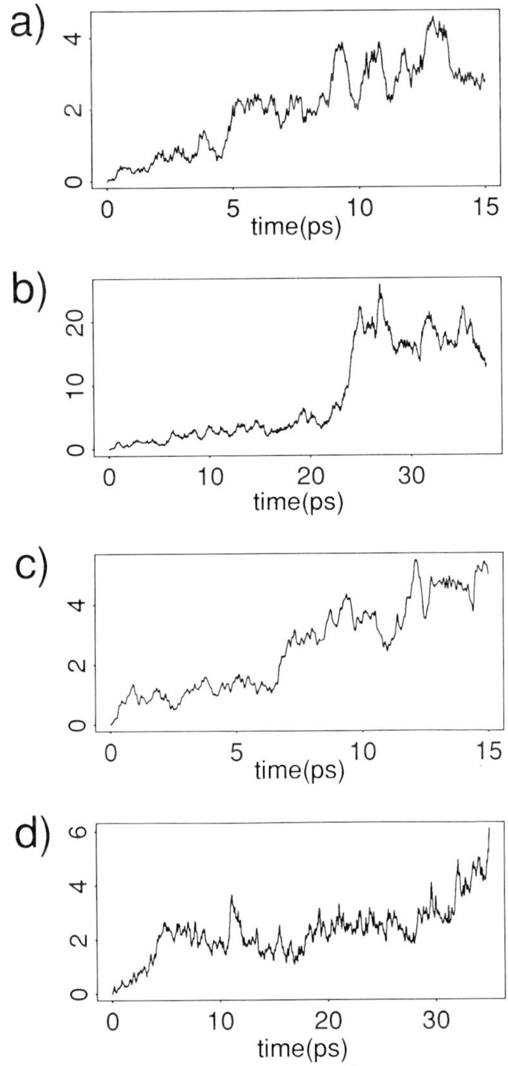

Figure 3.7 Mean-square displacements for (a) magnesium ions at 315 K, (b) at 350 K, and (c) sodium cations at 315 K and (d) at 350 K. Only at 350 K is significant cation diffusion observed. (Notice that the plots have different scales).

this paper will inspire more state-of-the-art QENS experiments, and believe that models derived from our simulation results could be valuable in their interpretation. This should alleviate some of the ambiguities of previous experiments which resulted from the lack of information on the processes involved.

The simulations do not yield accurate values for any cation diffusion rates, because those processes occur too slowly to reproduce on the timescale available to MD simulation. This is in accordance with the observations that cation diffusion constants are around two orders of magnitude lower than that of the water.

References

Allen, M. P. and Tildesley, D. J. (1987) *Computer Simulation of Liquids*, Clarendon Press, Oxford.

Beeman, D. (1976) Some multistep methods for use in molecular dynamics calculations. *Journal of Computational Physics*, **20,** 130–9.

Bounds, D. G. (1985) A molecular dynamics study of the structure of water around the ions Li^+, Na^+, K^+, Ca^{2+}, Ni^{2+} and Cl^-. *Molecular Physics*, **54**(6), 1335–55.

Calvet, R. and Chaussidon, J. (1969) Diffusion des cations compensateurs dans la montmorillonite aux faibles hydrations, in *Proceedings of the International Clay Conference*, (ed. L. Heller), Israel Universities' press, Jerusalem, pp. 635–47.

Calvet, R. and Mamy, J. (1971) Sur la nature des charges responsables de la conductivité électrique des argiles. *Compte Rendu Académie des Sciences, Paris*, **27D3,** 1251–3.

Curtis, L. A., Halley, J. W., Hautman, J. and Rahman, A. (1987) Nonadditivity of *ab initio* pair potentials for molecular dynamics of multivalent transition metal ions in water. *Journal of Chemical Physics*, **86**(4), 2319–27.

Fripiat, J. J. (1980) The application of NMR to the study of clay minerals, in *Advanced Chemical Methods of Soil and Clay Minerals Research*, vol. C(63), Reidel, Dordrecht, for NATO, pp. 245–315.

Hall, P. L., Ross, D. K., Tuck, J. J. and Hayes, M. H. B. (1979) Neutron scattering studies of the dynamics of interlamellar water in montmorillonite and vermiculite, in *Proceedings of the International Clay Conference*, Oxford, 1978, (eds M. M. Mortland and V. C. Farmer), Elsevier, Amsterdam, pp. 121–30.

Lai, T. M. and Mortland, M. (1968) Cation diffusion in clay minerals: II. Orientation effects. *Clays and Clay Minerals*, **16,** 129–36.

Matsuoka, O., Clementi, E. and Yoshimine, M. (1976) CI study of the water dimer potential surface. *Journal of Chemical Physics*, **64**(4), 1351–61.

Mooney, R. W., Keenan, A. G. and Wood, L. A. (1952) Absorption of water vapour by montmorillonite. II. Effect of exchangeable cations and lattice swelling as measured by X-ray diffraction. *Journal of the American Chemical Society*, **74**, 1371–4.

Neilson, G. W. and Enderby, J. E. (1970) Neutron and X-ray diffraction studies of concentrated aqueous electrolyte solutions. *Annual Reports on the Progress of Chemistry Section C, Physical Chemistry*, **76**, 185–220.

Nye, P. H. (1979) Diffusion of ions and uncharged solutes in soils and soil clays. *Advances in Agronomy*, **31**, 225–72.

Parinello, M. and Rahman, A. (1981) Polymorphic transitions in single crystals: A new molecular dynamics method. *Journal of Applied Physics*, **52**(12), 7182–90.

Refson, K. (1985) Molecular dynamics simulation of solid *n*-butane. *Physica*, **131B**, 256–66.

Ross, D. K. and Hall, P. L. (1980) Neutron scattering methods of investigating clay systems, in *Advanced Chemical Methods of Soil and Clay Minerals Research*, (eds J. W. Stucki and W. L. Banwart), vol. C(63), Reidel, Dordrecht for NATO, pp. 93–168.

Skipper, N. T., Refson, K. and McConnell, J. D. C. (1991) Computer simulation of interlayer water in 2:1 clays. *Journal of Chemical Physics*, **94**(11), 7434–45.

Slade, R. C. T., Barker, J. and Hirst, P. R. (1987) Conduction and diffusion in exchanged montmorillonite clays. *Solid State Ionics*, **24**, 289–95.

Tuck, J. J., Hall, P. L., Hayes, M. H. B., Ross, D. K. and Hayter, J. B. (1985) Quasi-elastic neutron scattering studies of intercalated molecules in charge-deficient layer silicates. II. *Journal of the Chemical Society, Faraday Transactions 1*, **81**, 833–46.

Tuck, J. J., Hall, P. L., Hayes, M. H. B., Ross, D. K. and Poinsignon, C. (1984) Quasi-elastic neutron scattering studies of the dynamics of intercalated molecules in charge-deficient layer silicates. I. *Journal of the Chemical Society, Faraday Transactions 1*, **80**, 309–24.

PART TWO
Natural Processes of Diagenesis and Weathering

CHAPTER FOUR

The composition of authigenic clay minerals in recent sediments: links to the supply of unstable reactants

A. C. Aplin

4.1 Introduction

The minerals and X-ray amorphous materials (mineraloids) which are eventually deposited in marine and lacustrine sediments are largely the products of continental weathering and sea floor hydrothermal activity. They are compositionally diverse and form under chemical conditions which are often very different to those typical of the sediments which they comprise. Such a diverse mineral assemblage is inherently unstable and progresses towards equilibrium through a series of diagenetic reactions (Garrels, 1986).

Clay mineral diagenesis begins within centimetres of the sediment–water interface. However, although pore-water data show that aluminosilicates dissolve and reform in all types of marine sediments (e.g. Hurd, 1973; Mackin and Aller, 1984a; 1984b), it is often difficult to detect authigenic clay minerals against a rapidly accumulating background of detrital aluminosilicates. The best places to study the early formation of authigenic clay minerals are, therefore: (1) in slowly accumulating sediments, and (2) at points where extremely unstable materials are supplied rapidly (hydrothermal systems). A wide variety of authigenic clays has been described from such sediments. The twin aims of this paper are to summarize and then rationalize the chemical and mineralogical variability exhibited by these authigenic phases.

4.2 Mineralogical and chemical trends

4.2.1 *Mineralogy*

The wide variety of authigenic clay minerals recognized in surface marine and lacustrine sediments is listed in Table 4.1 (in addition to references in Table 4.1

Geochemistry of Clay-Pore Fluid Interactions. Edited by D. A. C. Manning, P. L. Hall and C. R. Hughes. Published in 1993 by Chapman & Hall, London. ISBN 0 412 48980 5

Table 4.1 Authigenic clay minerals in recent marine and lacustrine sediments

Group	Species	Model environment	Example	References
Smectite	Fe(II)/Mg-rich	Si, Mg-rich anoxic hydrothermal	Red Sea	Cole (1988)
Smectite	Fe(II)/Fe(III)-rich	Si, Fe-rich anoxic/oxic hydrothermal	Red Sea	Badaut *et al.* (1985) Cole (1988)
Smectite	Nontronite	Si, Fe-rich oxic hydrothermal	Galapagos Mounds	Corliss *et al.* (1978)
			Famous Area, Mid-Atlantic Ridge	Hoffert *et al.* (1978)
		Si, Fe-rich oxic lacustrine	Lake Chad	Pedro *et al.* (1978)
Smectite	Fe(III)-rich	Si, Fe-rich oxic pelagic	Bauer Deep	Cole (1985)
Smectite	Fe(III)/Al-rich	Si, Fe, Al-rich oxic pelagic	Central Pacific	Hein *et al.* (1979) Aplin (1983)
Smectite	Mg-rich	Si, Mg-rich evaporative	Lake Chad	Tardy *et al.* (1974) Gac *et al.* (1977)
?Smectite	'Glauconitic smectite'	Si, Fe-rich sub-oxic continental shelf edge	Gulf of Guinea	Odin (1988b)
?Mica	'Glauconitic mica'	Si, Fe-rich sub-oxic continental shelf edge	Gulf of Guinea	Odin (1988b)
Chlorite	Al-rich	Al-rich, anoxic continental shelf	East China Sea	Mackin and Aller (1984a)
Kaolin–Serpentine	Odinite	Fe, Al-rich sub-oxic continental shelf	Senegalese continental shelf	Odin and Masse (1988)
?	Phyllite C	Si, Fe, Al-rich sub-oxic continental shelf	Senegalese continental shelf	Odin and Masse (1988)
Pyrophyllite–Talc	Talc	Si, Mg-rich high temperature hydrothermal	Red Sea	Zierenberg and Shanks (1983)
Palygorskite–Sepiolite	Palygorskite, Sepiolite	Low/moderate temperature alteration of volcanic debris (?)	Atlantic fracture zones	Bowles *et al.* (1971)

see: Bischoff, 1972; Elderfield, 1976; Kastner, 1981; Odin and Matter, 1981; Cole and Shaw, 1983; McMurtry *et al.*, 1983; Odin, 1988a; 1990). Chlorite has been reported in older literature (Rohrlich *et al.*, 1969; Swindale and Fan, 1967), but its precise mineralogy was often unspecified. Furthermore, more recent work has cast doubt on whether chlorite does actually form in surficial sediments (Odin, 1988a; 1990).

Most of the minerals listed in Table 4.1 are extensively described in standard texts (Brindley and Brown, 1984; Newman, 1987; Chamley, 1989). Exceptions are the 'green marine clays': glauconitic minerals, odinite and phyllite C, which have been the subject of extensive work by Odin and his co-workers (see Odin, 1988a). Odin has defined two marine facies, both of which are characterized by the occurrence of green grains, but green grains which have different mineralogical and chemical compositions. The *glaucony* facies is characterized by glauconitic minerals whilst the *verdine* facies is characterized by odinite and phyllite C. Glauconitic minerals exhibit a spectrum of chemical and physical properties between dioctahedral smectitic and micaceous (glauconite *sensu strictu*) end-members and may be thought of in the same way as mixed-layer illite–smectite. Odinite is a recently defined ferric-magnesian, dioctahedral–trioctahedral 1:1 serpentine-like mineral with a 7 Å basal spacing (Odin, 1990). Previously, this mineral was known as 'phyllite V' (Odin, 1985), but phyllite V is now thought to be a mixture of odinite, a ferric chlorite, a 7–14 Å mixed-layer clay and a pyrophyllite-like clay (Odin, 1990). The precise nature of phyllite C is enigmatic but it has properties which are intermediate between those of a smectite and a swelling chlorite (see Odin, 1988a, for greater detail).

Until Odin's investigations, berthierine (a 7 Å trioctahedral serpentine mineral) was thought to be the green mineral which gives the verdine facies its name (Porrenga, 1967). However, Odin (1988a, and references therein) suggests that berthierine is rarely found in recent sediments. The most compelling evidence for this is that the mineral found in 'verdinized' recent sediments (odinite) is greatly enriched in Fe(III) compared with berthierine, in which iron occurs as Fe(II).

4.2.2 Geographical distribution of minerals

It is striking that each group or species of authigenic clay mineral occurs in specific geographical regions of the modern ocean (Table 4.1; Figure 4.1). Odinite and phyllite C are characteristic of sediments accumulating between 10 and 60 m water depth on tropical continental shelves. Glauconitic minerals occur mainly between 60 and 500 m water depth at latitudes between 0 and 60°. Both glauconitic minerals and odinite occur as pellets, or replacing sand-sized minerals, or on hardgrounds, implying that both form preferentially in environments which are at least periodically subjected to wave or current activity.

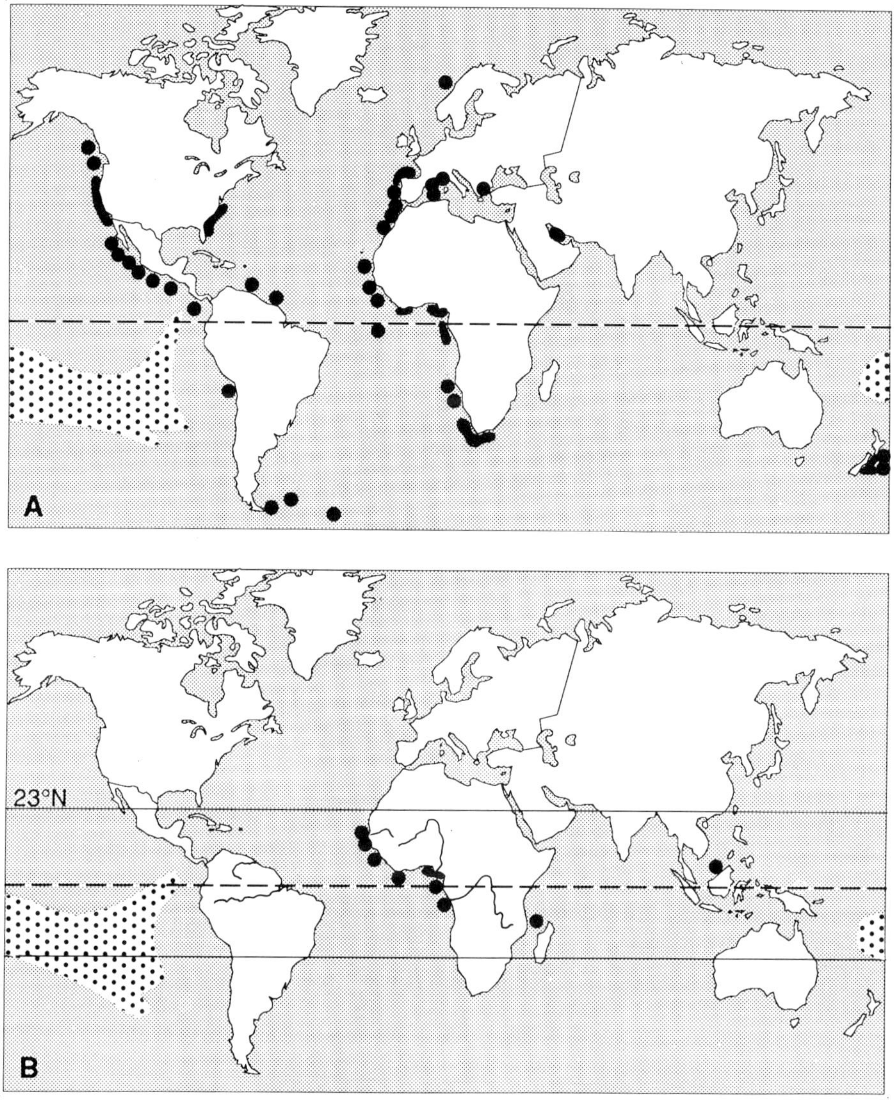

Figure 4.1 Distribution of (A) authigenic glauconitic minerals (glaucony facies), (B) odinite (verdine facies) and smectite (speckled) in marine surface sediments. Modified from Chamley (1989).

Smectite is the major authigenic clay mineral in pelagic sediments, where hydrothermal solutions debouch into the ocean, and in lacustrine sediments (Table 4.1). Authigenic palygorskite and sepiolite typically occur in evaporative environments (alkaline lakes and perimarine environments).

4.2.3 *Chemistry*

Representative chemical analyses of early diagenetic clay minerals are given in Table 4.2 and are displayed as a series of triangular diagrams in Figure 4.2. Specific mineral genera occupy particular regions of the seafloor, as do individual species. One example of this is smectite in sediments from the Pacific Ocean, where there is geographical uniformity of composition with clear species distributions (Figure 4.3; Tables 4.1 and 4.2). Variations in the aluminium and iron contents of the smectites are especially striking and, based on the available data, appear to be related to the proximity to centres of hydrothermal activity. On the Galapagos spreading centre, where iron and silica-rich hydrothermal solutions bleed on to the ocean floor, Fe(III) smectite (nontronite) dominates (McMurtry *et al.*, 1983). In the Bauer Deep, where hydrothermally derived iron oxides from the nearby East Pacific Rise are ponded, smectite is somewhat more aluminous, but is still dominated by ferric iron (Cole, 1985). In the Central North and South Pacific, which are unaffected by hydrothermal activity, smectite contains more aluminium than iron (Hein *et al.*, 1979). All the smectites are dioctahedral and contain no ferrous iron.

Trioctahedral smectites are rather rare in recent sediments. Magnesium-rich varieties occur in areas where the magnesium content of ambient waters is enhanced by evaporation (saline lakes; Tardy *et al.*, 1974) or hydrothermal activity (Red Sea; Cole, 1988), whilst Fe(II)-rich smectites have been isolated from sediments underlying the anoxic brine pools in the Red Sea (Badaut *et al.*, 1985; Cole, 1988).

Glauconitic minerals can be distinguished from early diagenetic smectites by their slightly lower SiO_2 contents and the occurrence of small amounts of ferrous iron and high concentrations of potassium. Odinite and phyllite C are compositionally distinct both from early diagenetic smectites and glauconitic minerals. They are depleted in SiO_2, enriched in Al_2O_3 and contain very significant amounts of both MgO and FeO (Figure 4.2).

4.3 Controls on mineralogy and chemistry

When they are first deposited, marine and lacustrine muds comprise a complex mixture of crystalline and amorphous phases, and organic matter. It is impossible for all these phases to be simultaneously in equilibrium with the aqueous phase. Diagenetic reactions are thus stages by which the mineral mixture progresses towards thermodynamic stability. These reactions initially involve the least stable, most soluble, often X-ray amorphous phases (Garrels, 1986). Characterizing these unstable phases, and defining their sources and distribution in ocean basins, is central to any understanding of the controls on the mineralogy and chemistry of early diagenetic clay minerals.

Table 4.2 Major oxide compositions of authigenic clay minerals in recent marine and lacustrine sediments

Group	Species	Example	SiO_2	Al_2O_3	Fe_2O_3	FeO	MgO	K_2O	Na_2O	CaO	Reference
Smectite	Fe(II)/Mg-rich	Red Sea	45.1	4.8	8.0	n.d.	22.9	0.0	3.4	0.8	Cole (1988)
Smectite	Fe(II)/Fe(III)-rich	Red Sea	43.2	2.4	33.0	n.d.	1.8	0.2	3.5	0.9	Cole (1988)
Smectite	Nontronite	Galapagos Mounds	51.6	0.1	26.5	n.d.	3.2	3.3	0.1	0.3	McMurtry *et al.* (1983)
Smectite	Fe(III)-rich	Bauer Deep	53.3	5.4	19.2	n.d.	4.8	0.6	1.4	0.3	Cole (1985)
Smectite	Al/Fe(III)-rich	Central Pacific	49.7	15.2	10.9	n.d.	4.1	1.5	2.7	0.8	This paper
Smectite	Mg-rich	Lake Chad	57.1	5.9	2.8	n.d.	17.9	0.2	0.9	0.2	Tardy *et al.* (1974)
Smectite (?)	'Glauconitic smectite'	Gulf of Guinea	46.9	6.9	21.2	0.9	3.6	3.4	0.3	1.4	Odin (1988b)
Mica (?)	'Glauconitic mica'	Galician Margin	43.4	6.0	21.3	1.5	3.5	8.2	0.2	0.7	Odin and Lamboy (1988)
Kaolin–Serpentine	Odinite	Senegalese shelf	34.3	9.2	21.4	6.5	13.0	0.5	0.2	0.0	Odin and Masse (1988)
?	Phyllite C	Senegalese shelf	41.9	9.9	17.4	4.1	11.2	0.5	0.0	0.0	Odin and Masse (1988)
Pyrophyllite–Talc	Talc	Red Sea	49.5	3.1	0.2	3.9	25.9	0.1	2.2	0.2	Zierenberg and Shanks (1983)
Palygorskite–Sepiolite	Palygorskite	Marianas volcanics	59	11	4	n.d.	9	1	n.d.	1	Desprairies (1981)
Palygorskite–Sepiolite	Sepiolite	Mid-Atlantic Ridge	58.6	0.4	0.2	0.05	25.4	n.d.	n.d.	0.3	Hathaway and Sachs (1965)

n.d. = not determined. Where there are no data for FeO, total iron is given as Fe_2O_3.
All data normalized to 85% total oxides.

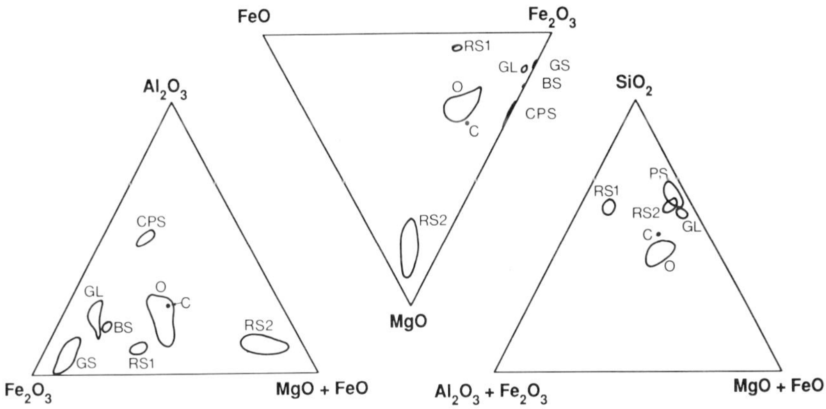

Figure 4.2 Triangular diagrams displaying the known range of chemical composition of authigenic smectite, odinite, glauconitic minerals and phyllite C in marine surface sediments. Data from McMurtry *et al.* (1983); Hein *et al.* (1979); Cole (1985; 1988); Odin (1988a) and from this paper. CPS = Central Pacific smectite; GL = glauconitic minerals; BS = Bauer Deep smectite; GS = Galapagos Mounds smectite; O = odinite; C = phyllite C; PS = Galapagos smectite + Bauer smectite + Central Pacific smectite; RS1 = Fe(II)/Mg-rich smectite, Red Sea; RS2 = Mg-rich smectite, Red Sea.

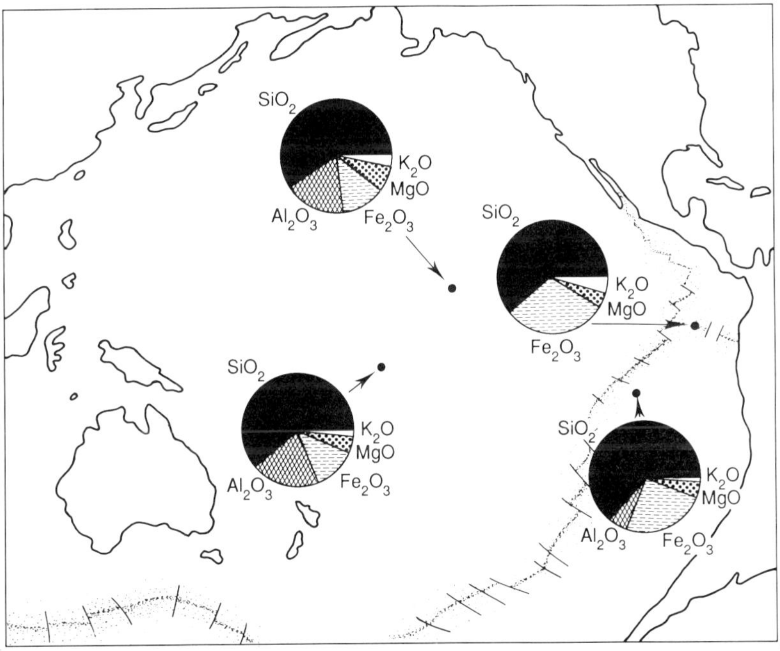

Figure 4.3 Geographical trends in the major oxide composition of authigenic smectite isolated from surface sediments of the Pacific Ocean. Data from McMurtry *et al.* (1983); Hein *et al.* (1979); Cole (1985) and from this paper.

4.3.1 Sources of reactive components

Early diagenetic clay minerals may be considered as a mixture of the oxides and hydroxides of silicon, aluminium, iron, magnesium and potassium. Of these, magnesium and potassium generally occur in high concentrations in marine systems and are only likely to exert a controlling influence on the nature of authigenic clay minerals where their concentrations are altered by evaporation or hydrothermal activity. In contrast, silicon, aluminium and iron are only sparingly soluble in seawater and occur mainly as solid phases: as aluminosilicates (Si, Al, Fe), oxyhydroxides (Fe, Al), and biogenic opal and quartz (Si). The availability of these components in the marine environment may be crucial, in many sediments, in determining the nature and distribution of early diagenetic clay minerals.

Key reactants are delivered to the ocean via two main routes: by hydrothermal activity, mainly at plate margins (Degens and Ross, 1969; Edmond *et al.*, 1982), and from the continents, by rivers and as atmospheric dust (Martin and Meybeck, 1979; Milliman and Meade, 1983; Uematsu *et al.*, 1983). Seafloor hot springs can be considered as point sources of silicon and iron plus, in the Red Sea, of magnesium. Iron is transported as Fe(II) and precipitates sequentially as sulphides, silicates and oxides as the solutions cool and are oxidized by contact with seawater (Corliss *et al.*, 1978). Where buoyant hydrothermal plumes eject material high above the ocean floor, deep ocean currents disperse the fine-grained iron oxides. This is most clearly seen in the eastern Pacific, where iron oxide-rich sediments accumulate regionally along the hydrothermally active East Pacific Rise, and in the adjacent Bauer Deep (Bostrom and Peterson, 1966; Dymond, 1981; Edmond *et al.*, 1982). We can think of these sediments as ones in which an additional flux of hydrothermally derived iron (and manganese) oxides is superimposed upon the normal Pacific pelagic background sedimentation of detrital minerals, oxyhydroxides and organic matter.

Materials delivered to the ocean through rivers and the atmosphere are the products of weathering on the continental surface. In its most simple expression, weathering can be considered as the titration of atmospheric CO_2 (carbonic acid) by cation-bearing silicates:

$$\text{cation-bearing aluminosilicate} + \text{carbonic acid} \rightarrow$$
$$\text{aluminosilicate residue} + \text{cations} + \text{silica} + \text{bicarbonate} + \text{aluminium.}$$

Increasingly intense chemical weathering leads to the progressive transformation and loss of cation-bearing silicates (ferromagnesian minerals, feldspars, mica, illite, smectite and hydroxy-interlayer smectite, vermiculite) from the soil profile and the corresponding enrichment of their reaction products: kaolinite, gibbsite and iron-manganese oxides.

Fluvial particulates are an integrated record of weathering in those parts of the catchment which supply sediment to the river system. In general, minerals resulting from intense chemical weathering are concentrated in rivers draining humid, tropical areas, whereas less stable minerals are relatively enriched in rivers draining cooler catchments. For example, minerals in the St. Lawrence are dominated by mica/illite, chlorite, feldspars and quartz, reflecting cool-climate weathering in the Canadian interior (Figure 4.4). In contrast, the clay mineralogy of the tropical Niger River is dominated by kaolinite, and includes gibbsite (Konta, 1985).

It is important to remember that rivers which enter the ocean within tropical zones do not necessarily carry a suite of minerals derived from weathering in hot and humid climates. This is illustrated by the Ganges, which carries large amounts of mica, chlorite and feldspar, most of which are derived by physical weathering in the Himalaya (Figure 4.4).

The chemical composition of fluvial suspensates also reflects the nature of the weathering process. Increased chemical weathering leaves an aluminium + iron-oxide-enriched residue, culminating in the formation of laterites (Persons, 1970). Suspensates in rivers draining humid, tropical climatic zones (like the Niger and Orinoco) are correspondingly enriched in aluminium and depleted in sodium, potassium and calcium compared with rivers draining cooler and/or drier catchments (Konta, 1985). And, although suspensates in tropical rivers may not show an absolute enrichment in iron, it is likely that the nature of the iron residues will be very different from those in rivers draining cool climatic zones. Labile oxides and oxyhydroxides will be the dominant ferroan phases in the tropical rivers, with a greater proportion of relatively refractory iron-bearing silicates (like chlorite) in rivers carrying sediment derived from weathering in cool climates.

Many of the minerals found in marine sediments (for example quartz, chlorite, mica, smectite and feldspars) persist beyond early diagenesis and are deeply buried. They may not be thermodynamically stable in marine sediments but their rate of reaction is sufficiently slow that they do not participate in the formation of early diagenetic, authigenic clays. However, a significant proportion of fluvial suspensates are effectively X-ray amorphous. Figure 4.5 shows X-ray diffractograms of suspensates from the Nile and Orinoco rivers (Konta, 1985). Peaks representing crystalline phases are perched on a broad background indicative of poorly crystalline material. Such material is poorly characterized but includes degraded aluminosilicates, iron, manganese and aluminium oxyhydroxides, opal, and organic matter. Its importance lies not only in its apparent abundance, but also in its likely chemical reactivity. X-ray amorphous material is amongst the first to disappear during mud and mudstone diagenesis (Foscolos and Powell, 1980), and is almost certainly an important feedstock for the growth of early diagenetic clay minerals.

Rivers, like hot springs, are point sources of key reactive components. And, like the hot springs, much of the material delivered to the ocean by rivers is dumped close to source: in the nearshore environment. Within estuaries, reactive iron and aluminium compounds flocculate with organic matter and are deposited in underlying sediments (Sholkovitz, 1976).

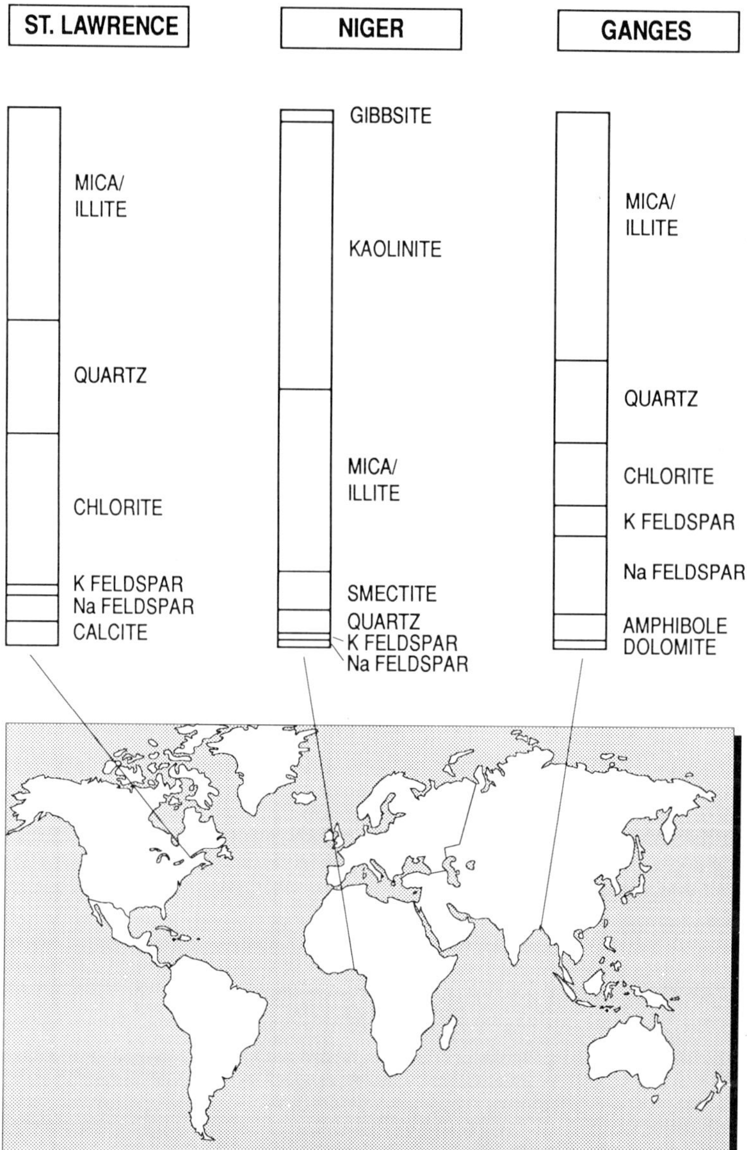

Figure 4.4 Semi-quantitative mineralogy of crystalline suspensates from the Ganges, St. Lawrence and Niger rivers. Data from Konta (1985). Rivers deriving sediment from hot, humid climates (Niger) carry suspensates which are depleted in cation-bearing silicates and enriched in aluminium-rich phases like kaolinite and gibbsite.

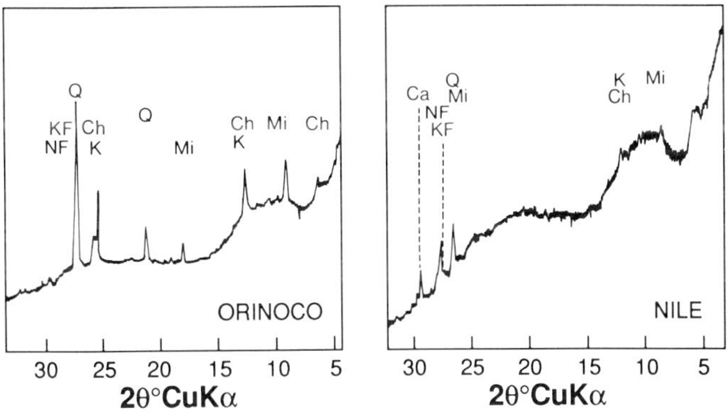

Figure 4.5 X-ray diffractograms of suspensates from the Nile and Orinoco rivers (data from Konta, 1985).

Material transported to the ocean via the atmosphere is rather poorly characterized but is likely to be compositionally similar to that delivered by rivers. In particular, it can form an important component of pelagic sediments (Blank *et al.*, 1985).

Additional key reactants are produced by biological activity within the marine (and lacustrine) environment. The most important of these are organic matter and opaline silica. The flux of organic matter to the sediment–water interface depends on its rate of production in surface waters and its rate of destruction during vertical transport through the water column. High fluxes of organic matter are thus favoured by high productivity and/or shallow waters, and tend to occur on continental shelves and underneath upwelling systems (Suess, 1980).

Opal forms the skeleton of diatoms and radiolaria and is produced throughout the oceans, and in lakes. Opal-rich sediments accumulate under those areas of the pelagic ocean where productivity is relatively high but where carbonate is not accumulating (Lisitzin, 1972). However, its high concentration in such sediments probably reflects its high rate of accumulation relative to other components, rather than an especially high absolute rate of production. In many sediments, particularly nearshore sediments, its relative abundance is diminished by the high accumulation rates of other sedimentary components.

4.3.2 Compositional links between reactants and products

Figure 4.6 illustrates the basic thesis of this paper: that a link exists between: (1) the formation of particular clay mineral species in particular regions of the world ocean, and (2) the relative rates of supply of a diverse suite of poorly crystalline and dissolved reactant phases (also see Table 4.1).

Smectites form in environments where reactive silica, either as opal or in hydrothermal solutions, is relatively abundant. The particular species of

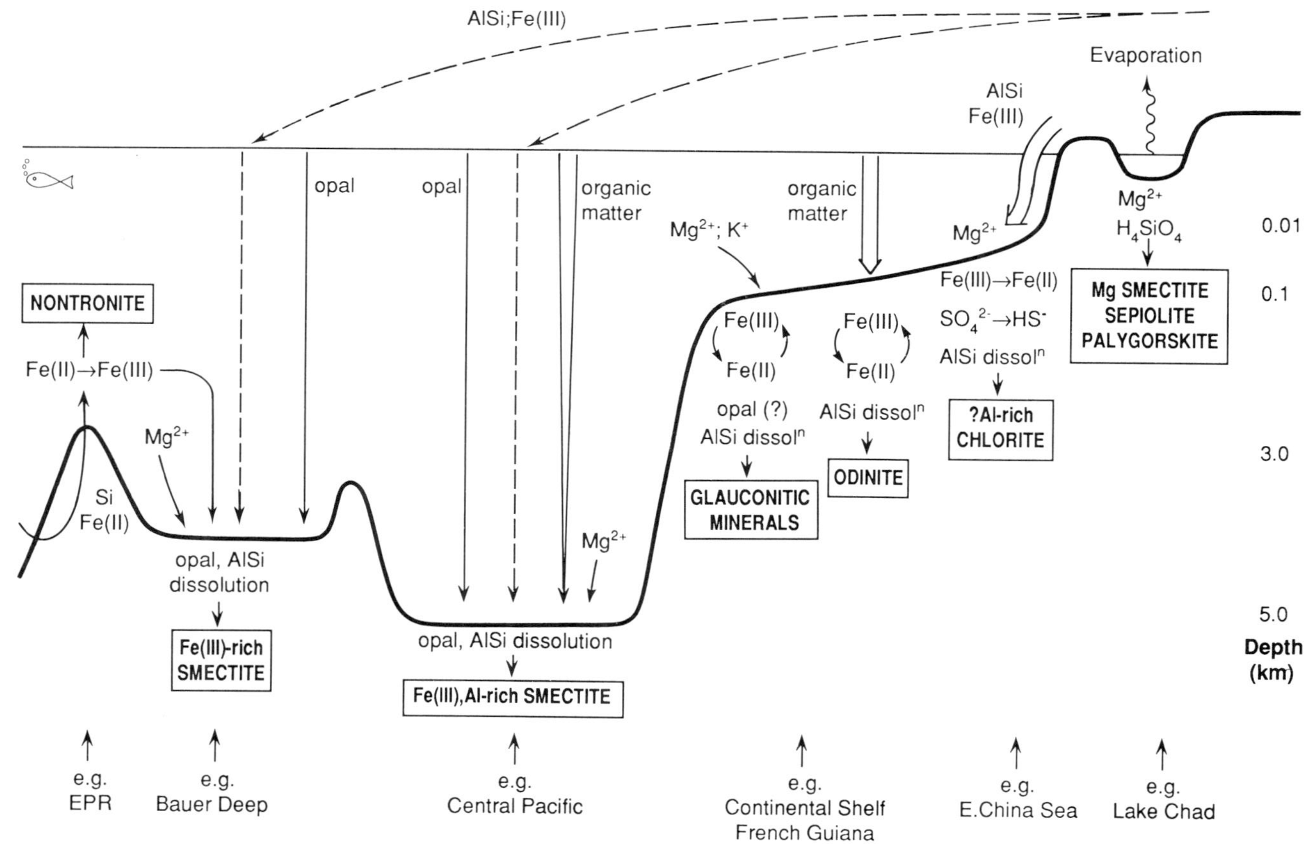

Figure 4.6 Proposed links between; (1) the supply of reactive materials (opal, degraded aluminosilicates (AlSi), iron and aluminium oxyhydroxides, organic matter), and (2) the mineralogy of early diagenetic clays.

authigenic smectite depends on the composition of reactants supplied along with the silica. Smectites forming in the Atlantis II Deep in the Red Sea provide a good example of this. There, hot (>200°C), anoxic brines are discharged from the underlying basement and are ponded beneath oxygenated seawater in naturally occurring depressions on the seafloor (Degens and Ross, 1969). The brines contain high concentrations of silica, reduced iron and (often) magnesium. Petrographic, chemical and isotopic data show that distinct species of smectite form as the hydrothermal waters are cooled and oxidized by contact with overlying seawater (Cole, 1983; 1988; Zierenberg and Shanks, 1983; Badaut *et al.*, 1985). Trioctahedral, magnesium-rich smectite (Fe(II)-bearing saponite) and talc form in anoxic, high-temperature (>150°C) conditions close to hydrothermal vents. Removal of magnesium from solution and partial oxidation of Fe(II) to Fe(III) leads to the formation of a mixed trioctahedral–dioctahedral iron-rich smectite at temperatures between about 80 and 140°C. Finally, when solutions have cooled to about 80°C and iron occurs mainly as Fe(III), dioctahedral nontronite forms.

The Red Sea is in two respects a rather unusual seafloor hydrothermal system. First, the hydrothermal fluids often contain magnesium; secondly, anoxic conditions occur above the sediment–water interface. At mid-ocean ridges, it is more typical for hydrothermal solutions to be depleted in magnesium and to debouch directly into oxygenated seawater (Edmond *et al.*, 1982). These solutions are also greatly enriched in iron relative to aluminium (Michard *et al.*, 1984). In these systems reduced iron is precipitated mainly as sulphides and magnesium-smectites are absent. Any iron which is not precipitated as sulphide is rapidly converted to oxyhydroxides and either combines with silica to form nontronite (as in the Galapagos hydrothermal mounds; Corliss *et al.*, 1978), or is dispersed by ocean currents.

In the Pacific Ocean, hydrothermally derived iron oxides occur in sediments accumulating along the East Pacific Rise and adjacent Bauer Deep. These deep-water sediments are oxic and, in addition to hydrothermally derived material, also contain abundant opal and lesser amounts of terrestrially derived aluminosilicates. Authigenic Fe(III)-rich smectite is common and contains small but significant amounts of aluminium (Cole, 1985; Table 4.2). Within pelagic sediments in the Central Pacific, away from the influence of hydrothermal activity, an Al–Fe(III) smectite forms. Like the Bauer Deep sediments, these sediments are both oxic and opaline, but contain relatively more terrestrially derived aluminosilicates and fewer (if any) hydrothermally derived iron oxyhydroxides (Aplin and Cronan, 1985).

Lake sediments also provide good examples of the links that exist between the composition of detrital reactants and authigenic smectites. Studies of sediments in Lake Chad have shown that Fe(III) smectite (nontronite) is the main authigenic clay close to sources of freshwater input, and that magnesium-smectite (stevensite) dominates the mineral assemblage in the distal, evaporative parts of the lake (Tardy *et al.*, 1974; Gac *et al.*, 1977; Pedro *et al.*, 1978).

Draining a tropical area, iron oxides are likely to be a significant constituent of material delivered to the lake. These oxides are sedimented close to river mouths and are converted to nontronite by the addition of silica, perhaps from dissolving opal (Pedro *et al.*, 1978). There are clear parallels here with the marine environment. In the evaporative parts of the lake, the input of detrital material is low, and evaporation leads to the increase of Mg^{2+}, silica and pH in lake waters (Gac *et al.*, 1977). Here, magnesium-smectite forms, perhaps as a two-stage process involving: (1) the precipitation of brucite ($Mg(OH)_2$), and (2) the adsorption of dissolved silica and conversion of brucite to smectite (Tardy *et al.*, 1974).

Terrestrially derived aluminosilicate debris, iron and aluminium oxyhydroxides and organic matter are all supplied rapidly to nearshore sediments. In general, the authigenic clays forming in these sediments (odinite, phyllite C and perhaps, chlorite) are more aluminous than deep-sea smectites, reflecting the relatively greater supply of reactive alumina compared with silica. The formation of particular mineral species seems to be controlled by the chemical composition of the terrestrial input plus the redox status of the sediments (Figure 4.6; Table 4.1). In sediments accumulating rapidly in the East China Sea, Mackin and Aller (1984a) suggested that an aluminous dioctahedral chlorite might be the main authigenic clay mineral. The detrital phases in these sediments are derived from rivers draining the Chinese mainland and are of average chemical maturity (Li *et al.*, 1984). Pore-water concentrations of dissolved iron are low (Mackin and Aller, 1984a). In these sediments, where the supply of reactive aluminium is high relative to silica and iron, aluminous chlorite may be the dominant authigenic clay. The same logic can be extended to argue that in sulphidic, nearshore sediments, reduced iron would be precipitated as pyrite with the result that aluminous clay minerals would form preferentially.

Aluminium and iron oxyhydroxides are major weathering products in tropical climates and are the apparent feedstock for the formation of aluminium- and/or iron-rich odinite and phyllite C on shallow, tropical continental shelves. These minerals define a particular redox niche because they contain both ferrous and ferric iron. In surface sediments organic matter is oxidized by bacterially catalysed reactions involving the sequential reduction of oxygen (oxic zone), manganese oxides, nitrate and iron oxides (sub-oxic zone) and sulphate (anoxic zone; Froelich *et al.*, 1979). As they contain both ferrous and ferric iron, odinite and phyllite C must form close to the oxic/sub-oxic redox boundary. Furthermore, specific redox conditions must be maintained for long periods in order for the clay minerals to develop. Such conditions must be maintained by a delicate balance between the supply of organic matter, via biological productivity, and the supply of oxygen by current and wave activity. In many muddy nearshore sediments, the sub-oxic zone is thin and sulphate reduction is the dominant oxidation process. In these sediments iron is rapidly stripped from pore waters by reaction with sulphide. Odinite and phyllite C form in sediments in which a significant supply rate of organic matter is balanced by a supply of oxygen from current activity. Both mineral species form in microenvironments (pellets, intragranular cracks) in which there is restricted exchange of solutes with overlying seawater. These may be sub-oxic niches in an otherwise oxic environment.

Both compositionally and geographically, glauconitic minerals are intermediate between Fe(III) smectite and odinite (Figures 4.1, 4.2 and 4.6; Tables 4.1 and 4.2). Compared with smectite, glauconitic minerals are somewhat depleted in silicon and contain small but significant amounts of Fe(II). Compared with odinite, they are enriched in silicon and depleted in Fe(II). The formation of glauconitic minerals is favoured in sediments which are: (1) marginally or occasionally sub-oxic (moderate supply of organic matter), and (2) deposited away from point sources of aluminium-rich silicates and oxyhydroxides. Like odinite, glauconitic minerals often form within pellets and microfossil tests which may be sub-oxic microenvironments in a generally oxic environment.

4.4 Involvement of the aqueous phase

It seems that the composition of early diagenetic clay minerals is significantly influenced by the composition of the most soluble, reactive components delivered to the sediment. However, at low temperatures the energetics of solid-state reactions are rarely favourable. It is more likely that the reactions by which unstable precursors are converted to authigenic clay minerals proceed through the aqueous phase. Accordingly, we now explore the mechanism(s) by which authigenic clays form in the early diagenetic environment, focussing on the role of pore waters.

A wide variety of common clay minerals has been rapidly synthesized in the laboratory (Harder, 1972; 1978; Huang *et al.*, 1986; Decarreau *et al.*, 1987). Of special relevance here is Harder's synthesis of dioctahedral and trioctahedral smectites, and berthierine, by the low-temperature ageing of magnesium, aluminium and iron oxyhydroxides in solutions containing low concentrations of dissolved silica. Harder proposed that the oxyhydroxides adsorbed silica from solution and were slowly converted to smectite. Cole (1985) used SEM evidence to argue that the abundant smectite in the Bauer Deep formed in a similar way. In this case silica was derived from the dissolution of opaline skeletal tests, whilst iron oxyhydroxides were supplied by hydrothermal activity on the adjacent East Pacific Rise. A similar mechanism was proposed to account for smectite formed authigenically in the North Pacific (Hein *et al.*, 1979).

Determining the mechanism by which smectite forms in South Pacific sediments is complicated by the presence of volcanic debris. Since South Pacific sediments contain more smectite than any others in the world ocean, and since smectite is known to be a low-temperature alteration product of volcanic ash, the high smectite contents of South Pacific sediments have generally been ascribed to the alteration of volcanic debris (e.g. Griffin *et al.*, 1968). However, there is good evidence that much of the smectite formed via other routes. Because Quaternary volcanics in the region are titanium-rich (McBirney and Aoki, 1968; Natland, 1980), the titanium content of surface sediments is proportional to the abundance of volcanic debris. However, the proportion of smectite in the smaller than 2 μm sediment fraction is inversely related to the titanium content of the sediment (Figure 4.7). Rather, there is a strong relation between the P_2O_5 and smectite contents of surface sediments (Figure 4.7). As the P_2O_5 content of pelagic sediments is thought to be inversely proportional to their sedimentation rate (Bischoff *et al.*, 1979), this implies that smectite is enriched in more slowly accumulating sediments. Thus, the common occurrence of smectite in sediments which are both slowly accumulating and relatively depleted in volcanic debris suggests that a very significant fraction of the smectite is formed by mechanisms other than the alteration of volcanic debris. The most probable mechanism is that suggested for other pelagic Pacific sediments: the combination of silica (from dissolving opaline tests) with iron (and aluminium) oxyhydroxides. This suggestion is supported by (1) microscopic and chemical evidence for the presence of the relevant reactants (Aplin, 1983; Aplin and Cronan, 1985), and (2) the chemical composition of smectites in South and North Central Pacific sediments, which is much more homogeneous than that of smectites formed by the oxic alteration of volcanic debris (Table 4.3).

Comparison of the chemical composition of coexisting sediment oxyhydroxides and smectites gives additional insight into the mechanism by which smectite forms. If smectite formed by the simple adsorption of silica by

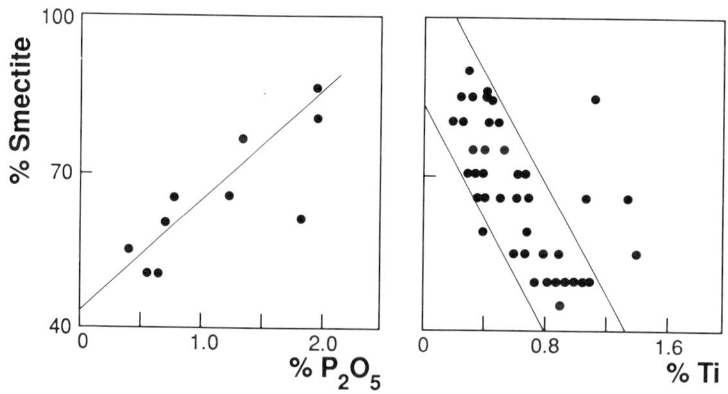

Figure 4.7 Proportion of smectite in the smaller than 2 μm fraction of sediments from the Central South Pacific plotted against the P_2O_5 and Ti contents of bulk sediments. Data from Aplin (1983). P_2O_5 is inversely proportional to sedimentation rate (Bischoff *et al.*, 1979) and titanium is proportional to volcanic debris. Smectite is most enriched in those slowly accumulating sediments which are relatively depleted in volcanic debris.

Table 4.3 Major oxide composition of smaller than 0.5 μm fractions of smectite-rich Central Pacific sediments and altered basalts

Sample	Location	%S	%I	%K+C	SiO_2	Al_2O_3	Fe_2O_3	MgO	CaO	Na_2O	K_2O	TiO_2	MnO
Sediments													
M70-33	6°S 172°W	95	3	2	44.8	13.3	10.6	3.5	1.3	2.6	1.3	0.8	0.1
M70-34	4°S 169°W	98	1	1	49.3	12.0	7.7	3.7	1.0	1.5	1.0	0.5	0.1
Cl-13	6°S 159°W	95	0	5	42.6	13.0	8.3	3.8	6.7	2.5	1.1	0.7	0.3
K3-1	7°S 149°W	90	4	6	45.9	13.2	8.8	3.9	3.4	2.3	0.9	0.6	0.2
Altered basalts													
051 2/7	13°S 151°W	95	2	2	42.9	6.9	17.2*	3.5	0.7	2.0	2.7	2.3*	0.1
060 2/4	13°N 167°W	98	2	0	42.6	14.8	7.4	2.6	4.7†	2.1	1.4	0.8	0.0
063 8/9	16°N 168°N	100	0	0	44.8	14.4	12.2	2.6	0.2	2.2	2.6	1.0	0.1

*Contaminated with Fe–Ti oxides.
†Contaminated with francolite.
All samples leached with citrate-dithionite-bicarbonate. S = smectite, I = illite, K + C = kaolinite + chlorite

oxyhydroxides followed by a solid-state transformation to smectite, one might expect the ratios of metals in smectite and oxyhydroxide to be similar. Data for sediments from the Bauer Deep and the South Central Pacific indicate that this is not the case. In particular, the Al/Fe ratios of oxyhydroxides (defined as material soluble in hydroxylamine hydrochloride) in Bauer Deep and Central Pacific sediments are greater than those in coexisting smectites (Figure 4.8). This suggests that the formation of smectite is a reaction in which the aqueous phase plays a key role.

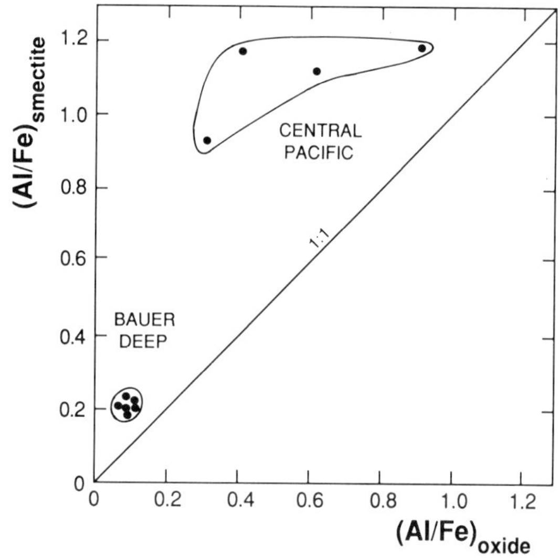

Figure 4.8 Cross-plot of the weight ratios of aluminium and iron in smectites and oxyhydroxides coexisting in sediments from the Central Pacific and Bauer Deep. Data from Varnavas (1979); Cole (1982); Aplin and Cronan (1985) and from this paper.

Limited oxygen isotope data support the assertion that smectite authigenesis involves the aqueous phase. Cole (1985) has shown that smectite in Bauer Deep sediments is in isotopic equilibrium with ambient pore waters. At low temperatures, oxygen isotopic exchange or isotopic shifts will only occur during chemical reactions; that is, when cation–oxygen bonds are broken (O'Neil, 1986). If smectite had formed by the solid-state transformation of silica-bearing oxyhydroxides, then the smectite would have inherited, at least partly, the isotopic composition of the oxyhydroxide. By analogy with oxides in ferromanganese nodules, the isotopic composition of the oxyhydroxides probably falls between +10 and +15‰ (Savin, 1980). This compares with +29‰ for Bauer Deep smectite, compatible with formation from seawater at 3°C (Cole, 1985). The smectite is fully equilibrated with ambient pore fluids, implying that it formed by a dissolution–reprecipitation mechanism, not solid-state transformation. The differing chemical compositions of precursor oxyhydroxides and authigenic smectite further suggest that aluminium is preferentially incorporated into the silicate structure during this process.

If clay authigenesis involves the aqueous phase, then the chemical composition of associated pore waters should reflect that process. What are required, ideally, are pore-water data for all the major clay-forming components (silica, aluminium, magnesium and iron) from sediments in which authigenic clay minerals have been chemically and mineralogically characterized. What actually exist are data for some of the key species in pore waters where authigenic clays have not been characterized. Nevertheless available data suggest that aluminosilicates *are* reactive in seawater, and that clay authigenesis involves the aqueous phase.

Profiles of dissolved silica have been determined in a wide variety of marine sediments. All the profiles show similar features, with low concentrations in surface sediments increasing to constant values within a metre of the sediment–water interface (Hurd, 1973; Mackin and Aller, 1984a; 1984b; Figure 4.9). Maximum values vary from sediment to sediment but are not related to sediment type or depositional environment. Maximum values are also much greater than the equilibrium solubility of quartz, but much lower than the equilibrium solubility of amorphous silica (Lindsay, 1979). Silica is not buffered by quartz or opal, but more likely by an aluminosilicate. This is supported both by laboratory and field evidence. First, in experiments involving the dissolution of amorphous silica, Iler (1973) showed that the solubility of silica in aluminium-bearing solutions was lower than its solubility in pure water. Presumably, this reflected the formation of aluminosilicates. Secondly, some of the pore waters analysed by Hurd (1973) were extracted from pelagic sediments in which biogenic opal was dissolving and in which authigenic smectite was forming.

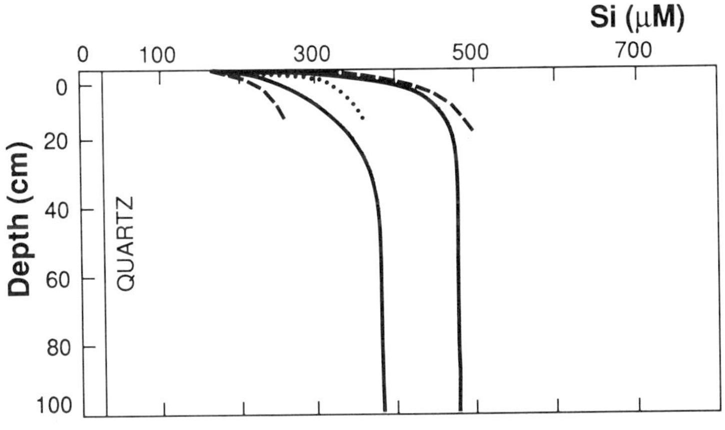

Figure 4.9 Profiles of dissolved silica in recent sediments. Data from Hurd (1973); Mackin and Aller (1984a; 1984b). Solid line: pelagic sediments from the Central Pacific; dashed line: shelf muds from the East China Sea; dotted line: estuarine mud from Mud Bay, eastern USA.

Unfortunately, there are no aluminium data from the pore waters of any sediments where authigenic clays have been fully characterized. The most complete set of pore-water aluminium data is probably that collected by Mackin and Aller (1984a; 1984b) from rapidly accumulating, nearshore and estuarine sediments. In the East China Sea, Mackin and Aller (1984a) found maximum aluminium concentrations in overlying seawater, declining downwards into the sediment pore waters. Silica showed the opposite trend, increasing with depth into the sediment (Figure 4.10). Mackin and Aller (1984a) interpreted these trends as indicating control of dissolved silica and aluminium by equilibrium with an authigenic clay mineral. In such rapidly accumulating sediments, it was impossible to identify any authigenic phase, but statistical analysis of the pore-water data suggested that the clay might have been an aluminium-rich dioctahedral chlorite.

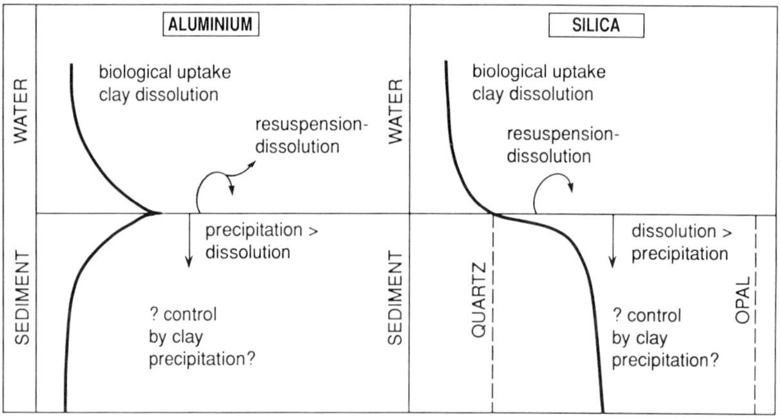

Figure 4.10 The cycles of aluminium and silica in recent muds (based on data in Mackin and Aller, 1984a).

Another interpretation of Mackin and Aller's (1984a) data would be that, rather than indicating true equilibrium between pore water and solid phases, a steady state exists between the dissolution of reactants (degraded aluminosilicates) and the precipitation of products (authigenic clay). Enhanced concentrations of aluminium occur in waters overlying both nearshore and pelagic sediments, suggesting that aluminosilicates are being dissolved (Mackin and Aller, 1984a; Orians and Bruland, 1986). Dissolution may continue within the sediment column, but if the rate of precipitation is greater than the rate of dissolution, concentrations of dissolved aluminium will decline. In contrast, increased levels of dissolved silica in pore waters suggest that the rate of dissolution of reactive silicon-bearing phases (opal, degraded aluminosilicates) is initially greater than the rate of uptake of silica into authigenic clays (Figure 4.10).

Although there is good evidence that clay mineral authigenesis involves aqueous reactants, the mechanisms by which the clays form are poorly constrained and may be different for different mineral species. In some cases aqueous reactants may nucleate to form an insoluble product, possibly a colloid. In a second step the colloid polymerizes and is reorganized into a crystalline lattice. This mechanism may apply to mineral species whose constituent components are reasonably soluble (Mg; Fe(II); Si). A second mechanism would involve the nucleation of an aqueous reactant on to a pre-existing solid phase. Again, a second step would involve polymerization and reorganization to yield a crystalline lattice. This mechanism may be pertinent for mineral species which contain a highly insoluble component such as Fe(III). Nevertheless, isotopic data suggest that if this mechanism is realistic, the polymerization step must involve dissolution and reprecipitation (see above).

The above discussion is restricted to processes which occur within a few centimetres of the sediment–water interface. Further dissolution–reprecipitation reactions involving clay minerals occur throughout the diagenetic zone (e.g. Hower *et al.*, 1976; Yeh and Savin, 1977). Some of these reactions occur close to the sediment–water interface and involve mineral species which are themselves early diagenetic products. For example, Odin and Fullagar (1988) have argued that the spectrum of glauconitic minerals on the seafloor represents an evolutionary trend from a potassium-poor glauconitic smectite to a potassium-rich glauconitic mica, whilst Odin (1990) suggests that odinite may convert to a mixture of ferric chlorite, a 7–14 Å mixed-layer phase and a pyrophyllite-like clay within a few thousand years. These reactions may be thought of as a second stage along a long diagenetic path by which terrestrial weathering products and hydrothermal precipitates are ultimately converted to low-grade metamorphic greenschists.

4.5 Summary

Although authigenic clay minerals probably form in most marine and lacustrine surface sediments, they are readily characterized only in slowly accumulating sediments and close to hydrothermal vents.

Available data suggest that authigenic clays display remarkable chemical and mineralogical coherence within ocean and lake basins. Furthermore, the chemical composition of the authigenic clays appears to reflect that of the least stable phases delivered to the sediment surface: degraded aluminosilicates, iron and aluminium oxyhydroxides, biogenic opal and organic matter. The balance of supply of these compounds is in turn controlled by: (1) their delivery to the ocean or lake from fluvial, aeolian and hydrothermal sources, (2) their mechanical and chemical sorting within the water column, and (3) biological activity in the water column (Figure 4.6). Climate plays an important role in that it controls the composition of the weathering products supplied to the river systems.

Experimental work demonstrates that clay minerals can form rapidly, given the necessary reactants. For example, Harder (1972; 1978) has shown that

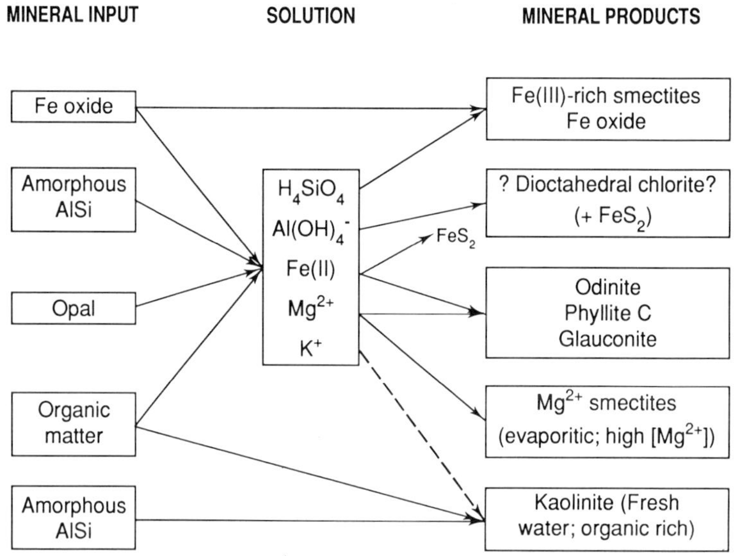

Figure 4.11 Possible routes by which reactive materials may be converted to more stable authigenic clays by reactions involving the aqueous phase. In general, the composition of the clay mineral depends on: (1) the composition of the most reactive material delivered to the sediment, and (2) the relative rates at which each of those materials are dissolved into the sediment pore waters.

smectites and chlorites can form at low temperature by the ageing of metal hydroxides in silica-bearing solutions. Although parallels may be drawn to the formation of clay minerals in the early diagenetic environment, chemical, isotopic and pore-water data all suggest that in natural systems clay-forming reactions proceed via the aqueous phase. Two reaction pathways may be proposed: one involves the dissolution of relatively soluble, unstable mineraloids (aluminosilicates, opal, iron oxyhydroxides under reducing conditions) and the subsequent precipitation of less soluble, more stable clay minerals directly from pore water. A second also starts with the dissolution of unstable aluminium- and silicon-bearing phases, but involves an intermediate stage in which dissolved silica, aluminium and magnesium are adsorbed by iron and aluminium oxyhydroxides. Subsequent 'ageing' of the oxyhydroxides, involving dissolution and reprecipitation, finally results in the formation of recognizable clay minerals. Crucially, the composition of reactants (unstable mineraloids) and products (authigenic clays) are linked through dissolution into, and precipitation from, the aqueous phase. These ideas are summarized in Figure 4.11.

Acknowledgements

The ideas presented in this paper germinated at Imperial College and matured at the Centre de Recherches Pétrographiques et Géochimiques (CRPG) in

Nancy and the BP Research Centre. Funds provided by NERC and the Royal Society during this time are gratefully acknowledged. I particularly thank Tom Cole for keen discussions over the years, and Harry Shaw, Tom Cole and Chas Curtis for constructive reviews of the manuscript.

References

Aplin, A. C. (1983) *The Geochemistry and Environment of Deposition of some Ferromanganese Deposits from the South Equatorial Pacific*, PhD Thesis, University of London.

Aplin, A. C. and Cronan, D. S. (1985) Ferromanganese oxide deposits from the Central Pacific Ocean, II. Nodules and associated sediments. *Geochimica et Cosmochimica Acta*, **49**, 437–51.

Badaut, D., Besson, G., Decarreau, A. and Rautureau, R. (1985) Occurrence of a ferrous, trioctahedral smectite in recent sediments of Atlantis II Deep, Red Sea. *Clay Minerals*, **20**, 389–404.

Bischoff, J. L. (1972) A ferroan nontronite from the Red Sea geothermal system. *Clays and Clay Minerals*, **20**, 217–23.

Bischoff, J. L., Heath, G. R. and Leinen, M. (1979) Geochemistry of deep-sea sediments from the Pacific manganese nodule province: DOMES sites A, B, and C, in *Marine Geology of the Pacific Manganese Nodule Province*, (eds J. L. Bischoff and D. Z. Piper), Plenum Press, New York, pp. 397–436.

Blank, M., Leinen, M. and Prospero, J. M. (1985) Major Asian aeolian inputs indicated by the mineralogy of aerosols and sediments in the western North Pacific. *Nature*, **314**, 84–6.

Bostrom, K. and Peterson, M. N. A. (1966) Precipitates from hydrothermal exhalations on the East Pacific Rise. *Economic Geology*, **61**, 1258–65.

Bowles, F. A., Angino, E. A., Hosterman, J. W. and Galle, O. K. (1971) Precipitation of deep-sea palygorskite and sepiolite. *Earth and Planetary Science Letters*, **11**, 324–32.

Brindley, G. W. and Brown, G. (eds) (1984) *Crystal Structures of Clay Minerals and their X-ray Identification*, Monograph Number 5, Mineralogical Society, London.

Chamley, H. (1989) *Clay Sedimentology*, Springer-Verlag, Berlin.

Cole, T. G. (1982) *Mineralogy and Geochemistry of Metalliferous Sediments from the Bauer Deep, Southeast Pacific, and the Atlantis II Deep, Red Sea*, PhD Thesis, University of London.

Cole, T. G. (1983) Oxygen isotope geothermometry and origin of smectites in the Atlantis II Deep, Red Sea. *Earth and Planetary Science Letters*, **66**, 166–76.

Cole, T. G. (1985) Composition, oxygen isotope geochemistry, and origin of smectite in the metalliferous sediments of the Bauer Deep, southeast Pacific. *Geochimica et Cosmochimica Acta*, **49**, 221–35.

Cole, T. G. (1988) The nature and origin of smectite in the Atlantis II Deep, Red Sea. *Canadian Mineralogist*, **26**, 755–63.

Cole, T. G. and Shaw, H. F. (1983) The nature and origin of authigenic smectites in some recent marine sediments. *Clay Minerals*, **18**, 239–52.

Corliss, J. B., Lyle, M., Dymond, J. and Crane, K. (1978) The chemistry of hydrothermal mounds near the Galapagos Rift. *Earth and Planetary Science Letters*, **40**, 12–24.

Decarreau, A., Bonnin, D., Badaut-Trauth, D., Couty, R. and Kaiser, P. (1987) Synthesis and crystallogenesis of ferric smectite by evolution of Si–Fe coprecipitates in oxidizing conditions. *Clay Minerals*, **22**, 207–23.

Degens, E. T. and Ross, D. A. (eds) (1969) *Hot Brines and Recent Heavy Metal Deposits in the Red Sea*, Springer-Verlag, Heidelberg.

Desprairies, A. (1981) Authigenic minerals in volcanogenic sediments cored during Deep Sea Drilling Project leg 60, in *Initial Reports of the Deep Sea Drilling Project*, vol. 60, (eds D. M. Hussong, and S. Uyeda), United States Government Printing Office, Washington, DC, pp. 455–66.

Dymond, J. (1981) Geochemistry of Nazca Plate sediments: An evaluation of hydrothermal, biogenic, detrital, and hydrogenous sources. *Geological Society of America, Memoir*, **154**, 133–74.

Edmond, J. M., Von Damm, K. L., McDuff, R. E. and Measures, C. I. (1982) Chemistry of hot springs on the East Pacific Rise and their effluent dispersal. *Nature*, **297**, 187–91.

Elderfield, H. (1976) Hydrogenous material in marine sediments, excluding manganese nodules, in *Chemical Oceanography*, vol. 5, (eds J. P. Riley and R. Chester), Academic Press, New York, pp. 137–215.

Foscolos, A. E. and Powell, T. G. (1980) Mineralogical and geochemical transformation of clays during catagenesis and their relation to oil generation, in *Facts and Principles of World Petroleum Occurrence*, (ed. A. D. Miall), Memoir 6, Canadian Society of Petroleum Geologists, Calgary, pp. 153–72.

Froelich, P. N., Klinkhammer, G. P., Bender, M. L., Luedtke, N. A., Heath, G. R., Cullen, D., Dauphin, P., Hammond, D., Hartman, B. and Maynard, V. (1979) Early oxidation of organic matter in pelagic sediments of the eastern equatorial Atlantic: suboxic diagenesis. *Geochimica et Cosmochimica Acta*, **43**, 1075–90.

Gac, J. Y., Droubi, A., Fritz, B. and Tardy, Y. (1977) Geochemical behaviour of silica and magnesium during the evaporation of waters in Lake Chad. *Chemical Geology*, **19**, 215–28.

Garrels, R. M. (1986) Sediment cycling and diagenesis. *United States Geological Survey Bulletin*, **1578**, 1–11.

Griffin, J. J., Windom, H. and Goldberg, E. D. (1968) The distribution of clay minerals in the World Ocean. *Deep-Sea Research*, **15**, 433–59.

Harder, H. (1972) The role of magnesium in the formation of smectite minerals. *Chemical Geology*, **10**, 31–9.

Harder, H. (1978) Synthesis of iron layer silicate minerals under natural conditions. *Clays and Clay Minerals*, **26**, 65–72.

Hathaway, J. C. and Sachs, P. L. (1965) Sepiolite and clinoptilolite from the mid-Atlantic ridge. *American Mineralogist*, **50**, 852–67.

Hein, J. R., Yeh, H. W. and Alexander, E. (1979). Origin of Fe-rich montmorillonite from the Mn nodule belt of the North Equatorial Pacific. *Clays and Clay Minerals*, **27**, 185–94.

Hoffert, M., Perseil, A., Hekinian, R., Choukroune, P. and Needham, H. D. (1978) Hydrothermal deposits sampled by diving saucer in transform fault 'A' near 37°N on the mid-Atlantic Ridge, Famous area. *Oceanologica Acta*, **1**, 73–86.

Hower, J., Eslinger, E. A., Hower, M. E. and Perry, E. A. (1976) Mechanism of burial metamorphism in argillacous sediment: 1. Mineralogical and chemical evidence. *Geological Society of America Bulletin*, **87**, 725–37.

Huang, W. L., Bishop, A. M. and Brown, R. W. (1986) The effect of fluid/rock ratio on feldspar dissolution and illite formation under reservoir conditions. *Clay Minerals*, **21**, 585–601.

Hurd, D. C. (1973) Interactions of biogenic opal, sediment and seawater in the central equatorial Pacific. *Geochimica et Cosmochimica Acta*, **37**, 2257–82.

Iler, R. K. (1973) Effect of adsorbed alumina on the solubility of amorphous silica in water. *Journal of Colloid and Interface Science*, **43**, 399–408.

Kastner, M. (1981) Authigenic silicates in deep-sea sediments: formation and diagenesis, in *The Sea*, vol. 7, (ed. C. Emiliani), J. Wiley & Sons, Chichester, 915–80.

Konta, J. (1985) Mineralogy and chemical maturity of suspended matter in major rivers sampled under the SCOPE/UNEP project. *Mitteilung Geologisch-Palaontologisches Institut, Universitat Hamburg, SCOPE/UNEP Sonderband Heft*, **58**, 569–92.

Li, Y.-H., Teraoka, H., Yang, T.-S. and Chen, J.-S. (1984) The elemental composition of suspended particles from the Yellow and Yangtse rivers. *Geochimica et Cosmochimica Acta*, **48**, 1561–4.

Lindsay, W. L. (1979) *Chemical Equilibria in Soils*, J. Wiley & Sons, Chichester.

Lisitzin, A. P. (1972) Sedimentation in the world ocean. *Society of Economic Paleontologists and Mineralogists, Special Publication*, **17**, 218 pp.

Mackin, J. E. and Aller, R. C. (1984a) Dissolved Al in sediments and waters from the East China Sea: Implications for authigenic mineral formation. *Geochimica et Cosmochimica Acta*, **48**, 281–97.

Mackin, J. E. and Aller, R. C. (1984b) Diagenesis of dissolved aluminium in organic-rich estuarine sediments. *Geochimica et Cosmochimica Acta*, **48**, 298–313.

Martin, J. M. and Meybeck, M. (1979) Elemental mass balance of material carried by major world rivers. *Marine Chemistry*, **7**, 173–206.

McBirney, A. R. and Aoki, K. (1968) Petrology of the island of Tahiti. *Geological Society of America, Memoir*, **116**, 523–56.

McMurtry, G. M., Wang, C. H. and Yeh, H. W. (1983) Chemical and isotopic investigations into the origin of clay minerals from the Galapagos hydrothermal mounds field. *Geochimica et Cosmochimica Acta*, **47**, 475–89.

Michard, G., Albarede, F., Michard, A., Minster, J.-F., Charlou, J.-L. and Tan, N. (1984) Chemistry of solutions from the 13°N East Pacific Rise hydrothermal site. *Earth and Planetary Science Letters*, **67**, 297–307.

Milliman, J. D. and Meade, R. H. (1983) Worldwide delivery of river sediment to the oceans. *Journal of Geology*, **91**, 1–21.

Natland, J. H. (1980) The progression of volcanism in the Samoan linear volcanic chain. *American Journal of Science*, **280**, 709–25.

Newman, A. C. D. (ed.) (1987) *Chemistry of Clays and Clay Minerals*, Monograph Number 6, Mineralogical Society, London.

Odin, G. S. (1985) La verdine, facies granulaire vert, marin et cotier, distinct de la glauconie: distribution actuelle et composition. *Comptes Rendues de l'Academie des Sciences, Paris*, **301**, 105–8.

Odin, G. S. (ed.) (1988a) *Developments in Sedimentology*, vol. 45, *Green Marine Clays*, Elsevier, Amsterdam.

Odin, G. S. (1988b) Glaucony from the Gulf of Guinea, in *Developments in Sedimentology*, vol. 45, *Green Marine Clays*, (ed. G. S. Odin), Elsevier, Amsterdam, pp. 225–47.

Odin, G. S. (1990) Clay mineral formation at the continent–ocean boundary: the verdine facies. *Clay Minerals*, **25**, 477–83.

Odin, G. S. and Fullagar, P. D. (1988) Geological significance of the glaucony facies, in *Developments in Sedimentology*, vol. 45, *Green Marine Clays*, (ed. G. S. Odin), Elsevier, Amsterdam, pp. 295–332.

Odin, G. S. and Lamboy, M. (1988) Glaucony from the margin off northwestern Spain, in *Developments in Sedimentology*, vol. 45, *Green Marine Clays*, (ed. G. S. Odin), Elsevier, Amsterdam, pp. 249–75.

Odin, G. S. and Masse, J. P. (1988) The verdine facies from the Senegalese continental shelf, in *Developments in Sedimentology*, vol. 45, *Green Marine Clays*, (ed. G. S. Odin), Elsevier, Amsterdam, pp. 83–104.

Odin, G. S. and Matter, A. (1981) De glauconiarum origine. *Sedimentology*, **28**, 611–41.

O'Neil, J. R. (1986) Preservation of H, C, and O isotopic ratios in the low temperature environment, in *Stable Isotope Geochemistry of Low Temperature Fluids*, (ed. T. K. Kyser), *Short Course*, No. 13, Mineralogical Association of Canada, Nepean, pp. 85–128.

Orians, K. J. and Bruland, K. W. (1986) The biogeochemistry of aluminum in the Pacific Ocean. *Earth and Planetary Science Letters*, **78**, 397–410.

Pedro, G., Carmouze, J. P. and Velde, B. (1978) Peloidal nontronite formation in recent sediments of Lake Chad. *Chemical Geology*, **23**, 139–49.

Persons, B. S. (1970) *Laterite: Genesis, Location, Use*, Plenum Press, New York.

Porrenga, D. H. (1967) Glauconite and chamosite as depth indicators in the marine environment. *Marine Geology*, **5**, 495–501.

Rohrlich, V., Price, N. B. and Calvert, S. E. (1969) Chamosite in the recent sediments of Loch Etive. *Journal of Sedimentary Petrology*, **39**, 624–31.

Savin, S. M. (1980) Oxygen and hydrogen isotope effects in low-temperature mineral–water interactions, in *Handbook of Environmental Geochemistry*, vol. 1, *The Terrestrial Environment*, (eds P. Fritz and J.-Ch. Fontes), Elsevier, Amsterdam, pp. 283–327.

Sholkovitz, E. R. (1976) Flocculation of dissolved organic and inorganic matter during the mixing of river water and seawater. *Geochimica et Cosmochimica Acta*, **40**, 831–45.

Suess, E. (1980) Particulate organic carbon flux in the ocean-surface productivity and oxygen utilization. *Nature*, **288**, 260–3.

Swindale, L. D. and Fan, P.-F. (1967) Transformation of gibbsite to chlorite in ocean bottom sediments. *Science*, **157**, 799–800.

Tardy, Y., Cheverry, C. and Fritz, B. (1974) Néoformation d'une argile magnésienne dans les dépressions interdunaires du lac Tchad. Application aux domaines de stabilité des phyllosilicates alumineux, magnésiens et ferrifères. *Comptes Rendues de l'Academie des Sciences, Paris*, **278**, 1999–2002.

Uematsu, M., Duce, R. A., Prospero, J. M., Chen, L., Merrill, J. T. and McDonald, R. L. (1983) Transport of mineral aerosol from Asia over the North Pacific Ocean. *Journal of Geophysical Research*, **88**, 5343–52.

Varnavas, S. P. (1979) *Geochemical Investigations on Sediments from the Eastern Pacific*, PhD Thesis, University of London.

Yeh, H. W. and Savin, S. M. (1977) Mechanism of burial metamorphism of argillaceous sediments: 3. O-isotope evidence. *Geological Society of America Bulletin*, **88**, 1321–30.

Zierenberg, R. A. and Shanks, W. C. III (1983). Mineralogy and geochemistry of epigenetic features in metalliferous sediment, Atlantis II Deep, Red Sea. *Economic Geology*, **78**, 57–72.

CHAPTER FIVE

Diagenesis in Jurassic mudrocks of the North Sea

J. M. Huggett and H. F. Shaw

5.1 Introduction

The Jurassic contains the most important hydrocarbon reservoirs and source rocks in the central and northern North Sea. The mudrocks, in addition to being important source rocks also act as cap rocks to many of the reservoirs. The Jurassic is most productive in the Central and Viking grabens and the Moray Firth Basin where the source rocks have achieved favourable maturation states (Cornford, 1990).

The Lower Jurassic, which consists predominantly of marine argillaceous sediments has only a limited distribution in the North Sea basins. The Lower Jurassic reaches its maximum thickness of about 900 m in the Sole Pit Trough (Kent, 1980) and the eastern Norwegian–Danish Basin (Michelson, 1978). In contrast the Middle Jurassic is dominated by sandstones and mudrocks deposited in non-marine to paralic environments. In the extreme north and some areas of the south (North Sea) marine conditions persisted throughout the Middle Jurassic. Within the Middle Jurassic mudrocks occur interbedded with the sandstones, but towards the top of the interval the mudrocks begin to predominate (e.g. the Heather Formation). The Upper Jurassic sees a return to marine conditions in the North Sea Basin and the deposition of argillaceous, organic-rich sediments (e.g. the Kimmeridge Clay) (Brown, 1990).

To date most studies of Jurassic mudrocks from the North Sea have been based entirely, or almost entirely, on interpretations of bulk mineralogical and geochemical data (e.g. Pearson *et al.*, 1982; 1983). Recent studies, employing back-scattered electron microscopy (e.g. Primmer and Shaw, 1987) and transmission electron microscopy (e.g. Glasmann *et al.*, 1989), have shown the heterogeneous nature of these mudrocks and the complexity of the diagenetic reactions that have occurred within them.

The present study examines both new petrological data and previously published data from North Sea Jurassic mudrocks, in particular the Kimmeridge Clay Formation (KCF) (samples from the northern North Sea, the Viking Graben and the outer Moray Firth) and the Heather Formation (samples from the northern North Sea and outer Moray Firth). The object is to examine the wide variety, both laterally, and between formations, of the

Geochemistry of Clay-Pore Fluid Interactions. Edited by D. A. C. Manning, P. L. Hall and C. R. Hughes. Published in 1993 by Chapman & Hall, London. ISBN 0 412 48980 5

depositional and diagenetic characteristics. It is not intended to be treated as an exhaustive study, more an illustrated review of data from several sources, to present an overview of North Sea Jurassic mudrocks.

5.2 Materials and methods

The materials studied were core and cuttings samples of Jurassic mudrocks from the Viking and Central Grabens, the northern North Sea, the outer Moray Firth, and the central and northern Norwegian continental shelf. All samples were examined by standard X-ray diffraction (XRD) techniques. Analyses were made of random powder mounts for whole-rock mineral identification, and of orientated smaller than 2 μm size fractions for clay mineral identification. Illite–smectite compositions were determined by the method of Reynolds and Hower (1970). Semi-quantitative mineral percentages were determined from the whole-rock data, using peak area weightings determined from measurement of known quartz/mineral standard mixtures.

Samples were also examined in back-scattered electron imaging (BEI) mode in either a JEOL 733 scanning electron microscope/superprobe or a Hitachi S2500 scanning electron microscope, both with energy-dispersive X-ray analytical facilities (EDS). Quantitative EDS analyses were also carried out on illite and illite–smectite phases in some samples. The application of BEI in the SEM to petrological studies has been discussed by Lloyd and Hall (1981) and to studies of sedimentary rocks in particular by White *et al.* (1984). To summarize, the proportion of back-scattered electrons in the signal is a function of the mean atomic number of the mineral being imaged. Hence, the higher the mean atomic number of the mineral the brighter it appears in the image, enabling particles of different composition to be differentiated. The BEI contrast is maximized through the use of polished surfaces, and for this study polished thin sections or polished blocks with a conductive coating of carbon were used. Authigenic clays were identified in BEI images through their petrographic characteristics; these are described in Section 5.3.

Scanning transmission electron microscope (STEM) analysis was carried out on several KCF samples using a JEOL 200FX microscope. In the STEM mode authigenic quartz was identified through euhedral crystal terminations.

5.3 Results

X-ray diffraction and BEI analysis have shown that in the North Sea the Jurassic mudrocks are extremely variable in composition. Illite, illite–smectite, kaolinite and chlorite are generally all present, but in highly variable proportions. Illite–smectite is a minor component of the KCF samples examined, and is generally ordered with less than 20% smectite. Smectite was only encountered in relatively shallow-buried KCF samples from the northern Norwegian shelf. This localized distribution suggests a local, possibly volcanic source for the smectite. Not uncommonly, samples from similar depths but different wells contain widely different proportions of illite and illite–smectite. This may reflect variations in either diagenetic history or in original detrital composition; without detailed knowledge of burial and thermal histories which of these is the most significant cannot be determined.

X-ray diffraction data from several previous studies of Jurassic mudrocks from the North Sea have shown that the illite–smectites tend to be illite-rich ordered varieties (e.g. Pearson and Small, 1988). However, there are exceptions to this with smectite-rich random illite–smectites being found in the Upper Jurassic mudrocks (Primmer and Shaw, 1987). Hurst (1982) also reported the presence of smectite-rich random illite–smectites in Middle and Lower Jurassic mudrocks from Brora, north-east Scotland. Some typical KCF clay analyses from various North Sea locations at similar burial depths are shown in Table 5.1, which illustrate the variability in reported illite–smectite compositions from these Upper Jurassic mudrocks. These data also indicate the variable nature of the assemblages found in the KCF mudrocks. Some samples are predominantly composed of illite, illite–smectite or kaolinite, whilst others contain similar proportions of these clay phases. In all the samples chlorite is only found in relatively minor proportions. X-ray diffraction data from various sources show that illite, ordered illite–smectite and kaolinite also predominate in the Middle and Lower Jurassic mudrocks.

Table 5.1 Some 'typical' KCF clay analyses

Composition	Magnus (3000 m)	Brae (3650 m)	E. Shetland Basin (3400 m)	Gullfaks (3000 m)
% illite	5–20	30–70	25	10–15
% I/S	20–75	25–45	25	10–15
% kaolinite	20–60	5–30	35	60–70
% chlorite	5	5	15	n.d.
% Sm in I/S	30–45	20–70	10–35	20–40

n.d. = not detected

The SEM data provide evidence of neo-formed illite-rich illite–smectites and illite infilling cavities in partially dissolved K-feldspars (Figure 5.1), replacing muscovite (Figure 5.2) and also within the matrix of the mudrocks (Figure 5.3). Some KCF samples contain aggregates of pure ordered illite–smectites with an organic coating; these may be mineralized faecal pellets (Figure 5.4). However, the illite–smectites are also present mixed with illite in the matrix, often making it difficult to determine whether it is detrital or authigenic in orgin. Microprobe analyses of illite–smectites and illites that could be differentiated by their petrographic characteristics as being detrital or authigenic show that these illite–smectites also differ in composition. This is illustrated in a generalized

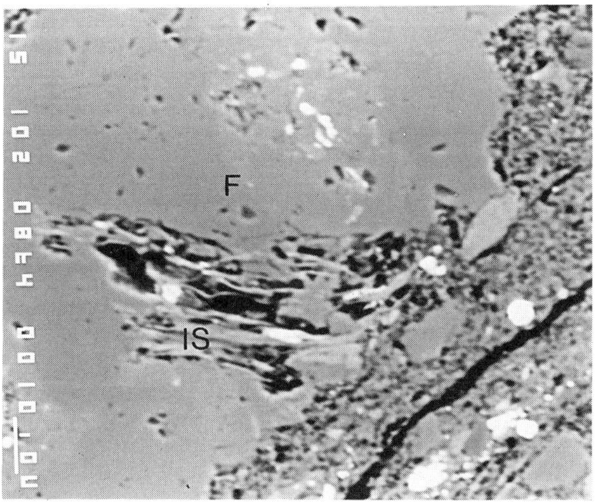

Figure 5.1 Neo-formed illite-rich illite–smectite (IS) in partially dissolved K-feldspar (F). (KCF, Brae area.)

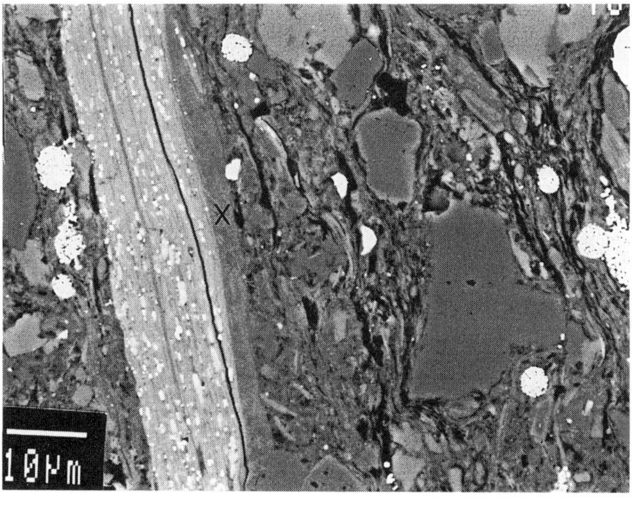

Figure 5.2 Illite or illite–smectite (X) (the analysis is uncertain, coming from so close to a large muscovite grain) replacing phengitic mica. (Fulmar Formation, Central Graben.)

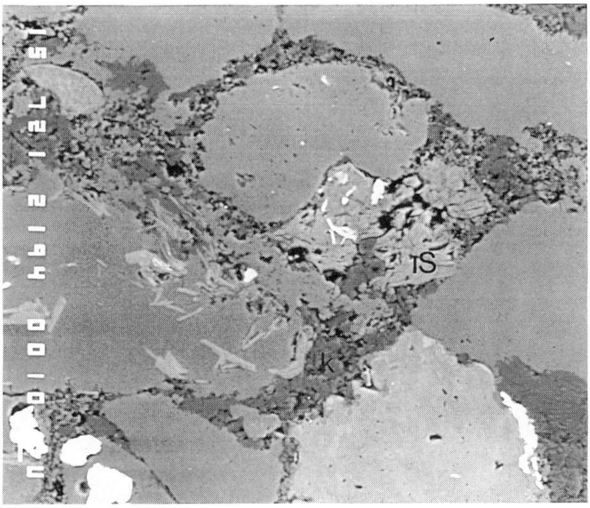

Figure 5.3 Neo-formed illite-rich illite–smectite (IS) within matrix. Kaolinite (K) in the matrix could be either a neo-formed pore filling or a replacement of another mineral; the origin cannot be clearly identified on the basis of the textures observed. (KCF, Brae area.)

Figure 5.4 Aggregate of pure illite–smectite (IS), with a complex internal structure and an organic coating (bottom right-hand corner). Possibly this aggregate is a mineralized fecal pellet. 'Shallow-buried' sample from Northern Norwegian continental shelf.

fashion in Figure 5.5, which uses data from numerous different examples to show the overall compositional ranges. The authigenic ordered illite–smectite aggregates in some 'shallow-buried' mudrocks have a higher K^+ content than the matrix illite–smectite in the same samples (Figure 5.4; Table 5.2: samples 2/1 and 2/2), yet only one illite–smectite phase was recognized from the XRD data (e.g. KCF sample 2 in Table 5.2). Overall the K^+ values are lower and Na^+ values are higher than would be expected for illite, especially where no illite–smectite has been detected by XRD analysis. This suggests that either the illite is K^+ depleted or that the clay being analysed is illite–smectite with a greater proportion of smectite than was calculated from the XRD data. The SEM and XRD data would, therefore, seem to be at odds.

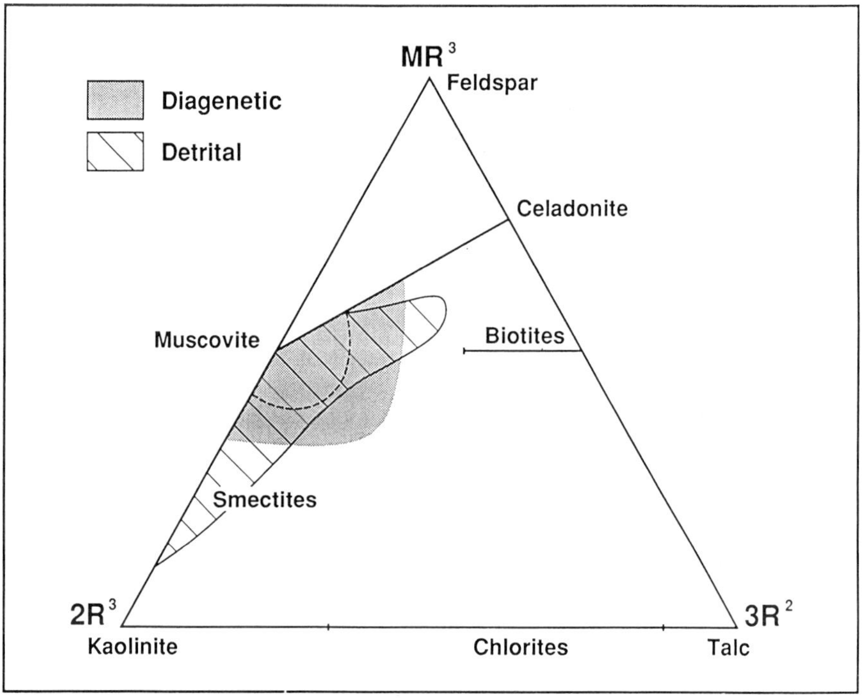

Figure 5.5 Velde plot showing variations in detrital and diagenetic illite–smectite phases found in KCF mudrock (unpublished data of Primmer and Shaw).

Kaolinite is ubiquitous and often authigenic, occurring as a replacement of chlorite and muscovite, as cavity infills (Figure 5.6), and, less commonly, in the matrix as booklets which have grown perpendicular to the detrital clay fabric (Figure 5.7). It was identified in all the formations analysed in amounts typically varying from a few percent to 20%. In one KCF sample with an unusually high organic matter content, 60% of the inorganic material is books of apparently authigenic kaolinite, with the remainder being quartz and pyrite (Figure 5.8).

Table 5.2 Typical superprobe data for matrix clays from selected samples

% oxide	KCF									Heather Form		Fulmar Form
	Magnus Field			Brae		N. Norway				Thistle Field		Clyde Field
	Sample 32	Sample 33	Sample 45	Sample 1	Sample 2	Sample 2/1	Sample 2/2	Sample 50	Sample 18	Sample 11	Sample 12	Sample 57
Na_2O	0.81	1.69	0.16	0.26	1.93	0.52	0.76	4.59	0.54	0.57	6.98	0.60
MgO	2.56	2.01	0.71	1.33	0.80	3.43	4.32	3.62	1.69	1.70	1.10	1.22
Al_2O_3	24.96	28.46	24.01	28.09	29.51	25.04	19.23	20.46	26.59	26.77	29.23	32.77
SiO_2	52.59	48.47	61.66	52.02	44.20	50.35	52.17	53.67	50.75	51.10	45.92	50.19
K_2O	5.33	4.94	3.74	7.98	9.26	6.88	5.18	3.51	6.06	6.108	3.33	3.73
CaO	0.23	0.11	0.09	0.00	0.08	0.00	0.00	0.96	0.21	0.211	0.04	0.40
TiO_2	0.29	0.27	0.31	0.17	0.53	0.33	0.30	0.46	1.28	1.291	0.23	0.32
Cr_2O_3	0.00	0.06	0.31	0.00	0.00	0.00	0.12	0.36	0.00	0.00	0.00	0.00
MnO	0.00	0.00	0.00	0.04	0.00	0.00	0.00	0.00	0.06	0.062	0.00	0.01
FeO	2.01	2.99	0.68	2.50	2.69	6.06	6.52	2.26	3.26	3.28	1.56	2.17
Total	88.77	93.92	91.67	92.39	88.98	92.60	88.60	89.90	90.44	91.09	88.40	91.40
% illite	4	11	34	—	—	—	16	13	9	16	17	17
% illite–smectite	50	12	0	—	—	—	11	0	8	7	14	2
% smectite	0	0	0	—	—	—	0	17	0	0	0	0
% kaolinite	11	33	0	—	—	—	19	17	20	17	20	2

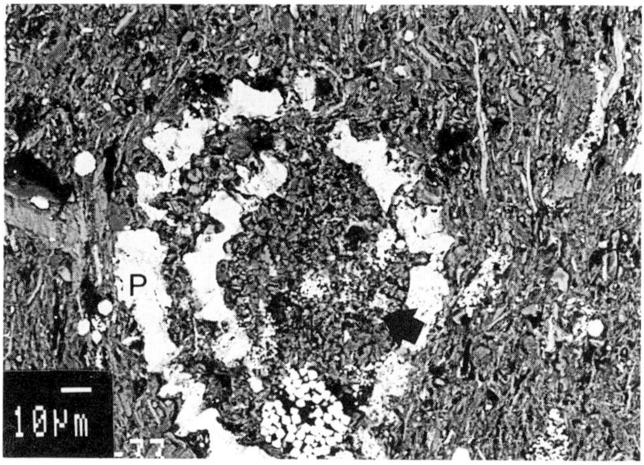

Figure 5.6 Authigenic kaolinite (arrow) infilling a pyritized (P) microfossil test. (Brent Group, Thistle Field, Northern North Sea.)

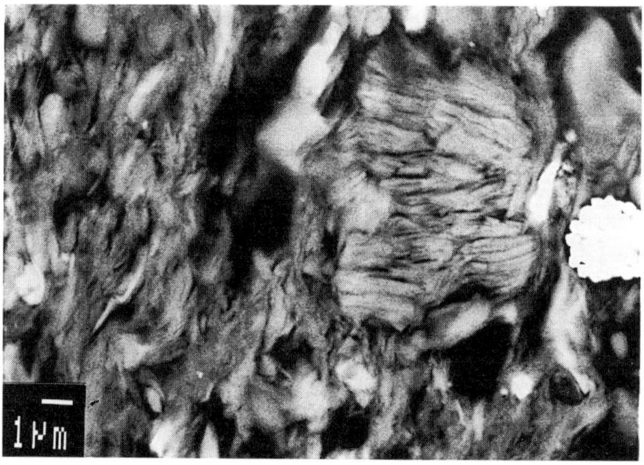

Figure 5.7 Authigenic kaolinite booklet in illite/illite–smectite matrix. The kaolinite has grown perpendicular to the preferred orientation of the detrital matrix. (Brent Group, Thistle Field, Northern North Sea.)

Figure 5.8 Organic-rich mudrock with a predominantly kaolinite matrix (K). Authigenic kaolinite has infilled a pyritized (P) microfossil test. (Hekkingen Formation, northern Norwegian continental shelf.)

Minor chlorite occurs as clay-grade to silt-grade particles in the matrix of less than half the KCF samples examined, and is generally only present in minor amounts in the Middle and Upper Jurassic mudrocks. Two exceptional samples, one from the KCF, the other from the Humber Group (outer Moray Firth) mudrocks have a very fine-grained and well-compacted matrix consisting entirely of kaolinite and chlorite (Figure 5.9). In these instances both clays are probably detrital, perhaps from a depositional environment not generally encountered in the North Sea Jurassic mudrocks, or from a different source area. From the limited information available on depositional environments (most samples are believed to be either shallow shelf, lagoonal, deep-water hemipelagic or turbidite deposits) it does not appear, in general, that variations in the clay mineralogy can be related to differing depositional environments.

Quartz content varies from as little as 10% to as much as 60% (estimates based on semi-quantitative XRD analysis), and does not appear to be related to depositional environment. In some samples the grains are 'flattened' and have irregular outlines (Figure 5.10): this implies that extensive dissolution has occurred, and as there often appears to be no widespread reprecipitation of the silicon in any form it must be assumed that it has been removed from the mudrock. In other samples there is evidence of both dissolution and precipitation of microquartz cements and quartz overgrowths (Figures 5.11 and 5.12). However, other samples of mudrocks show well-preserved quartz grains with

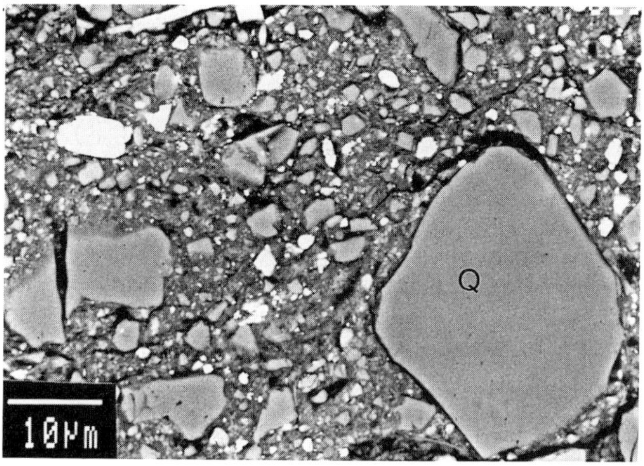

Figure 5.9 Rounded, unleached quartz grains (Q) in a matrix of kaolinite and chlorite. The largest of the white grains is baryte, the rest are pyrite. This sample is unusual in that it contains no organic matter. (Humber Group, Moray Firth Basin.)

Figure 5.10 Dissolution of quartz grains producing elongate grains (e.g. arrow). (Middle Jurassic, Brae area.)

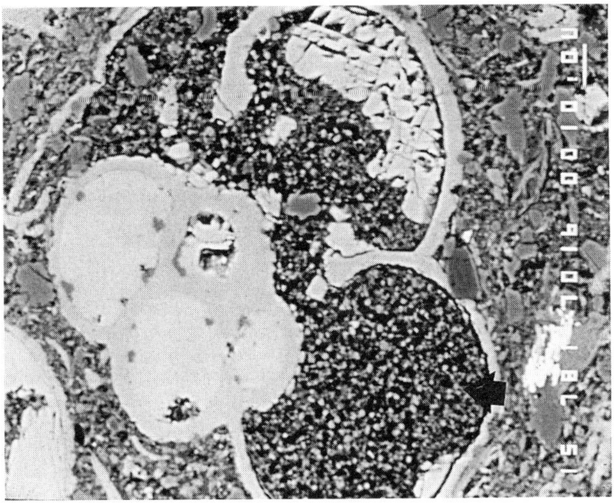

Figure 5.11 Microquartz (arrow) cement in a calcite microfossil test. (KCF, Brae area.)

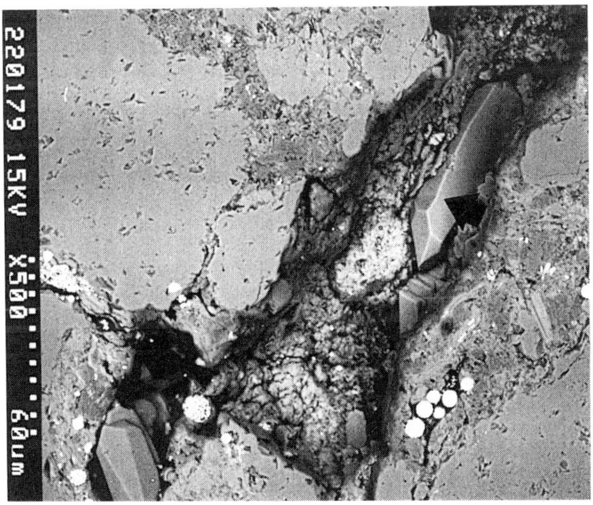

Figure 5.12 Quartz overgrowths (arrow) in Middle Jurassic mudrocks (Brae area). The relief on the quartz is seen because this is an SEI image superposed upon a BEI image. (Courtesy of P. J. Greenwood.)

no evidence of diagenetic alteration (e.g. Figure 5.9). Quartz dissolution appears to correlate with organic matter: out of ten samples where dissolution was positively identified, eight contain organic matter visible in BEI micrographs, and out of seven samples with no organic matter only two showed evidence of dissolution of quartz. Scanning transmission electron microscope images of the KCF formation confirm that many quartz grains are flattened. However, this technique also revealed the presence of some clay-grade quartz crystals with euhedral crystal terminations, believed to be authigenic (Figure 5.13).

Figure 5.13 Scanning transmission electron microscope image of quartz (q) in illitic clay matrix (KCF, Magnus Field). The central grain appears to have overgrowths at either end: p = pyrite; c = clay.

The feldspar content of the mudrocks is highly variable. In the KCF samples from the northern North Sea, the Viking Graben and the outer Moray Firth visual estimates and XRD semi-quantitative analyses of total feldspar rarely exceed a few percent, whilst in the KCF formation from northern offshore Norway there is up to 12% K-feldspar, and 3% albite (XRD determination). The high feldspar content in the northern Norwegian KCF Formation may, like the smectite, be related to a volcanic source. The feldspar content of the Heather samples is similarly variable. Mudrocks from the Fulmar Formation (Central Graben) have typically about 10% total feldspar, with variable amounts of replacement of albite by K-feldspar or vice versa (Figure 5.14). K-feldspar also shows evidence of dissolution, with or without accompanying neo-formed clay phases (Figures 5.1 and 5.15, respectively).

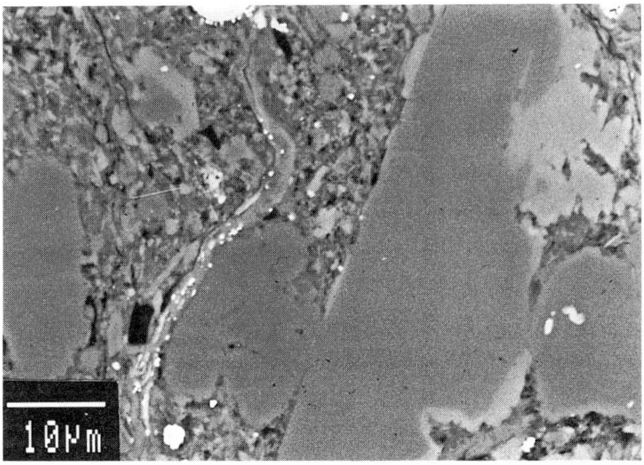

Figure 5.14 Grain containing both K-feldspar (light grey margin) and albite (darker grey centre). Which has replaced which is not clear from this or any other BSE micrographs of the mudrocks sampled. (Fulmar Formation, Viking Graben.)

Carbonate minerals are present in the Jurassic mudrocks as calcite microfossil tests and fragments plus cements that are usually calcite but also occasionally dolomite (Figure 5.16), siderite and ankerite (Figure 5.17). Pyrite is present in all samples examined, occurring as an early, pre-compactional framboidal cement, and as a post-compactional euhedral cement. Some microfossils have been completely pyritized.

5.4 Discussion

X-ray diffraction data from previous studies have been used to suggest that there has been progressive illitization of smectite and illite–smectite with increased burial in the North Sea basins, as found in the Texas Gulf Coast Tertiary mudrocks (Pearson *et al.*, 1982; 1983; Pearson and Small, 1988; Hansen and Lindgreen, 1989). The ordered illite-rich illite–smectite in the Jurassic mudrocks has been interpreted as the end-product of this progressive illitization.

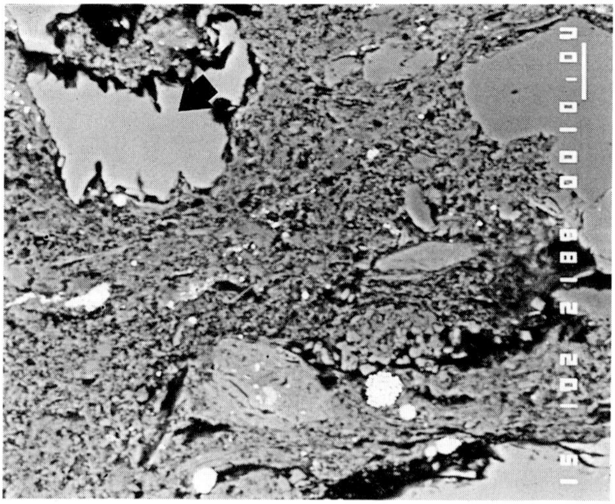

Figure 5.15 Partially dissolved feldspar (arrow). (Northern Norwegian continental shelf.)

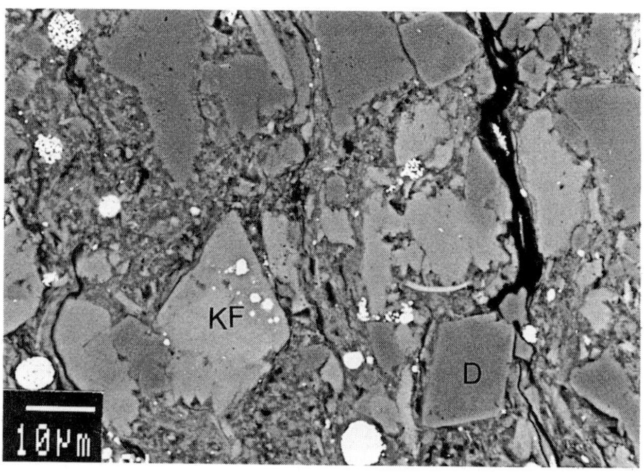

Figure 5.16 Dolomite cement (D) replacing matrix and K-feldspar (KF). (Fulmar Formation, Viking Graben.)

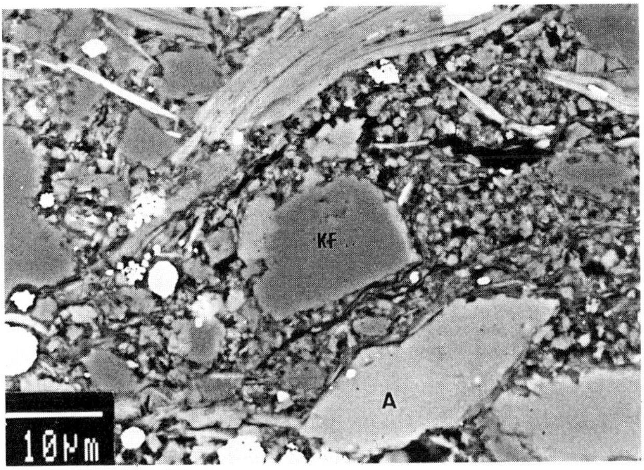

Figure 5.17 Ankerite cement (A), kaolinite-mica stacks and K-feldspar (KF) in illite/illite–smectite matrix. (KCF, northern Norwegian continental shelf.)

If the illite–smectites were predominantly diagenetic phases formed by transformation of earlier smectites and smectite-rich illite–smectite one would have expected more evidence of well-formed diagenetic illite–smectites within the matrix of the mudrocks than is found during SEM analyses. Also if the ordered illite–smectites identified from XRD analyses represent the end-product of transformation illitization reactions, it might be expected that they would show similar microprobe compositions, whereas they show wide variability (Figure 5.5). This variability does not simply reflect the varying proportions of illite and smectite, as can be seen by comparing the XRD data with the analyses (Table 5.2).

It would seem that the XRD analyses of the illite–smectites in the Jurassic mudrocks represents an 'averaging' of the analysis of illite–smectites that are both diagenetic (neo-formed) and detrital in origin. The XRD analysis could also reflect the mixing together of illites and illite–smectites of these different origins during sample preparation. The resultant illite–smectite composition determined from the XRD data could thus be an artefact as suggested by Nadeau *et al.* (1984) and Ahn and Peacor (1986) in their studies of interparticle diffraction effects, though not necessarily simply by stacking of 'fundamental' particles, but possibly by stacking of illite and illite–smectite particles, cleaved or intact.

We would, therefore, conclude that the illite–smectite compositions of the Jurassic mudrocks in the North Sea do not simply represent the end-products of illitization reactions similar to those of the Texas Gulf Coast Model. Rather the illite–smectites are both detrital and diagenetic in origin and the SEM data suggest that the diagenetic illite–smectites identified in the matrix and in tests are neo-formed, although transformation origins cannot be entirely excluded.

Comparison of samples rich in organic matter with those depleted in it reveals that authigenic kaolinite, pyrite and possibly quartz dissolution are characteristically associated with high organic matter content. Reaction of organic matter through sulphate-reducing microbial methanogenesis (Berner, 1984) releases HS^-, which reacts with any available Fe^{2+} to precipitate pyrite. In organic-rich samples, organic matter undergoing diagenesis releases other soluble components which generally increase acidity (Carothers and Kharaka, 1978), though this effect could be modified by buffers such as carbonate phases. In a closed system phyllosilicates and K-feldspar are dissolved by these acids and consequently the concentration of silicon, aluminium, magnesium, iron and potassium in solution and pH all rise. However, the amount of feldspar and phyllosilicate dissolution would have to be considerable to outweigh the effect upon pH of organic acid formation during kerogen maturation. The solubility of quartz and other silicates may be enhanced through complexing with organic compounds (e.g. Taguchi *et al.*, 1987). This is corroborated by our observation that there appears to be more evidence of silica dissolution in organic-rich mudrocks.

Phase diagrams can be used to illustrate how the pore fluids may have evolved to produce the textures observed above. The phase diagram (Figure 5.18) for aluminium against pH (Warren, 1987) shows that, if sufficient K-feldspar and muscovite dissolution occurs with a lowering of pH, illite or illite-rich illite–smectite may precipitate as the pore fluids become saturated with respect to these phases (e.g. Figure 5.1). If the rate of K-feldspar and muscovite dissolution exceeds the rate of illite precipitation the kaolinite saturation line will be crossed because the fluid will evolve back towards a higher, more neutral pH and higher aluminium concentration. It would then be possible for both illite and kaolinite to precipitate (e.g. Figure 5.3). Alternatively limited muscovite and K-feldspar dissolution could result in a fluid being initially saturated with respect to kaolinite.

Addition of iron and magnesium (from biotite and detrital chlorite) into the pore fluid, or possibly evolved sea water could provide the necessary cations for chlorite precipitation. Iron and magnesium would also be available for formation of ankerite and dolomite. Precipitation of the ankerite and dolomite cements requires a shift towards a more alkaline pore fluid, implying that it post-dates silicate dissolution, and available carbonate in solution. This might have been achieved through reduction of iron minerals by organic matter (Curtis, 1977; 1987). Where dissolved sulphur was available it would have reacted with the reduced iron to form pyrite, and any carbonate cements would be iron poor. In sediments free from dissolved sulphur, or where all the available sulphur had already reacted to form pyrite, the reduced iron and the CO_2 would be incorporated into iron-rich carbonate.

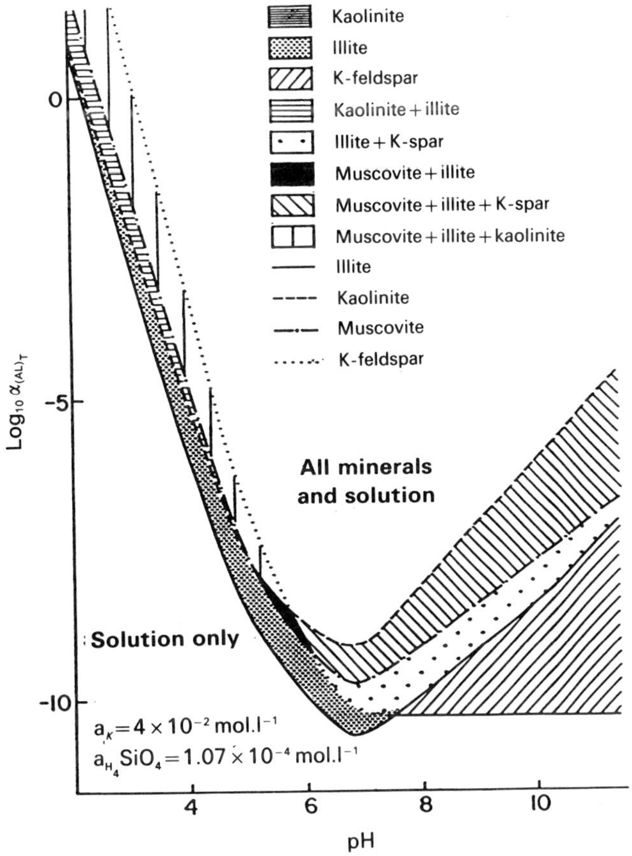

Figure 5.18 Solution-mineral equilibria in the Al_2O_3–SiO_2–H_2O system at $T = 298$ K, $P = 1$ bar and given activities of K^+ and H_4SiO_4. (After Warren, 1987.)

Thus, it is possible to interpret the conditions under which the various clay and non-clay minerals observed may have formed. These scenarios can, however, be criticized on the grounds that they assume a closed system. This may not always be the case, especially where there is evidence for removal of dissolved quartz or other components.

Astin and Evans (1990) have estimated that shales may typically lose 10–20% of their solid volume during diagenesis, primarily through quartz dissolution. Whilst diagenesis of mudrocks may involve a net export of silica and thus possibly provide sources for quartz cementation in adjacent sandstones, it must be borne in mind that there is also widespread evidence of quartz cementation in the mudrocks and that not all of the mobilized silica is exported from them. However, there appears to be more evidence of quartz dissolution than cementation in the mudrocks examined, though it is difficult to quantify any apparent net loss of silica from the mudrocks. There is also the problem as to why quartz should preferentially precipitate in sandstones. One reason why this may happen is that the clay coats on the quartz grains in the mudrocks will be much thicker and thus more likely to inhibit the formation of quartz overgrowths than in the sandstones. Another major problem regarding mudrocks as a possible source of silica for sandstone quartz cementation is the large volumes of fluid required to remove significant amounts of silica from the mudrocks into the sandstones. These volumes would be much greater than could be accounted for by loss of fluids during mudrock compaction (Bjørlykke, 1983). Movement of components out of mudrocks implies that they need not always be 'closed' systems. In an 'open' system mineral neo-formation may have been controlled by flow rates. Huang *et al.* (1986) demonstrated that flow rates may control whether illite or kaolinite precipitate in sandstones. Kaolinite precipitation may, therefore, have occurred exclusively during early diagenesis when flow rates were relatively high, and illite formation may have taken over as flow rates decreased with increasing compaction. Clearly in mudrocks there is a strong possibility that the flow rate will be slower than ionic diffusion. Consequently the dissolution of detrital minerals will control pore fluid composition and hence the diagenetic sequence and different diagenetic pathways may be followed simultaneously in adjacent sectors of the mudrock.

5.5 Summary

X-ray diffraction analyses in this study have shown that the Jurassic North Sea mudrocks, even at similar burial depths, have diverse mineralogical compositions. Back-scattered electron imaging electron microscopy has shown that this diversity is not limited to detrital mineralogy, but also applies to authigenic phases, and that these mudrocks are not the homogeneous sediments that XRD data alone might imply. Moreover, in the absence of SEM textural data it will often not be possible to adequately differentiate authigenic and detrital minerals, which is obviously imperative for any interpretation of diagenetic processes. For example, SEM analyses have shown the illite–smectites in Jurassic mudrocks to be both detrital and diagenetic in orgin and not simply the end-product of an illitization reaction similar to the Texas Gulf Coast Model as suggested by some previous studies of these mudrocks.

Overall the Jurassic mudrocks in the North Sea show a complex association of various diagenetic microenvironments in which the formation of clay and non-clay phases occurred in response to fluctuations in pore water and sediment compositions during diagenesis. The mudrocks exhibit aspects of multiphase diagenesis similar to those observed in sandstones, reflecting the changing response of the detrital minerals to changes in pore-water compositions, temperature and pressure during burial. Organic matter is thought to be an important factor in controlling diagenetic reactions in the mudrocks, particularly the precipitation of kaolinite, pyrite, ankerite and dolomite, and also possibly the dissolution of quartz.

The reactants required for these diagenetic reactions seem largely to be derived from the mudrocks themselves and consequently the diagenesis may be considered to take place predominantly in effectively 'closed' systems. These changes seem to occur in discrete stages rather than as gradual transitions and attempts can be made to produce paragenetic sequences for the diagenetic processes, similar to those used to represent sandstone diagenesis.

References

Ahn, J. H. and Peacor, D. R. (1986) Transmission and analytical electron microscopy of the smectite to illite transition. *Clays and Clay Minerals*, **34**, 156–64.

Astin, T. R. and Evans, I. J. (1990) Volume loss through silicate dissolution during shale diagenesis: implications for modelling the compaction and maturation histories of clastic sequences, in *International Sedimentological Congress Nottingham, Abstracts Volume*, pp. 29.

Berner, R. A. (1984) Sedimentary pyrite formation: an update. *Geochimica et Cosmochimica Acta*, **48**, 605–15.

Bjørlykke, K. (1983) Diagenetic relations in sandstones, in *Sediment Diagenesis, Proceedings NATO Advanced Studies Institute*, (eds A. Parker and B. W. Sellwood), Reidel, Dordrecht, pp. 169–214.

Brown, S. (1990) The Jurassic, in *Introduction to the Petroleum Geology of the North Sea*, (ed. K. W. Glennie), Blackwell Scientific Publications, Oxford, pp. 294–361.

Carothers, W. W. and Kharaka, Y. K. (1978) Aliphatic acid anions in oil-field waters—Implications for origin of natural gas. *American Association Petroleum Geologists Bulletin*, **62**, 2441–53.

Cornford, C. (1990) Source rocks and hydrocarbons of the North Sea, in *Introduction to the Petroleum Geology of the North Sea*, (ed. K. W. Glennie), Blackwell Scientific Publications, Oxford, pp. 219–54.

Curtis, C. D. (1977) Sedimentary geochemistry: environments and processes dominated by involvement of an aqueous phase. *Philosophical Transactions Royal Society London, Series A*, **286**, 353–72.

Curtis, C. D. (1987) Mineralogical consequences of organic matter degradation in sediments: Inorganic/organic diagenesis, in *Marine Clastic Sedimentology*, (eds J. K. Leggett and G. G. Zuffa), Graham and Trotman, London, pp. 108–23.

Glasmann, J. R., Larter, S., Briedis, N. A. and Lundegard, P. D. (1989) Shale diagenesis in the Bergen High area, North Sea. *Clays and Clay Minerals*, **37**, 97–112.

Hansen, P. L. and Lindgreen, H. (1989) Mixed layer diagenesis in the Upper Jurassic claystones from the North Sea and onshore Denmark. *Clays and Clay Minerals*, **24**, 197–214.

Huang, W. L., Bishop, A. M. and Brown, R. W. (1986) The effect of fluid/rock ratio on feldspar dissolution and illite formation under reservoir conditions. *Clay Minerals*, **21**, 585–602.

Hurst, A. (1982) The clay mineralogy of Jurassic shales from Brora, N.E. Scotland, in *International Clay Conference*, Bologna and Pavia, 1981, (eds H. Van Olphen and F. Veniale), *Developments in Sedimentology*, vol. 35, Elsevier, Amsterdam, pp. 677–684.

Kent, P. E. (1980) Subsidence and uplift in East Yorkshire and Lincolnshire: a double inversion. *Proceedings of the Yorkshire Geological Society*, **42**, 505–24.

Lloyd, M. G. and Hall, G. E. (1981) The SEM examination of geological samples with a semi-conductor back-scattered electron detector. *American Mineralogist*, **66**, 362–8.

Michelson, O. (1978) Stratigraphy and distribution of Jurassic deposits of the Norwegian–Danish Basin. *Geological Survey of Denmark Series B*, **2**, 28.

Nadeau, P. H., Wilson, M. J., McHardy, W. J. and Tait, J. M. (1984) Interparticle diffraction, a new concept for interstratified clays. *Clays Minerals*, **19**, 757–69.

Pearson, M. J. and Small, J. S. (1988) Illite–smectite diagenesis and palaeotemperatures in northern North Sea Quaternary to Mesozoic shale sequences. *Clay Minerals*, **23**, 109–32.

Pearson, M. J., Watkins, D. and Small, J. S. (1982) Clay diagenesis and organic maturation in northern North Sea sediments, in *International Clay Conference*, Bologna and Pavia, 1981, (eds H. Van Olphen and F. Veniale), *Developments in Sedimentology*, vol. 35, Elsevier, Amsterdam, pp. 665–75.

Pearson, M. J., Watkins, D., Pittion, J. L., Caston, D. and Small, J. S. (1983) Aspects of burial diagenesis, thermal maturation and palaeothermal history of an area in the South Viking Graben, North Sea, in Geological Society of London Special Publication, No. 12, *Petroleum Geochemistry and Exploration of Europe*, (ed. J. Brooks), Blackwell Scientific Publications, Oxford, pp. 161–73.

Primmer, T. J. and Shaw, H. F. (1987) Diagenesis in shales: evidence from back-scattered electron microscopy and electron microprobe analyses, in *Proceedings of the International Clay Conference*, Denver, 1985, (eds L. G. Schultz, H. Van Olphen and F. A. Mumpton), The Clay Minerals Society, Indiana, pp. 135–43.

Reynolds, R. C. and Hower, J. (1970) The nature of interlayering in mixed layer illite–montmorillonites. *Clays and Clay Minerals*, **18**, 25–36.

Taguchi, K., Hasegawa, K. and Suzuki, T. (1987) The relationship between silica minerals and organic matter diagenesis: its implications for the origin of oil. *Organic Geochemistry*, **13**, 97–108.

Warren, E. A. (1987) The application of a solution-mineral equilibrium model to the diagenesis of Carboniferous sandstones, Bothamsall oil field, East Midlands, England, in Geological Society of London Special Publication, No. 36, *Diagenesis of Sedimentary Sequences*, (ed. J. Marshall), Blackwell Scientific Publications, Oxford, pp. 55–69.

White, S. H., Shaw, H. F. and Huggett, J. M. (1984) The use of back-scattered electron imaging for the petrographic study of sandstones and shales. *Journal of Sedimentary Petrology*, **54**, 487–94.

CHAPTER SIX

Diagenetic pore fluid evolution in the Kimmeridge Clay Formation: from concretions to sandstone cements

I. C. Scotchman

6.1 Introduction

Sedimentation rate and burial depth have been demonstrated to be major controls on marine organic-rich mudrock diagenesis (Irwin *et al.*, 1977; Coleman *et al.*, 1979, Curtis, 1987; Burns and Baker, 1987), and in particular the evolution of their pore fluids during burial and diagenesis (Curtis and Coleman, 1986; Scotchman, 1991).

Depth-related reactions based on the oxidation of the organic matter content of the mudstones have profound effects on the composition of the mudrock pore fluids, particularly during early diagenesis. The sequence of these reactions is due, in the presence of excess organic matter, to the thermodynamics of the individual reactions (Froelich *et al.*, 1979) and the oxygenation level of the pore waters (Coleman, 1985). During early diagenesis, sedimentation rate appears to be the overall control on the sequence of reaction zones and their relative importance and thickness (Scotchman, 1989).

With increasing sedimentation rate, at shallow burial depths, reactions change from microbially-mediated iron reduction (FeR) through sulphate reduction (SR) to methanogenesis (Me) (Curtis, 1978; 1987) (Figure 6.1). At burial depths greater than 1 km and generally associated with high sedimentation rates, thermally-mediated decarboxylation (D) reactions occur, continuing to depths of 2.5–3 km when catagenic hydrocarbon generation reactions become important. Generation of CO_2 and organic acids in the lower part of the D zone appear important in the generation of secondary porosity in sandstone reservoirs (Hayes, 1979; Schmidt and McDonald, 1979; Al-Shaieb and Shelton, 1981; Franks and Forester, 1984; Surdam and Crossey, 1985a; Curtis, 1987). The production of acidic pore fluids in the D zone appears to be related to the mobilization of silica and its precipitation as quartz and clay

Geochemist of Clay-Pore Fluid Interactions, Edited by D. A. C. Manning, P. L. Hall and C. R. Hughes Published in 1993 by Chapman & Hall, London. ISBN 0 412 48980 5

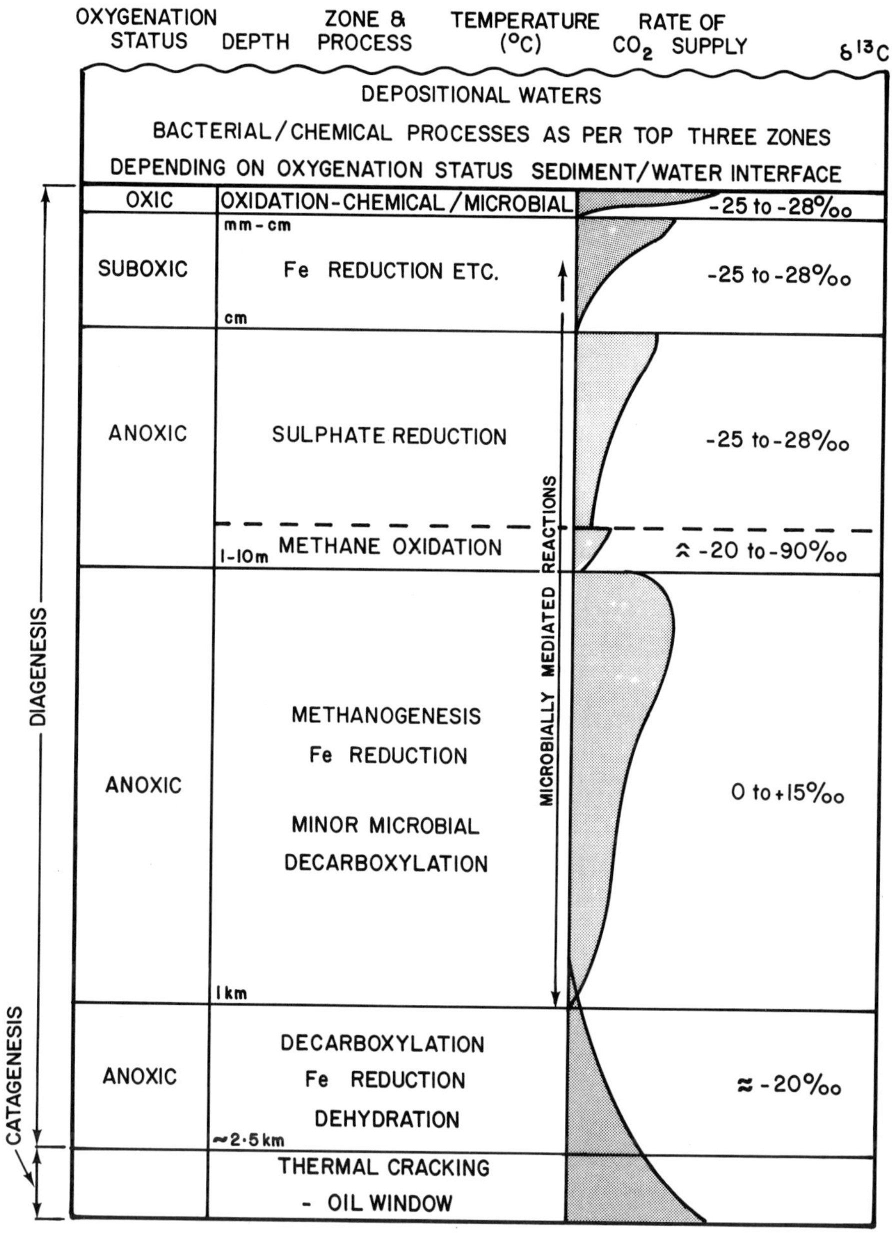

Figure 6.1 Depth-related diagenetic reaction zones based on the degradation of organic matter showing carbonate by-product $\delta^{13}C$ composition. (After Curtis, 1978.)

cements (largely kaolinite and illite) in sandstone reservoirs, often just prior to filling by hydrocarbons (Hancock and Taylor, 1978; Glasmann *et al.*, 1989a; Scotchman *et al.*, 1989).

Finally, at depths of 2–3 km, clay diagenetic reactions such as the illitization of smectite (Hower *et al.*, 1976, Pearson *et al.*, 1982; Pearson and Small, 1988) and conversion of kaolinite to chlorite (Curtis *et al.*, 1985) are taking place, releasing chemically-bound water and cations into the pore fluids. There is, therefore, from studies of both modern and ancient organic-rich mudrocks, a progression of geochemical reactions related to organic matter degradation, controlled by the influences of sedimentation rate and burial depth (Figure 6.1).

The applicability of the diagenetic reaction zone scheme to the organic-rich mudstones of the Upper Jurassic Kimmeridge Clay Formation (KCF) has been demonstrated in the type-section at Kimmeridge Bay, Dorset, southern England (Irwin *et al.*, 1977), in the formation as a whole in the onshore UK (Scotchman, 1989) and in part in the North Sea Central Graben (Scotchman, 1988). Sedimentation rate and burial depth appear to be major controls on the diagenetic reactions (Scotchman, 1989) resulting in a variable and complex evolution of pore fluid composition. Results of studies of diagenetic carbonates from the KCF from the North Sea graben system are presented and data from the onshore sequence (Scotchman, 1991)* are reviewed to ascertain the evolutionary history of pore fluid composition within the formation and determine controlling factors.

6.2 Geological setting

The KCF organic-rich mudstone sequence was deposited by a widespread epicontinental sea across much of north-west Europe, during the end Jurassic transgression (see Ziegler, 1988, Plate 13; Figure 6.2). Depositional setting varied widely, from slow deposition and shallow burial of the basin margin shelf and swell areas to rapid deposition and deep burial in basinal areas, particularly in the North Sea graben system. Details of the depositional setting and burial history of the onshore UK section are discussed in Scotchman (1989).

Offshore, the KCF comprises a sequence of organic-rich mudrocks with dolomitic carbonate bands throughout much of the North Sea graben system (Brown, 1990; Vollset and Doré, 1984). Though core material from only very limited sections has been studied, electric log data suggest this to be representative of much of the section (Scotchman, unpublished data; Norwegian Petroleum Directorate, 1977). In the North Sea, sedimentation rates were

*Full tabulated data are available as a supplementary publication No. SUP 15002 available (upon payment) from British Library Document Supply Centre, Boston Spa, West Yorkshire LS3 7BQ, UK, or upon request from the author.

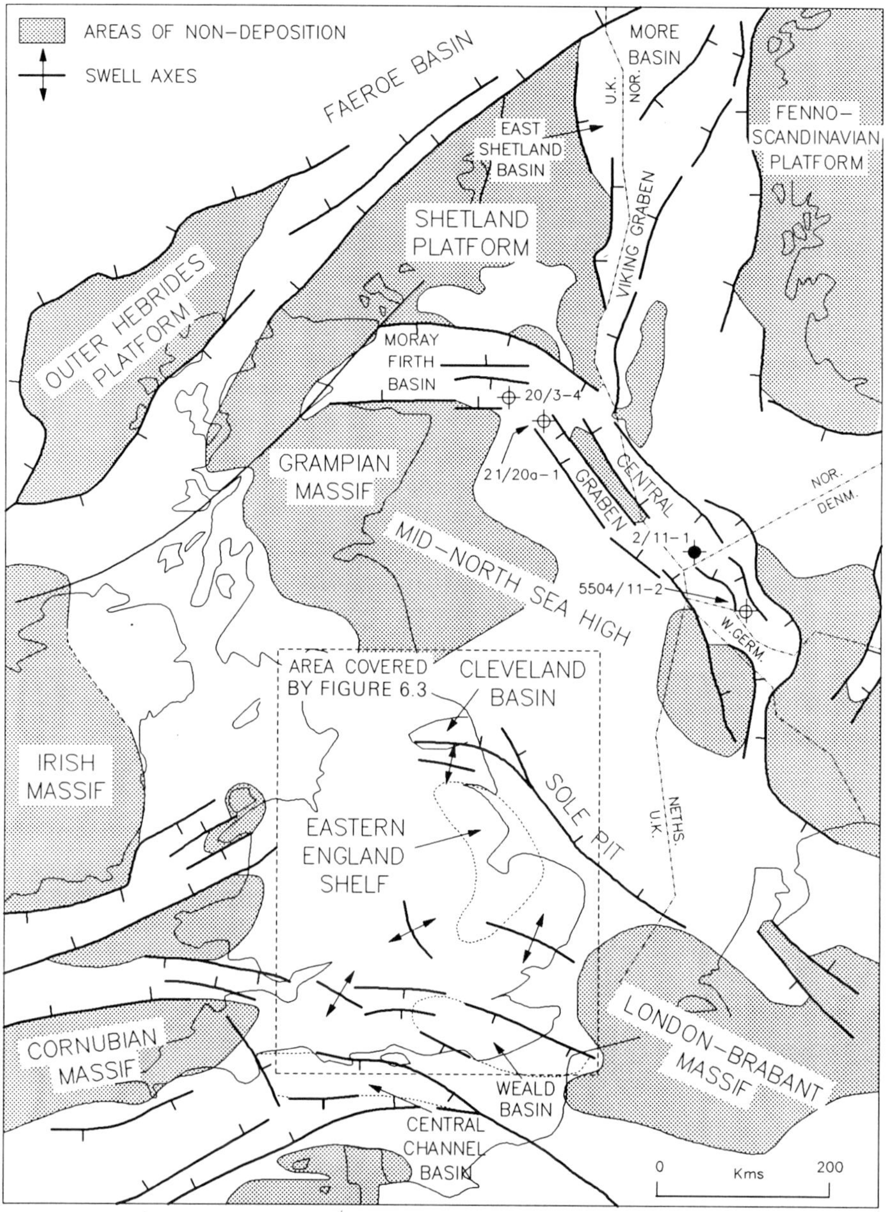

Figure 6.2 Distribution of the Upper Jurassic Kimmeridge Clay Formation in the UK and North Sea area showing sample locations.

highest in the central areas of the Central and Viking Grabens where thicknesses of the whole Upper Jurassic shale sequence reach 500–1000 m, while burial depths, currently at maximum, exceed 4–5 km. In basinal areas flanking the graben, such as the East Shetland Basin or the Outer Moray Firth, KCF thicknesses are of the order of 500 m with burial attaining 3–4 km. Sedimentation rates are difficult to estimate in the offshore sections due to lack of detailed biostratigraphical zonation comparable to that of the onshore section. It should be noted that the 'Kimmeridge Clay Formation' of the North Sea graben system has a much larger chronostratigraphic span (Oxfordian to Ryazanian) than the Kimmeridgian-aged Dorset type-section (Rawson and Riley, 1982; Doré *et al.*, 1985).

6.3 Sampling and analytical techniques

Two sample sets were used in this study; concretions from the onshore UK outcrop/sub-crop reported in Scotchman (1991) (Figure 6.3) and concretions and diagenetic carbonates from core in three wells in the Outer Moray Firth and Central Graben areas of the North Sea (Figure 6.2). Data from an additional well in the Norwegian Central Graben are also included (Scotchman, 1988).

Analytical techniques for the onshore sample set are reported in Scotchman (1991) and new analyses were made in a similar manner with sampling of concretion bodies and cements using a dentist's drill. Carbonate mineralogy was determined by standard X-ray diffraction (XRD) and staining techniques. Stable isotope ratios were measured on a Finnigan mass-spectrometer at Amoco Production Co.'s Tulsa Research Center and are reported in standard ‰ notation relative to the PDB standard. Oxygen isotopic analyses of silicate cements were made by Krueger Enterprises of Cambridge, Massachusetts, USA and are reported in ‰ relative to the SMOW standard. Pore fluid $\delta^{18}O$ values are also reported in ‰ relative to SMOW. Chemical analysis of the carbonate cements by electron microprobe was performed by Dr T. R. Astin at the Postgraduate Research Institute for Sedimentology, University of Reading.

6.4 Results

6.4.1 Onshore concretions

Data from the onshore UK sample set, fully discussed in Scotchman (1991), are summarized in Figures 6.4 and 6.5. The samples show a wide range of isotopic compositions, $\delta^{18}C$ ranging from -23.3 to 0.6‰ and $\delta^{18}O$ from -14.5 to 1.1‰. Standard staining and XRD techniques show the majority of onshore concretions to comprise non-ferroan calcite cement in the concretion matrix and first-generation septarian cement while second and any later generations of cement are ferroan calcite. Exceptions are concretion D179.95

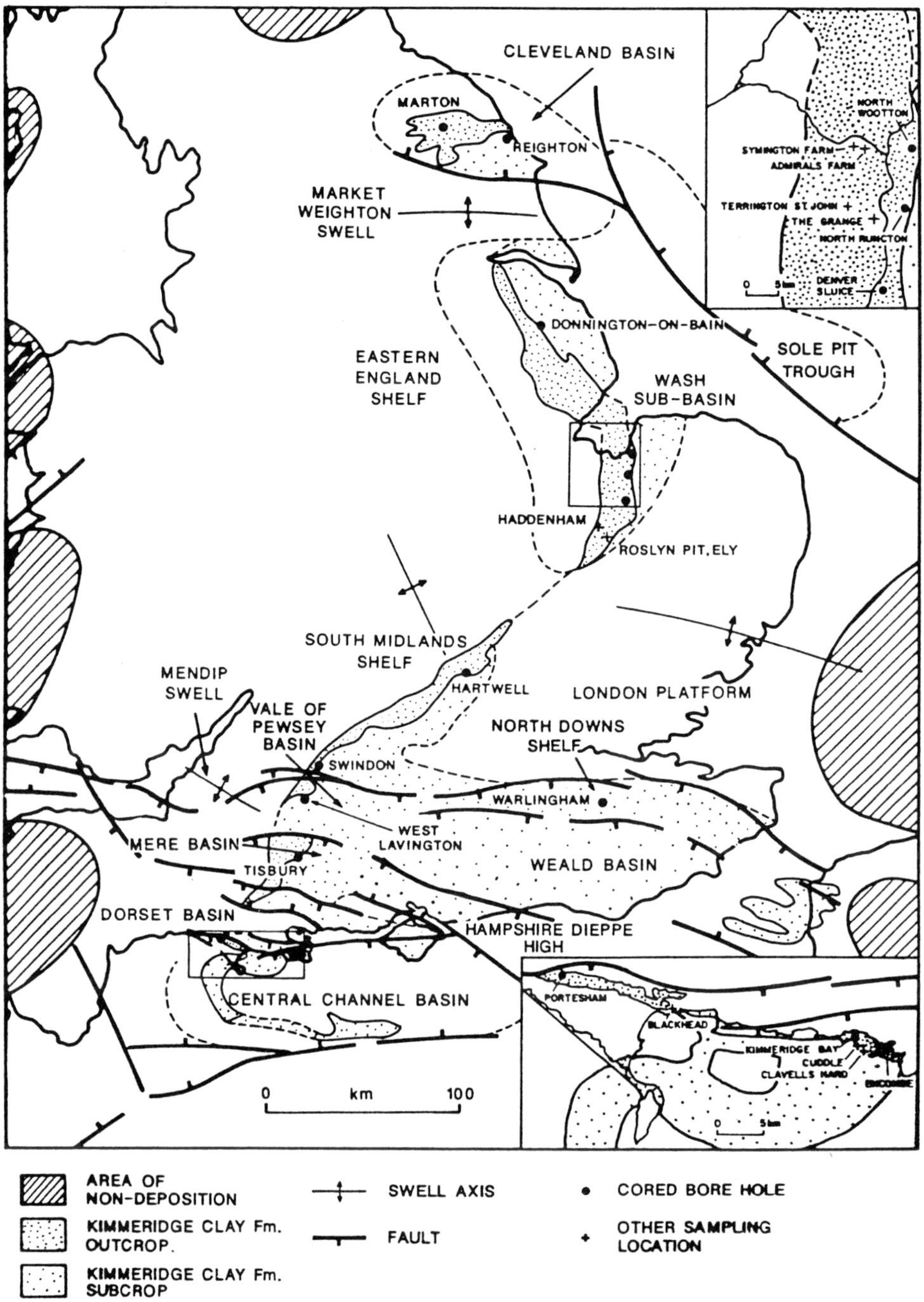

Figure 6.3 Kimmeridge Clay Formation distribution and sampling locations in southern and eastern England. (From Scotchman, 1991.)

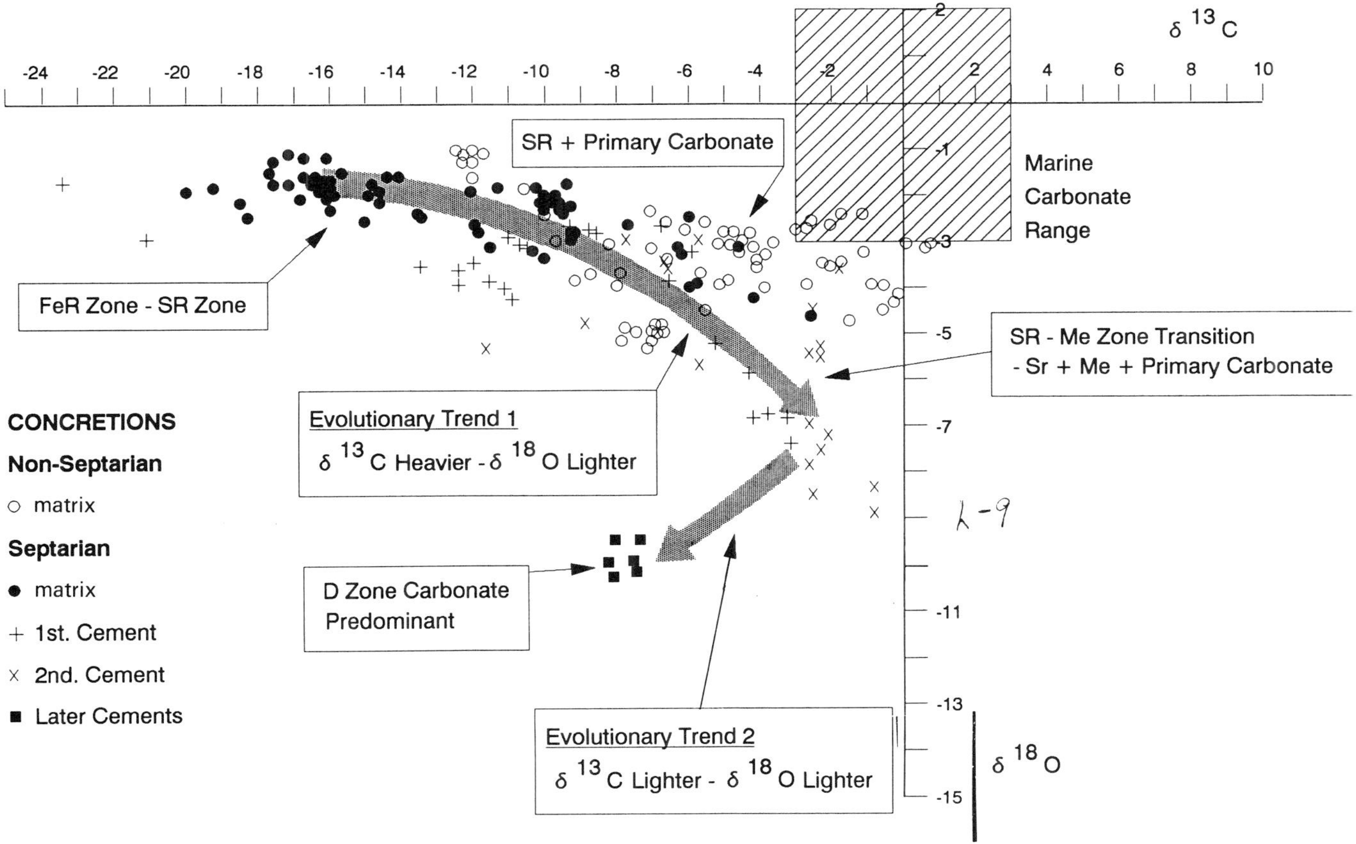

Figure 6.4 Isotopic composition of early diagenetic concretions from the onshore section. (Modified from Scotchman, 1991.)

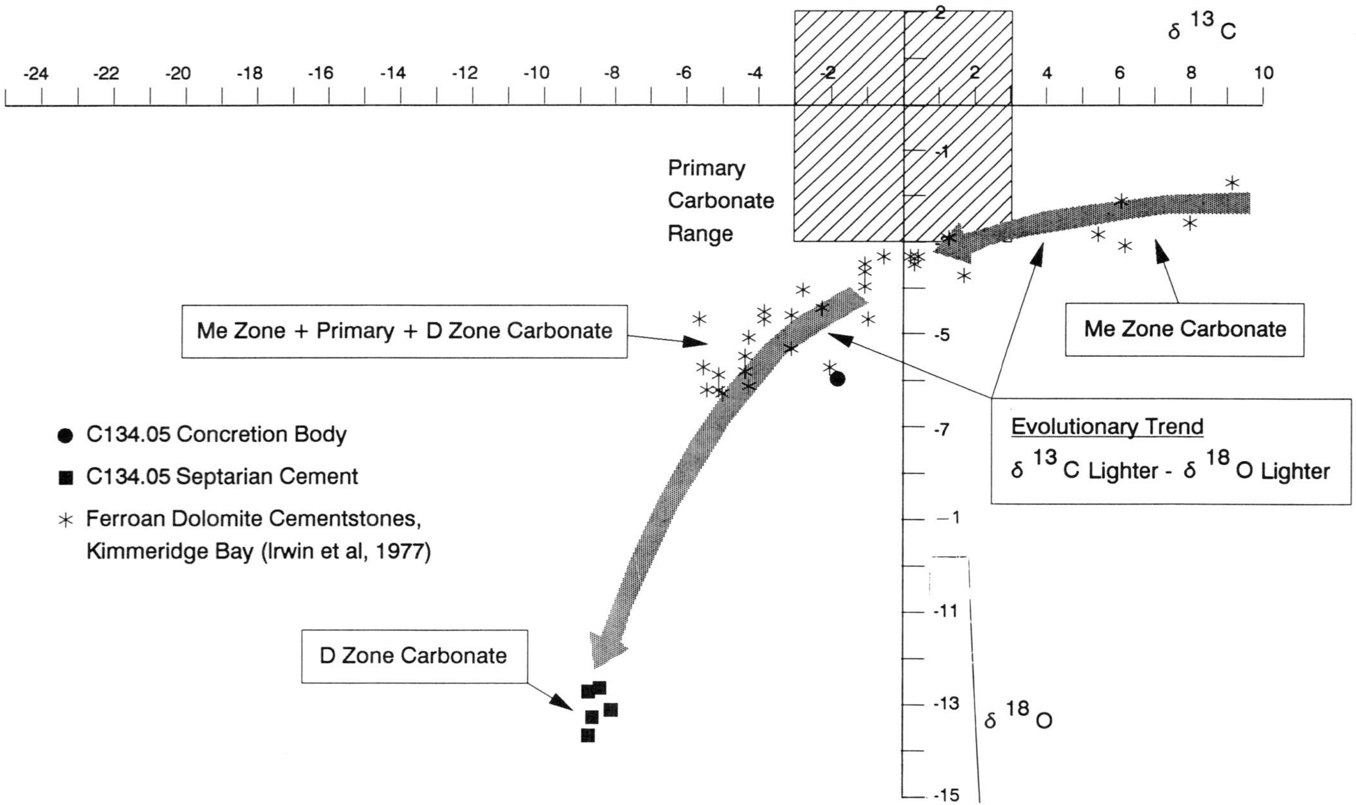

Figure 6.5 Isotopic composition of later diagenetic carbonates from the onshore section. (Modified from Scotchman, 1991.)

Table 6.1 Geochemical data.

Sample	Description	Mineralogy	Carbonate	Chemistry			$\delta^{13}C$	$\delta^{18}O$	
			Ca (wt%)	Mg (wt%)	Fe (wt%)	Mn (ppm)	(‰)	(‰ PDB)	(‰ SMOW)
Well, 20/3-4									
10 318′4″	Conc. matrix	Dol.	—	—	—	—	+1.0	−3.9	—
10 318′6″–10 321′6″									
A	Conc. matrix	Dol.	—	—	—	—	+7.4	−2.8	—
B	Dk. sept. cmt	Dol.	—	—	—	—	+9.9	+1.7	—
C	Grey sept. cmt	Dol.	—	—	—	—	−0.8	−8.6	—
D	White sept. cmt	Dol. + Qtz	—	—	—	—	+1.2	−9.9	—
E	Clear coarse cmt	Qtz	—	—	—	—	—	—	+26.3, +26.5
F	Clear coarse cmt	Qtz	—	—	—	—	—	—	+21.1
10 318′11″	Conc. matrix	Dol?	—	—	—	—	+9.0	+1.0	—
10 321′0″	Conc. matrix	Dol.	—	—	—	—	+9.0	+1.1	—
10 322′2″	Conc. matrix	Dol.	—	—	—	—	−4.0	−6.5	—
10 323′9″	Conc. matrix	Dol.	—	—	—	—	0.0	−4.6	—
10 323′10″	Conc. matrix	Dol.	—	—	—	—	+6.5	−1.1	—
10 323′10″	Conc. matrix	Dol.	—	—	—	—	+6.1	−1.5	—
10 324′0″–10 326′6″									
A	Conc. matrix	Dol.	52.0	47.7	0.3	200	+8.0	−0.2	—
B	Dk. sept. cmt	Dol.	53.0	46.7	0.3	0	—	—	—
			54.2	45.0	0.8	0	—	—	—
			54.7	44.7	0.6	0	+8.8	−1.2	—
			54.9	45.0	0.6	0			
			54.6	44.7	0.7	100			
			53.2	43.6	3.2	200			
C	2nd sept. cmt	Saddle dol.	54.3	33.5	12.2	400			
			53.9	33.9	12.2	300	+6.1	−0.9	—
			53.4	32.0	14.5	300			

Table 6.1 *(contd)*

Sample	Description	Mineralogy	Carbonate	Chemistry			$\delta^{13}C$	$\delta^{18}O$	
			Ca (wt%)	Mg (wt%)	Fe (wt%)	Mn (ppm)	(‰)	(‰ PDB)	(‰ SMOW)
	Sept. cmt.	Qtz	—	—	—	—	—	—	—
	3rd sept. cmt.	Saddle dol.	53.7	39.4	6.9	0			
			53.7	37.8	8.6	300	—	—	—
			54.2	40.4	5.4	300			
	4th sept. cmt.	Saddle dol.	53.5	35.1	11.4	400	—	—	—
			52.9	35.0	12.1	400			
			52.8	36.2	11.0	300			
10 325′0″	Conc. matrix	Dol.	—	—	—	—	−4.2	−5.8	—
10 326′0″	Conc. matrix	Dol.	—	—	—	—	+8.7	+0.8	—
10 327′2″	Conc. matrix	Dol.	—	—	—	—	−4.9	−6.8	—
10 327′7″	Conc. matrix	Dol.	—	—	—	—	+3.7	−4.1	—
Well 21/20a-1									
13 296′	Conc. matrix	Dol.	—	—	—	—	−3.0	−7.7	—
13 630′3″	Conc. matrix	Dol.	—	—	—	—	+2.1	−3.5	—
13 631′	Conc. matrix	Dol.	52.0	45.2	2.7	400	—	—	—
			51.6	45.2	3.2	300			
	1st sept. cmt.	Saddle dol.	55.1	33.1	11.8	400	—	—	—
			56.8	32.4	10.8	200			
			56.8	31.3	11.8	300			
			57.3	29.6	13.1	200			
			56.0	31.3	12.6	200			
			56.0	29.3	14.7	200			
			55.6	29.5	14.9	300			
			54.4	29.9	15.7	300			
			56.4	29.2	14.4	400			

Table 6.1 *(contd)*

Sample	Description	Mineralogy	Carbonate	Chemistry			$\delta^{13}C$	$\delta^{18}O$	
			Ca (wt%)	Mg (wt%)	Fe (wt%)	Mn (ppm)	(‰)	(‰ PDB)	(‰ SMOW)
	2nd sept. cmt.	Saddle dol.	53.4	28.8	17.8	600			
			53.3	28.6	18.2	900			
			52.8	28.7	18.5	500			
			53.1	28.7	18.3	700			
			53.6	28.4	18.0	600			
			52.4	29.5	18.1	600			
			52.5	29.7	17.8	500			
			52.6	28.0	19.3	300			
13 633′0″	Conc. matrix	Dol.	—	—	—	—	−2.2	−5.5	—
13 634′0″	Conc. matrix	Dol.	—	—	—	—	−1.5	−4.1	—
13 635′0″	Conc. matrix	Dol.	—	—	—	—	−2.5	−6.6	—
13 652′8″	Conc. matrix	Dol.	—	—	—	—	−2.1	−10.0	—
13 653′2″	Conc. matrix	Dol.	—	—	—	—	−1.6	−4.7	—
Denmark 5504/11-2									
9948′	Mst. matrix	Dol.	—	—	—	—	+0.8	−5.7	—

Conc. = concretion. Sept. = septarian. Cmt. = cement. Dk. = dark. Dol. = dolomite. Qtz = quartz. Mst. = mudstone.

Table 6.2 Maturity data.

Sample	TOC (wt%)	TMAX (°C)	PI*	Elemental ratios H/C	O/C	V.R. (%Ro)	Hopane†	Sterane‡	Tri-aromatic sterane§
Well 20/3-4									
10 301′8″	3.6	426	0.09	1.12	0.08	0.29	0.58	0.28	0.04
10 324′7″	2.7	436	0.07	1.09	0.09	0.29	0.59	0.28	0.05
10 340′	3.6	431	0.08	1.18	0.07	0.29	0.56	0.30	0.06
Well 21/20a-1									
13 231′	4.5	442	0.26	0.71	0.05	0.63	0.60	0.55	0.28
13 300′6″	4.1	431	0.38	0.67	0.07	0.67	0.57	0.62	0.68
13 625′	4.1	445	0.38	0.72	0.06	0.82	0.50	0.67	0.79
13 652′8″	1.4	468	0.27	0.63	0.05	0.93	0.56	0.58	0.68
Well NOCS 2/11-1									
12 676′	7.3	427	0.09	—	—	0.33	0.57	0.50	0.10
12 717′	10.8	431	0.07	1.10	0.07	0.31	0.56	0.53	0.10
12 720′	1.7	427	0.28	1.02	0.09	0.32	0.60	0.49	0.06

*Productivity Index $S_1/(S_1+S_2)$.
†C_{31} 22S/(22S+22R) ratio.
‡C_{29} 20S/(20S+20R) ratio.
§$C_{20}+C_{21}/(C_{20}+C_{21}+C_{26}S+C_{26}R+C_{27}S+C_{28}S+C_{27}R+C_{28}R)$ cracking ratio (Lewan *et al.*, 1986).

from Donington-on-Bain Borehole and the pyritic nodule KC from the Kimmeridge Bay section where the cements are entirely ferroan calcite. Ferroan dolomites occur in the cementstone bands of Kimmeridge Bay (Irwin *et al.*, 1977; Irwin, 1980; 1981) and in sample C134.05 from Marton Borehole.

6.4.2 Offshore concretions

These concretions are of broadly equivalent Middle Volgian age and occur at 10,318′6″ (3145.1 m) and 10,324′ (3146.9 m) drilled depth in Amoco well 20/3–4 (Kelly Bushing Elevation (KBE)=77′, 23.5 m) and at 13,631′ (4154.7 m) and 13,652′ (4161.1 m) drilled depth in Amoco well 21/20a-1 (KBE=82′, 25.0 m). These wells are located in the Outer Moray Firth, well 20/3–4 being situated on a structural high, downdip of the Ettrick Oilfield, while well 21/20a-1 lies in a downflank, basinal setting in the West Forties Basin. Geochemical analysis of core samples adjacent to the concretions indicates the KCF to be early mature in wells 20/3–4 and NOCS 2/11–1 and well into oil window, beyond peak oil, in well 21/20a–1 (Table 6.1 and 6.2). The Upper Jurassic section in this area is overpressured (Buhrig, 1989), overpressure exceeding 6000 psi in well 21/20a-1.

The concretions are septarian and the cements are predominantly dolomitic with saddle dolomite becoming important in later cements. Unusually, silica in the form of both massive quartz and chalcedony occurs as late cements in concretions from well 20/3–4. Elsewhere in the North Sea Central Graben, dolomitic, non-septarian diagenetic carbonates occur in cemented intervals in Norwegian well (NOCS) 2/11-1 (Scotchman, 1988) and in Danish Central Graben well 5504/11-2.

Isotopic composition is highly variable in the offshore samples, both between successive cements and between individual concretions, with $\delta^{13}C$ ranging from $+4.9$ to $-10.0‰$ and $\delta^{18}O$ from -10.0 to $+1.7‰$ (Figure 6.6). In particular very light $\delta^{13}C$ values of the -25 to $-10‰$ range typically found in concretions from the onshore shallow burial sequences of KCF are not seen, suggesting a deeper and later diagenetic regime in the offshore section. Oxygen isotopes in the two late generations of quartz cement in sample 10,318′6″ from well 20/3-4 give $\delta^{18}O$ SMOW values of $+21.1$ and $+26.4‰$.

Microprobe analysis of the offshore samples shows them to have a complex, but predominantly ferroan, sequence of cements. The matrix cements in the 21/20a-1 concretion are of ferroan dolomite while in the concretions from 20/3-4 the matrix is of dolomite.

Septarian cements comprise a first generation of dolomitic cement in the concretions from well 20/3-4 (10,324′5″) (Figure 6.7), associated with up to four phases of fracturing, followed by a thin saddle dolomite cement. A zone of quartz cement associated with a further episode of fracturing is followed by two generations of iron-rich saddle dolomite. The shallower concretion from well 20/3-4 (10,318′6″) shows a different sequence of septarian cements (Figure 6.8), comprising a first generation of dolomite with two episodes of fracturing,

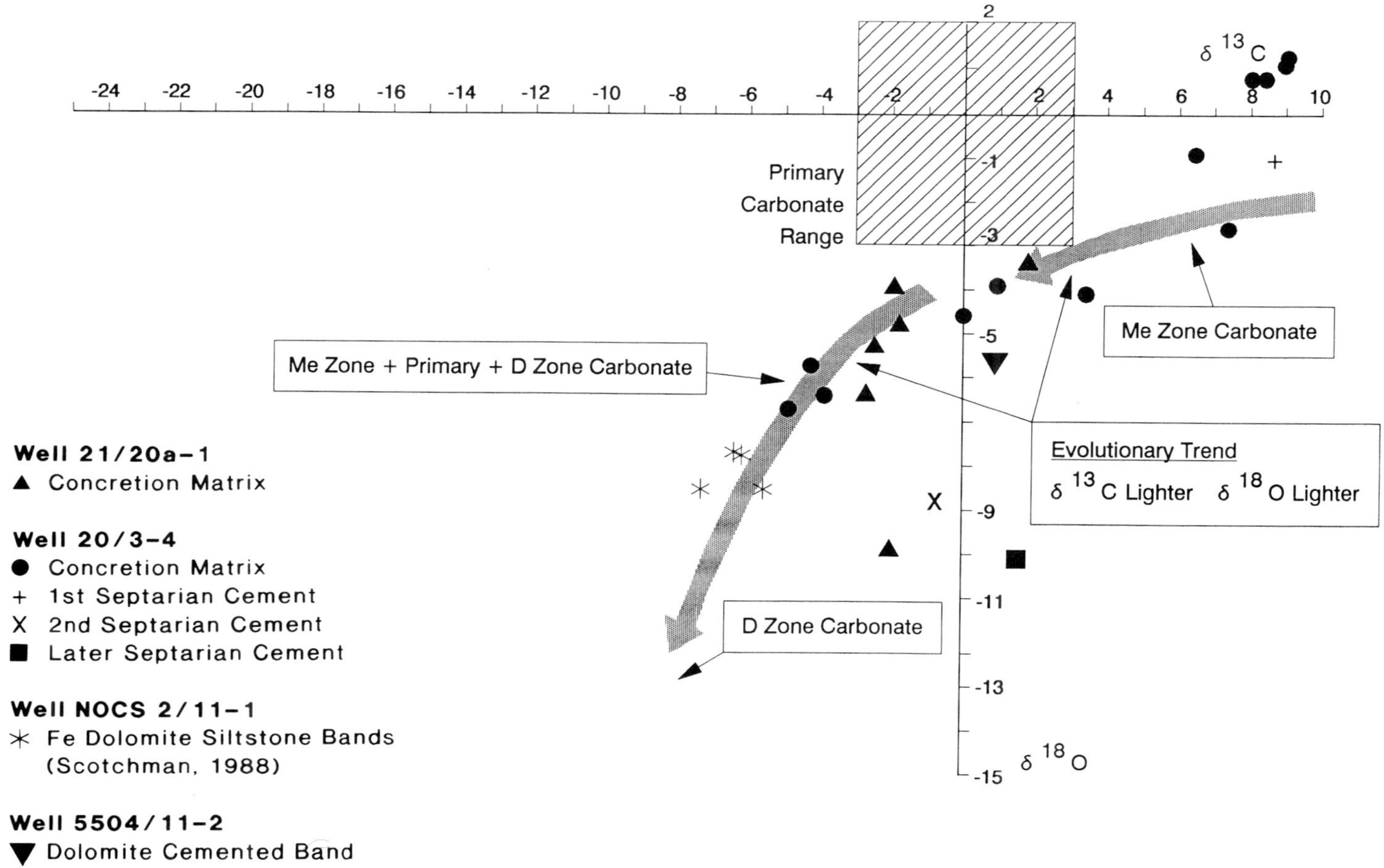

Figure 6.6 Isotopic composition of offshore concretions and diagenetic carbonates.

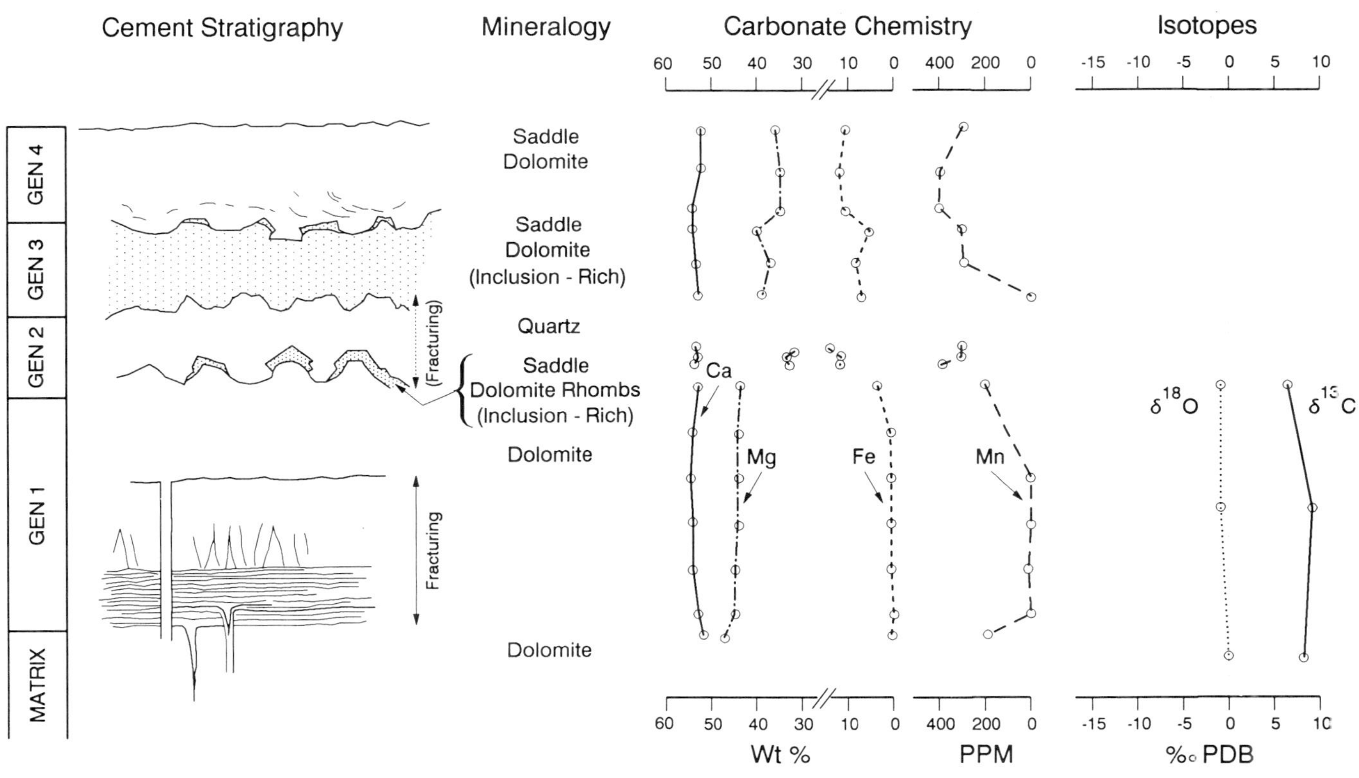

Figure 6.7 Geochemistry of septarian cements in lower concretion from well 20/3–4 at 10 324′ 5″.

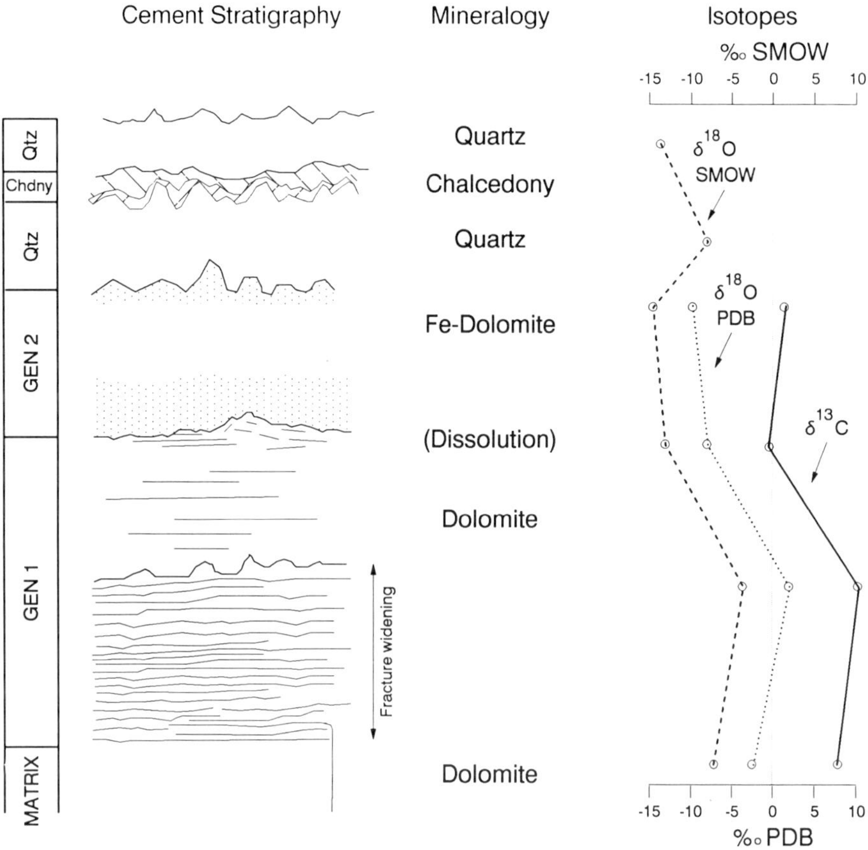

Figure 6.8 Geochemistry of septarian cement sequence in upper concretion from well 20/3–4 at 10 318′ 6″.

ending with a dissolution surface. A ferroan dolomite cement is followed by two generations of coarsely crystalline quartz cement separated by a powdery white chalcedony zone.

The carbonate concretion matrix and the septarian cements both show trends to lighter $\delta^{13}C$ and $\delta^{18}O$, starting with a similar isotopic composition. However, the successive septarian cements become more rapidly depleted in oxygen than the successive matrix cements. $\delta^{13}C$ of the septarian cements also becomes heavier and increases from -0.8 to $+1.2‰$ (Figure 6.6). Carbonate precipitation then ceased, being followed by the sequence of silicate cements.

The upper concretion from well 21/20a-1 (13,631′) contains two generations of iron-rich saddle dolomite septarian cement, separated by a dissolution zone (Figure 6.9). The dolomitic matrix cements of the concretions from well 21/20a-1 also show a lightening $\delta^{18}C$ and $\delta^{18}O$ isotopic trend with low positive $\delta^{13}C$ values and $\delta^{18}O$ in the range -3.5 to $-5.0‰$ in the earliest cements. The

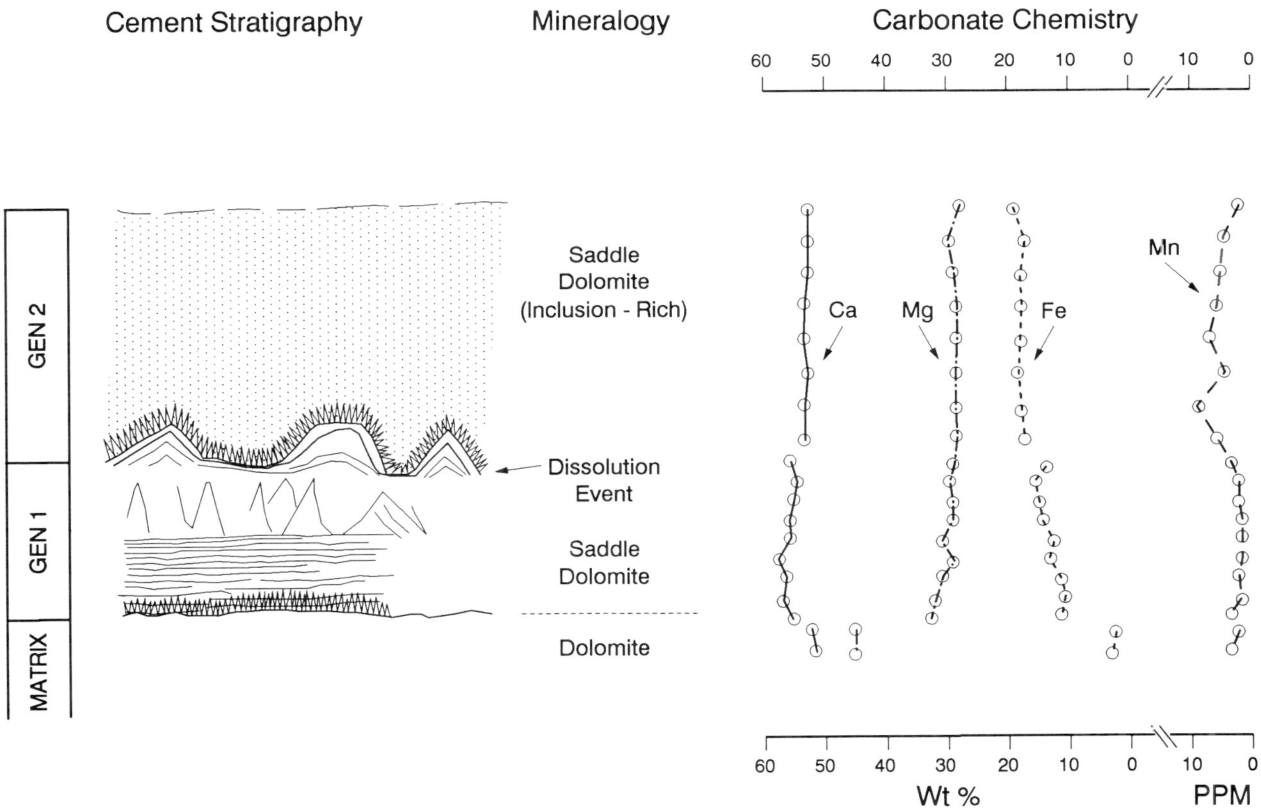

Figure 6.9 Geochemistry of septarian cement sequence from well 21/20a–1 concretion at 13 631′.

later cements have $\delta^{13}C$ and $\delta^{18}O$ values of -2.1 to $-3.0‰$ and -6.6 to $-10.0‰$, respectively.

The offshore concretions show a trend of increasing iron and decreasing magnesium through successive septarian cements, starting from the concretion matrix (Figure 6.9), the switch to silicate deposition marking the end of the trend. Calcium remains relatively constant.

The cemented intervals in NOCS well 2/11-1 comprise ferroan dolomites with $\delta^{13}C$ of -5.8 to $-7.3‰$ and $\delta^{18}O$ of -7.7 to $-8.5‰$ (Scotchman, 1988). The similar dolomite-cemented interval from Danish well 5504/11-2 has an isotopic composition of $+0.8‰$ $\delta^{13}C$ and $-5.7‰$ $\delta^{18}O$.

6.5 Discussion

6.5.1 Pore fluid evolution in the onshore section

Textural, isotopic and geochemical data from concretions of the shallow buried onshore section suggest growth soon after burial with a sedimentation-rate-controlled pore fluid evolutionary sequence. Pore waters were initially marine within the organic-rich sediments which contain excess iron (Davey and Curtis, 1989; J. H. S. Macquaker, personal communication). The earliest reactions occurred in the condensed, low sedimentation rate, swell sequences. FeR zone iron-rich calcites were precipitated from sub-oxic pore waters with light $\delta^{13}C$ (-20 to $-23‰$) and near zero ‰ $\delta^{18}O$ indicating shallow, low-temperature precipitation.

SR zone conditions were prevalent over much of the basin as the first diagenetic zone, except in the rapidly subsiding basinal areas where these reactions occurred in the bottom waters. Light carbon (-15 to $-16‰$ $\delta^{13}C$) associated with marine, near zero $\delta^{18}O$ (-1 to $-2‰$) characterized the SR zone pore waters, non-ferroan calcite and pyrite being precipitated. The calcites were non-ferroan due to the preferential partitioning of iron into pyrite, forming an iron 'sink'. Pore-water evolution through the SR zone was characterized by the addition of increasing amounts of heavier carbonates, largely from remobilization of primary carbonates from shell and coccolithic material in the sediments, and the depletion of oxygen. While a satisfactory explanation for the lightening of $\delta^{18}O$ in the SR zone has yet to be put forward, suggested causes include mixing with meteoric water (Hudson, 1978), precipitation of diagenetic minerals (Irwin *et al.*, 1977) and ultrafiltration by clay minerals (Coplen and Hanshaw, 1973). The latter has not been demonstrated beyond laboratory scale (Benzel and Graf, 1984) while mixing with depleted meteoric waters can be discounted at shallow depths beneath marine waters.

Precipitation of diagenetic minerals, principally carbonate, appears the likely cause of oxygen depletion, as suggested by Sass *et al.* (1990). They propose that under the increasingly restricted pore-water circulation conditions at the base of the SR zone, precipitation of carbonate as a by-product of organic matter oxidation causes effective pore-water oxygen depletion.

In parts of the shelf area with low sedimentation rate and burial, pore-water evolution ceased at the base of the SR zone with no further organic-matter-related reactions (Astin and Scotchman, 1988). The SR zone signature of light $\delta^{13}C$ non-ferroan calcite is, therefore, carried to considerable depth, beyond the 10 m usually associated with the zone (Curtis 1978; 1987), although increasing dilution by marine carbon occurs. Cation concentrations vary along the SR zone evolutionary trend, magnesium decreasing as the influence of marine pore waters declines while iron increases. The increase in iron causes the calcite composition to change from non-ferroan to ferroan in the late SR zone concretions, such as those from Kimmeridge Bay. This reflects the waning influence of sulphate reduction and the attendant pyrite iron 'sink'. The relative timing of growth of concretions within the SR zone can be determined by their position along this pore-water evolution curve. This is seen in pyritic nodules and concretions from Kimmeridge Bay where ferroan calcite with $\delta^{13}C$ values of $-5‰$ indicate late formation near the base of the SR zone.

The succeeding phase of pore-water evolution is exhibited by the basinal facies dolomitic cementstones at Kimmeridge Bay. Here Me zone conditions developed just below the sediment–water interface where the pore waters were marine but depleted in sulphate. With the absence of pyrite precipitation, no iron 'sink' was available so the carbonates were iron-rich and dolomitic, the magnesium probably being derived from seawater, due to close proximity of the sediment–water interface. Me zone carbonates are typified by a heavy $\delta^{13}C$ of $+5$ to $+15‰$, values up to $+16‰$ being recorded in dolomitic concretions from the Monterey Formation in California (Garrison *et al.*, 1984), and a marine $\delta^{18}O$ (typically -2 to $-3‰$: Irwin *et al.*, 1977).

At Kimmeridge Bay, ferroan dolomite cementstones precipitated in the Me zone, the later cements showing a trend of rapidly lightening $\delta^{13}C$ with relatively constant $\delta^{18}O$. Initially, lightening of $\delta^{13}C$ appears to have been due to the addition of near zero ‰ marine carbonate before a more rapid change of $\delta^{13}C$ in the later cements due to increasing amounts of light (-6 to $-20‰$ $\delta^{13}C$) D zone carbonate.

The rate of change of $\delta^{13}C$ decreases through the Me–D zone transition while that of $\delta^{18}O$ rapidly increases, $\delta^{18}O$ becoming much lighter in the later cements. This suggests near zero ‰ $\delta^{13}C$ marine carbonate, from dissolution of shell and coccolithic material, formed a considerable proportion of the dissolved carbonate in the pore waters. The lack of very light $\delta^{13}C$ values probably reflects both the buffering effect of the marine carbonate pool and an early D zone carbonate having a $\delta^{13}C$ composition of about $-6‰$ (Whiticar *et al.*, 1986). The increasingly negative $\delta^{18}O$ values suggest precipitation at elevated temperatures from relatively heavy basinal pore-waters (Scotchman, 1991).

However, the possibility of a minor component of depleted meteoric water cannot be totally discounted.

The final phase of pore-water evolution in the onshore section is shown by Marton concretion C134.05 and by late septarian cements in several other concretions. These show both $\delta^{13}C$ and $\delta^{18}O$ becoming increasingly negative, suggesting a D zone carbonate source. Pore waters were enriched in magnesium and iron although magnesium declines through successive cements. The carbonates are, therefore, strongly ferroan and the lightening $\delta^{13}C$ trend suggests that buffering by marine carbonate was waning, the values of -7.9 to $-8.6‰$ indicating that the later D zone carbonate was lighter with a $\delta^{13}C$ of -10 to $-20‰$ (Irwin *et al.*, 1977). Values of $\delta^{18}O$ range from -8 to $-14‰$ in the late cements of these concretions, suggesting precipitation either from depleted waters or at high temperatures.

In several of the concretions, such as samples TA, TB and P30A, only slight lightening of $\delta^{13}C$ is associated with the dramatic $\delta^{18}O$ lightening in the late cements. This suggests that the light $\delta^{18}O$ values are due to precipitation from depleted pore waters and, in the relatively shallow burial conditions of the shelf areas, depletion from an increasing component of meteoric water is possible (Hudson, 1978; Thyne and Boles, 1989). However, Rukin (1988), in his study of Lias fibrous calcite veins, attributes a similar dramatic decrease in $\delta^{18}O$ to temperature increase associated with overpressure development rather than to any significant charge in pore-water composition. This appears a more tenable explanation, as overpressuring during rapid burial plays a key role in the development of septarian fractures and cements in concretions (Astin, 1986; Astin and Scotchman, 1988). In other samples the lightening $\delta^{18}O$ is mirrored by a lightening $\delta^{13}C$ trend (samples D179.95 and C134.05), indicative of an increasing D zone influence with precipitation at elevated temperatures.

Pore fluid evolution in the onshore section can, therefore, be demonstrated to be dependent on basinal setting with sedimentation rate being the major control (Figure 6.10). Burial rate and maximum burial depth are also important parameters; the development of septaria and cement infills in concretions can be related to periods of rapid burial punctuated by relative still-stands or uplifts (Figure 6.11). In contrast, deeper continuous burial resulted in development of Me–D zone carbonates as cementstones, their common dewatering and fracturing structures (Irwin, 1980; Leddra *et al.*, 1987) attesting to rapid burial conditions and development of overpressuring.

6.5.2 Pore fluid evolution: offshore sections

In contrast to the onshore section, the record of pore fluid evolution in the KCF samples from the Outer Moray Firth and Central Graben areas begins in the Me zone, a situation comparable to the onshore basinal sections (Figure 6.12). SR zone conditions, which were not conducive to carbonate precipitation, appear to have occurred in the top few centimetres of the sediments and

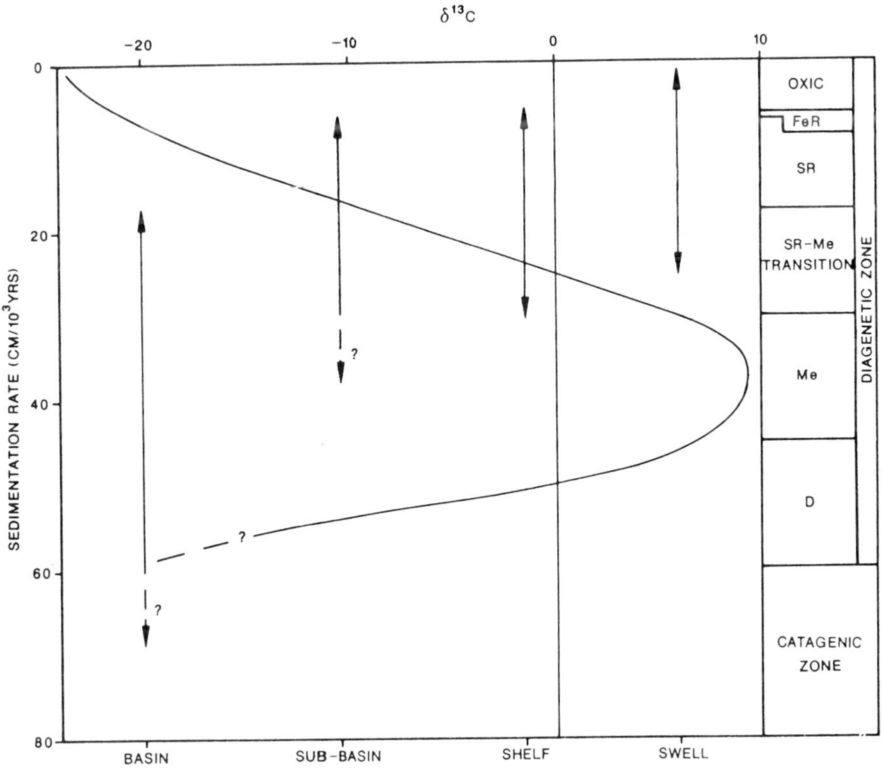

Figure 6.10 Diagram illustrating sedimentation rate control on the $\delta^{13}C$ (‰) composition of carbonate cements in the onshore KCF. (From Scotchman, 1991.)

probably extended up into the overlying water column. Me zone conditions are suggested by the heavy $\delta^{13}C$ values, while the earliest matrix cements with heavy $\delta^{18}O$ values around +1‰ indicate precipitation from either marine pore waters ($\delta^{18}O$ SMOW = −1.2‰, Shackleton and Kennett, 1975) at low temperature or from heavier basinal pore waters ($\delta^{18}O$ SMOW = +3‰) at 45°C, using the dolomite fractionation curves of Sellwood *et al.* (1989, their Figure 9). By analogy with the present day Gulf of California, where Me zone conditions are developed 0.45–0.6 m below the sediment–water interface (Goldhaber and Kaplan, 1980), low-temperature precipitation from marine pore waters appears likely, with the influence of basinal pore waters becoming more important in later cements due to the updip expulsion of compactional waters.

The concretions generally follow a similar Me zone to D zone isotopic evolution trend to the Kimmeridge Bay cementstones (Irwin *et al.*, 1977; Irwin, 1980) with a relatively faster change in $\delta^{13}C$ to lighter values compared with $\delta^{18}O$ (Figure 6.5). This suggests that the relatively lighter carbon was of marine origin, precipitated from marine pore waters at gradually increasing temperatures, as indicated by the lightening $\delta^{18}O$ composition. Burial rates appear to

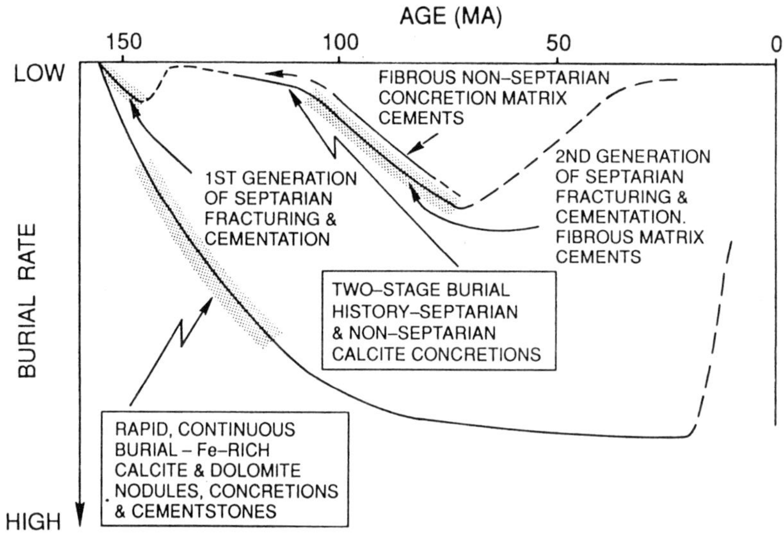

Figure 6.11 Concretion development in relation to geological history, onshore KCF. (From Scotchman, 1991.)

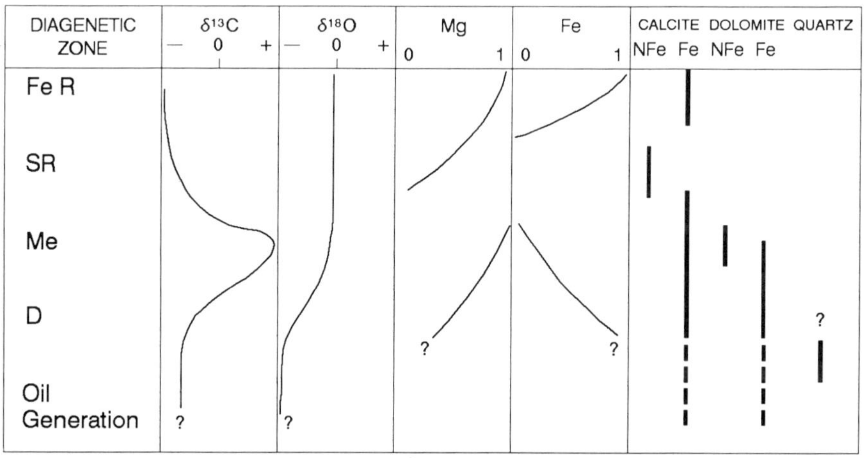

Figure 6.12 Summary of diagenetic pore fluid evolution in the KCF.

have been higher and more continuous in the North Sea grabens than in the onshore basins, as shown by the Me–D zone carbonates being septarian concretions rather than the laterally extensive cementstones which require long-lived hiatuses to develop.

The later matrix and septarian cements show a rapid lightening of $\delta^{18}O$ while $\delta^{13}C$ shows little change, suggesting a large reservoir of marine carbonate which overwhelmed the predominant light D zone carbonate being generated at this time. D zone acidic pore waters have been suggested as the cause of carbonate dissolution in many sandstones (Schmidt and McDonald, 1979; Burley, 1986; Curtis, 1987), providing a ready source of marine carbonate. The rapid change in $\delta^{18}O$ suggests both higher temperature precipitation and the increasing importance of heavier basinal pore fluids, probably of about $+3‰$ $\delta^{18}O$ SMOW as reported from present-day North Sea pore waters (Lønøy *et al.*, 1986; Egeberg and Aagaard, 1989).

The saddle dolomites, forming the later carbonate cements, are indicative of high-temperature precipitation, published data suggesting precipitation at 60–120°C from isotopic and fluid inclusion evidence (Radke and Mathis, 1980; Gregg, 1983; Woronick and Land, 1985; Gawthorpe, 1987; Sellwood *et al.*, 1987; 1989). Iron shows a steady increase through the cements in the sample from well 21/20a-1 while magnesium slowly decreases before reaching a plateau. Manganese increases through the first-generation saddle dolomite, then decreases through the second. The pore waters precipitating these late saddle dolomite cements were thus iron rich but relatively poor in magnesium and manganese.

Clay mineral reactions are potential sources of cations during deep burial (Kahle, 1965; McHargue and Price, 1982; Gawthorpe, 1987). In particular the transformation of smectite through mixed-layer illite–smectite to illite has been well documented (e.g. Burst, 1969; Perry and Hower, 1970; Foscolos and Kodama, 1974; Hower *et al.*, 1976; Pearson *et al.*, 1982; Pearson and Small, 1988). The smectite illitization reaction appears capable of releasing large amounts of cations, particularly Ca^{2+}, Mg^{2+}, Fe^{2+} and Si^{4+}, and water into the pore fluid system (Boles and Franks, 1979; Perry and Hower, 1972; Hower *et al.*, 1976). Mass balance calculatios on the Gulf Coast USA Tertiary section indicate the reaction is more than capable of accounting for the volumes of dolomite and ankerite cements within the Tertiary sandstones (Boles, 1978; Boles and Franks, 1979). Ordered illite–smectites have been reported from both the onshore and offshore KCF sections (Scotchman, 1987; Dypvik, 1983; Pearson *et al.*, 1982; 1983; Pearson and Small, 1988; Glasmann *et al.*, 1989b) and authigenic illite cements are commonplace in many North Sea sandstones (e.g. Glasmann *et al.*, 1989a; 1989c; Blanche and Whitaker 1978; Liewig *et al.*, 1987; Thomas, 1986; Jourdan *et al.*, 1987) suggesting the illitization reaction as a potential source of cations. The water released by the reaction is ^{18}O enriched (Suchecki and Land, 1983) with $\delta^{18}O$ SMOW of $+6‰$; Fisher and Land (1986) suggesting a range of -3 to $+8‰$ $\delta^{18}O$ SMOW from the Gulf Coast Tertiary Wilcox sandstones. Heavy $\delta^{18}O$ values are characteristic of the formation waters at depth within the Gulf Coast Tertiary section (Land and Prezbindowski, 1981; Woronick and Land, 1985) and other basins (Clayton *et al.*, 1966) including the North Sea (Bjørlykke *et al.*, 1986; Lønøy *et al.*, 1986; Egeberg and Aagaard, 1989). While isotopically heavy brines derived from evaporites appear to be an important constituent of the deep pore fluids in all of these basins, illitization of smectite is seen as an important source of ^{18}O enriched fluids during deep burial.

However, recent work on the clay diagenesis of the KCF and other Jurassic shales in the onshore UK and the North Sea has questioned the importance of smectite illitization during burial with the recognition of the large detrital ordered illite–smectite and illite content of many mudrocks. Source areas in the North Sea were very variable both laterally and vertically within the section, the variable chemical compositions making determination of real burial diagenetic sequences difficult (Primmer and Shaw, 1987). This is particularly so in the onshore section where ordered illite–smectites are largely of detrital origin (Scotchman, unpublished work), while mixed-layer kaolinite–smectite is common (J. H. S. Macquaker, personal communication). Thus, the Gulf Coast Tertiary Model cannot be simply applied to the KCF of the North Sea: the effects of the highly variable chemical compositions of the precursor smectite and kaolinite–smectite as well as detrital illite–smectite on the cations released during deep burial diagenesis require further study before mass-balance calculations can be attempted. Nevertheless, iron and to a lesser extent, magnesium, appear to be released by these reactions.

The detrital mineralogy should not be neglected as a cation source during later diagenesis. Modelling of the Me and D zone reactions shows increasing acidification of the pore waters (Curtis, 1978; 1983a; 1987), dependent on the availability of mineral oxidants such as Fe_2O_3 and of organic acids generated from kerogen in the D zone (Carothers and Kharaka, 1978; Surdam *et al.*, 1984; 1989). Appreciable amounts of detrital ferric iron have been shown to survive diagenesis through the SR, and Me zones into the D zone in the onshore KCF section (Davey and Curtis, 1989; J. H. S. Macquaker, personal communication) and be available as an oxidant in the D zone. The high availability of iron in the Me and D zones is shown by the highly ferroan cements precipitated.

Magnesium sources are more problematic as after initial high concentrations in the Me and early D zone carbonates, there is a general rapid decline. This suggests either a limited source or that iron is preferentially substituted into the dolomite crystal lattice leaving magnesium in the pore fluids. Widespread dissolution is associated with D zone conditions (Foscolos and Powell, 1979; 1980; Surdam *et al.*, 1984; Curtis, 1987; Gawthorpe, 1987), both in the mudrocks and in associated sandstones where widespread development of secondary porosity occurs (Schmidt and McDonald, 1979; Larese *et al.*, 1984). Detrital carbonate in the mudrocks is largely low-Mg calcite in abundant coccoliths and metastable high-Mg calcite and aragonite in shell debris. Dissolution would release large amounts of magnesium (and lesser amounts of strontium) into the pore fluid system. Also, secondary porosity development in sandstones has often been documented as being due to the dissolution of early-formed marine cements which are often magnesium-rich (e.g. Schmidt and McDonald, 1979; Burley, 1986), forming a further source of magnesium which would decline in time. Finally, the illitization of smectite may be a discrete, relatively short-lived event (Morton, 1985; Glasmann *et al.*, 1989b) suggesting a supply of cations, particularly Mg^{2+}, for a limited period of time.

6.6 Silicate cements

The late silicate cements in the 20/3-4 well concretions are unusual—no other septarian concretions with silicate cements have been reported. The deeper concretion (10,324′) contains a single quartz cement between earlier dolomite, saddle dolomite and later saddle dolomite cements. Pore fluid compositions appear to have oscillated, initially favouring carbonate precipitation as in most other concretions, then silicate precipitation, before finally reverting to carbonates. The shallower concretion (10 318′6″) shows complete change from carbonate to silicate precipitation. The two quartz cements decrease in $\delta^{18}O$ SMOW from $+26.4$ to $+21.1‰$ ($\pm 0.1‰$), while fluid inclusion analysis indicates precipitation temperatures of 85–95°C (Figure 6.13), giving a pore-water $\delta^{18}O$ SMOW of -2 to $+4‰$.

Cements forming concretions are derived from the surrounding host mudrocks (Scotchman, 1989). The change from carbonate to quartz cements suggests that mudrocks were undergoing diagenetic reactions releasing silica with a pore fluid environment favouring quartz rather than carbonate precipitation. As previously discussed, illitization of smectite is capable of generating excess silica and, from the data of Pearson and Small (1988), the reaction was taking place off-structure at depths greater than 6000′ (1830 m). Assuming a $30°C/km^{-1}$ geothermal gradient, the quartz cements from well 20/3-4 were precipitated below 2 km burial depth (Figure 6.13).

Organic acids are a major component of porefluids in oilfields (Carothers and Kharaka, 1978; 1980; Surdam *et al.*, 1984; Crossey, 1985; MacGowan and Surdam, 1988). Hydrous pyrolysis studies have shown that these can be generated in large amounts from kerogen in hydrocarbon source rocks under D zone conditions, prior to the onset of hydrocarbon generation (Kawamura and Kaplan, 1987; Eglinton *et al.*, 1987; Barth *et al.*, 1988; 1989). As well as increasing acidity of the pore fluids and causing carbonate dissolution, difunctional organic acids such as oxalic acid are able to complex silica and Al^{3+}, greatly increasing the dissolution and transport of quartz and aluminosilicate minerals (Curtis, 1983b; 1987; Surdam and Crossey, 1985b; Bennett and Siegel, 1987; Surdam *et al.*, 1989). Concentrations of organic acids are greatest in oilfield waters in the 80–100°C temperature range (Figure 6.13): above this range decarboxylation of first difunctional and, later, monofunctional acids occurs (Crossey, 1985; MacGowan and Surdam, 1988; Surdam *et al.*, 1989). This temperature range is broadly coincident with the illitization of smectite in the northern North Sea (Dypvik, 1983; Pearson *et al.*, 1983) and, therefore, pore fluids in a hydrocarbon source rock are likely to be enriched in both organic acids and the by-products of illitization (silica and cations) (Figure 6.13).

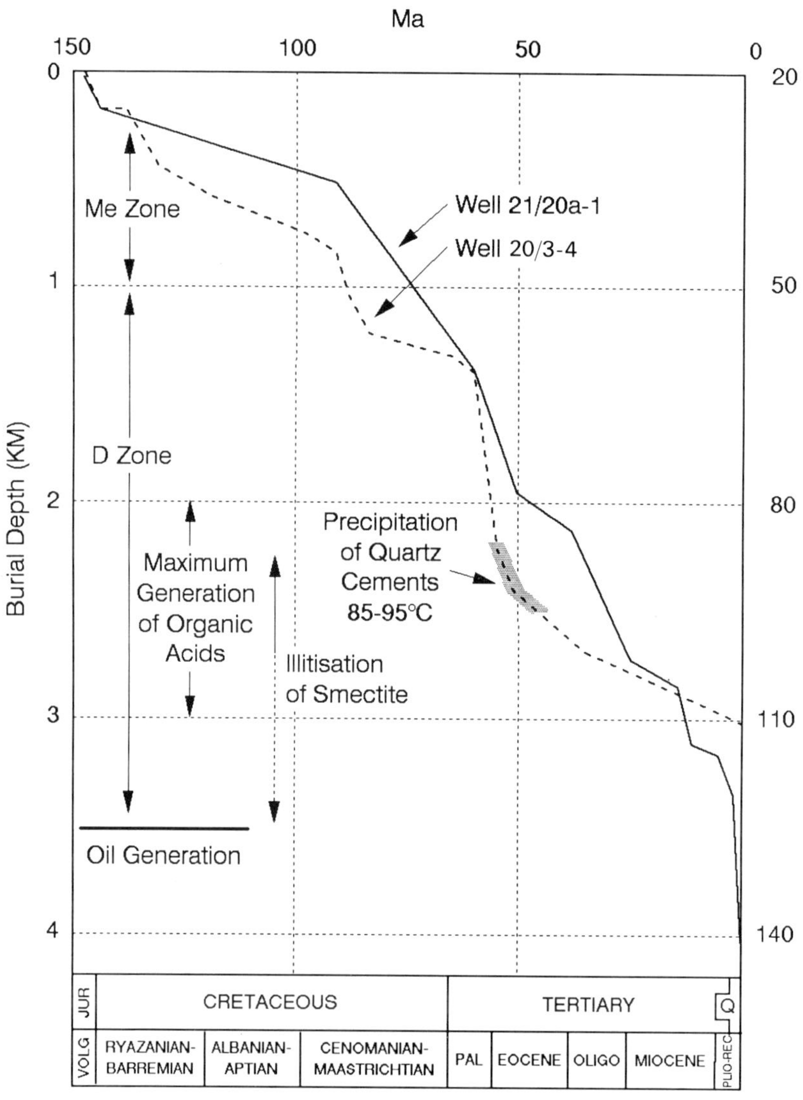

Figure 6.13 Burial history curves for wells 21/20a–1 and 20/3–4 showing concretion development relative to geological history. Data on organic maturity from Scotchman (unpublished); Me and D zones (Curtis, 1978); illitization of smectite (Pearson and Small, 1988) and organic acids (Surdam *et al.*, 1989).

Local redox conditions, dependent on the balance of Fe (III) reduction to acid-generating reactions such as decarboxylation, will determine precipitation or dissolution of minerals (Curtis, 1987). Merely by their existence, concretions can be inferred to be sites of preferential precipitation with only limited dissolution, implying mostly alkaline conditions during early diagenesis (SR to early D zones). The change to silica cements suggests an increase in acidity in the pore fluids, which, with increasing temperature, favoured carbonate dissolution. The balance from silicate to carbonate precipitation can be tipped by increasing alkalinity, due either to the decarboxylation of organic acids with increasing temperature (Surdam *et al.*, 1989) or to increased Fe (III) reduction (Curtis, 1987), resulting in silica and, ultimately, carbonate cements being precipitated.

6.6.1 Links to sandstone cementation

Having demonstrated the presence of excess silica in the pore fluids of the deeply buried KCF, albeit in a limited amount, by the presence of silica cements in the concretions, it is tempting to make a tentative link to sandstone cementation. Late-stage carbonate cements of D zone composition are common in North Sea sandstone reservoirs (Larese *et al.*, 1984; Kantorowicz, 1985) and, perhaps more importantly, many reservoir sandstones on structures adjacent to KCF source 'kitchens' show a period of often intense cementation by quartz and clay minerals immediately prior to oil migration (e.g. Gluyas, 1985; Burley, 1986; Jourdan *et al.*, 1987; Scotchman *et al.*, 1989). Circumstantial evidence from basin modelling suggests this is related to the development of D zone conditions in the downdip source rock facies, with hot silica-laden pore fluids being expelled updip into the reservoir-bearing structures where kaolinite, quartz and, finally, illite cements were precipitated prior to hydrocarbon emplacement. A strong link between the later stages of diagenesis in the mudrocks and sandstone cementation is therefore suggested, with the mudrocks supplying much of the silica and carbonate cements in the sandstones. Other workers, however, report mudrocks as being closed, isochemical systems during diagenesis (Hower *et al.*, 1976) with any silica generated being precipitated as microquartz in the mudrock matrix (Primer and Shaw, 1987; I. J. Evans, personal communication, 1990). Therefore, while the link between mudrock and sandstone diagenesis remains unproven, there is much evidence to implicate mudrocks as major players in the sandstone diagenesis story (Curtis, 1983a; 1983b; 1987). In particular, the diagenetic evolutionary sequence preserved in the concretions demonstrates that mudrocks generated pore fluids carrying dissolved species capable of producing the observed diagenetic sequences in sandstone reservoirs. Only proof of large-scale expulsion and migration of these pore fluids, providing the physical link between mudrocks and sandstones, is required.

6.7 Conclusions

1. Pore fluid evolution in the KCF is dominated by diagenetic reactions based on the degradation of organic matter (Figure 6.12).
2. The succession and duration of these organic matter-based reaction zones, under conditions of excess organic matter, is dependent on sedimentation rate and burial depth.

3. The low sedimentation rate/shallow burial shelf areas of the onshore section are dominated by SR zone conditions while the Me and D zone conditions were attained only under the more rapid sedimentation rates and deeper burial of the basins (Figure 6.12).
4. From the samples studied and by analogy with the basinal areas of the onshore section, the offshore basins of the North Sea rift system are characterized by Me and, predominantly, D zone conditions with high sedimentation rates coupled with rapid, deep burial.
5. Generation of organic acids under later D zone conditions, coincident with the illitization of smectite, with consequent mobilization of silica, appears to characterize the deep burial of the offshore basins.
6. Late-stage quartz cements indicate generation or, at least, migration of silica-bearing pore fluids through the less deeply buried section in well 20/3-4, suggesting a link between mudrock diagenesis and sandstone cementation (Curtis, 1978; 1983b; 1987).

Acknowledgements

Part of this work was supported by a NERC studentship at the University of Sheffield and the Institute of Geological Sciences, Geochemistry Division, London (now British Geological Survey). Professor C. D. Curtis and M. L. Coleman are thanked for their support and encouragement, while an anonymous reviewer is thanked for helpful criticism. Additional analyses were performed at Amoco's Research Center, Tulsa, Oklahoma, USA, where Paul Wagner, Mike Smith and John Wheeler are thanked, and by Dr Tim Astin at the Postgraduate Research Institute, University of Reading. Amoco are thanked for permission to publish and for provision of word processing and draughting facilities.

References

Al-Shaieb, Z. and Shelton, J. W. (1981) Migration of hydrocarbons and secondary porosity in sandstones. *American Association of Petroleum Geologists Bulletin*, **65**, 2433–6.

Astin, T. R. (1986) Septarian crack formation in carbonate concretions from shales and mudstones. *Clay Minerals*, **21**, 617–32.

Astin, T. R. and Scotchman, I. C. (1988) The diagenetic history of some septarian concretions from the Kimmeridge Clay, England. *Sedimentology*, **35**, 349–68.

Barth, T., Borgund, A. E. and Hopland, A. L. (1989) Generation of organic compounds by hydrous pyrolysis of Kimmeridge Oil shale – Bulk results and activation energy calculations. *Organic Geochemistry*, **14**, 69–76.

Barth, T., Borgund, A. E., Hopland, A. L. and Grave, A. (1988) Volatile organic acids produced during kerogen maturation – Amounts, composition and role in migration of oil, in *Advances in Organic Geochemistry 1987*, (eds L. Mattavelli and L. Novelli), *Organic Geochemistry*, **13**, 461–5.

Bennett, P. and Siegel, D. I. (1987) Increased solubility of quartz in water due to complexing by organic compounds. *Nature*, **326**, 684–6.

Benzel, W. M. and Graf, D. L. (1984) Studies of smectite membrane behaviour: importance of layer thickness and fabric in experiments at 20°C. *Geochimica et Cosmochimica Acta*, **48**, 1169–88.

Bjørlykke, K., Aagaard, P., Dypvik, H., Hastings, D. S. and Harper, A. S. (1986) Diagenesis and reservoir properties of Jurassic sandstones from the Haltenbanken area, offshore mid Norway, in *Habitat of Hydrocarbons on the Norwegian Continental Shelf*, (eds A. M. Spencer *et al.*), Graham and Trotman, London, pp. 275–86.

Blanche, J. B. and Whitaker, J. H. McD. (1978) Diagenesis of part of the Brent Sand Formation (Middle Jurassic) of the northern North Sea Basin. *Journal of the Geological Society of London*, **135**, 73–82.

Boles, J. R. (1978) Active ankerite cementation in the subsurface Eocene of Southwest Texas. *Contributions to Mineralogy and Petrology*, **68**, 13–22.

Boles, J. R. and Franks, S. G. (1979) Clay diagenesis in Wilcox sandstones of Texas: implications of smectite diagenesis on sandstone cementation. *Journal of Sedimentary Petrology*, **49**, 55–70.

Brown, S. (1990) Jurassic, in *Introduction to the Petroleum Geology of the North Sea*, 3rd edn, (ed. K. W. Glennie), Blackwell Scientific Publications, Oxford, pp. 219–54.

Buhrig, C. (1989) Geopressured Jurassic reservoirs in the Viking Graben: modelling and geological significance. *Marine and Petroleum Geology*, **6**, 31–48.

Burley, S. D. (1986) The development and destruction of porosity within Upper Jurassic reservoir sandstones of the Piper and Tartan oilfields, Outer Moray Firth, North Sea. *Clay Minerals*, **21**, 649–94.

Burns, S. J. and Baker, P. A. (1987) A geochemical study of dolomite in the Monterey Formation, California. *Journal of Sedimentary Petrology*, **57**, 128–39.

Burst, J. F. (1969) Diagenesis of Gulf Coast clayey sediments and its possible relation to petroleum migration. *American Association of Petroleum Geologists Bulletin*, **53**, 73–93.

Carothers, W. W. and Kharaka, Y. K. (1978) Aliphatic acid anions in oil-field waters—implications for origin of natural gas. *American Association of Petroleum Geologists Bulletin*, **62**, 2441–53.

Carothers, W. W. and Kharaka, Y. K. (1980) Stable carbon isotopes of HCO_3^- in oil-field waters—implications for the origin of CO_2. *Geochimica et Cosmochimica Acta*, **44**, 323–32.

Clayton, R. N., Friedman, I., Graf, D. L., Mayenda, T. K., Meents, W. F. and Shimp, N. F. (1966) The origin of saline formation waters: 1. Isotopic composition. *Journal of Geophysical Research*, **71**, 3869–82.

Coleman, M. L. (1985) Geochemistry of diagenetic non-silicate minerals: kinetic considerations. *Philosophical Transactions of the Royal Society of London*, Series A, **315**, 39–56.

Coleman, M. L., Curtis, C. D. and Irwin, H. (1979) Burial rate: a key to reservoir potential. *World Oil*, **188**, 83–92.

Coplen, T. B. and Hanshaw, B. B. (1973) Ultrafiltration by a compacted clay membrane. 1: Oxygen and hydrogen fractionation. *Geochimica et Cosmochimica Acta*, **37**, 2295–310.

Crossey, L. J. (1985) *The Origin and Role of Water-soluble Organic Compounds in Clastic Diagenetic Systems*, PhD Dissertation, University of Wyoming, Laramie, Wyoming.

Curtis, C. D. (1978) Possible links between sandstone diagenesis and depth-related geochemical reactions occurring in enclosing mudstones. *Journal of the Geological Society of London*, **135**, 107–17.

Curtis, C. D. (1983a) Geochemistry of porosity enhancement and reduction in clastic sediments, in *Petroleum Geochemistry and Exploration of Europe*, (ed. J. Brooks), Geological Society of London Special Publication, No. 12, Blackwell Scientific Publications, Oxford, pp. 113–26.

Curtis, C. D. (1983b) Link between aluminium mobility and destruction of secondary porosity. *American Association of Petroleum Geologists Bulletin*, **67**, 390–3.

Curtis, C. D. (1987) Inorganic geochemistry and petroleum exploration, in *Advances in Petroleum Geochemistry*, vol. 2, (eds J. Brooks and D. Welte), Academic Press, London, pp. 91–140.

Curtis, C. D. and Coleman, M. L. (1986) Controls on the precipitation of early diagenetic calcite, dolomite and siderite concretions in complex depositional sequences, in *Roles of Organic Matter in Sediment Diagenesis*, (ed. D. L. Gautier), *Society of Economic Paleontologists and Mineralogists. Special Publication*, **38**, 23–33.

Curtis, C. D., Hughes, C. R., Whiteman, J. A. and Whittle, C. K. (1985) Compositional variations within some sedimentary chlorites and some comments on their origin. *Mineralogical Magazine*, **49**, 375–86.

Davey, R. and Curtis, C. D. (1989) Mössbauer and chemical investigations of mudrocks. *Clay Minerals*, **24**, 53–65.

Doré, A. G., Vollset, J. and Hamar, G. P. (1985) Correlation of the offshore sequences referred to the Kimmeridge Clay Formation – relevance to the Norwegian Sector, in *Petroleum Geochemistry in Exploration of the Norwegian Shelf*, (eds B. M. Thomas *et al.*), Graham and Trotman, London, pp. 27–38.

Dypvik, H. (1983) Clay mineral transformations in Tertiary and Mesozoic sediments from the North Sea. *American Association of Petroleum Geologists Bulletin*, **67**, 160–5.

Egeberg, P. K. and Aagaard, P., (1989) Origin and evolution of formation waters from oil fields on the Norwegian shelf. *Applied Geochemistry*, **4**, 131–42.

Eglinton, T. I., Curtis, C. D. and Rowlands, S. J. (1987) Generation of water-soluble organic acids from kerogen during hydrous pyrolysis: implications for porosity development. *Mineralogical Magazine*, **51**, 495–503.

Fisher, R. S. and Land, L. S. (1986) Diagenetic history of Eocene Wilcox sandstones, South-Central Texas. *Geochimica et Cosmochimica Acta*, **50**, 551–61.

Foscolos, A. E. and Kodama, H. (1974) Diagenesis of clay minerals from Lower Cretaceous shales of northeastern British Columbia. *Clays and Clay Minerals*, **22**, 319–35.

Foscolos, A. E. and Powell, T. G. (1979) Catagenesis in shales and occurrence of authigenic clays in sandstones, North Sabine H-49 Well, Canadian Arctic Islands. *Canadian Society of Petroleum Geologists*, **16**, 1309–14.

Foscolos, A. E. and Powell, T. G. (1980) Mineralogical and geochemical transformations of clays during catagenesis and their relations to oil generation, in *Facts and Principles of World Petroleum Occurrences*, (ed. A. D. Miall), Memoir 6 of the Canadian Society of Petroleum Geologists, Calgary, pp. 153–72.

Franks, S. G. and Forester, R. W. (1984) Relationships among secondary porosity, pore-fluid chemistry and carbon dioxide, Texas Gulf Coast, in *Clastic Diagenesis*, (eds D. A. McDonald and R. C. Surdam), *Memoirs, American Association of Petroleum Geologists*, **37**, 63–80.

Froelich, P. N., Klinkhammer, G. P., Bender, M. L., Luedtke, N. A., Heath, G. R., Cullen, D. and Dauphin, P. (1979) Early oxidation of organic matter in pelagic sediments of the eastern equatorial Atlantic. Suboxic diagenesis. *Geochimica et Cosmochimica Acta*, **43**, 1075–90.

Garrison, R. E., Kastner, M. and Zenger, D. H. (eds) (1984) Dolomites of the Monterey Formation and other organic-rich units. *Publication of the Society of Economic Paleontologists and Mineralogists* (*Pacific Section*), No. 41, SEPM, Tulsa.

Gawthorpe, R. L. (1987). Burial dolomitization and porosity development in a mixed carbonate-clastic sequence: an example from the Bowland Basin, northern England. *Sedimentology*, **34**, 533–48.

Glasmann, J. R., Larter, S., Briedis, N. A. and Lundegard, P. D. (1989b) Shale diagenesis in the Bergen High area, North Sea. *Clays and Clay Minerals*, **37**, 97–112.

Glasmann, J. R., Clark, R. A., Larter, S., Briedis, N. A. and Lundegard, P. D. (1989c) Diagenesis and hydrocarbon accumulation, Brent Sandstone (Jurassic), Bergen High area, North Sea. *American Association of Petroleum Geologists Bulletin*, **73**, 1341–60.

Glasmann, J. R., Lundegard, P. D., Clark, R. A., Penny, B. and Collins, I. (1989a) Isotopic evidence for the history of fluid migration and diagenesis: Brent Sandstone, Heather Field, North Sea. *Clay Minerals*, **24**, 255–84.

Gluyas, J. G. (1985) Reduction and prediction of reservoir potential, Jurassic, North Sea. *Philosophical Transactions of the Royal Society of London*, Series A, **315**, 187–202.

Goldhaber, M. and Kaplan, I. R. (1980) Mechanisms of sulfur incorporation and isotope fractionation during early diagenesis in sediments of the Gulf of California. *Marine Chemistry*, **9**, 95–144.

Gregg, J. M. (1983) On the formation and occurrence of saddle dolomite – Discussion. *Journal of Sedimentary Petrology*, **53**, 1025.

Hancock, N. J. and Taylor, A. M. (1978) Clay mineral diagenesis and oil migration in the Middle Jurassic Brent Sand Formation. *Journal of the Geological Society of London*, **135**, 69–72.

Hayes, J. B. (1979) Sandstone diagenesis: the hole truth, in *Aspects of Diagenesis*, (eds P. A. Scholle and P. R. Schluger), *Society of Economic Paleontologists and Mineralogists, Special Publication*, **26**, 127–39.

Hower, J., Eslinger, E., Hower, E. M. and Perry, E. A. (1976) Mechanism of burial metamorphism of argillaceous sediments: 1. Mineralogical and chemical evidence. *Bulletin of the Geological Society of America*, **87**, 725–37.

Hudson, J. D. (1978) Concretions, isotopes, and the diagenetic history of the Oxford Clay (Jurassic) of Central England. *Sedimentology*, **25**, 339–70.

Irwin, H. (1980) Early diagenetic carbonate precipitation and pore fluid migration in the Kimmeridge Clay of Dorset, England. *Sedimentology*, **27**, 577–91.

Irwin, H. (1981) On calcic dolomite-ankerite from the Kimmeridge Clay. *Mineralogical Magazine*, **44**, 105–7.

Irwin, H., Curtis, C. D. and Coleman, M. (1977) Isotopic evidence for source of diagenetic carbonates formed during burial of organic rich sediments. *Nature*, **269**, 209–13.

Jourdan, A., Thomas, M., Brevart, O., Robson, P., Sommer, F. and Sullivan, M. (1987) Diagenesis as the control of Brent sandstone reservoir properties in the greater Alwyn Area (East Shetland Basin), in *Petroleum Geology of North West Europe*, (eds J. Brooks and K. W. Glennie), Graham and Trotman, London, pp. 951–62.

Kahle, C. F. (1965) Possible roles of clay minerals in the formation of dolomite. *Journal of Sedimentary Petrology*, **35**, 448–53.

Kantorowicz, J. D. (1985) The origin of authigenic ankerite from the Ninian Field, U.K. North Sea. *Nature*, **315**, 214–16.

Kawamura, K. and Kaplan, I. R. (1987) Dicarboxylic acids generated by thermal alternation of kerogen and humic acids. *Geochimica et Cosmochimica Acta*, **51**, 3201–8.

Land, L. S. and Prezbindowski, D. R. (1981) Origin and evolution of saline formation water, Lower Cretaceous carbonates, south central Texas, USA. *Journal of Hydrology*, **54**, 51–74.

Larese, R. E., Haskell, N. L., Prezbindowski, D. R. and Beju, D. (1984) Sedimentologic and diagenetic controls on porosity development in selected Jurassic sandstone specimens from the Norwegian and North Seas, Norway—an overview, in *Petroleum Geology of the Northern European Margin*, (ed. A. M. Spencer), Graham and Trotman, London, pp. 81–95.

Leddra, M. J., Yassir, N. A., Jones, C. and Jones, M. E. (1987) Anomalous compression structures formed during diagenesis of a dolostone at Kimmeridge Bay, Dorset. *Proceedings of the Geologists' Association*, **98**, 145–55.

Lewan, M. D., Bjorøy M. and Dolcater, D. L. (1986) Effects of thermal maturation on steroid hydrocarbons as determined by hydrous pyrolysis of Phosphoria Retort shale. *Geochimica et Cosmochimica Acta*, **50**, 1977–87.

Liewig, N., Clauer, N. and Sommer, F. (1987) Rb–Sr and K–Ar dating of clay diagenesis in Jurassic sandstone oil reservoir, North Sea. *American Association of Petroleum Geologists Bulletin*, **71**, 1467–74.

Lønøy, A., Akselsen, J. and Rønning, K. (1986) Diagenesis of a deeply buried sandstone reservoir: Hild Field, northern North Sea. *Clay Minerals*, **21**, 711–34.

MacGowan, D. and Surdam, R. C. (1988) Difunctional carboxylic acid anions in oil-field waters. *Organic Geochemistry*, **12**, 245–59.

McHargue, T. R. and Price, R. C. (1982) Dolomite from clay in argillaceous shale-associated marine carbonates. *Journal of Sedimentary Petrology*, **52**, 873–86.

Morton, J. P. (1985) Rb–Sr evidence for punctuated illite/smectite diagenesis in the Oligocene Frio Formation, Texas Gulf Coast. *Geological Society of American Bulletin*, **96**, 114–22.

Norwegian Petroleum Directorate (1977) Lithology wells no. 2/11-1 and no. 2/8-1. *Norwegian Petroleum Directorate, Paper* 7, Stavanger.

Pearson, M. J. and Small, J. S. (1988) Illite–smectite diagenesis and palaeotemperatures in Northern North Sea Quaternary to Mesozoic shale sequences. *Clay Minerals*, **23**, 109–32.

Pearson, M. J., Watkins, D., Pittion, J. L., Caston, D. and Small, J. S. (1983) Aspects of burial diagenesis, organic maturation and palaeothermal history of an area in the South Viking Graben, North Sea, in *Petroleum Geochemistry and Exploration of Europe*, (ed. J. Brooks), Geological Society of London Special Publication, No. 12, Blackwell Scientific Publications, Oxford, pp. 161–73.

Pearson, M. J., Watkins, D. and Small, J. S. (1982) Clay diagenesis and organic maturation in northern North Sea sediments, in *International Clay Conference*, Bologna and Pavia, 1981, (eds H. Van Olphen and F. Veniale), *Developments in Sedimentology* vol. 35, Elsevier, Amsterdam, 665–75.

Perry, E. A. and Hower, J. (1970) Burial diagenesis in Gulf Coast pelitic sediments. *Clays and Clay Minerals*, **18**, 165–78.

Perry, E. A. and Hower, J. (1972) Late stage dehydration in deeply buried pelitic sediments. *American Association of Petroleum Geologists Bulletin*, **56**, 2013–21.

Primmer, T. J. and Shaw, H. F. (1987) Diagenesis in shales: evidence from backscattered electron microscopy and electron probe analyses, in *Proceedings of the International Clay Conference*, Denver, 1985, (eds L. G. Schultz, H. Van Olphen and F. A. Mumpton), Clay Minerals Society, Bloomington, Indiana, pp. 135–43.

Radke, B. M. and Mathis, R. L. (1980) On the formation and occurrence of saddle dolomite. *Journal of Sedimentary Petrology*, **50**, 1149–68.

Rawson, P. F. and Riley, L. A. (1982) Latest Jurassic–Early Cretaceous events and the 'Late Cimmerian Unconformity' in North Sea area. *American Association of Petroleum Geologists Bulletin*, **66**, 2628–48.

Rukin, N. (1988) Fibrous calcite and overpressuring in mudrocks, in *Abstracts, 27th Annual Meeting of the British Sedimentological Research Group*, Cambridge, 18–21 December 1988.

Sass, E., Bein, A. and Almogi-Labin, A. (1990) Oxygen isotopic composition of diagenetic calcite in reducing environments: An alternative approach, in *Abstracts, 13th International Sedimentological Congress Proceedings, Sediments 1990*, Nottingham, England, 26–31 August 1990.

Schmidt, V. and McDonald, D. A. (1979) The role of secondary porosity in the course of sandstone diagenesis, in *Aspects of Diagenesis*, (eds P. A. Scholle and P. R. Schluger), *Society of Economic Paleontologists and Mineralogists, Special Publication*, **26**, 175–208.

Scotchman, I. C. (1987) Relationship between clay diagenesis and organic maturation in the Kimmeridge Clay Formation, onshore UK, in *Petroleum Geology of North West Europe*, (eds J. Brooks and K. W. Glennie), Graham and Trotman, London, pp. 251–62.

Scotchman, I. C. (1988) Diagenesis and primary migration in Upper Jurassic claystone source rocks in North Sea: Discussion. *American Association of Petroleum Geologists Bulletin*, **72**, 1423–5.

Scotchman, I. C. (1989) Diagenesis of the Kimmeridge Clay Formation, onshore UK. *Journal of the Geological Society of London*, **146**, 285–303.
Scotchman, I. C. (1991) The geochemistry of concretions from the Kimmeridge Clay Formation of southern and eastern England. *Sedimentology*, **38**, 79–106.
Scotchman, I. C., Johnes, L. H. and Miller, R. S. (1989) Clay diagenesis and oil migration in Brent Group sandstones of NW Hutton Field, UK North Sea. *Clay Minerals*, **24**, 339–74.
Sellwood, B. W., Scott, J., James, B., Evans, R. and Marshall, J. (1987) Regional significance of 'dedolomitization' in Great Oolite reservoir facies of Southern England, in *Petroleum Geology of North West Europe*, (eds J. Brooks and K. W. Glennie), Graham and Trotman, London, pp. 1129–37.
Sellwood, B. W., Shepherd, T. J., Evans, M. R. and James, B. (1989) Origin of late cements in oolitic reservoir facies: a fluid inclusion and isotopic study (Mid-Jurassic, Southern England). *Sedimentary Geology*, **61**, 223–37.
Shackleton, N. J. and Kennett, J. P. (1975) Palaeotemperature history of the Cenozoic and the initiation of Antarctic glaciation: oxygen analysis in DSDP Sites 277, 279, and 281, in *Initial Reports of the Deep-Sea Drilling Project 24*, (eds J. P. Kenneth and R. E. Houtz), US Government Printing Office, Washington, pp. 743–55.
Sucheki, R. K. and Land, L. S. (1983) Isotopic geochemistry of burial metamorphosed volcanogenic sediments, Great Valley sequence, northern California. *Geochimica et Cosmochimica Acta*, **47**, 1487–99.
Surdam, R. C., Boese, S. W. and Crossey, L. J. (1984) The chemistry of secondary porosity, in *Clastic Diagenesis*, (eds D. A. McDonald and R. C. Surdam), *Memoirs, American Association of Petroleum Geologists*, **37**, 129–49.
Surdam, R. C. and Crossey, L. J. (1985a) Organic–inorganic reactions during progressive burial: key to porosity and permeability enhancement and preservation. *Philosophical Transactions of the Royal Society of London*, Series A, **315**, 135–56.
Surdam, R. C. and Crossey, L. J. (1985b) Mechanisms of organic/inorganic interactions in sandstone/shale sequences, in *Relationships of Organic Matter and Mineral Diagenesis*, (eds D. L. Gautier, Y. K. Kharaka and R. C. Surdam), *Society of Economic Paleontologists and Mineralogists Short Course Notes*, No.17, SEPM, Tulsa, pp. 177–232.
Surdam, R. C., Crossey, L. J., Hagen, E. S. and Heasler, H. P. (1989) Organic–inorganic interactions and sandstone diagenesis. *American Association of Petroleum Geologists Bulletin*, **73**, 1–23.

Thomas, M. (1986) Diagenetic sequences and K/Ar dating in Jurassic sandstones, Central Viking Graben: effects on reservoir properties. *Clay Minerals*, **21**, 695–710.
Thyne, G. and Boles, J. (1989) Isotopic evidence for origin of the Moeraki Septarian Concretions, New Zealand. *Journal of Sedimentary Petrology*, **59**, 272–9.

Vollset, J. and Doré, A. G. (1984) A revised Triassic and Jurassic lithostratigraphic nomenclature for the Norwegian North Sea. *Bulletin of the Norwegian Petroleum Directorate*, **3**, 53 pp.

Whiticar, M. J., Faber, E. and Shoell, M. (1986) Biogenic methane formation in marine and freshwater environments: CO_2 reduction vs. acetate fermentation-isotope evidence. *Geochimica et Cosmochimica Acta*, **50**, 693–709.
Woronick, R. E. and Land, L. S. (1985) Late burial diagenesis, Lower Cretaceous Pearsall and Lower Glan Rose Formations, South Texas, in *Carbonate Cements*, (eds N. Schneidermann and P. M. Harris), *Society of Economic Paleontologists and Mineralogists, Special Publication*, **36**, 265–75.

Ziegler, P. A. (1988) Evolution of the Arctic–North Atlantic and the Western Tethys. *Memoirs, American Association of Petroleum Geologists*, **43**, 198 pp.

PART THREE
Experimental Studies of Diagenesis and Weathering

CHAPTER SEVEN

The stability of experimentally grown clay minerals: implications for modelling the stability of neo-formed diagenetic clay minerals

T. J. Primmer, E. A. Warren, B. K. Sharma and M. P. Atkins

7.1 Introduction

The prediction of authigenic clay mineral occurrence and distribution in potential oil-bearing reservoir sandstones requires a clear understanding of the stabilities of authigenic clay minerals in terms of the range of temperatures and compositions from which different clay minerals will precipitate. Few attempts at such prediction have been published, not least because of the paucity of experimentally determined thermodynamic data on clay minerals upon which to base predictions. The experimental data that do exist (e.g. Routson and Kittrick, 1971; May *et al.*, 1986; Sass *et al.*, 1987) must be considered very carefully as they have been based on work with materials which were not formed by the same processes or under the same conditions as authigenic clay minerals found in sandstones. In the absence of reliable experimental data, past efforts to model clay mineral stability have relied on a variety of schemes aimed at estimating appropriate thermodynamic data. Of these, the simplest approach has been direct use of data for stoichiometric end-member phyllosilicates (e.g. micas) to proxy for non-stoichiometric (2:1) clay minerals (e.g. illites and smectites, Bjørkum and Gjelsvik, 1988). Others have considered a summation of individual free energies of oxide components in a model composition to describe the stability of non-stoichiometric clay minerals (e.g. Tardy and Garrels, 1974; Nriagu, 1975; Chermak and Rimstidt, 1989). In each of these schemes the free energies of the constituents were derived from experimentally determined thermodynamic data for stoichiometric phyllosilicates. More elaborate approaches have also been devised which treat non-stoichiometric clay minerals as solid solutions (Stoessell, 1979; 1981; Tardy and Fritz, 1981; Aagaard and Helgeson, 1983). However, reliance has again been placed on using existing or arbitrarily modified data for pure end-member phyllosilicates as data sources.

Geochemistry of Clay-Pore Fluid Interactions. Edited by D. A. C. Manning, P. L. Hall and C. R. Hughes. Published in 1993 by Chapman & Hall, London. ISBN 0 412 48980 5

As yet, no experimentally determined thermodynamic properties have been published on demonstrably neo-formed clay minerals typical of those observed in sandstones. Similarly, none of the schemes described above have been closely appraised by investigations of clay mineral stability in laboratory synthesis experiments.

Fundamental to our investigations is the observation that most authigenic clay minerals in sandstones appear to have precipitated directly from solution rather than by transformation of a precursor (as is often the presumption in soils and mudrocks, for example). At the outset, experiments were conducted to mimic natural processes such that clay minerals were precipitated directly from solution at different temperatures and over a range of solution conditions. Experiments were undertaken broadly according to the methods devised by Guven *et al.* (1982) and Huang *et al.* (1986) and were devised to investigate the stability relationships of kaolinite, illite and smectite in the Na_2O–K_2O–Al_2O_3–SiO_2–H_2O system. Albite was mixed with acidified KCl solutions of different molarity containing dissolved SiO_2 and run at 150 and 200°C for periods of up to several weeks. The chief merit of this approach is that by acidifying the solution dissolution of albite releases aluminium into solution. It is necessary for aluminium to be mobile (albeit over micron-scale distances) in order for clay minerals to be precipitated. This is directly analogous to natural systems, where aluminium mobility over short distances is an essential pre-requisite for the precipitation of authigenic clay minerals as neo-formed phases in sandstones (Warren, 1987).

The nature and composition of solubility-controlling phases were determined from analysis of the quench solutions and microanalysis of precipitated solids and used with the solution analyses to construct isobaric, isothermal activity diagrams. Values of log K_{sp} for precipitated phases were calculated from the experimental data and used in a simple ideal solid-solution model to 'unmix' values for log K_{sp} of the respective end-member phyllosilicates. These values were then compared with published, internally consistent data (SUPCRT; Helgeson *et al.*, 1978) for the two stoichiometric end-member phyllosilicates muscovite and pyrophyllite, in order to assess whether data on such end-member phyllosilicates are appropriate for predicting the stability of authigenic clay minerals in diagenetically altered sandstone reservoirs.

7.2 Materials and methods

7.2.1 *Materials*

Following the work of Huang *et al.* (1986), soda feldspar BCS (British Chemical Standards) No. 375 was used as the starting material. All the experiments described were conducted using screened feldspar (particle size range/ BET surface area for different experiments were 5–125 $\mu m/0.4\,m^2.g^{-1}$ and

10–53 μm/0.31 $m^2.g^{-1}$). Coarser material was removed by sieving and fines were removed by conventional sedimentation techniques (Tanner and Jackson, 1947).

Starting solutions (0.1 M and 1 M KCl) were prepared by adding sufficient silica in solution (from a stock solution of DuPont Ludox AS40 (10,000 ppm)) to produce 80 ml of solution containing either 50 or 150 ppm dissolved silica. The pH was adjusted to 2.0 by adding dilute (*c.* 2 M) hydrochloric acid dropwise to the mixture.

7.2.2 Experimental approach

Experiments were conducted in batches using 150 cm^3 type 316 stainless steel cylindrical pressure vessels. These were fitted with PTFE inner sleeves to prevent iron leaching from the walls under the initial acidic conditions. Starting solutions were mixed with the feldspar to produce fluid/mineral ratios of either 10:1 or 5:1. The filled pressure vessels were then sealed and placed in a convection oven set at the appropriate temperature. Vessels were rotated as they came up to temperature to ensure good mixing of solution and feldspar. Internal pressure was not monitored as it was assumed to have been controlled by the vapour pressure of solutions at the respective temperatures investigated.

Individual experiments were left to run in each batch for periods of just a few hours (to investigate the rate of dissolution and pH change) to several weeks. Runs were terminated by rapid quenching of reaction vessels in cold water. pH was measured using a combination electrode and Beckman pH meter immediately after filtration of solid products from the solution. Each run solution was then quantitatively analysed for dissolved silica and sodium by conventional flame atomic adsorption methods using a Varian AA1275 spectrophotometer, calibrated using standard solutions with a similar matrix. Semi-quantitative analysis of aluminium in solution (i.e. to see if [Al] exceeded 5 ppm) was also performed to monitor the degree of aluminium mobility during initial feldspar dissolution.

Element concentrations from solution analyses at 25°C were converted to element activities at 150 and 200°C using a geochemical model, Gt (Bethke, personal communication), similar to EQ3/6 (Wolery, 1986). Corrections were made for ionic association using the model of Harvie *et al.* (1984), and for temperature using the 1986 version of the Lawrence Livermore database. These values, listed in Table 7.1, were used for the construction of activity diagrams (see below).

Solid run products were washed thoroughly in distilled water to remove salts and air-dried at room temperature. Fractions of each sample were screened by SEM in a JEOL JSM-840A machine equipped with a Link AN10000 energy dispersive X-ray analyser to examine the extent of crystalline growths with clay-like morphologies and chemistry. A fine size fraction (<0.5 μm) was separated from the run products by ultrasonic disaggregation

Table 7.1 Experimental data

Run no.	Temperature (°C)	Time (h)	Initial KCl conc. (M)	Initial silica conc. (ppm)	Final* $\log (a_{K^+}/a_{H^+})$	Final* $\log a (SiO_2)$	Observed mineral paragenesis
A4	200	314	1.0	150	5.06	−2.21	Kaolinite–smectite
B5/4	200	122	1.0	150	2.79	−2.08	Kaolinite–boehmite
B5/5	200	334	1.0	150	3.10	−2.19	Illite–kaolinite–boehmite
B6/2	200	169	1.0	150	5.31	−2.12	Illite
B6/3	200	334	1.0	150	5.27	−2.34	Illite
B10/2	200	289	1.0	150	3.99	−2.34	Illite–kaolinite
B13/1	200	262	1.0	150	3.99	−2.51	Kaolinite
B13/3	200	262	1.0	150	4.38	−2.54	Illite–kaolinite
B16/3	200	214	1.0	150	4.93	−2.44	Illite
B18/1	200	356	0.1	50	3.88	−2.29	Illite–kaolinite
B18/2	200	356	0.1	150	3.70	−2.26	Illite–kaolinite
B18/3	200	356	1.0	50	4.09	−2.28	Illite–kaolinite
B18/4†	200	356	1.0	150	1.30	−1.59	Kaolinite
B18/5	200	356	1.0	150	5.12	−2.36	Illite
B18/6	200	356	1.0	150	5.06	−2.36	Illite
19/1	150	478	0.1	50	4.57	−2.63	Illite–kaolinite
19/2	150	478	0.1	150	4.53	−2.52	Illite–kaolinite–smectite
19/3	150	478	1.0	50	5.56	−2.73	Illite–smectite
19/4	150	168	1.0	150	5.50	−2.52	Illite–kaolinite–smectite
19/5	150	316	1.0	150	5.48	−2.52	Illite–kaolinite–smectite
19/6	150	478	1.0	150	5.51	−2.52	Illite–kaolinite–smectite

*Corrected for ionic strength and temperature.
†Starting pH = 1.0.

and centrifugation and used as a slurry to sediment directly on to glass slides for XRD analysis. Products were characterized using Ni-filtered CuKα radiation. Fine-grained feldspar remnants (low albite) served as an internal standard for fixing peak positions accurately. More dilute suspensions of selected samples were also prepared for dispersion on to 3 mm diameter (250 mesh) Cu TEM grids. These were examined in a JEOL-2000FX TEM similarly equipped with a Link AN10000 energy-dispersive X-ray analyser for spatially resolved quantitative analyses of the major elements in the synthesized run products. This arrangement was calibrated according to the methods of Cliff and Lorimer (1975), using thin foils of well-characterized, mainly phyllosilicate standards (e.g. biotite, phlogopite, muscovite and sodalite) similar to those described by Hughes *et al.* (1990).

7.3 Results

7.3.1 *Reaction progress*

By varying the fluid/mineral ratio it was possible to control the evolution of the reaction in much the same way as Huang *et al.* (1986). In all experiments, initial feldspar dissolution (under acidic conditions) released aluminium and other components into solution. This was followed within a few hours by a decrease in the amount of aluminium in solution and either titration of the solution towards more neutral conditions as H^+ was consumed in experiments with low fluid/mineral ratios or maintenance of acidic conditions in experiments with high fluid/mineral ratios (Figure 7.1). In both cases, crystalline precipitates were observed in the SEM. Subsequent XRD investigation also confirmed that the run products were similar to those obtained by Huang *et al.* (1986);

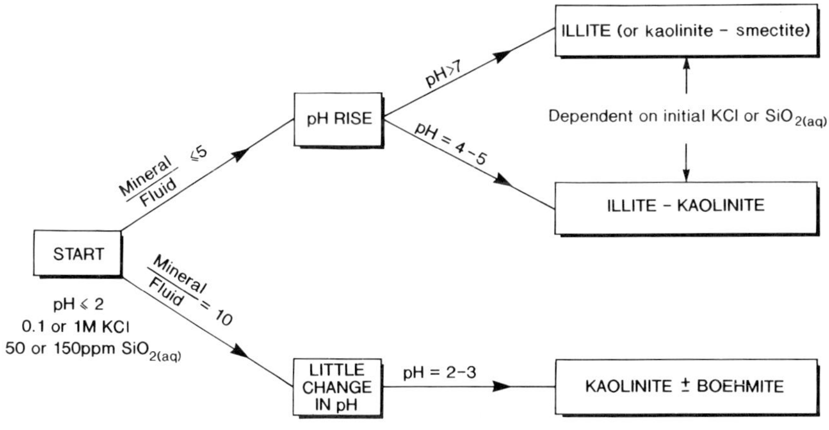

Figure 7.1 Flow diagram to show how final solution composition and precipitated products were controlled by starting composition and mineral/fluid ratio of experimental charge.

illite precipitation in solutions titrated to near neutral conditions and kaolinite precipitation in solutions which remained acidic.

7.3.2 Characterization of solid products

Closer investigation of the solid products from each run showed that distinct mineral parageneses were developed as a function of temperature, starting solution and final solution composition. These are shown in Table 7.1. Phase identification was based on the following XRD criteria:

Illite	10 Å maxima from air-dried samples; *no* shift in peak position upon glycolation (although some modification in profile was observed).
Smectite	Broad maxima between 12 and 15 Å from air-dried samples which shifted to 17 Å and sharpened upon glycolation.
Kaolinite	7 Å maxima from air-dried samples; no change in peak position or profile upon glycolation.
Boehmite	6.1 Å maxima from air-dried samples; no change in peak position or profile upon glycolation.

Characteristic morphologies of crystalline products as seen in the SEM are shown in Figure 7.2; qualitative EDS analysis helped confirm mineral identity. Illite nucleates tangentially to the feldspar substrate but grows away from the surface in curved, rounded platelets occasionally with stumpy pseudohexagonal projections (Figure 7.2(a)). Smectite similarly nucleates tangentially to the feldspar but continues to grow as a characteristic thin, continuous, poorly crystalline grain coating (Figure 7.2(b)). Kaolinite on the other hand appears to nucleate and grow with a fairly consistent hexagonal habit perpendicular to the feldspar surface (Figure 7.2(c)) whereas boehmite precipitated as subhedral bladed lath-shaped bundles of crystals (Figure 7.2(d)).

Quantitative chemical analysis of these phases by TEM shows that the kaolinite composition, as expected, is invariant. In contrast, illite composition from seven selected runs (Table 7.2) conducted at both 150 and 200°C varies considerably. Unfortunately quantitative analysis of the products was hindered by their extremely thin, fine-grained nature. This rendered the mineral susceptible to damage during exposure to the beam, resulting in very low EDS X-ray counts and relatively few reliable analyses were obtained. Table 7.2 and Figure 7.3 show that for all single-grain analyses which conform to the 2:1 dioctahedral mica-type structural formula (in which octahedral site occupancy varies from 3.9 to 4.1) there is a range in aluminium for silicon substitution which directly corresponds to a variation in interlayer K^+ content. Moreover, the trend appears to be governed by the associated mineral paragenesis (rather than temperature), in that illite which precipitated in the presence of kaolinite

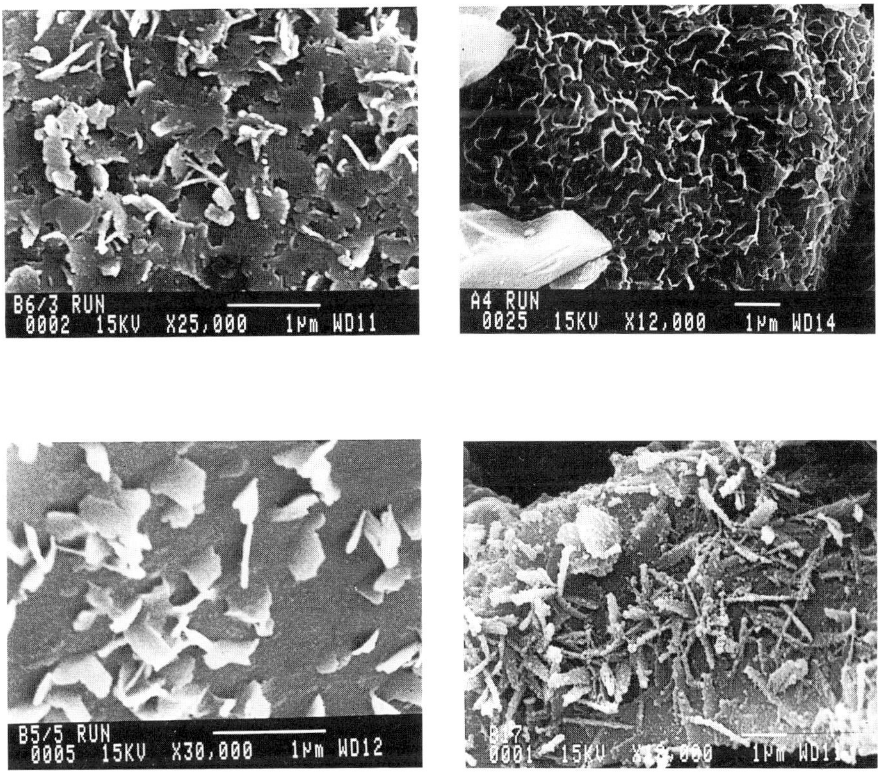

Figure 7.2 Scanning electron microscope photomicrographs showing examples of precipitated products, (a) illite, (b) smectite, (c) kaolinite and (d) boehmite (note mottled, uneven surface texture is a fine-grained coating of precipitated KCl).

shows a much narrower range in composition (0.8–1.4 K^+ per $O_{20}(OH)_4$) than illite which did not (0.9–1.9 K^+ per $O_{20}(OH)_4$).

7.4 Discussion

7.4.1 Construction of activity diagrams

In order to represent the five-component system Na_2O–K_2O–Al_2O_3–SiO_2–H_2O in two dimensions on a standard isobaric, isothermal activity diagram (Garrels and Christ, 1965) a few simplifying assumptions must be made. As the sodium and aluminium released by feldspar dissolution both act conservatively (sodium remaining in solution and aluminium being totally consumed within precipitating phases) they can both be eliminated as immobile components. By assuming that charge balance is also maintained the system is reduced to the pseudobinary (K^+/H^+)–SiO_2 system. Corrected solution

Table 7.2 Analytical TEM data for illite and smectite precipitated in this study

Run and analysis no.	Si	Al(IV)	Al(VI)	Fe(III)	Octahedral occupancy	K
Run 6/3						
1	7.15	0.85	3.82	0.09	3.91	1.12
2	6.45	1.55	3.78	0.18	3.96	1.67
3	6.40	1.60	3.82	0.28	4.10	1.33
4	6.21	1.79	3.75	0.27	4.02	1.74
Run 16/2						
5	6.37	1.63	3.70	0.24	3.94	1.82
Run 18/3						
6	6.78	1.22	3.94		3.94	1.41
7	6.63	1.37	3.76	0.28	4.04	1.26
8	6.81	1.19	3.79	0.27	4.06	1.00
9	6.34	1.36	3.73	0.29	4.02	1.31
Run 18/5						
10	6.58	1.42	3.76	0.18	3.94	1.60
11	6.87	1.13	3.78	0.28	4.06	0.95
12	6.64	1.36	3.72	0.23	3.95	1.56
13	6.68	1.32	3.80	0.20	4.00	1.31
Run 19/2						
14	6.50	1.50	3.81	0.24	4.05	1.36
Run 19/3						
15	6.10	1.90	3.73	0.25	3.98	1.97
16	7.40	0.60	3.84	0.18	4.02	0.55
17	6.90	1.10	4.01		4.01	1.09
18	6.83	1.17	3.87	0.11	3.98	1.24
Run 19/6						
19	7.43	0.57	3.76	0.15	3.91	0.85

activities (Table 7.1) for experiments undertaken at 200°C are shown plotted thus in Figure 7.4.

There is no compelling evidence to suggest that any of the experiments represented on the activity diagrams have reached equilibrium. However, most of the data group linearly according to observed parageneses (Figure 7.4). By analogy to similar constructions made by Sass *et al.* (1987), the Phase Rule decrees that an equilibrium paragenesis of the two minerals observed (+ solution) is represented by a line of univariant equilibrium. This certainly appears to be the case for experiments which precipitated both kaolinite and illite. However, data from experiments which only precipitated illite also plot linearly even though the parageneses are a phase short of assemblages representing univariant equilibrium. Figure 7.3 shows that this illite has a much wider range in composition than illite associated with kaolinite precipitation. Such a spread cannot (given the criteria for an acceptable analysis above) be

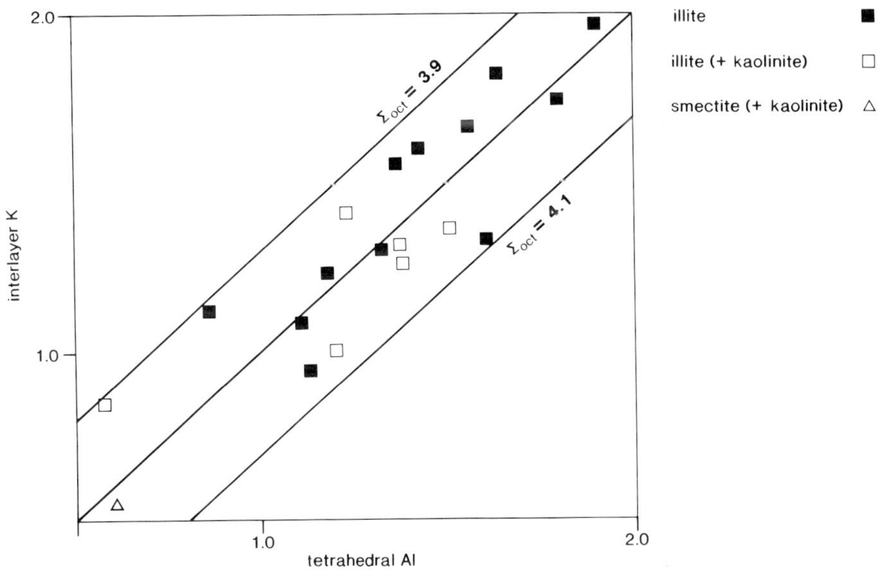

Figure 7.3 Variation in composition of precipitated 2:1 clay minerals shown in terms of interlayer K^+ content and degree of tetrahedral aluminium substitution per $O_{20}(OH)_4$.

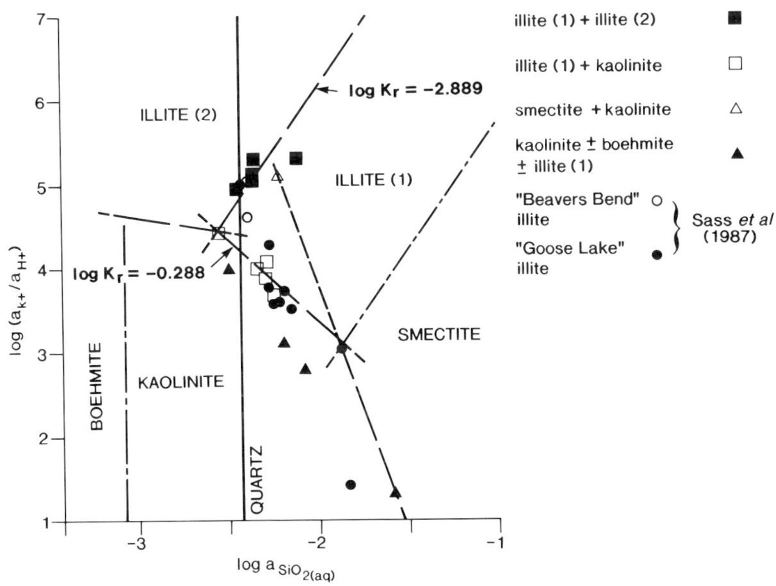

Figure 7.4 Fully interpreted activity diagram for data from experiments at 200°C showing calculated values of log K_r for two lines of univariant equilibrium and metastable extension of assumed kaolinite–smectite equilibrium (see text). Also included are appropriate data for 'Beavers Bend' and 'Goose Lake' illite from Sass *et al.* (1987).

due to intrasample analytical error. Warren and Ransom (1992) show that the error envelope for potassium content is ± 0.3 atoms per unit cell given a 10% uncertainty in the calibration factors used in analytical TEM. This is smaller than the range in illite composition observed, indicating that the range in compositions shown in Figure 7.3 must be real.

To investigate this further, a hierarchical clustering technique was performed to see if any structure could be determined in the spread of data shown in Figure 7.3 where illite was the only phase precipitated. Clustering was performed by calculating a 'distance' coefficient between each pair of analyses based on the variability of interlayer potassium and tetrahedral aluminium contents of each individual analysis (Davis, 1986). A low 'distance' indicates the analyses are similar whereas a high 'distance' indicates dissimilarity. Pairs of analyses which are most similar are merged iteratively to produce a dendrogram (Figure 7.5). This shows that two clusters of data exist at the point of dissimilarity where discrimination between variation due to analytical uncertainty and real variation due to compositional differences within the suite of analyses can be assured. This implies that in the reduced system shown in Figure 7.4 the corresponding solution data define a line of univariant equilibrium between illites of two different compositions (+ solution).

Intersections of the lines of univariant equilibria must represent invariant points. None of the solution compositions plot very near the projected positions of the invariant points, indicating that most of the results must represent metastable equilibria. Solutions failed to reach equilibrium with

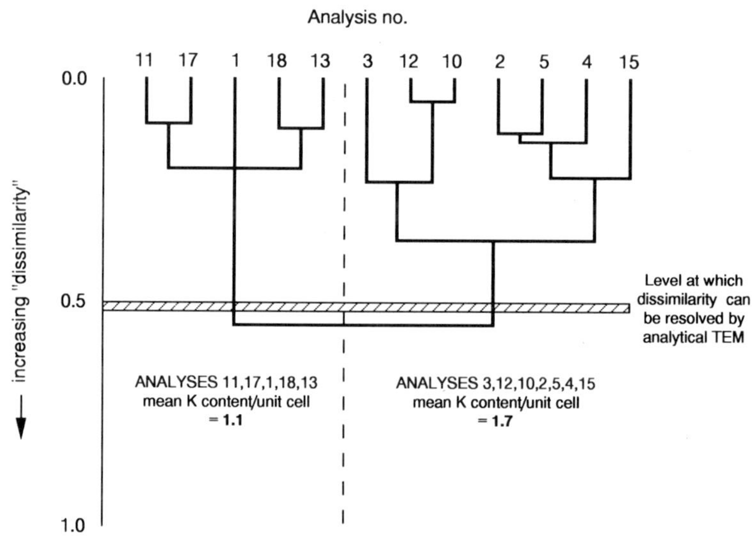

Figure 7.5 Dendrogram illustrating the bimodal character in the analytical TEM data for precipitated 2:1 clay minerals.

respect to the stable SiO_2 polymorph (e.g. precipitation of quartz was not observed); this was perhaps due to the relatively short duration of experiments. However, relatively rapid equilibration of the a_K^+/a_H^+ ratio enabled metastable univariant equilibria to be investigated over a range of different silica activities (mainly supersaturated with respect to quartz).

7.4.2 *Phase stability relationships*

The same expressions for equilibria between phases as defined by Sass *et al.* (1987) are used here. Equilibrium between illite (or smectite) and kaolinite is given by the equation:

$$\underset{\text{illite (or smectite)}}{K_xAl_2(Al_xSi_{4-x})O_{10}(OH)_2} + xH^+ + (1+x/2)H_2O = \frac{2+x}{2}\underset{\text{kaolinite}}{Al_2Si_2O_5(OH)_4} + xK^+ + (2-2x)SiO_2(aq).$$

The equilibrium constant for this reaction can be obtained from the equation:

$$\log(a_{K^+}/a_{H^+}) = \frac{2x-2}{x}\log a_{SiO_2} + \frac{1}{x}\log K_r. \quad (1)$$

Equilibrium between two illites of different compositions (where $y > x$) is given by the equation:

$$\underset{\text{illite (1)}}{K_xAl_2(Al_xSi_{4-x})O_{10}(OH)_2} + \left(\frac{2+x}{2+y}y - x\right)K^+ = \frac{2+x}{2+y}\underset{\text{illite (2)}}{K_yAl_2(Al_ySi_{4-y})O_{10}(OH)_2} + \left(\frac{2+x}{2+y}y - x\right)H^+ + \left[(4-x) - \frac{(2+x)}{(2+y)}(4-y)\right]SiO_2(aq).$$

The equilibrium constant for this reaction can be obtained from the equation:

$$\log(a_{K^+}/a_{H^+}) = 3\log a_{SiO_2} - \frac{\log K_r}{\left(\frac{2+x}{2+y}y - x\right)}. \quad (2)$$

Values for the equilibrium constants ($\log K_r$) for the different reaction equilibria above can be easily calculated from the slope and intercept of lines representing equations (1) and (2) on any appropriate activity diagram. Linear regression through points considered to represent equilibrium between illite (denoted illite (1)) and kaolinite at 200°C yields a line of slope $= -1.953$ (Table 7.2). The composition of illite (1) and log K_r can be calculated from equation (1). Note the value of 0.51 per $O_{10}(OH)_2$ (half-cell) for 'x' compares favourably with the restricted range in illite composition in parageneses associated with kaolinite determined by quantitative analytical TEM (0.4–0.7 K^+ per $O_{10}(OH)_2$).

Data for smectite–kaolinite equilibria were treated in the same way. Unfortunately, only one run at 200°C resulted in the metastable precipitation of a smectite–kaolinite assemblage. No corresponding univariant line could, therefore, be inserted with any degree of confidence, the line shown in Figure 7.4 being inferred from the composition of the phase precipitated at 150°C.

Lines representing equilibria between two illites (or between illite and smectite) are always constrained to have a slope of 3 (see equation (2) above). Neither phase composition can be determined independently without prior knowledge of the log K_r at the appropriate temperature. However, calculated values of 'x' for illite (1) and appropriate quantitative TEM data for the value of 'y' in the more potassic illite (denoted illite (2)) do allow log K_r to be calculated for equilibrium between illite(1) and illite(2) at 200°C. The completed activity diagram at 200°C is shown in Figure 7.4; the remaining phase boundaries were included in accordance with the principles of Schreinemaker.

Data from Sass *et al.* (1987) for appropriate experiments at 200°C are also included in Figure 7.4. Their equilibrium solubility compositions show a close disposition to data trends from this study. Closer investigation of the starting 'illites' on which they made their solubility investigations showed them to be heterogeneous mixtures of up to three different phases. In the course of their experiments they had no independent means of identifying which of these phases controlled solubility. Thus, it is interesting to note that their assertions concerning the nature and composition of the solubility controlling phases (the 'mica-, illite- or smectite-like' phases contained in their starting 'illite') appear to be in close agreement with both the calculated and analysed compositions of the phases precipitated in this study (Table 7.3).

Table 7.3 Analysed and calculated compositions of illite and smectite

Mineral species	ATEM	K^+ per $O_{10}(OH)_2$ calculated from univariant equilibria	
		This study	Sass *et al.* (1987)
Smectite	0.27	0.29	0.24
Illite (1)*	0.55	0.51	0.67
Illite (2)†	0.85	0.87	0.90

*Taken as equivalent to the 'illite' component of Sass *et al.*
†Taken as equivalent to the 'mica' component of Sass *et al.*

A similar isothermal, isobaric construction was made on the data obtained from experiments undertaken at 150°C (Figure 7.6). In this case, lines of univariant equilibria were constrained for illite (2)–kaolinite and smectite–kaolinite equilibria from experimental data. These were linked using a line of similar slope to that calculated from experiments at 200°C to define illite (1)–kaolinite stability.

The calculated values of log K_r shown in Table 7.4 were then used to determine, where possible, the stabilities of the different non-stoichiometric 2:1 clay mineral species precipitated (i.e. the smectite and illites (1) and (2)). The different equilibrium expressions for each mineral at the respective temperature for which data are available and the calculated log K_{sp} are shown below. Note that each expression requires knowledge of kaolinite solubility at some stage in the calculation for values of log K_{sp} to be determined. In order to be consistent with later analysis, data for kaolinite were taken from the SUPCRT data set (Helgeson *et al.*, 1978).

At 200°C, to calculate log $K_{sp\,[\text{illite (1)}]}$:

$$\log K_{sp\,[\text{illite (1)}]} = \log K_{r\,[\text{illite (1)–kaolinite}]} + \frac{(2+x)}{x} \log K_{sp\,[\text{kaolinite}]}. \qquad (3)$$

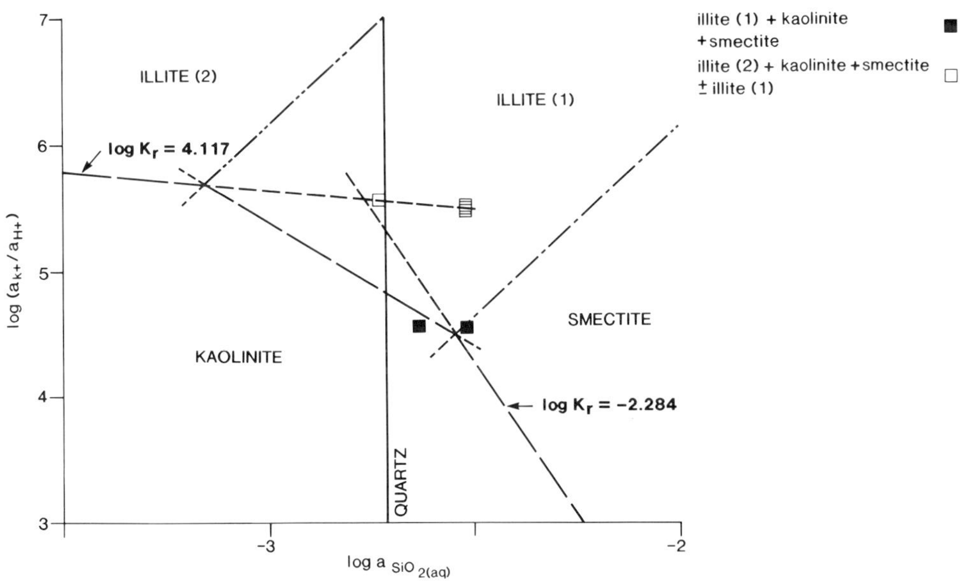

Figure 7.6 Similarly interpreted activity diagram for data from experiments at 150°C also showing calculated values of log K_r for two lines of univariant equilibrium based on observed metastable reaction products.

Table 7.4 Calculated values of log K_r for selected univariant equilibria

Equilibrium reaction	200°C				150°C			
	Slope	x	y*	log K_r	Slope	x	y	log K_r
Illite (1) –kaolinite	−1.95	0.51		−0.29				
Illite (2) –kaolinite					−0.30		0.87	4.12
Smectite –kaolinite					−4.83	0.29		−2.28
Illite (1) –illite (2)	3	0.51	0.85	−2.89				

*Average of ATEM analyses of illite (2) at 200°C.

Using data for kaolinite from Helgeson *et al.*, and values of log K_r from Table 7.2 and x from Table 7.4:

$$\log K_{\text{sp [illite (1)]}} = -2.22.$$

Similarly, to calculate log $K_{\text{sp [illite (2)]}}$:

$$\log K_{\text{sp [illite (2)]}} = \frac{\log K_{\text{sp [illite (1)]}}}{(2+x)/(2+y)} - \frac{\log K_{\text{r [illite (1)–illite (2)]}}}{(2+x)/(2+y)}. \tag{4}$$

Using data for kaolinite from Helgeson *et al.*, and values of log K_r from Table 7.2 and x and y from Table 7.4:

$$\log K_{\text{sp [illite (2)]}} = 0.76.$$

Similar equations can be used to calculate values of log K_{sp} from experimentally derived data at 150°C, for example:

$$\log K_{\text{sp [illite (2)]}} = 4.12,$$

$$\log K_{\text{sp [smectite]}} = -2.06.$$

As pointed out by Lippmann (1982), these experimentally determined solubility products are of clays with non-stoichiometric elemental compositions and should be regarded as solid solutions. Consequently the data have been recalculated in terms of solubility products of stoichiometric ideal end-members akin to the approach of Tardy and Fritz (1981). For an ideal solid solution they assume:

$$\log K_{\text{sp [clay mineral]}} = X_i \log K^i_{sp} + X_i \log X_i,$$

where X_i is the mole fraction of the ith end-member and K^i_{sp} is the corresponding solubility product.

In the highly constrained system investigated in these experiments, the clay mineral compositions can be considered in terms of just two ideal stoichiometric end-members, pyrophyllite and muscovite. As there are only two end-members considered, it is possible to determine values of $\log K^i_{sp}$ for both end-members by simultaneously solving expressions for each value of $\log K_{sp}$ determined at each temperature for the experimentally precipitated 2:1 clay minerals. By having sufficient data at both 150 and 200°C, we can also make a preliminary assessment of the relative variations in calculated values of $\log K^i_{sp}$ with temperature.

Calculated values of $\log K^i_{sp}$ for the muscovite and pyrophyllite end-members are shown in Table 7.5 and plotted on Figure 7.7 along with corresponding self-consistent literature data (Helgeson *et al.*, 1978) and the empirical values used in the solid solution model (Fritz, 1981), which were derived at 25°C and extrapolated to higher temperatures using the same Maier–Kelly power functions as Helgeson *et al.* (1978). The value of $\log K^i_{sp}$ for pyrophyllite at 25°C given by Fritz (1981) is actually taken from Helgeson *et al.* (1978). When this is extrapolated to higher temperatures, values of $\log K^i_{sp}$ for pyrophyllite are within 3.5 kJ.mol^{-1} of the experimentally determined values at both 150 and 200°C. Conversely, the extrapolated values of $\log K^i_{sp}$ for muscovite given by Fritz are markedly different (by $\geqslant$15.8 kJ.mol^{-1}) from the data set of Helgeson *et al.* (1978). The same is true of the experimental data which are less stable (by up to *c.* 26 kJ.mol^{-1}) than the values of $\log k^i_{sp}$ obtained from the Helgeson *et al.* data set. The net result is that, within reasonable experimental error (*c.* 10 kJ.mol^{-1}), the experimental data and the extrapolated values of Fritz for the muscovite end-member at 150 and 200°C appear to be similar. Clearly the new data, derived experimentally, are in agreement with those previously estimated by Fritz (1981) for the end-member phyllosilicates using by Tardy and Fritz (1981) in their solid solution model.

Table 7.5 Comparison of $\log K_{sp}$ for stoichiometric phyllosilicate end-members

Mineral species	Calculated from experimental data (this study)	Extrapolation of Fritz's empirical values at 25°C	From SUPCRT data set of Helgeson *et al.* (1978)
Muscovite			
150°C	6.00	4.74	2.79
200°C	2.20	1.78	−0.08
Pyrophyllite			
150°C	−4.98	−4.63	−4.56
200°C	−6.21	−6.12	−5.99

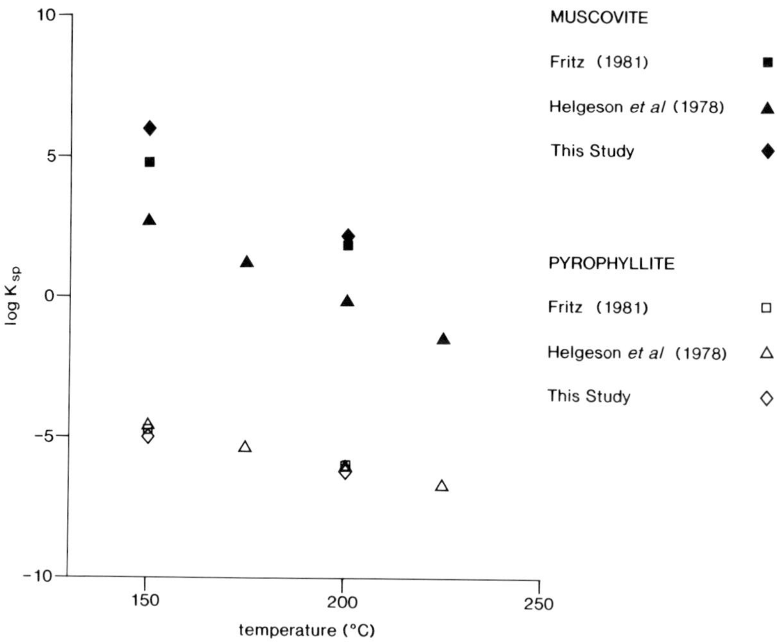

Figure 7.7 Log K^i_{sp} values for end-member phyllosilicates calculated from experimental data in this study compared with the extrapolated values of Fritz (1981) and those calculated from the SUPCRT data set of Helgeson *et al.* (1978).

7.5 Implications for modelling the stability of 2:1 clay minerals

This work suggests that, on current evidence, solid solution modelling is a simple and robust approach to modelling the stability of neo-formed clay minerals at diagenetic temperatures. However, the data presented reinforce the picture that such neo-formed clay minerals are not ideal solid solutions of ideal end-members. The experimentally determined value of log K^i_{sp} for the muscovite end-member shows that this is the case, being the equivalent of 21–26 kJ.mol^{-1} less stable in free energy terms than the value given by the internally consistent SUPCRT data set of Helgeson *et al.* (1978). Literature data suggest that neo-formed 2:1 clay minerals are 1 M or 1 M_d polytypes (e.g. Guven, 1974), whereas data for muscovite in the Helgeson *et al.* data set have been derived on the $2M_1$ polytype. Aagaard and Helgeson (1983) suggested that there was a free energy difference in the order of 8.36 kJ.mol^{-1} between the disordered 1 M_d and the more stable 2 M_1 polytype. Data presented here suggest it may in fact be greater than this or that there may be a degree of non-ideal mixing, particularly with respect to the muscovite end-member. The clue to this may be in the recognition of three chemically distinct 2:1 clay mineral compositions in the phases precipitated experimentally (i.e. smectite, illite (1) and illite (2)) and the fact that any two of them are capable of defining lines of univariant reaction equilibria.

The suggestion that non-ideal mixing may occur should not detract from the usefulness of the ideal solid solution approach. What is important to emphasize is that almost all the other models discussed earlier incorporate the use of thermodynamic data based on 2 M_1 muscovite and must, by definition, be in greater error than the solid solution models of Tardy and Fritz (1981) and to a lesser extent Aagaard and Helgeson (1983).

In the light of the work discussed here there is a clear requirement for thermodynamic data on different mica polytypes and further experimental work to determine the applicability of ideal solid solution modelling beyond the simple binary system investigated. It is envisaged that work on the trioctahedral Mg^{2+} talc–phlogopite system would be most beneficial as it appears that Mg^{2+} clay minerals are, like the Al^{3+} system investigated here, less stable than their ideal end-member phyllosilicate analogues (according to the empirical values given by Fritz, 1981). Such knowledge would make more accurate prediction of 'phengite-type' substitutions on the stability of diagenetic 2:1 clay minerals a possibility.

Acknowledgements

We wish to thank the British Petroleum Company PLC for permission to publish this paper and Colin Hughes (University of Manchester) for the loan of his analytical TEM standards.

References

Aagaard, P. and Helgeson, H. C. (1983) Activity/composition relations among silicates and aqueous solutions: II. Chemical and thermodynamic consequences of ideal mixing of atoms on homological sites in montmorillonites, illites and mixed-layer clays. *Clays and Clay Minerals*, **31**, 207–17.

Bjørkum, P. A. and Gjelsvik, N. (1988) An isochemical model for formation of authigenic kaolinite, K-feldspar and illite in sediments. *Journal of Sedimentary Petrology*, **58**, 506–11.

Chermak, J. A. and Rimstidt, J. D. (1989) Estimating the thermodynamic properties (ΔG_f° and ΔH_f°) of silicate minerals at 298 K from the sum of polyhedral contributions. *American Mineralogist*, **74**, 1023–31.

Cliff, G. and Lorimer, G. W. (1975) The quantitative analysis of thin films. *Journal of Microscopy*, **103**, 203–4.

Davis, J. C. (1986) *Statistics and Data Analysis in Geology*, 2nd edn, John Wiley, New York.

Fritz, B. (1981) Etude thermodynamique et modèlisation des réactions hydrothermale et diagénétiques. *Mémoire Sciences Géologiques, Strasbourg*, **65**, 197 pp.

Garrels, R. M. and Christ, C. L. (1965) *Solutions, Minerals & Equilibria*, Harper and Row, New York.

Guven, N. (1974) Lath-shaped units in fine-grained micas and smectites. *Clays and Clay Minerals*, **22**, 385–90.

Guven, N., Lafon, G. M. and Lee, L. J. (1982) Experimental hydrothermal alteration of albite to clays: preliminary results, in *Proceedings of the 7th International Clay Conference*, Bologna and

Pavia, 1981, (eds H. Van Olphen and F. Veniale), *Developments in Sedimentology*, vol. 35, Elsevier, Amsterdam, pp. 495–511.

Harvie, C. E., Møller, N. and Weare, J. H. (1984) The prediction of mineral solubilities in natural waters: The Na–K–Mg–Ca–Cl–SO_4–OH–HCO_3–CO_3–CO_2–H_2O system to high ionic strength at 25°C. *Geochimica et Cosmochimica Acta*, **48**, 723–51.

Helgeson, H. C., Delaney, J. M., Nesbitt, H. W. and Bird, D. K. (1978) Summary and critique of the thermodynamic properties of rock-forming minerals. *American Journal of Science*, **278A**, 1–229.

Huang, W. L., Bishop, A. M. and Brown, R. W. (1986) The effect of fluid/rock ratio on feldspar dissolution and illite formation under reservoir conditions. *Clay Minerals*, **21**, 585–601.

Hughes, C. R., Curtis, C. D., Whiteman, J. A., Heping, S., Whittle, C. K. and Ireland, B. J. (1990) Selected applications of analytical electron microscopy in clay mineralogy, in *CMS Workshop Lectures*, vol. 2, *Electron-Optical Methods in Clay Science*, (eds. I. D. R. MacKinnon and F. A. Mumpton), Clay Minerals Society, Evergreen, Colorado, pp. 69–88.

Lippmann, F. (1982) The thermodynamic status of clay minerals, in *Proceedings of the 7th International Clay Conference*, Bologna and Pavia, 1981, (eds H. Van Olphen and F. Veniale), *Developments in Sedimentology*, vol. 35, Elsevier, Amsterdam, pp. 475–85.

May, H. M., Kinniburgh, D. G., Helmke, P. A. and Jackson, M. L. (1986) Aqueous dissolution, solubilities and thermodynamic stabilities of common aluminosilicate clay minerals: kaolinite and smectites. *Geochimica et Cosmochimica Acta*, **50**, 1667–77.

Nriagu, J. O. (1975) Thermodynamic approximations for clay minerals. *American Mineralogist*, **60**, 834–9.

Routson, R. C. and Kittrick, J. A. (1971) Illite solubility. *Proceedings of the Soil Science Society of America*, **35**, 714–18.

Sass, B, M., Rosenberg, P. E. and Kittrick, J. A. (1987) The stability of illite/smectite during diagenesis: an experimental study. *Geochimica et Cosmochimica Acta*, **51**, 2103–15.

Stoessell, R. K. (1979) A regular solution site-mixing model for illites. *Geochimica et Cosmochimica Acta*, **43**, 495–544.

Stoessell, R. K. (1981) Refinements in a site-mixing model for illites: local electrostatic balance and the quasi-chemical approximation. *Geochimica et Cosmochimica Acta*, **45**, 1733–41.

Tanner, C. B. and Jackson, M. L. (1947) Nomographs of sedimentation times for soil particles under gravity or centrifugal acceleration. *Proceedings of the Soil Science Society of America*, **12**, 60–5.

Tardy, Y. and Fritz, B. (1981) An ideal solid solution model for calculating solubility of clay minerals. *Clay Minerals*, **16**, 361–73.

Tardy, Y. and Garrels, R. M. (1974) A method of estimating the Gibbs free energies of formation of layer silicates. *Geochimica et Cosmochimica Acta*, **38**, 1101–16.

Warren, E. A. (1987) The application of a solution–mineral equilibrium model to the diagenesis of Carboniferous sandstones, Bothamsall oilfield, East Midlands, England, in *Diagenesis of Sedimentary Sequences*, (ed. J. D. Marshall), Geological Society of London Special Publication, No. 36, Blackwell Scientific Publications, Oxford, pp. 55–69.

Warren, E. A. and Ransom, B. (1992) The influence of analytical error upon the interpretation of chemical variations in clay minerals. *Clay Minerals* (in press).

Wolery, T. J. (1986) *Calculation of Chemical Equilibrium Between Aqueous Solution and Minerals: the EQ3/6 Software Package. Lawrence Livermore Laboratory Publication* UCRL-52658, Lawrence Livermore National Laboratory, University of California, Livermore, California.

CHAPTER EIGHT

Laboratory reproduction of morphological variation in petroleum reservoir clays: monitoring of fluid composition during illite precipitation

J. S. Small and D. A. C. Manning

8.1 Introduction

The processes involved in precipitation of authigenic reservoir clays to date have mainly been elucidated from petrographic studies, including radiometric dating, which yield detailed knowledge of the paragenesis and absolute timing of diagenetic phases. Indirect knowledge of the origin of fluids associated with a particular diagenetic event can also be obtained from various isotopic studies of diagenetic phases. Further evaluation of the reactions involved in clay precipitation requires a more detailed knowledge of the equilibrium pore fluid chemistry and its evolution during dissolution and precipitation. The collection of uncontaminated formation water from petroleum boreholes is at present technically prohibitive. In addition it is not always clear what reaction history the present-day fluid has experienced. This study directly monitors fluid composition during clay precipitation in laboratory experiments. In this way the stability of clay minerals can be determined, and their rate of formation monitored by changes in fluid composition.

Among the authigenic clays illite has perhaps the most detrimental effect on reservoir properties because of the development of a delicate fibrous morphology (Stalder, 1973; Wilson and Pittman, 1977; Sommer, 1978; Guven *et al.*, 1980; Rossel, 1982; McHardy *et al.*, 1982). Despite the widespread occurrence of fibrous illite, knowledge of the controls of illite morphology derived from petrographic studies is virtually non-existent (Macchi, 1987). Huang (1990) has synthesized realistic fibrous illites, under conditions of high silica activity, above quartz saturation and suggests that under these

Geochemistry of Clay-Pore Fluid Interactions. Edited by D. A. C. Manning, P. L. Hall and C. R. Hughes. Published in 1993 by Chapman & Hall, London. ISBN 0 412 48980 5

conditions they are formed more slowly than illite plates. Preliminary experiments (Small *et al.*, 1992a) have already shown that highly realistic clay morphologies can be produced by the inclusion of a sandstone substrate, and that fibrous illites can be formed very rapidly in experiments containing oxalate. The study reported here examines in detail the formation of illite from oxalate-bearing solutions and monitors inorganic and organic solutes during precipitation.

8.2 Methods

8.2.1 Experimental approach

This study adopts an approach where the fluid composition associated with the precipitation of reservoir-type illites is determined in experiments where equilibrium is approached by precipitation of illite. Oversaturated fluid conditions with respect to illite are established by the initial dissolution of amorphous alumina and silica in various potassium-bearing solutions (Table 8.1). Previous experiments have shown that illites comparable in composition and morphology to reservoir clays can be precipitated following dissolution of both mineral phases (e.g. feldspar, kaolinite) and amorphous gels (Small *et al.*, 1992a). To promote nucleation of illites a quartz substrate is included in the experiments in the form of a sandstone core.

Table 8.1 Experimental run conditions and product summary.

Experiment	Starting materials	Time (days)	Products
Exploratory illite precipitation experiments (PTFE apparatus, 75 MPa, 250°C)			
B1	Al_2O_3, 0.1M $K_2(COO)_2$	13	Zeolites, fibrous illite (low Al)
B2	Al_2O_3, 0.2M $K_2(COO)_2$	13	Zeolites, fibrous illite (low Al)
B7	Al_2O_3, SiO_2, 0.5M KCl	12	Scarce platy illite, kaolinite
B22	Al_2O_3, SiO_2, 0.01M $H_2(COO)_2$ 0.1M KCl	12	Lath illite, lath kaolinite,
B23	Al_2O_3, SiO_2, 0.01M $H_2(COO)_2$ 0.1M KCl	3	Lath illite, lath kaolinite, Al-oxyhydroxide
B24	Al_2O_3, SiO_2, 0.01M $H_2(COO)_2$ 0.1M KCl	1	Lath illite, lath kaolinite, Al-oxyhydroxide
B25	Al_2O_3, SiO_2, 0.1M KCl	3	Poor reaction no ppt.
B29	Al_2O_3, SiO_2, 0.01M $K_2(COO)_2$	3	Fibrous/lath illite
B30	Al_2O_3, SiO_2, 0.01M $K_2(COO)_2$	1	Fibrous/lath illite
Fluid-sampling experiments (gold-titanium cell, 50 MPa, 250°C)			
R1	Al_2O_3, SiO_2, 0.01M $H_2(COO)_2$ 0.1M KCl	26	Lath illite, lath kaolinite, lath Al-oxyhydroxide
R2	Al_2O_3, SiO_2, 0.01M $K_2(COO)_2$	10	Fibrous/lath illite
R3	Al_2O_3, SiO_2, 0.01M $K_2(COO)_2$	2	Fibrous/lath illite
R4	Al_2O_3, SiO_2, 0.1M KCl	12	Poor reaction no ppt.

8.2.2 Experiment configuration

Two types of experiment have been performed:

1. Exploratory experiments where the objective was to examine solid-phase products, and to identify reaction conditions which have a significant effect on illite morphology.
2. Fluid-sampling experiments where the fluid composition associated with illite precipitation could be monitored.

The exploratory experiments were performed in a 250 ml PTFE reaction vessel, enclosed in a custom-made 5 cm ID stainless steel pressure vessel. Exploratory experiments were performed at 250°C and 750 bar (75 MPa). Fluid-sampling experiments were carried out in a Dickson-type fluid-sampling autoclave supplied by Core-Test Systems Inc., Mountain View, California. A 200 ml gold reaction cell with titanium closure, sampling tube and sampling valve were used, identical to that described by Seyfried *et al.* (1987). The fluid sampling experiments were all performed at 250°C and 500 bar (50 MPa). The same arrangement of starting materials was used in both types of apparatus, and is illustrated in Figure 8.1. For the exploratory experiments a temperature gradient of 10°C was maintained across the reaction vessel; this provides a means of convective circulation of fluid which effectively transports solutes from the source material at the bottom of the vessel to the sandstone core,

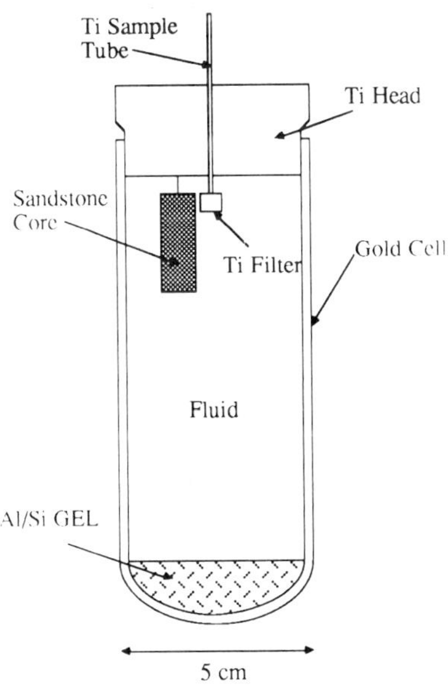

Figure 8.1 Diagrammatic arrangement of starting materials in a fluid-sampling gold/titanium reaction vessel; a similar configuration was used in the exploratory experiments performed in a PTFE vessel.

where the majority of precipitation occurs. The fluid-sampling apparatus was rocked continuously through an angle of 120°, thus ensuring mixing of fluid and equilibration of temperature. One experiment was performed in a static position, with inversion of the apparatus only during sampling; this was in order to be consistent as possible with the previous exploratory experiments.

8.2.2.1 MATERIALS

The amorphous alumina and silica gel starting material of composition 50/50 mol % Al/Si was prepared by the method of Hamilton and Henderson (1968). The surface area of the sintered gel determined by the BET method, using nitrogen, was $0.4\,m^2.g^{-1}$. Starting solutions used are summarized in Table 8.1, and were made up from analytical grade reagents. The sandstone substrate used was a sample of Penrith Sandstone, a high-porosity, high-permeability authigenic clay-free quartz-arenite from Stoneraise Quarry, Cumbria. This sandstone was chosen for its absence of authigenic clays, so that synthetic clays could be unambiguously identified. The major cement phase is quartz producing well-developed overgrowths. Both fresh quartz crystal faces and abraded grain surfaces are present and serve as a means of studying substrate surface effects on nucleation of clays. Petrographic and XRD study has shown that small quantities of feldspar (5%) are present in the sample together with lithic fragments (2%), which will contribute some detrital clay. Infiltrated/detrital clay and iron hydroxide/haematite coatings are present on the grain surfaces of abraded grains.

8.2.2.2 ANALYSIS OF SAMPLES

The morphology of precipitated clays on the surface of the sandstone core were examined in detail by secondary electron SEM and their composition characterized qualitatively from energy-dispersive X-ray spectra. X-ray diffraction was used to examine the post-run gel starting material. A fine fraction of dispersed clay from the outside of the sandstone core was prepared by ultrasonic disaggregation of a portion of the core in 5 ml of deionized water followed by settling for 5 min to produce an approximately smaller than 2 μm clay fraction. An orientated XRD specimen of the fine fraction was prepared, to identify clay species. A diluted drop of the fine clay fraction was dried on a carbon film mounted on a Cu TEM grid for analytical TEM (ATEM) investigation to obtain accurate single-grain analyses of clay particles, and to relate these to clay morphology. Analytical TEM analyses were made using a Phillips 400T TEM/STEM with EDAX energy-dispersive X-ray analysis. The X-ray spectra were collected from electron-transparent clay particles at 100 kV for 100 s and processed by the method of Cliff and Lorimer (1975). Calibration factors were determined from independently analysed macroscopic micas.

Samples of 8 ml volume fluid were recovered at intervals, measured from the start of furnace heating, and passed through a 0.2 μm syringe filter. A 1 ml aliquot was immediately diluted ten times in deionized water, to avoid precipitation of silica and alumina on standing prior to analysis, although in practice no precipitate was observed. The pH was determined immediately on the undiluted aliquot, under a nitrogen atmosphere, with a Microelectrodes Inc. 1 drop electrode calibrated to pH 4 and 7. The major soluble elements of interest in this study are silicon, aluminium and potassium. Silicon and potassium were determined on both aliquots by ICP optical emission spectroscopy (ICPOES); in addition some potassium analyses were checked by flame atomic absorption (AA) spectroscopy. Aluminium was determined in the undiluted aliquot by ICP down to a detection limit of $0.2\,mg.l^{-1}$; concentrations below this were analysed by graphite furnace AA spectroscopy. Trace elements iron, magnesium, calcium, and sodium were analysed on the undiluted aliquot by ICP. Anions were analysed by ion chromatography, using a Dionex AS5A anion separator column. Anions of interest were oxalate, acetate and formate. All analyses were performed within 3 days of sampling; samples were stored at 5°C to avoid bacterial breakdown of the organics. Freezing of the sample was avoided, since this resulted in precipitation of silica. Fluid analyses are presented in Table 8.2.

8.3 Exploratory precipitation of illite morphotypes

A series of experiments was performed in the PTFE apparatus to identify the conditions in which illite and other clay minerals could be precipitated in a fluid-dominated vessel and which could be subsequently repeated in the fluid-sampling apparatus. The experiments provide information on the solid-phase products, and the post-run fluid composition. However, the latter is often of limited quality, owing to the long quench time of around 3 h before recovery of the fluid. The reported values of *in situ* ($\log(a_{K^+}/a_{H^+})$ for the exploratory experiments estimated using EQ3/6, are considered to have a total error of around ± 1 log unit. In many cases short run times have been used which yield valuable information about the nature and timing of the initial precipitation of phases. The results of these experiments are presented in detail in Small *et al.* (1992b), and are summarized here.

8.3.1 Precipitation of clays in KCl solutions

Experiments with initial solutions of KCl produced significant precipitation of clays only after 12 days of reaction with relatively concentrated solutions (0.5 M KCl). An example of this clay precipitate (Figure 8.2(a)) shows euhedral clay plates precipitated on the surface of the sandstone core. Analytical TEM analysis showed that individual clay particles had a composition intermediate between kaolinite and smectite, consistent with interlayering of these two components. In more dilute KCl (0.05 M) only trace precipitation of rod- and lath-shaped clays was produced. Similar experiments of short duration (3 days) produced no evidence of precipitation. Small *et al.* (1992b) show that, using much lower fluid/rock ratios (approximately 1.0) in a small-volume apparatus, platy illites with a chemical composition indistinguishable from reservoir illite can be precipitated in the pore space of a sandstone substrate following dissolution of amorphous gels in KCl solution between 150 and 350°C. The present, large-volume, experimental arrangement (Figure 8.1) represents a much slower reaction regime for a given set of chemical conditions.

8.3.2 *Precipitation of clays in oxalate-bearing solutions*

A series of experiments was performed with initial solutions containing oxalate. The rationale behind the inclusion of this carboxylic acid anion being that it has been suggested to be a complexing ligand for aqueous aluminium, with the potential for enhancing dissolution of feldspar under diagenetic conditions (Surdam *et al.*, 1984). Such enhanced solubility of aluminium would be expected to have an effect on clay precipitation experiments. Three groups of experiments have been performed delineated by a_{K^+}/a_{H^+} activity ratios which cover a range of clay stabilities.

8.3.2.1 PRECIPITATION IN THE ILLITE STABILITY FIELD

Two experiments with post-run fluid compositions falling close to the illite field were attained by reaction of an initial solution of 0.01 M potassium-oxalate, with an initial pH of 7.0 (B29, B30, Table 8.1). In both experiments a remarkable heavy precipitate formed on the sandstone core, which is illustrated in Figures 8.2(b) and (c). The precipitate is composed of interlocking laths 1–2 μm in length (Figure 8.2(c)) and forms a layer of the order of 20 μm thickness. Analytical TEM analysis of dispersed clay laths indicates a mean formula of $Si_{3.32}Al_{2.20}Fe_{0.04}K_{0.96}O_{10}(OH)_2$, which is comparable to ATEM analyses of authigenic reservoir illite (Warren and Curtis, 1989), although it is noted that these synthetic illites have high potassium contents. X-ray diffraction analysis of the original amorphous starting material showed, by the presence of 10, 5 and 3.3. Å basal spacings, that it had crystallized to illite.

The fluids recovered from the quenched experiments B29 and B30 have an estimated *in situ* $\log(a_{K^+}/a_{H^+})$ of around 4.5–5.5. Compared with the $\log(a_{K^+}/a_{H^+})$ vs. $\log(a_{SiO_2})$ phase diagram at 250°C, based on Sass *et al.* (1987), and EQ3/6, these $\log(a_{K^+}/a_{H^+})$ values are within experimental error of the value for K-feldspar–illite equilibrium ($\log(a_{K^+}/a_{H^+})=4.2$), and illite–muscovite equilibrium ($\log(a_{K^+}/a_{H^+})=4.0$), at quartz saturation. They are also consistent with the value $\log(a_{K^+}/a_{H^+})=5.5$ for muscovite/microcline/quartz at 250°C recently published by Aja *et al.* (1991). Experiments B29 and B30 thus represent conditions at the upper $\log(a_{K^+}/a_{H^+})$ limit of illite stability. Ion chromatography showed that in experiment B29 no oxalate was detected, but that 40% of the original oxalate had decayed to formate. The remaining oxalate is assumed to have decomposed to gaseous products, most probably CO_2.

Table 8.2 Fluid analyses of experiments R1, R2, R3 and R4 (50 MPa, 250°C)

Experiment	Inital fluids	Solid phase	Duration (days)
R1	0.01 M oxalic acid, 0.1M KCl	Amorphous Al/Si gel	26
R2	0.01 M K-oxalate	Amorphous Al/Si gel	10
R3	0.01 M K-oxalate	Amorphous Al/Si gel	2
R4	0.1 M KCl	Amorphous Al/Si gel	12

Sample	Time (h)*	Temp (°C)	Si ($mg.l^{-1}$)	Al ($mg.l^{-1}$)	K ($mg.l^{-1}$)	Fe ($mg.l^{-1}$)	Mg ($mg.l^{-1}$)	Ca ($mg.l^{-1}$)	Na ($mg.l^{-1}$)	Meas. pH	*In situ* pH	$\log(a_{SiO_2})$	$\log(a_{K^+}/a_{H^+})$	Oxalate ($mg.l^{-1}$)	Formate ($mg.l^{-1}$)	Acetate ($mg.l^{-1}$)
R1a	2.5	110	21.3	83.00	3778	4.10	0.78	1.87	20.20	2.52	2.46	−3.12	1.26	780	0	0
R1b	8.5	250	120.6	4.16	3883	5.40	0.80	1.55	8.80	3.08	3.32	−2.36	2.07	0	170	113
R1c	24.5	250	297.3	3.90	3853	n.a.	n.a.	n.a.	n.a.	3.09	3.04	−1.97	1.78	0	30	33
R1d	50.0	250	318.2	1.92	3969	n.a.	n.a.	n.a.	n.a.	3.20	3.10	−1.94	1.86	0	0	0
R1e	96.0	250	395.6	1.35	3917	14.00	1.41	1.28	n.a.	3.46	3.19	−1.84	1.95	n.a.	n.a.	n.a.
R1f	145.0	250	404.0	2.19	3746	13.70	1.68	1.36	n.a.	3.61	3.22	−1.83	1.96	n.a.	n.a.	n.a.
R1g	220.0	250	406.3	1.55	3631	12.34	1.66	1.34	1.11	3.62	3.27	−1.83	2.03	n.a.	n.a.	n.a.
R1h	299.0	250	371.0	0.37	3654	8.98	1.60	1.26	0.98	3.76	3.45	−1.86	2.19	n.a.	n.a.	n.a.
R1i	389.0	250	300.0	0.95	3573	5.31	1.80	2.10	1.10	3.96	3.59	−1.95	2.32	n.a.	n.a.	n.a.
R1j	480.0	250	250.7	0.67	3603	3.66	1.72	1.37	1.03	4.22	3.77	−2.03	2.50	n.a.	n.a.	n.a.
R1k	625.0	250	211.4	0.28	3473	4.27	1.82	1.58	1.69	4.03	3.73	−2.11	2.46	n.a.	n.a.	n.a.
R2in†	0.0	25	2.6	8.96	644	0.08	0.00	0.08	0.00	7.45	—	—	—	800	0	0
R2a	1.8	100	4.3	12.41	651	0.08	0.12	0.09	0.00	7.61	—	—	—	755	0	0
R2b	3.5	175	30.4	25.98	624	0.04	0.14	0.00	0.00	7.40	9.52	−2.98	7.63	849	0	0
R2c	7.0	250	113.5	30.42	534	0.00	0.04	0.00	0.00	6.92	8.60	−2.39	6.63	172	265	4
R2d	24.0	250	200.7	2.85	269	0.00	0.01	0.02	0.00	6.08	7.55	−2.15	5.31	11	192	6
R2e	48.5	250	168.1	1.23	183	0.00	0.01	0.04	0.00	6.07	7.22	−2.23	4.82	11	87	8
R2f	72.0	250	164.2	1.52	156	0.00	0.00	0.00	0.00	6.01	7.11	−2.23	4.65	7	46	5
R2g	97.3	250	166.3	1.80	132	0.00	0.00	0.00	0.00	6.24	7.01	−2.23	4.48	26	31	4
R2h	168.3	250	182.9	1.77	104	0.00	0.00	0.00	0.00	6.22	6.89	−2.19	4.26	18	28	4
R2i	240.5	250	196.3	1.62	85	0.00	0.00	0.00	0.00	6.36	6.77	−2.16	4.07			

Table 8.2 *(contd)*

Experiment	Inital fluids	Solid phase	Duration (days)
R1	0.01 M oxalic acid, 0.1M KCl	Amorphous Al/Si gel	26
R2	0.01 M K-oxalate	Amorphous Al/Si gel	10
R3	0.01 M K-oxalate	Amorphous Al/Si gel	2
R4	0.1 M KCl	Amorphous Al/Si gel	12

Sample	Time (h)*	Temp (°C)	Si (mg.l^{-1})	Al (mg.l^{-1})	K (mg.l^{-1})	Fe (mg.l^{-1})	Mg (mg.l^{-1})	Ca (mg.l^{-1})	Na (mg.l^{-1})	Meas. pH	*In situ* pH	$\log(a_{SiO_2})$	$\log(a_{K^+}/(a_{H^+})$	Oxalate (mg.l^{-1})	Formate (mg.l^{-1})	Acetate (mg.l^{-1})
R3in†	0.0	25	0.1	0.00	742	0.16	0.01	0.15	0.00	6.94	—	—	—	805	0	0
R3a	0.5	40	1.5	2.84	725	0.15	0.08	0.13	0.00	7.19	—	—	—	810	0	0
R3b	3.5	100	4.3	7.59	692	0.11	0.09	0.16	0.00	7.89	10.42	−3.84	8.59	811	0	0
R3c	4.5	180	25.8	20.04	646	0.13	0.11	0.20	0.01	7.55	9.50	−3.05	7.63	803	0	0
R3d	7.8	250	110.4	29.80	550	0.12	0.07	0.00	0.26	7.26	8.15	−2.41	6.73	313	228	0
R3e	12.0	250	177.3	34.90	462	0.00	0.00	0.00	0.13	7.11	8.20	−2.20	6.18	6	397	0
R3f	16.0	250	192.4	19.80	403	0.00	0.00	0.00	0.10	6.83	8.03	−2.18	5.95	0	363	0
R3g	20.0	250	204.0	10.22	407	0.00	0.00	0.00	0.00	6.99	8.05	−2.14	5.98	0	352	0
R3h	24.0	250	202.0	6.91	368	0.00	0.00	0.00	0.00	6.68	7.94	−2.14	5.83	0	343	0
R3i	29.0	250	206.9	4.42	333	0.00	0.00	0.00	0.00	6.65	7.83	−2.13	5.68	0	322	0
R3j	47.5	250	195.5	2.24	250	0.00	0.00	0.00	0.00	6.56	7.52	−2.16	5.25	0	235	0
R4in†	0.0	25	1.0	0.00	3809	0.00	0.11	0.52	0.63	5.50	—	—	—	n.a.	n.a.	n.a.
R4a	1.5	100	2.3	0.00	3818	0.00	0.18	0.43	0.54	6.59	6.05	−4.08	4.89	n.a.	n.a.	n.a.
R4b	3.5	192	27.0	0.03	3846	0.00	0.21	0.60	0.56	5.69	5.33	−3.01	4.08	n.a.	n.a.	n.a.
R4c	7.5	250	141.4	0.66	3837	0.03	0.25	1.32	0.98	4.66	4.22	−2.28	2.97	n.a.	n.a.	n.a.
R4d	23.0	250	222.4	1.07	3823	0.44	0.43	1.60	0.96	3.67	3.62	−2.09	2.37	n.a.	n.a.	n.a.
R4e	54.0	250	221.2	0.44	3880	0.83	0.55	0.75	0.87	4.23	4.04	−2.09	2.79	n.a.	n.a.	n.a.
R4f	120.0	250	232.9	0.29	3817	2.60	0.66	0.85	0.96	4.22	3.92	−2.07	2.67	n.a.	n.a.	n.a.
R4g	144.0	250	235.5	0.31	3914	3.57	0.72	1.25	1.21	4.30	3.87	−2.06	2.63	n.a.	n.a.	n.a.
R4h	288.0	250	237.0	0.31	3908	9.35	0.82	0.63	0.70	4.48	3.69	−2.06	2.45	n.a.	n.a.	n.a.

n.a. = not analysed.
*Sample time recorded from start of heating.
†Sample taken just before heating started.

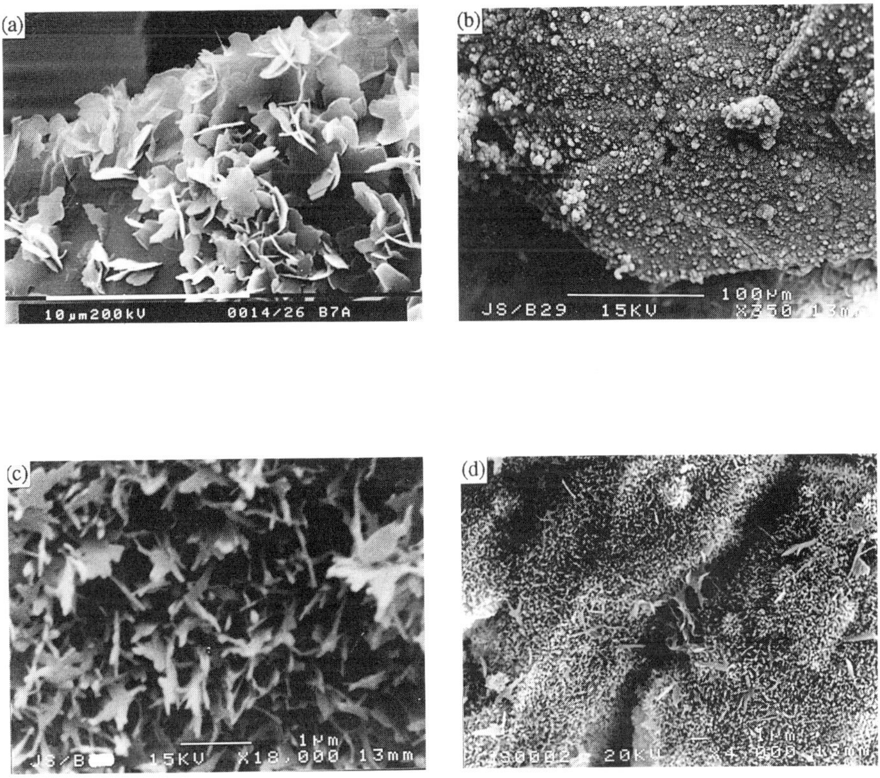

Figure 8.2 Scanning electron microscope photomicrographs of illites and other secondary precipitates formed in the exploratory experiments. (a) Euhedral platy kaolinite–smectite nucleated on the surface of the quartz sandstone substrate. Formed following dissolution of amorphous Al/Si gel in 0.5M KCl at 250°C and 75 MPa for 12 days. (b) Low-magnification view of the surface of sandstone substrate showing extensive precipitation of illite following initial dissolution of Al/Si gel in 0.01M potassium oxalate at 250°C and 75 MPa. This very dense precipitate formed within the first 24 h of reaction. (c) Detail of (b) showing lath/fibrous nature of illite. (d) Lath/fibrous precipitate formed in initially acidic solution of 0.01 M oxalic acid and 0.1 M KCl at 250°C and 75 MPa. Analytical TEM (Figure 8.4) shows that the composition of these laths varies considerably and includes illite, kaolinite, and pure aluminium-oxyhydroxide.

8.3.2.2 PRECIPITATION FROM INITIAL ACIDIC SOLUTIONS

Three experiments (B22, B23, B24, Table 8.1) were performed with an initial solution of 0.01 M oxalic acid and 0.1 M KCl, with run times of 12, 3 and 1 days. A very extensive precipitate of lath-shaped clays was produced in all three experiments. An example of that formed after 12 days is shown in Figure 8.2(d), and is typical of many examples of natural reservoir illites. The experiments of shorter duration produced identical precipitates indicating that the main phase of precipitation has occurred within the first 24 h of the experiment.

Analytical TEM analyses of clay laths recovered from the surface of the core substrate show a wide range of compositions (Figure 8.3) which include illite, kaolinite, and aluminium-rich phases intermediate between pure aluminium (boehmite) and kaolinite. In the 1 and 3 day experiments illites and the aluminium-rich phase are present, whilst in the 12 day experiment only kaolinites and illites are identified. X-ray diffraction analyses of the crystallized gel identified kaolinite and boehmite in all three experiments.

Fluids recovered from the experiments are all oversaturated with respect to quartz and have $\log(a_{K^+}/a_{H^+})$ of around 2–3, which, according to the phase diagram at 250°C (see Sass *et al.*, 1987), coincides with conditions of kaolinite and smectite stability. Considering experimental and calculational errors, these data may represent illite–smectite or kaolinite–smectite equilibria. The presence of an aluminium-rich precipitate in the two short-duration runs suggests that this phase precipitated initially, when the silica activity was below that required for kaolinite stability. After 12 days this early-formed phase appears to have been replaced by kaolinite and illite. All three phases have a characteristic lath-shaped morphology; this could conceivably result from an isomorphous replacement of the early-formed aluminium phase. Ion chromatography revealed that in all three experiments oxalate had completely broken down, and no formate was detected. Associated with this breakdown there was an increase in quench pH from 2 to 4.

8.3.2.3 PRECIPITATION IN THE ZEOLITE STABILITY FIELD

Experiments with a more concentrated initial solution of 0.1 M potassium-oxalate (B1 and B2, Table 8.1) and an initial pH of 8 produced very extensive

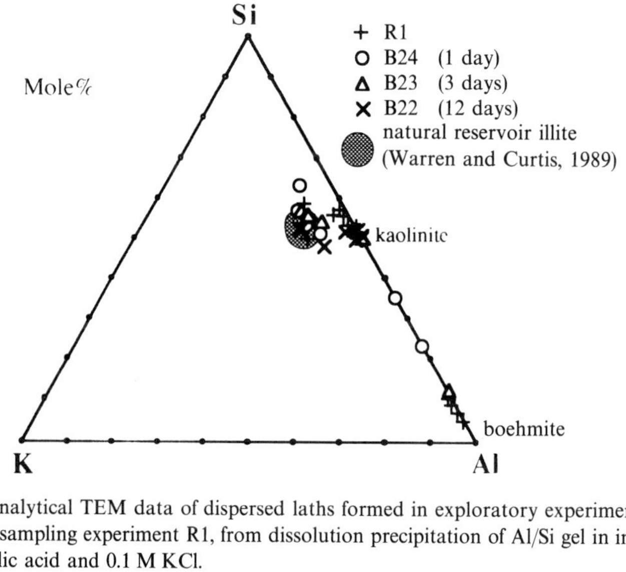

Figure 8.3 Analytical TEM data of dispersed laths formed in exploratory experiments B22, B23, B24 and fluid-sampling experiment R1, from dissolution precipitation of Al/Si gel in initial solution of 0.01 M oxalic acid and 0.1 M KCl.

precipitation; however, in this case, mainly zeolite phases precipitated. Among the precipitated zeolites was a fibrous clay mineral identified on the basis of electron diffraction and XRD analysis as illite. Analytical TEM analyses of single fibres are highly unusual in that they contain only 1.7 mol of Al/11 oxygen formula unit, which is insufficient to assign to any layer silicate formula. It is assumed that these aluminium-deficient particles are mixtures of illite and a silica-rich phase, possibly a zeolite (mordenite), and may represent an unstable phase analogous to the early aluminium-rich phases formed under initially acidic conditions. The post-run fluids from these experiments have $\log(a_{K^+}/a_{H^+})$ around 7.5, 3 log units above the field of illite stability (see Aja *et al.*, 1991) and represent conditions in which zeolites (mordenite) and feldspar are stable.

8.4 Fluid sampling experiments

8.4.1 Objectives

The previous experiments show that substantial precipitation of illite can be produced in the initial presence of oxalate. Using the same experimental configuration in the absence of oxalate only platy clays are produced at a much slower rate. Rapid precipitation appears to be a major factor in controlling the morphology of illites. The occurrence of unstable mixed-phase composition precipitates in experiments which have not attained equilibrium may also be of significance to the control of clay morphology. These earlier experiments have been repeated using a fluid-sampling apparatus with the following objectives in mind:

1. To monitor fluid composition in order to assess the rate of approach of steady-state conditions.
2. To determine the saturation state of various potential unstable phases during the approach to equilibrium, and comparison with observed secondary phases.
3. To determine the rate of decay of oxalate, and its influence on clay precipitation.
4. To determine the stability of illite, and other authigenic clay minerals from steady-state fluid composition.

8.4.2 Calculation of in situ *activities*

The following discussion of results uses measured fluid composition and pH at 25°C to assess mineral stabilities at 250°C and 500 bar. Activities of aqueous species including pH are calculated for 250°C using the EQ3/6 software package 3245R111 (Wolery, 1983) using the steam-saturated data file (data 0.3245R54feb. 88). A Debye–Hückel activity correction was applied, ionic strengths of all solutions being less than 0.1. The *in situ* pH of the experiments is controlled partly by clay precipitation reactions but also by dissociation reactions of oxalic acid and its breakdown products, formic acid and carbon dioxide. EQ3 was run at 25°C to charge-balance the measured fluid analyses, by adjustment of HCO_3^-. An assumed concentration of $CO_2(aq)$ at 250°C was then calculated to maintain a mass balance of carbon, the measured pH being made after dissolved CO_2 was evolved. The *in situ* pH at 250°C was then calculated by EQ6 using a dissociation constant (K_{diss}) for formic acid estimated by Shock and Helgeson (1990), and by back-reacting the calculated mass of CO_2 evolved during sampling. The dissociation constants of oxalic acid at elevated temperatures are poorly constrained; oxalate is, however, completely broken down in all but one sample taken at 250°C. In this sample and in those taken during heating, K_{diss} for formic acid was used for the first dissociation of oxalic acid to provide an approximate pH correction; this would probably produce an error of around ± 0.25 pH units.

Errors in $\log(a_{K^+}/a_{H^+})$ and $\log(a_{SiO_2})$ at *in situ* P and T are a combination of sampling/analytical and activity calculation errors. For the internal comparison of concentration data only the sampling/analytical error is important, which is estimated to be $\pm 5\%$ for silicon, aluminium and potassium concentrations. For comparison with other studies, activity calculation errors are important and are estimated to be ± 0.4 log units for $\log(a_{K^+}/a_{H^+})$ and ± 0.05 log units for $\log(a_{SiO_2})$.

8.4.3 Precipitation under initially acidic conditions

Experiment R1 (Table 8.1) was a repeat of experiments B22, B23 and B24 monitoring fluid composition at 11 intervals with a total run time of 620 h (26 days). The sampling interval was shortest during the first 100 h, which included the heating of the apparatus.

8.4.3.1 FLUID ANALYSES

The major solute variables are plotted in Figure 8.4. None of the plots show the attainment of a steady state, although an approach to equilibrium is apparent. Silicon concentrations initially rise rapidly to around 400 ppm after 100 h, which is in excess of quartz saturation, but below amorphous silica saturation (585 ppm). Subsequently silicon concentrations progressively decrease, and at the termination of the experiment approach quartz saturation at 180 ppm. Aluminium concentration is at a maximum in the first sample taken after 2 h at a temperature of 100°C. In the next sample, taken at 250°C after 8 h, the aluminium concentration has dropped by over ten times. Subsequent samples show that the aluminium concentration decreases steadily by a further order of magnitude during the remainder of the experiment. Both quench and *in situ* pH show a general inverse relationship to aluminium concentration, with an initial sharp rise followed by a more gentle increase to an *in situ* pH of 3.5. Potassium

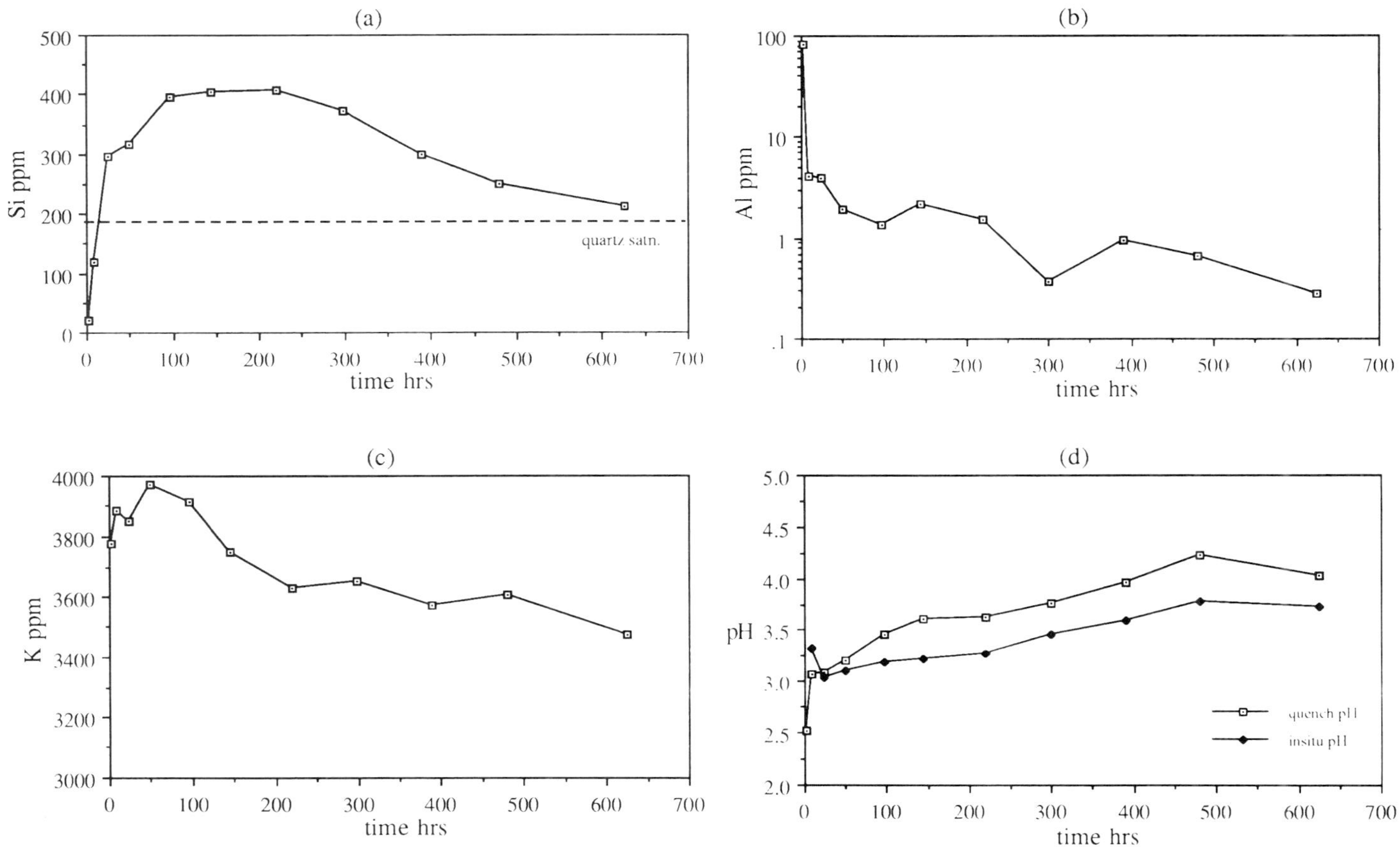

Figure 8.4 Major solute concentrations, quench pH, and calculated *in situ* pH, in fluid-sampling experiment R1, 250°C, initial solution 0.01 M oxalic acid, 0.1 M KCl.

concentration shows a decrease after 100 h, with a total loss of 300 ppm K from the fluid.

Figure 8.5 shows that oxalate is highly unstable in the experiment, decomposing above 100°C after only 2 h heating. Formate and acetate are intermediate products of oxalate breakdown, but are themselves totally decomposed after 50 h. The rapid breakdown of oxalate is considered responsible for the initial increase in pH and drop in aluminium concentration between 2 and 8 h after the start of the experiment.

8.4.3.2 SECONDARY PHASES

Scanning electron microscope examination of solid-phase products from R1 revealed the formation of a very extensive lath-shaped precipitate on the surface of the sandstone core (Figure 8.6(a)). Analytical TEM analyses of dispersed laths are illustrated in Figure 8.3 which shows the presence of illite, kaolinite, and a mixed aluminium-rich phase all having a lath-shaped morphology. The Al/Si gel starting material formed a consolidated cake at the bottom of the gold cell. X-ray diffraction analysis showed that it had crystallized to kaolinite and illite–smectite. The secondary precipitates in this fluid-sampling experiment closely resemble those found in the exploratory experiments B22, B23, and B24 and suggest that kaolinite and illite have formed by replacement of an early formed aluminium oxyhydroxide.

8.4.3.3 MINERAL STABILITY

Fluid analyses recalculated as $\log(a_{K^+}/a_{H^+})$ and $\log(a_{SiO_2})$ at 250°C are plotted on the activity–activity diagram in Figure 8.7. This plot shows that the fluid composition passes through different mineral stability fields on its approach to equilibrium. The first sample (Table 8.2, R1a), which plots in the boehmite stability field of the 250°C phase diagram, cannot be compared directly with

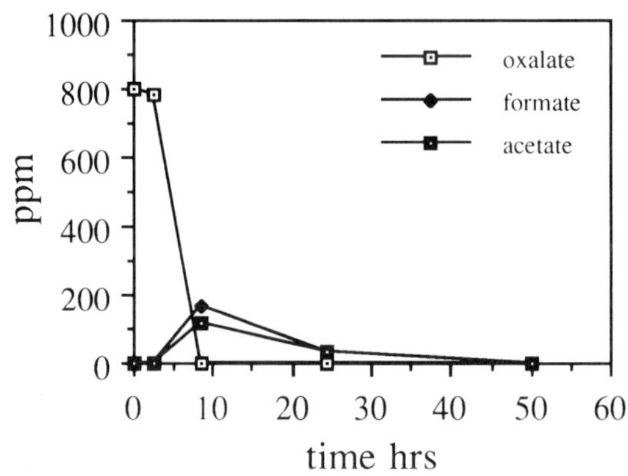

Figure 8.5 Decay of oxalate, and formation of acetate and formate during first 50 h of experiment R1.

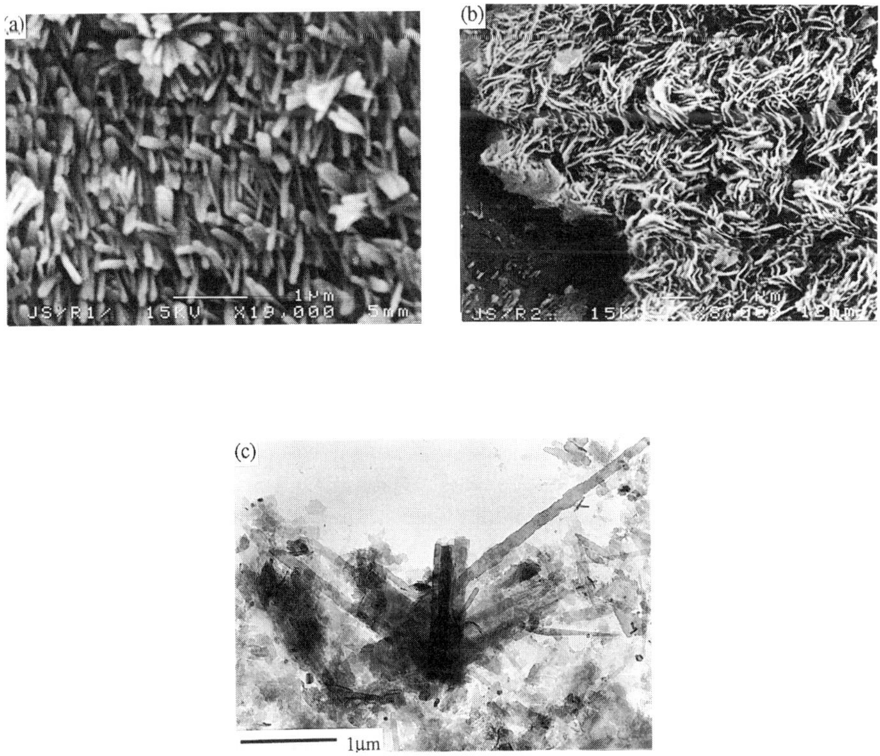

Figure 8.6 Scanning electron microscope and TEM photomicrographs of precipitates formed in fluid-sampling experiments. (a) Lath-shaped illite, kaolinite and aluminium-oxyhydroxide formed in experiment R1, all three composition precipitates have an identical lath morphology. (b) Pseudo-platy illite precipitate formed in initial solution of 0.01 M potassium oxalate in experiment R2. (c) Transmission EM photomicrograph of dispersed illite precipitate formed in experiment R2. Individual particles have a predominantly lath and ribbon morphology, typical of fibrous reservoir illite. In this experiment illite laths appear to have clumped together to form the platy texture illustrated in (b).

others taken later in the same experiment, since it was sampled at 100°C. Composition R1a falls in the kaolinite field at 100°C, but since temperature was rising rapidly it is likely that boehmite would be stable at some stage during heating. Figure 8.4(b) shows that substantial aluminium precipitation occurred after this first sampling. The precipitation of the observed aluminium-rich phase and kaolinite are, therefore, likely to have occurred in the first 8 h, during the heating phase of the experiment. The measured drop in aluminium concentration of 76 ppm is equivalent to the precipitation of 42 mg of boehmite or 91 mg of kaolinite in the reaction vessel.

Fluid analyses of samples taken at 250°C can be compared with the phase diagram; these compositions pass through the kaolinite and smectite fields and at the termination of the experiment lie within experimental error of the

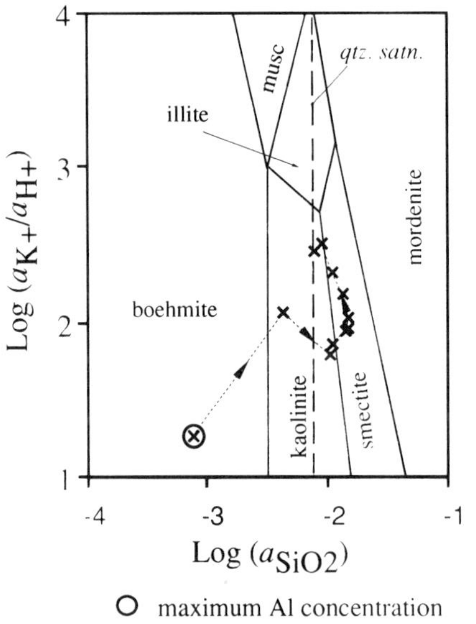

Figure 8.7 $\text{Log}(a_{K^+}/a_{H^+})$ vs. $\log(a_{SiO_2}^+)$ activity–activity diagram for 250°C and saturated vapour pressure showing pathway of fluid composition during experiment R1. The phase diagram is constructed for 250°C and saturated vapour pressure using equilibrium constants for illite–smectite, kaolinite–smectite, and illite–kaolinite calculated from data in Sass *et al.* (1987), where the following formulae are assumed, illite: $K_{0.67}Al_{2.67}Si_{3.33}O_{10}(OH)_2$, smectite: $K_{0.24}Al_{2.24}Si_{3.76}O_{10}(OH)_2$. The mordenite stability field is defined by a solubility constant for K-mordenite provided in the EQ3/6 database (Wolery, 1983). Remaining data were taken from Robie *et al.* (1978) and Helgeson *et al.* (1978). Circled point represents maximum in aluminium concentration, arrows indicate direction of fluid-composition pathway.

illite–kaolinite phase boundary defined by Sass *et al.* (1987). The final $\log(a_{K^+}/a_{H^+})$ of 2.46 is in excellent agreement with that for muscovite–kaolinite stability at this temperature (Hemley, 1959; Chermak and Rimstidt, 1990). Modelling of the final fluid composition is thus partially consistent with the mineralogical observations, with illite and kaolinite being identified. The aluminium-rich phase, however, occurs metastably and $\log(a_{SiO_2})$ is slightly above quartz saturation.

The rise in $\log(a_{SiO_2})$ during the first 100 h of the experiment results from the dissolution of amorphous silica gel starting material, producing quartz supersaturated conditions. The subsequent fall in $\log(a_{SiO_2})$ must be controlled by the precipitation of secondary phases. Figure 8.7 shows that the fall in $\log(a_{SiO_2})$ is accompanied by a large increase in $\log(a_{K^+}/a_{H^+})$. Potassium concentration also shows a decrease after 100 h indicating precipitation of a potassium-bearing phase. The fluid composition pathway during this phase of silica and potassium precipitation (Figure 8.7) is towards illite, which is observed as a precipitation product. The drop in potassium concentration during the experiment is equivalent to the precipitation of 0.1 g of illite within the vessel, which is reasonable since this includes illite formed within the crystallized gel. During the period of illite formation, aluminium concentrations fall from around 1 ppm to 0.3 ppm, which accounts for less than 1% of the loss of potassium by the formation of illite. The source of aluminium for illite formation in this case is more likely to be by alteration of early-precipitated aluminum rich phases. The identical lath-shaped morphologies of the aluminium rich phase, kaolinite and illite is further, circumstantial, evidence for solid-replacement of early-formed phases.

8.4.4 *Precipitation from alkaline K-oxalate*

Experiments R2 and R3 (Table 8.1) are repeats of exploratory experiments B29 and B30 using an initial solution of 0.01 M potassium-oxalate. R2 was run for a total time of 240 h whilst R3 examined in detail the first 50 h of the experiment. R3 was operated in a static mode, because a high proportion of the total run time was represented by heating, where the apparatus must remain static.

8.4.4.1 FLUID ANALYSES

Figure 8.8 illustrates the variation in major solutes, and shows that quartz saturation and steady-state conditions are established after about 50 h. The short experiment R3 defines a maximum in aluminium concentration of 35 ppm at 12 h, with apparent attainment of a steady state after 50 h. Aluminium concentrations in R3 are all slightly higher than equivalent samples in R2. Potassium concentration approaches a steady state after about 100 h, when approximately 80% of the K^+ has been precipitated; higher potassium concentrations are noted for R3 than R2. Quench pH shows rather erratic behaviour, although a general decrease in pH is noted. Calculated *in situ* pH shows a much smoother behaviour, and closely follows the potassium concentration curves.

The stability of the organic acid anions is illustrated in Figure 8.9. As in the previous experiment, R1, oxalate decomposes rapidly on reaching 250°C and is not detectable after 12 h. Formate is again recognized as a breakdown product; however, in this case it remains throughout the duration of the experiment. Greater amounts of formate are present in R3 than in R2. The presence of a significant concentration of formate in these two experiments produces a large difference in quench and calculated *in situ* pH. Smaller corrections are noted for R2 and the correction decreases with time as formate decays. The smaller corrections are a result of both lower formate concentrations and higher CO_2 concentrations, the latter effectively lowering the *in situ* pH.

The higher pH and potassium, aluminium and formate concentrations in R3 after 16 h most probably result from slower reaction kinetics in this run, in which the pressure vessel assembly remained stationary during the whole of the experiment.

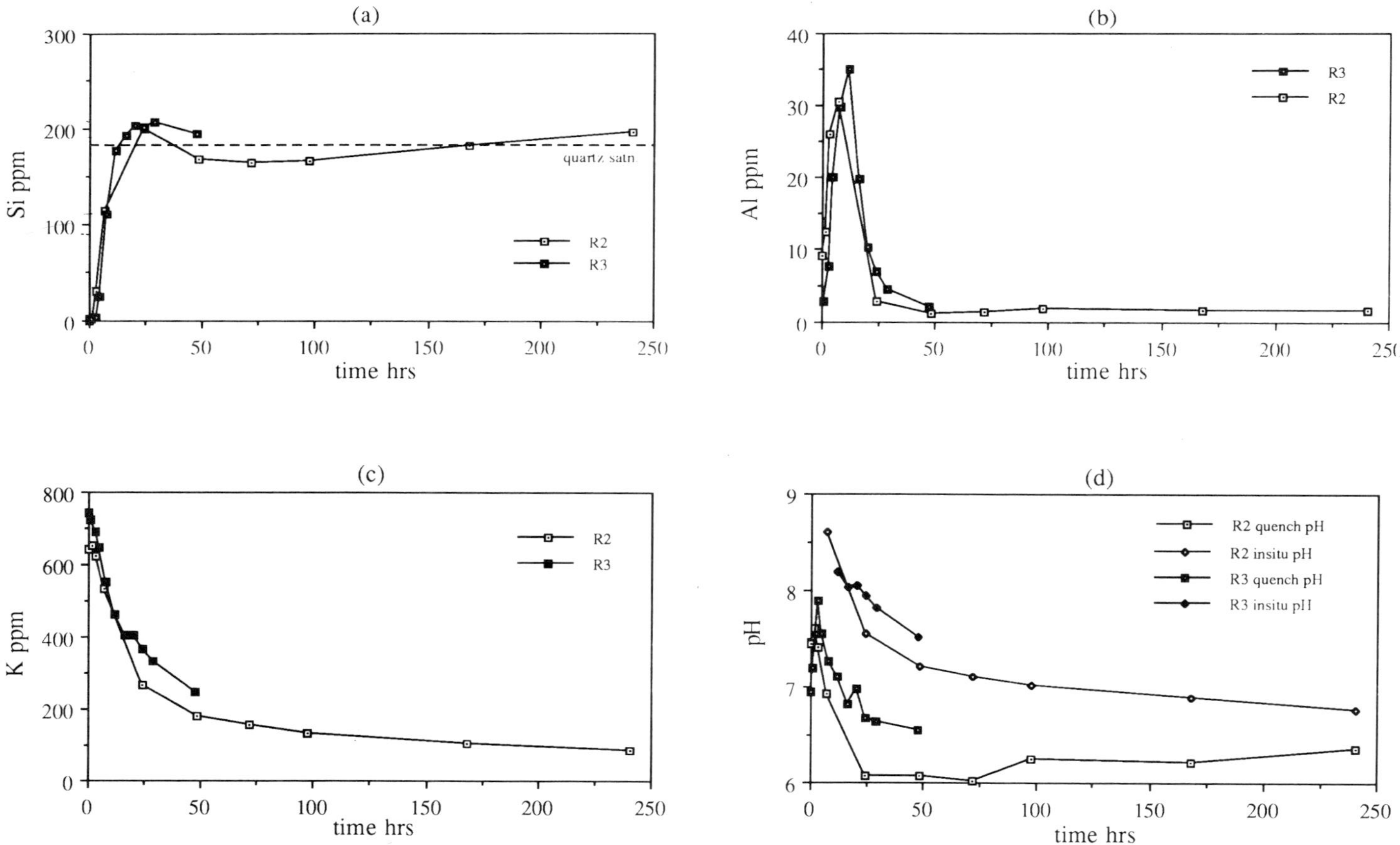

Figure 8.8 Major solute concentrations, quench pH, and calculated *in situ* pH, in fluid-sampling experiments R2 and R3, 250°C, initial solution 0.01 M potassium oxalate.

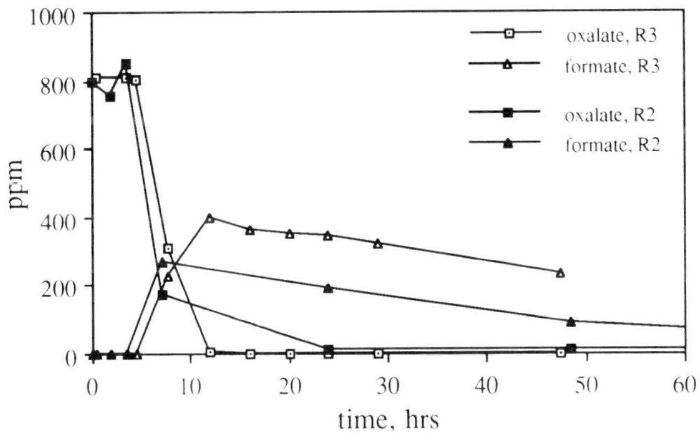

Figure 8.9 Decay of oxalate and production of formate in experiments R2 and R3 initially with 0.01 M potassium oxalate. Oxalate decay is slower than in experiment R1 under acidic conditions and formate remains for the duration of the experiment R2 (640 h).

8.4.4.2 SECONDARY PHASES

Scanning electron microscope examination of both experiments revealed that a very extensive clay precipitate had formed on the surface of the sandstone cores; this clay has a platy appearance (Figure 8.6(b)), which is in contrast to that formed in the exploratory experiments B29 and B30 (Figure 8.2(c)). Examination of dispersed clays by TEM, however, revealed that the clay plates are in fact aggregates of lath- and ribbon-type particles (Figure 8.6(c)). The rapid precipitation of fibrous particles is thus consistent between both sets of experiments. Analytical TEM analyses of fibrous particles from R2 give a mean formula of $K_{0.65}Si_{3.38}Al_{2.33}Fe_{0.19}Mg_{0.10}O_{10}(OH)_2$ which is consistent with published ATEM analyses of natural reservoir illites (Warren and Curtis, 1989; Macchi *et al.*, 1990). X-ray diffraction analysis of the gel material shows that it is also crystallized to illite.

8.4.4.3 MINERAL STABILITY

Figure 8.10 compares fluid analyses from R2 and R3 expressed as $\log(a_{K^+}/a_{H^+})$ and $\log(a_{SiO_2})$ with the phase diagram for 250°C. Together the two experiments define the pathway of fluid composition during illite precipitation. Initially, during heating and dissolution of the Al/Si gel, fluid compositions are in the muscovite and K-feldspar stability fields. In both experiments at the point of maximum aluminium concentration (defined in Figure 8.7) fluid composition occurs just below quartz saturation under conditions of K-feldspar stability. Subsequent points, which record the main phase of precipitation, temporarily move into the mordenite stability field, and then follow the K-mordenite–K-feldspar field boundary. At the same time (a_{K^+}/a_{H^+}) decreases until the final steady-state conditions are attained at the K-feldspar–illite phase boundary. The only significant secondary phase observed in these two experiments is

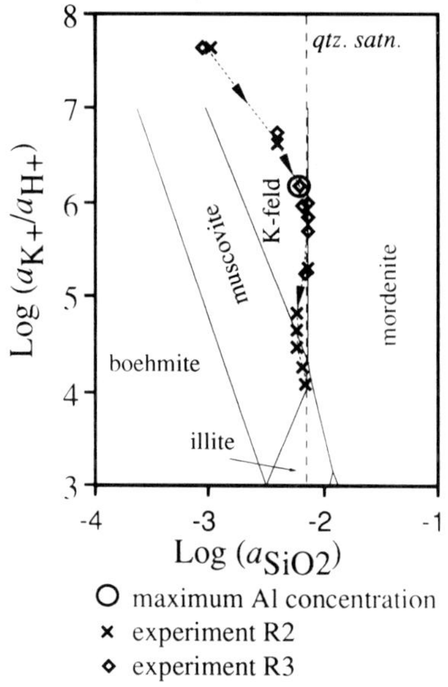

Figure 8.10 Log(a_{K^+}/a_{H^+}) vs. log(a_{SiO_2}) activity–activity diagram for 250°C and saturated vapour pressure showing pathway of fluid composition during experiment R2 and R3. Construction, and legend of phase diagram as in Figure 8.7.

fibrous illite; thus, the final fluid composition and the model prediction are in excellent agreement. The pathway of fluid-composition migration and the timing of decrease in potassium and aluminium concentrations, however, indicates that fibrous illite has precipitated metastably in the K-feldspar and mordenite stability fields.

8.4.5 Precipitation from KCl solution

Experiment R4 was performed with an initial solution of 0.1 M KCl as a control for comparison with the previous fluid-sampling experiments which contained oxalate. No secondary phases were identified, either by SEM examination of the sandstone core, or by XRD analysis of the amorphous gel starting material.

Fluid analyses are shown in Figure 8.11. All major solutes show the attainment of steady-state conditions after 100 h. Silicon reaches a steady state above quartz saturation, presumably buffered metastably by amorphous silica. Aluminium concentration shows the attainment of a maximum of 1 ppm Al after 24 h, and reaches a steady state of 0.4 ppm. Clearly much less aluminium has been dissolved in this experiment than in the previous oxalate-bearing experiments; this is reflected in the absence of secondary clays in R4. Potassium concentration shows no change during the reaction, since no potassium-bearing phase precipitates. pH shows a large-scale decrease associated with the dissolution of the Al/Si gel.

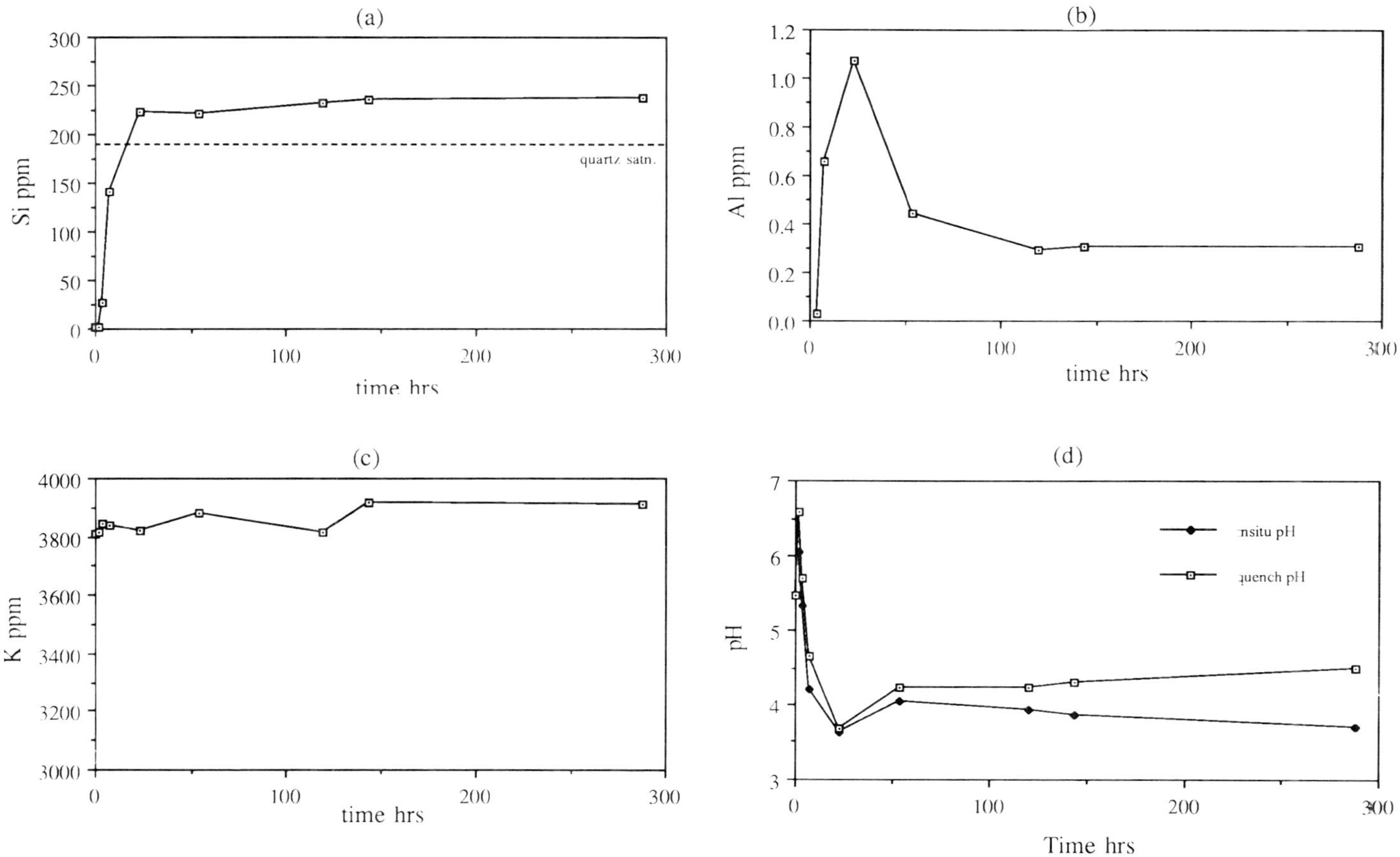

Figure 8.11 Major solute concentrations, quench pH, and calculated *in situ* pH, in fluid-sampling experiment R4, 250°C initial solution 0.1 M KCl.

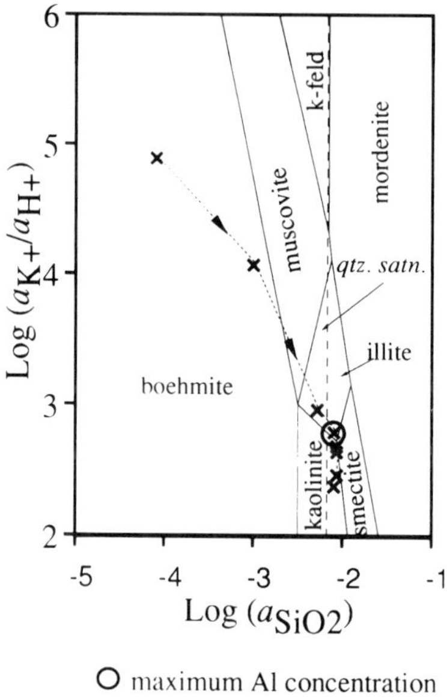

Figure 8.12 Log(a_{K^+}/a_{H^+}) vs. log(a_{K^+}/a_{H^+}) activity–activity diagram for 250°C and saturated vapour pressure showing pathway of fluid composition during experiment R4. Construction, of phase diagram as in Figure 8.7.

Plotting of fluid data on the log(a_{K^+}/a_{H^+}) vs. log(a_{SiO_2}) phase diagram (Figure 8.12) shows that steady-state conditions are established at log(a_{K^+}/a_{H^+}) of around 2.45 which is somewhat lower than for the predicted illite stability but identical to that established in R1. In R4 the maximum in aluminium concentration is attained in the illite field. The limited precipitation that is indicated by the small drop in aluminium concentration (Figure 8.11) is likely to have occurred in the illite and kaolinite fields.

8.5 Discussion

8.5.1 Control of illite morphology

Previous studies of illite–smectite and illite both during hydrothermal alteration and during burial diagenesis (Pollastro, 1985; Keller *et al.*, 1986; Inoue *et al.*, 1987; Glasmann *et al.*, 1989; Lanson and Champion, 1991) have recorded morphological changes which occur during the smectite–illite transformation. These studies have concluded that original crenulated, flaky smectite-rich clays are replaced by lath-shaped 1 M_d illite–smectite, and finally produce euhedral hexagonal 2 M illite. In contrast, fibrous and platy illites formed experimentally by dissolution-precipitation do not show any dependence on the illite–smectite to illite compositional series and are all exclusively 1 M_d polytype (Small *et al.*, 1992b). These synthetic illites indicate that a particular polytype does not have a particular morphology. The present experimental results identify some other controls on the morphology of illite formed by precipitation from solution.

This study demonstrates that the fibrous illite morphology is associated with rapid precipitation. These conditions are established in the experiments by the initial presence of oxalate, in either acidic or alkaline conditions. Typically, sandstone cores are coated in a 20 μm thick layer after 24 h of reaction. In the absence of oxalate, precipitation is virtually non-existent, with platy clays only being formed at a much slower rate in concentrated 0.5 M KCl solution. The fluid sampling experiments show that very high aluminium solubilities are associated with the fibrous illite-forming experiments, which are up to two orders of magnitude above that produced in the KCl-bearing experiment, in the absence of oxalate. From the initial rise and fall in aluminium concentration the timing of the major phase of precipitation can be determined. For initially acidic conditions (R1) the major phase of precipitation is completed after only 8 h, which is essentially the heating period of the experiment. For alkaline conditions, precipitation occurs slightly later, but is virtually completed after 24 h.

Another significant feature of some of the fibrous-illite-producing experiments is the formation of unstable phases with compositions of individual clay particles which are intermediate between the compositions of discrete mineral phases. These unstable phases are common in the initially acidic experiments where mixed boehmite–kaolinite lath-shaped phases are produced together with identically shaped illites. The early precipitation of boehmite and kaolinite and subsequent formation of illite is also predicted from the pathway of fluid composition during the experiments. The alteration of early-formed lath-shaped boehmite to kaolinite and illite is, therefore, indicated.

Under alkaline conditions in experiments using 0.1 M potassium oxalate, aluminium-deficient fibrous illites are produced, which compositionally represent a mixture of illite and zeolite (K-mordenite). Post-run fluid conditions of these experiments are in the feldspar and zeolite stability fields. In more dilute potassium oxalate (0.01 M) fibrous illites with a realistic stoichiometry produce a very extensive precipitate in experiments which appear to be at equilibrium. Interpretation of fluid analyses indicates that in these latter experiments illite must have precipitated metastably in the predicted K-feldspar and mordenite stability fields. The relationship between these two sets of potassium oxalate experiments is intriguing and suggests that fibrous illites have formed via an aluminium-deficient precursor which has precipitated metastably in the mordenite stability field at higher (a_{K^+}/a_{H^+}) and a_{SiO_2} than for illite stability.

In contrast, under a comparable experimental configuration, platy illite formation is much slower than for the fibrous formation. For platy illites, precipitation, defined by the decline in aluminium concentration, occurs in the illite and kaolinite fields. Under these conditions illite is completely neo-formed and so its morphology may simply be controlled by crystallography, the plate-like morphology reflecting its layer silicate structure.

Two distinct mechanisms of formation of fibrous illites have been recognized, which represent formation from initially acidic or alkaline conditions. Small *et al.* (1992b) show that platy illites can be produced, starting from acidic or alkaline solutions with low fluid/rock ratio; thus, a direct pH control of morphology can be ruled out. The main difference controlling platy versus fibrous growth demonstrated in this fluid-sampling study is the initial degree of supersaturation of aluminium, and the subsequent pathway of fluid migration during precipitation. Combined with the observed occurrence of unstable, mixed-composition fibrous phases, these results suggest that the fibrous morphology of reservoir illite could be inherited by means of an isomorphous replacement of precursor precipitates.

The presence of oxalate (or other organic acids) is not a prerequisite of fibrous illite formation since fibrous illites can be synthesized in the absence of oxalate (Huang, 1990; Small *et al.*, 1992b); in these cases fibrous illite is favoured by high a_{SiO_2} above quartz saturation. The illite system phase diagram (Figure 8.10) shows that fluids of high a_{SiO_2} would approach illite stability from the zeolite field (mordenite), and are thus essentially similar to experiments R2 and R3. Fibrous pyrophyllite is also recognized in experiments by Huang (1990); clearly this has similarities to the unstable aluminium-deficient illites (runs B1 and B2) and lath-shaped kaolinite and boehmite (runs B22, B23, B24, and R1) which appear in this study to be precursors of fibrous illite. The results of Huang (1990) are thus generally consistent with the present study indicating that fibrous illites are formed under a specific set of chemical conditions. Major differences in the rate of illite formation are evident. Huang (1990) states that fibrous illite is a late precipitate and that the extent of fibrous illite formation increases with time, probably resulting from the migration of fluid composition into a condition favouring fibrous illite formation. These differences in the rate of illite formation are presumed to result from timing of the migration of fluid composition into the fibrous illite stability field. In the absence of oxalate these will be driven by mineral precipitation which pre-dates illite formation, while in this study the inclusion of oxalate allows the critical conditions to be established at a much faster rate. The effect of precipitation rate on fibrous illite formation may, therefore, be secondary to the occurrence of critical fluid composition conditions favouring fibrous formation.

8.5.2 *Illite stability*

For illite precipitation experiments which attain a steady state the final fluid composition enables the stability of the precipitated illite to be determined. Of the four experiments performed, R2 reached steady-state conditions, and contains the mineral assemblage of illite/quartz. Examination of the $\log(a_{K^+}/a_{H^+})$ vs. $\log(a_{SiO_2})$ phase diagram (Figure 8.10) shows that the steady-state fluid composition falls very close to illite/muscovite stability predicted from equilibrium constants given by Sass *et al.* (1987). The thermodynamic status of illite and illite–smectite as being a multiphase mixture of high- and low-charged interlayers (Garrels, 1984; Sass *et al.*, 1987; Aja *et al.*, 1991) is supported by this result. The final fluid composition is also well within experimental error of feldspar–illite or mordenite–illite stability defined by equilibrium constants in the EQ3/6 database, although they are around 1 $\log(a_{K^+}/a_{H^+})$ unit below that recently determined by Aja *et al.* (1991) for muscovite–microcline equilibrium. Aluminium solubility data (see below) suggest that steady-state aluminium concentrations in R2 are buffered by K-feldspar or mordenite rather than illite or muscovite. The solubility-controlling phases in the present experiments are insufficiently constrained to further evaluate the discrepancies in illite–K-feldspar equilibria (Aja *et al.*, 1991) and further fluid-sampling experiments are warranted.

For experiment R1 steady-state conditions are approached, but equilibrium is not attained since boehmite persists, and the solution is supersaturated with respect to silica. Further reaction of boehmite and aqueous silica ($Si(OH)_4$) to produce kaolinite should not affect pH, and hence $\log(a_{K^+}/a_{H^+})$. Experiment R1 could, therefore, be considered to be in equilibrium with respect to $\log(a_{K^+}/a_{H^+})$. An identical steady-state $\log(a_{K^+}/a_{H^+})$ is attained in experiment R4 from high $\log(a_{K^+}/a_{H^+})$, and thus represents a reversibly defined point. The determined illite–kaolinite equilibrium fluid composition (Figure 8.7) agrees with the illite–kaolinite equilibrium defined by Sass *et al.* (1987) to within experimental error, and is very close (<0.01 $\log(a_{K^+}/a_{H^+})$ units) to the muscovite–kaolinite equilibrium defined by Hemley (1959) and Chermak and Rimstidt (1990).

Huang (1990) has proposed that the common occurrence of illite during late-stage diagenesis is a consequence of the strong buffering capability of the quartz–feldspar–kaolinite mineral assemblage, which buffers fluid composition in the illite stability field. Huang (1990) presents experimental evidence which shows that various fluid compositions initially outside the illite stability field move into the illite field upon reaction with the above assemblage. The present results from the fluid-sampling experiments also show this behaviour where starting with $\log(a_{K^+}/a_{H^+})$ either above or below the illite stability field the fluid composition migrates towards the illite field. In this case, however, the buffering assemblage is not present; this suggests that illite merely represents the most stable phase at quartz saturation, and that the appropriate illite-bearing assemblage forms depending on the composition of the system. In this study the pathway of fluid migration is accurately defined and shows that the fluid composition deviates on its route as intermediate phases (such as boehmite, kaolinite and mordenite) form.

8.5.3 Oxalate stability and its role in clay precipitation

The initial presence of the oxalate anion appears to be crucial to the precipitation of fibrous phases in these experiments. The behaviour of oxalate has two separate controls over clay formation: (1) establishing initial high aluminium concentrations, and (2) controlling the timing of precipitation.

The timing of the precipitation is a consequence of the decay of oxalate which is variable under different pH conditions. Under acidic conditions oxalate breaks down to gaseous products below 250°C, whereas in R2 and R3 under alkaline conditions formate remains at the end of the experiments. Lower formate concentrations are observed in R2 than in R3 and final pH is lower in R2 (Figure 8.9; Table 8.2). It is suspected that the differences in formate stability between R2 and R3 are a catalytic effect, possibly resulting from R3 being maintained in a static position, and that pH is in turn controlled by formate concentration. The major differences in oxalate breakdown between R1 and R2 are possibly a result of different, pH-dependent, decarboxylation mechanisms. The decarboxylation of oxalate and formate is dependent on the redox conditions of the experiments but is unconstrained in the present experiments. Gaseous decarboxylation products (CO_2, CO, CH_4, H_2) have not been analysed, and these could provide a constraint.

With regard to the initial high aluminium concentrations, the presence of oxalate could result in the attainment of high aluminium concentrations by the formation of an aluminium–oxalate complex with enhanced solubility over hydrated aluminium complexes (Surdam *et al.*, 1984). Alternatively oxalate may buffer pH to produce higher aluminium concentrations by the formation of charged species such as $Al(OH)_4^-$ and $Al(OH)_2^+$. Figure 8.13 compares directly the decay of oxalate, the maximum in aluminium concentration and the change in quench pH for R3. The decrease in oxalate above 180°C is mirrored by an immediate fall in quench pH; aluminium concentration, however, reaches a maximum 8 h after the initial decay of oxalate. Assuming

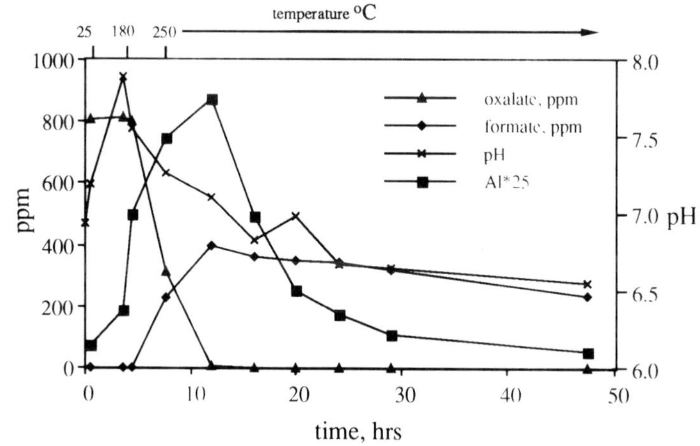

Figure 8.13 Comparison of the decay of oxalate, production of formate, pH, aluminium concentration and reaction temperature during experiment R3.

the aluminium–oxalate complex contains two oxalate anions for each aluminium atom (Surdam *et al.*, 1984) then there is sufficient oxalate present to account for measured aluminium concentrations up to 8 h. At 12 h, however, when the maximum aluminium concentration is reached, oxalate is virtually totally decomposed. After this point the aluminium concentration sharply declines.

To evaluate the effects of pH and oxalate complexation Figure 8.14 compares the dependence of measured total aluminium activity on *in situ* pH, with calculated aluminium solubilities of various phases derived from mineral solubility constants and Al^{3+} hydrolysis constants in the EQ3/6 database. The calculated mineral solubilities illustrate the dependence of aluminium solubility on pH, where under both acidic and alkaline conditions higher solubilities occur resulting from charged hydrated aluminium species. The gradient of the mineral solubility curves at the two extremes of pH is determined by the Al^{3+}/H^{+} ratio of the dissolution reaction, and the charge of the hydrated aluminium species. This effect is best illustrated for pH > 7 where the $Al(OH)_4^-$ species is dominant. The intercept of the solubility curves with pH = 0 is controlled by the solubility constant of the mineral, and for aluminosilicate phases is also dependent on the activities of components other than aluminium. The curves in Figures 8.14(a) and (b) have been calculated to be representative of SiO_2, K^+ and Mg^{2+} (for illite) activities in experiments R2 and R3 and R1 and R4, respectively. Changes in component activities and solubility constant merely offset the solubility curves along the $\log a_{(\mathrm{Al\ total})}$ axis. Given these constraints for the aqueous speciation model, and assuming equilibrium, the slope of measured solubility data should follow the gradient of the mineral curve controlling aluminium activity. If assumptions regarding the solubility constant and other component activities are correct the data should coincide exactly.

The data for R2 and R3 at 250°C (Figure 8.14(a)) define a linear trend in decreasing aluminium activity as *in situ* pH drops towards neutral conditions. The gradient of this trend is slightly greater than that of illite and muscovite solubility, and is comparable to the kaolinite and boehmite solubility curves. All measured aluminium concentrations are below that predicted for illite or muscovite solubility. This indicates that the variation in aluminium activity can be accounted for purely by the aqueous speciation model, and that complexing by organic species such as oxalate or formate does not need to be invoked. The analyses with pH > 8 are all above K-feldspar–mordenite saturation and represent the main phase of precipitation when fibrous illite formed metastably. The analyses from R2 at pH 7–7.5 define a trend with zero gradient slightly below the predicted K-feldspar–mordenite solubility curves. These latter points appear to represent equilibrium activities and suggest that aluminium activity is controlled by K-feldspar–mordenite rather than by illite.

For R1 (Figure 8.14(b)) the slope of the data at 250°C is consistent with the mineral solubility curves indicating no enhanced solubility above

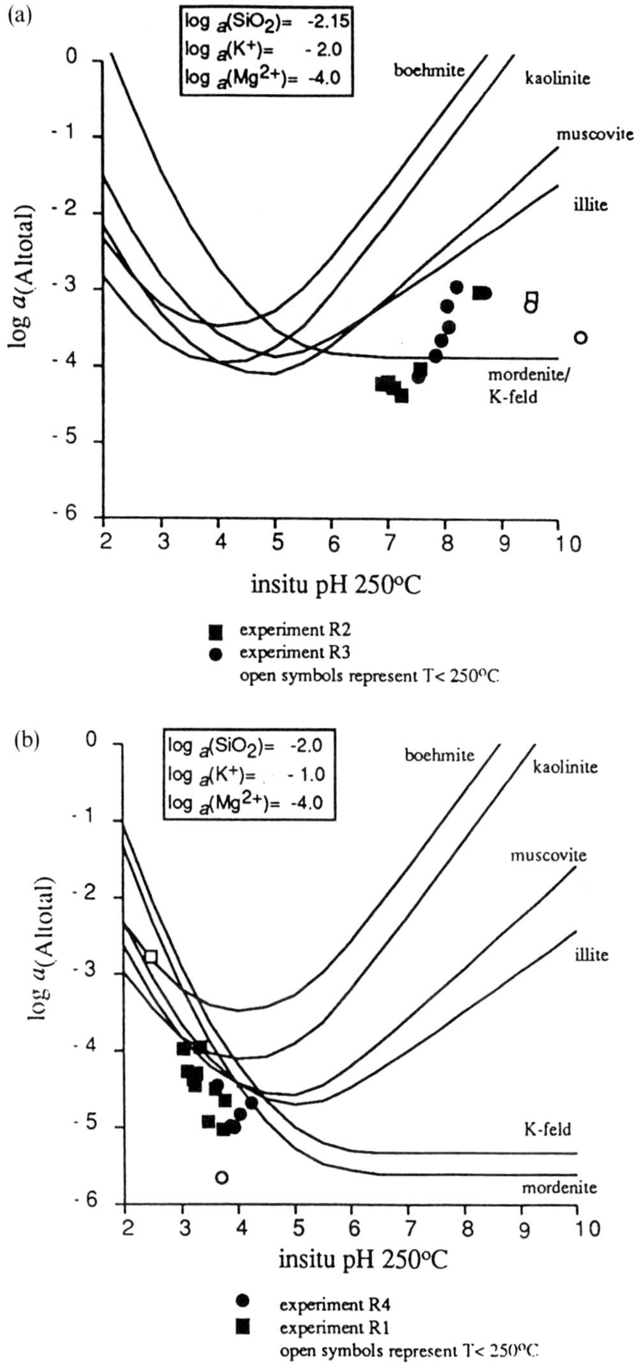

Figure 8.14 Dependence of total aluminium activity on pH for various phases, calculated from solubility constants and Al^{3+} hydrolysis constants in the EQ3/6 database, and comparison of measured aluminium activities. (a) Experiments R2 and R3, (b) experiments R1 and R4.

that predicted by the aqueous speciation model. Measured aluminium concentrations are slightly below the predicted curves; in this case illite–muscovite are likely to control aluminium activity. The single point with very high aluminium activity taken at 100°C is undersaturated with respect to boehmite at *in situ* pH and 100°C. There is thus no requirement for additional aluminium complexing by oxalate.

The results from the oxalate-free experiment R4 (Figure 8.14(b)) show that only low aluminium activities are produced in KCl solutions at pH 4–5, which are undersaturated with respect to all phases. The aluminium activities in R4 are, however, consistent with samples of equivalent pH from R1. Higher initial aluminium activities might be expected in R4, such as those in R1, R2 and R3, if the initial rise in aluminium activity is purely controlled by aqueous dissolution of amorphous alumina. Furrer and Stumm (1986) determined that oxalate considerably enhanced the dissolution of δ-alumina at 25°C. Their results show that below pH 3.5 at 25°C the rates of oxalate and aqueous dissolution are comparable, at around 20 $nmol.m^{-2}.h^{-1}$ but at pH 6 the rate of aqueous dissolution is reduced by a factor of 10. In this study aluminium solubility data at 25°C are available for sample R2 in (Table 8.2): the calculated dissolution rate of this sample is approximately 900 $nmol.m^{-2}.h^{-1}$; more than one order of magnitude above that of the maximum combined oxalate and aqueous dissolution rates measured by Furrer and Stumm (1986). In view of the findings of Furrer and Stumm (1986) and the very high aluminium dissolution rate observed in this study, it is most likely that oxalate-induced dissolution does contribute to the high initial aluminium activities observed in the presence of oxalate.

In summary, the measured aluminium activities in these illite precipitation experiments can be accounted for by pH variation and short-lived metastable conditions following initial dissolution of amorphous alumina. Oxalate is, however, considered to play an important role in controlling pH during its decay, which in turn results in rapid precipitation. In addition oxalate is likely to enhance dissolution rates which will raise initial aluminium activity. These results suggest that the principal control of aluminium solubility is pH, in this case varying in response to oxalate instability. The implication then is that other more abundant organic acids may also enhance fibrous illite formation by similarly acting as pH controls. The generation of carboxylic acids and other polar organics from petroleum source rocks may be important not only for the transport of aluminium and silicon from such sources but also for the control of clay precipitation as the organic anions break down following accumulation within a reservoir.

8.6 Conclusions

These experiments demonstrate that fibrous and lath-type illites can be precipitated extremely rapidly to produce clay cements realistic of those common in petroleum reservoirs in less than 24 h at 250°C. The initial presence of the oxalate anion is critical to this rapid dissolution and precipitation. In the absence of oxalate, in neutral conditions only minor dissolution and precipitation occurs.

Oxalate is unstable in these experiments and produces gaseous products (CO_2) in initially acidic conditions, and formate plus CO_2 in alkaline conditions. Oxalate is considered to control the rapid precipitation of fibrous illite by firstly enhancing the dissolution rate of alumina starting material and secondly, by control of pH during breakdown. There is, however, no requirement for an aqueous aluminium–oxalate complex to account for measured aluminium activities.

The changes in fluid composition during the early stages of the experiments control the formation of precursor precipitates. In acidic conditions boehmite and kaolinite form first, followed by illite. Under alkaline conditions aluminium-deficient precursor illites form metastably in the K-feldspar and mordenite stability fields. These precursor phases all have a distinctive fibrous morphology, and possibly control the development of fibrous illite formed by a topotactic replacement of the early-formed phases. In contrast, platy illites precipitate directly from solution in the illite–muscovite stability fields.

Illite stability estimated from these precipitation experiments is in reasonable agreement with predictions made from previous studies. Fibrous illite is, however, formed metastably outside the established illite stability field. This has implications for geochemical modelling of authigenic illite formation since existing illite solubility constants would be realistic of fluid equilibrated with reservoir-type illites, but would not be representative of the conditions in which such metastable fibrous illite formed.

Acknowledgements

This research was primarily supported by the Natural Environment Research Council by awards to Professor C. D. Curtis (D1/G1/189), D. L. Hamilton (GR3/6477), and D. A. C. Manning (GR3/6141 and GST/02/157). Some exploratory experiments reported were supported by The Geochem Group, Chester, with funding from Amoco UK Exploration, British Gas, Enterprise Oil, Mobil North Sea, and Shell UK. We are most grateful to Dr Wuu-Liang Huang for providing us with a preprint of his paper presented at the 27th annual meeting of the US Clays Minerals Society.

References

Aja, S. U., Rosenberg, P. E. and Kittrick, J. A. (1991) Illite equilibria in solutions: I. Phase relationships in the system K_2O–Al_2O_3–SiO_2–H_2O between 25 and 250°C. *Geochimica et Cosmochimica Acta*, **55**, 1353–64.

Chermak, J. A. and Rimstidt, J. D. (1990) The hydrothermal transformation rate of kaolinite to muscovite/illite. *Geochimica et Cosmochimica Acta*, **54**, 2979–90.

Cliff, G. and Lorimer, G. W. (1975) The quantitative analysis of thin specimens. *Journal of Microscopy*, **103**, 203–7.

Furrer, G. and Stumm, W. (1986) The coordination chemistry of weathering: I. Dissolution kinetics of δ–Al_2O_3 and BeO. *Geochimica et Cosmochimica Acta*, **50**, 1847–60.

Garrels, R. M. (1984) Montmorillonite/illite stability diagrams. *Clays and Clay Minerals*, **32**, 161–6.

Glasmann, J. R., Larter, S., Nowell, A. B. and Lundegard, P. D. (1989) Shale diagenesis in the Bergen High area, North Sea. *Clays and Clay Minerals*, **37**, 97–112.

Guven, N., Hower, W. F. and Davis, D. K. (1980) Nature of authigenic illites in sandstone reservoirs. *Journal of Sedimentary Petrology*, **45**, 761–6.

Hamilton, D. L. and Henderson, C. M. B. (1968) The preparation of silicate compositions by a gelling method. *Mineralogical Magazine*, **36**, 832–8.

Helgeson, H. C., Delaney, J. M., Nesbitt, H. W. and Bird, D. K. (1978) Summary and critique of the thermodynamic properties of rock-forming minerals. *American Journal of Science*, **278-A**, 1–229.

Hemley, J. J. (1959) Some mineralogical equilibria in the system K_2O–Al_2O_3–SiO_2–H_2O. *American Journal of Science*, **257**, 241–70.

Huang, W. L. (1990) Illitic clay formation during experimental diagenesis of arkoses, in *Symposium of Clays in Sandstones, Program and Abstracts 27th Annual Meeting*, Clay Mineral Society, Boulder, p. 62.

Inoue, A., Kohyama, N., Kitagawa, R., and Watanabe, T. (1987) Chemical and morphological evidence for the conversion of smectite to illite. *Clays and Clay Minerals*, **35**, 111–20.

Keller, W. D., Reynolds, R. C. and Inoue, A. (1986) Morphology of clay minerals in the smectite-to-illite conversion series by scanning electron microscopy. *Clays and Clay Minerals*, **34**, 187–97.

Lanson, B. and Champion, D. (1991) The I/S-to-illite reaction in the late stage of diagenesis. *American Journal of Science*, **291**, 473–506.

Macchi, L. (1987) A review of sandstone illite cements and aspects of their significance of hydrocarbon exploration and development. *Geological Journal*, **22**, 333–45.

Macchi, L., Curtis, C. D., Levison, A., Woodward, K. and Hughes, C. R. (1990) Chemistry, morphology and distribution of illites from Morecambe Gas Field, Irish Sea, offshore United Kingdom. *American Association of Petroleum Geologists Bulletin*, **74**, 296–308.

McHardy, W. J., Wilson, M. J. and Tait, J. M. (1982) Electron microscope and X-ray diffraction studies of filamentous illitic clay from sandstones of the Magnus Field. *Clay Minerals*, **17**, 23–39.

Pollastro, R. M. (1985) Mineralogical and morphological evidences for the formation of illite at the expense of illite/smectite. *Clays and Clay Minerals*, **33**, 265–74.

Robie, R. A., Hemingway, B. S. and Fisher, J. R. (1978) Thermodynamic properties of minerals and related substances at 298.15 K and 1 bar (10^5 Pascals) pressure and at higher temperatures. *U.S. Geological Survey, Bulletin*, **1452**, 456 pp.

Rossel, N. C. (1982) Clay mineral diagenesis in Rotliegend aeolian sandstones of the southern North Sea. *Clay Minerals*, **17**, 69–77.

Sass, B. M., Rosenberg, P. E. and Kittrick, J. A. (1987) The stability of illite/smectite during diagenesis: An experimental study. *Geochimica et Cosmochimica Acta*, **51**, 2103–15.

Seyfried, W. E., Janecky, D. R. and Berndt, H. E. (1987) Rocking autoclaves for hydrothermal experiments II. The flexible reaction-cell system, in *Hydrothermal Experimental Techniques*, (eds G. C. Ulmer and H. L. Barnes), John Wiley, New York, pp. 216–39.

Shock, E. L. and Helgeson, H. C. (1990) Calculation of the thermodynamic and transport properties of aqueous species at high pressures and temperatures: Standard partial molal properties of organic species. *Geochimica et Cosmochimica Acta*, **54**, 915–45.

Small, J. S., Hamilton, D. L. and Habesch, S. (1992a) Experimental simulation of clay mineral precipitation in reservoir sandstones 1: Techniques and examples. *Journal of Sedimentary Petrology*, **62**, 508–19.

Small, J. S., Hamilton, D. L. and Habesch, S. (1992b) Experimental simulation of clay mineral precipitation in reservoir sandstones 2: Mechanism of illite formation and controls on morphology. *Journal of Sedimentary Petrology*, **62**, 520–9.

Sommer, F. (1978) Diagenesis of Jurassic sandstones in the Viking Graben. *Journal of the Geological Society, London*, **125**, 63–7.

Stalder, P. J. (1973) Influence of crystallographic habit and aggregate structure of authigenic clay minerals on sandstone permeability. *Geologie en Mijnbouw*, **52**, 217–19.

Surdam, R. C., Boese, S. W. and Crossey, L. J. (1984) The chemistry of secondary porosity, in *Clastic Diagenesis*, (eds D. A. McDonald and R. C. Surdam), American Association of Petroleum Geologists, Tulsa, pp. 127–49.

Warren, E. A. and Curtis, C. D. (1989) The chemical composition of authigenic illite within two sandstone reservoirs as analysed by ATEM. *Clay Minerals*, **24**, 137–56.

Wilson, M. D. and Pittman, E. D. (1977) Authigenic clays in sandstones—recognition and influence on reservoir properties and palaeoenvironmental analysis. *Journal of Sedimentary Petrology*, **47**, 3–31.

Wolery, T. J. (1983) *EQ3NR, A Computer Program for Geochemical Aqueous Speciation–Solubility Calculations: User's Guide and Documentation. Lawrence Livermore Laboratory Publication* UCRL-53414, Lawrence Livermore National Laboratory, University of California, Livermore, California.

CHAPTER NINE

Experimental diagenesis of mudrocks

E. I. C. Rae, D. A. C. Manning and C. R. Hughes

9.1 Background

It has been suggested that mudrock pore fluids expelled during burial diagenesis may have a substantial influence on the properties of adjacent lithologies, including the development of secondary porosity or diagenetic cements (including ores as a particular example), and the composition and distribution of aquifer/reservoir pore fluids (hydrocarbons and brines). Fundamentally, mudrock pore fluids are poorly defined mixtures of aqueous and organic fluids within which a very wide spectrum of inorganic species are dissolved. These fluids may be responsible for the removal from mudrocks of material such as major cations, trace metals and polar and non-polar organic species, which are then available to be redistributed within a sedimentary basin. Mudrock pore-fluids may contain major and trace elements derived principally from mineral sources, and organic anions and polar species derived from their organic constituents. Major cations (Ca, Al, Si, Fe, Mg, K, etc.) potentially reprecipitate as common mineral cements such as clays, quartz or carbonates (Moncure *et al*, 1984; Crossey *et al.*, 1986; Stoessell and Pittman, 1990; Sullivan and McBride, 1991), while mudrock-derived trace metals such as lead and zinc may form Mississippi-Valley-type mineral deposits or recrystallize as disseminated sulphides (Beales and Jackson, 1966; Anderson and Macqueen, 1982; Kříbek, 1991). Both organic acid anions and CO_2 (derived from the decarboxylation of organic matter) potentially buffer pore fluid pH, thus influencing mineral dissolution or precipitation (Schmidt and McDonald, 1979a; 1979b; Curtis, 1983a; 1983b; Surdam *et al.*, 1989; Barth *et al.*, 1990). Organic acids may also act as ligands contributing to the transport of some or all of the cement-forming and mineralizing cations (Sverjensky, 1984; Giordano, 1985; Drummond and Palmer, 1986; Hennet *et al.*, 1988 and Yang *et al.*, 1989).

Although cited in explanations of many diagenetic phenomena, the compositions of mudrock pore fluids remain poorly known in detail. Analytical data for oilfield brines suggest some constraints, as some of their components may be derived from this source. Oilfield brines with ore-forming potential have

Geochemistry of Clay-Pore Fluid Interactions. Edited by D. A. C. Manning, P. L. Hall and C. R. Hughes. Published in 1993 by Chapman & Hall, London. ISBN 0 412 48980 5

been identified by Billings *et al.* (1969), and are thought broadly to resemble fluids responsible for Mississippi-Valley-type (MVT) mineralization (Sverjensky, 1984). Although chloride predominates in oilfield brines, organic anions are present in significant quantities (acetate up to 10 000 $\mu g.ml^{-1}$ with lesser concentrations of longer chain organics; Collins, 1974; MacGowan and Surdam, 1988; Kharaka *et al.*, 1986) and cannot be neglected in models in models of chemical speciation.

While compositional data are available from some oilfield brines and other formation waters, they are not necessarily simple to interpret. Fluids sampled from an aquifer or reservoir will have undergone reactions with aquifers and/or host rocks at any stage during their generation, migration and sampling. Sampling, especially, may involve the exsolution of dissolved components (mainly CH_4, H_2S and CO_2 but perhaps other short-chain organic species) which will affect the measured pH. Certain fluids will be comparatively immature, if sampled close to their source or before they equilibrate fully with their reservoir rocks, whilst others will be mature, in chemical equilibrium with their reservoir minerals. Little is known, however, about the processes involved in the evolution of aqueous pore fluids. Shock (1988, 1989) has discussed the metastability of organic acids under diagenetic conditions. Experimental studies of rates of decomposition of organic acids have also been carried out (Drummond and Palmer, 1986; Schleusener *et al.*, 1987, 1988; Crossey, 1991) since this process is believed to act as a control on the pH of pore fluids (Barth *et al.*, 1990; Cantrell *et al.*, 1990; see also discussion section later in this paper). Oilfield brines represent a complex cocktail of stable and metastable species and computer modelling of such solutions must take this into account if they are to be meaningful with reference to natural processes.

9.2 Experimental maturation of mudrocks

In order to understand better the behaviour of organic matter and base metal-rich mudrocks during diagenesis a series of artificial maturation experiments has been carried out under controlled conditions of high pressure (50 MPa; 500 bar) and at selected temperatures (150, 250 and 310°C). The experiments were carried out using Dickson-type fluid-sampling pressure vessels (Dickson *et al.*, 1963; Seyfried *et al.*, 1987) which permit fluid samples to be taken during long experiments without quenching, so that the composition of the fluid can be monitored during the course of the experiment. Samples were contained within a gold reaction cell (wall thickness 0.025 mm), with titanium closure, exit tube and sampling valve (Seyfried *et al.*, 1979); this combination of metals is believed to be least active catalytically (Palmer and Drummond, 1986). Oxygen fugacity was neither controlled nor monitored in the experiments, but because of the extremely low permeability of gold to hydrogen at the temperatures of interest (see Seward and Kishima, 1987) it can be assumed that external factors (such as reaction between the pressure vessel and water used as the confining pressure medium) had no influence on redox reactions within the sample. Limitations of this sample containment include the possibility that base metals such as copper might amalgamate with the gold, although again this is unlikely to be a major problem at these low temperatures (Seyfried *et al.*, 1987). Using this experimental apparatus, pressure and temperature are controlled independently, and are measured continuously. Pressure was controlled to 500 ± 10 bar, and temperatures to $\pm 5°C$; each experiment lasted up to 6 weeks.

The solid starting materials used in this study are Kimmeridge Blackstone Band (from the Dorset coastal section; 45% total organic carbon (TOC); e.g. Williams and Douglas, 1979) and Permian Marl Slate (British Coal borehole material from offshore County Durham; 14% TOC). Fluid starting materials were either distilled water or 0.5 M NaCl solution, with a fluid/rock mass ratio of approximately 12/1. Chemical analyses of unreacted starting materials are given in Table 9.1(a) and the minerals present (on the basis of XRD analysis) in Table 1(b).

9.2.1 *Experimental procedures*

Samples of the Kimmeridge Blackstone Band and the Marl Slate were ground in a 'TEMA' mill. Rock powders were used to maximize the surface area of the sample and hence the extent of reaction. No size separations were carried out, but grain size was approximately 10 μm or less. Mudstones were extracted with an azeotropic mixture of DCM:MeOH (dichloromethane:methanol; 93:7, boiling point 38°C) to remove bitumen already present so that any organic products seen were known to have been generated during the experiment. Characteristics of organic matter and extracted bitumen for unheated starting materials are given in Table 9.1(c) (see Section 9.3.2 for explanations of ratios). Distilled water used was boiled before use to exsolve dissolved CO_2. NaCl brine was prepared from AnalaR grade chemicals dissolved in distilled water. The levels of impurities found in both starting fluids were negligible compared with those found in the reacted fluids. The gold cells used were annealed before use to avoid problems of cracking and to ensure that the cell was sufficiently flexible to transmit the external pressure to the sample. Approximately 7 g of rock and 170 ml fluid were used in each experiment; gas absorption surface areas of the unreacted extracted Kimmeridge Blackstone Band and Marl Slate were 3.3 and 9.0 $m^2 g^{-1}$, respectively. Experiment durations varied but were of the order of 500–1000 h. A summary of the experimental conditions is given in Table 9.2.

Paired fluid samples were taken at intervals during the experiments. One sample was collected in a glass syringe and extracted with organic solvents to give an organic fraction (referred to as total soluble extract or 'TSE'). After extraction, the aqueous portion of this sample was acidified to pH 1 with concentrated nitric acid and re-extracted with DCM, for carboxylic acid

Table 9.1 Characteristics of unheated starting materials.
(a) Chemical composition

Composition (%)	Kimmeridge Blackstone Band				Marl Slate
	KU1+		KE1+		MS6
SiO_2	7.2±0.5		8.0±0.5		40.9
Al_2O_3	2.5		2.5		14.4
Fe_2O_3*	2.05		2.13		5.63
MgO	0.20		0.24		3.29
CaO	6.71		6.86		5.01
Na_2O	0.18		0.20		0.94
K_2O	0.41		0.38		3.39
TiO_2	0.11	0.11†	0.11	0.12†	0.54
P_2O_5	0.56	0.53†	0.58	0.56†	0.18
MnO	0.01		0.01		0.08
Ignition loss	76.8		75.6		–
CO_2	–		–		7.61
TOC	–		–		13.77
ΣS	–		–		3.62
Total	96.06		95.92		99.36
(ppm)					
Cr	41	35†	39	42†	475
Li	14	11†	13	14†	–
Ni	82	68†	75	80†	303
Co	7	6†	6	8†	147
Zn	79	77†	56	102†	28
Cu	50	44†	38	56†	138
V	58	55†	59	65†	1690
Pb	b.d.		–	–	141
Mo	118		–	–	431

*Fe as total iron. †Duplicate analyses.
–, not analysed; b.d., below detection limit.
+KU1 = unextracted Kimmeridge Blackstone Band.
+KE1 = extracted Kimmeridge Blackstone Band.

(b) Minerals present on the basis of XRD analysis

Kimmeridge Blackstone Band unheated		Marl Slate (MS6) unheated	
<2 μm‡	>2<8 μm	<2 μm	>2<8 μm
Mica	Calcite	Mica	Quartz
Kaolinite	Quartz	Kaolinite	Dolomite
Calcite	Mica	Quartz	Mica
	Kaolinite		Kaolinite
	Pyrite		(Pyrite)
			(?Apatite/ Siderite)

‡Indistinct peaks. ( ) trace only.

Table 9.1 (*contd*)

(c) Characteristics of organic matter from unheated Kimmeridge Blackstone Band (KEB) and Marl Slate (MS6). Ratios obtained from extracted bitumen, for explanations of ratios see text.

Shale sample	Kerogen type	pr/ph	pr/n–C_{17}	ph/n–C_{18}	CPI C_{20}–C_{30}
KEB	II	0.59	1.77	4.84	2.65
MS6	II	1.42	1.18	1.06	1.01

Table 9.2 Summary of experimental details.

Experiment name	Shale and fluid used	Experiment temperature (°C)	Experiment duration (h)
K5	Kimmeridge, distilled water	150	868
K3	Kimmeridge, distilled water	250	486
K9	Kimmeridge, distilled water	310	1033
K7	Kimmeridge, 0.5 M NaCl brine	310	506
M1	Marl Slate, distilled water	310	559
M2	Marl Slate, 0.5 M NaCl brine	310	627

determination by gas chromatography mass spectrometry (GCMS). The second sample was collected in a disposable polythene syringe, transferred to an acidified volumetric flask, weighed and made up to a standard volume. This sample was then available for determination of inorganic species present, using a combination of inductively coupled plasma spectroscopy, atomic adsorption spectroscopy and ion chromatography (anions only).

Unreacted and reacted solids were analysed by XRD using Cu Kα radiation, scanning at 1° per min. Selected samples were glycolated to confirm the presence of smectite. In order to obtain further compositional information from newly formed minerals, subsamples were examined by analytical transmission electron microscopy (ATEM) using a Philips 400T transmission electron microscope with an EDAX analytical system (Hughes *et al.*, 1990). Samples for analysis were prepared by suspension in distilled water, single drops of which were pipetted on to a coated copper grid to form dispersed mounts. When dry, these were carbon-coated to prevent charging in the microscope.

Following isolation by HF digestion, unheated and heated kerogens were analysed by Fourier Transform Infra-Red (FTIR) to examine the behaviour of substituent functional groups during experiments. Kerogens were mixed with KBr and formed into discs which were then scanned on a Hewlett Packard

1710 FTIR. Background scans were subtracted from the sample signal and corrections were made for variations in thickness of discs by manipulation of the computerized signal.

9.3 Results

The results obtained from fluid and solid products are presented separately in sections below. As a general observation, fluid samples taken during the experiments frequently formed emulsions during extraction with solvents and are inferred to be surfactant rich. Residual fluids (end fluids) frequently produced a precipitate on standing. This showed no reaction with acid and may consist of calcium salts of organic acids or be siliceous. Precipitation of calcium carbonates within sampling tubes is known to have occurred during sampling, providing constraints on fluid compositional data for calcium as well as a demonstration of scaling phenonema.

9.3.1 Fluids–inorganic considerations

Fluid compositional data are given in summary form in Table 9.3 for samples taken during experiments. The concentrations of certain species (notably silicon and calcium) were found to vary during cooling at the end of an experiment, which took about 24 h; during this time pressure was maintained to reduce the effects of carbonate precipitation. Because of this, the nature of the residual fluid obtained on quenching the experiment is not included in this discussion.

9.3.1.1 KIMMERIDGE EXPERIMENTS

At 150°C (K5, see Figure 9.1(a)), fluid produced is dominated by calcium (up to 400 $\mu g.ml^{-1}$) and silicon ($\sim 60\ \mu g.ml^{-1}$). Also detected in solution were magnesium ($\sim 8\ \mu g.ml^{-1}$), boron (1–2 $\mu g.ml^{-1}$) and strontium ($\sim 1\ \mu g.ml^{-1}$). About 1 $\mu g.ml^{-1}$ molybdenum is seen in samples taken throughout this experiment. At 250°C (K3, see Figure 9.1(b)), the fluid produced is dominated by silicon ($\sim 100\ \mu g.ml^{-1}$), sodium and potassium ($\sim 30\ \mu g.ml^{-1}$). Lesser concentrations of boron (1–2 $\mu g.ml^{-1}$) and strontium ($\sim 3\ \mu g.ml^{-1}$) are present. At 310°C with distilled water (K9, see Figure 9.1(c)), the solution is dominated by calcium ($\sim 80\ \mu g.ml^{-1}$) and silicon ($\sim 300\ \mu g.ml^{-1}$); potassium ($\sim 80\ \mu g.ml^{-1}$) and sodium (20–30 $\mu g.ml^{-1}$) are also present in significant concentrations. Many species show enhanced solubility in the brine experiment at 310°C (K7, see Figure 9.1(d)), for example potassium ($\sim 110\ \mu g.ml^{-1}$) and calcium ($\sim 500\ \mu g.ml^{-1}$). However, silicon solubility ($\sim 200\ \mu g.ml^{-1}$) is lower in this experiment than in the corresponding experiment using distilled water. Trace concentrations of lithium and boron were found.

Calcium concentration is inversely related to temperature (see Table 9.3). The behaviour of silica in experiments is shown in Figure 9.2. The dashed lines

Table 9.3 Summary of fluid compositions from Kimmeridge and Marl Slate heating experiments (in $\mu g.ml^{-1}$).

	Kimmeridge Blackstone Band									Marl Slate				
	150°C + H_2O		250°C + H_2O		310°C + H_2O		310°C + NaCl brine			310°C + H_2O		310°C + NaCl brine		
	Average	Range	Average	Range	Average	Range	Initial	Average	Range	Average	Range	Initial	Average	Range
Si	50	34–63	95	60–147	345	262–400	1.7 ± 0.5	214	196–240	429	425–438	b.d.	422	345–482
Ca	377	346–394	> 340	–	89.6	81.1–112.6	0.26	497	442–562	15.6	7.8–23.8	0.43	500	410–648
Na	20.3	18.3–21.9	31.9	30.0–34.6	26.4	24.6–28.3	*	*	11 000–7900	72.3	71.7–73.4	*	*	14 000–10 200
K	20.6	17.4–26.2	33.9	33.3–34.5	76.6	41.0–89.9	0.28	111.8	99.5–125.3	62.1	53.5–67.1	50.27	309.5	242–405
Mg	8.5	7.49–10.2	5.88	5.40–6.58	0.04	0.02–0.11	n.a.	0.02	0.00–0.04	0.07	0.06–0.09	n.a.	0.85	0.74–1.18
Sr	1.19	1.06–1.37	3.33	3.11–3.45	b.d.	–	n.a.	b.d.	–	0.07	0.06–0.09	n.a.	4.44	3.47–5.26
B	1.62	1.00–2.62	1.76	1.56–1.99	0.61	0.17–1.38	n.a.	0.55	0.30–0.81	3.35	3.09–3.70	n.a.	6.74	4.83–7.99
Mo	1.21	0.96–1.47	n.a.	–	b.d.	–	n.a.	b.d.	–	b.d.	–	n.a.	b.d.	–
Li	0.09	0.06–0.11	0.27	0.25–0.28	0.07	0.04–0.12	1.26	0.63	0.08–2.00	1.40	1.26–1.55	n.a.	3.74	3.06–4.28
pH†	6.65	6.54–6.80	5.78	5.56–5.95	6.89	6.69–6.94	n.a.	7.64	6.94–8.08	6.29	6.06–6.47	n.a.	6.83	6.40–7.20
Cl‡	556	–	n.a.	–	180	–	n.a.	14900	–	69	–	n.a.	16 800	–
SO_4‡	120	–	n.a.	–	169	–	n.a.	232	–	330	–	n.a.	200	–
$^3HCO_3$§	596	–	~530	–	137	–	n.a.	758	–	24	–	n.a.	765	–

Abbreviations used: n.a., not analysed; b.d., below detection limits.

Average = average value of samples taken once equilibrium value approached.

*Sodium lost steadily throughout experiments: range shows initial and final concentrations.

†Measured under N_2

‡Measured on fluid recovered after experiment quenched.

§Calculated from average calcium and magnesium concentrations. It was assumed that all calcium and magnesium were present as carbonate minerals which dissolved completely to give the bicarbonate concentration in this table. This was to give a default value for modelling and is not a valid assumption.

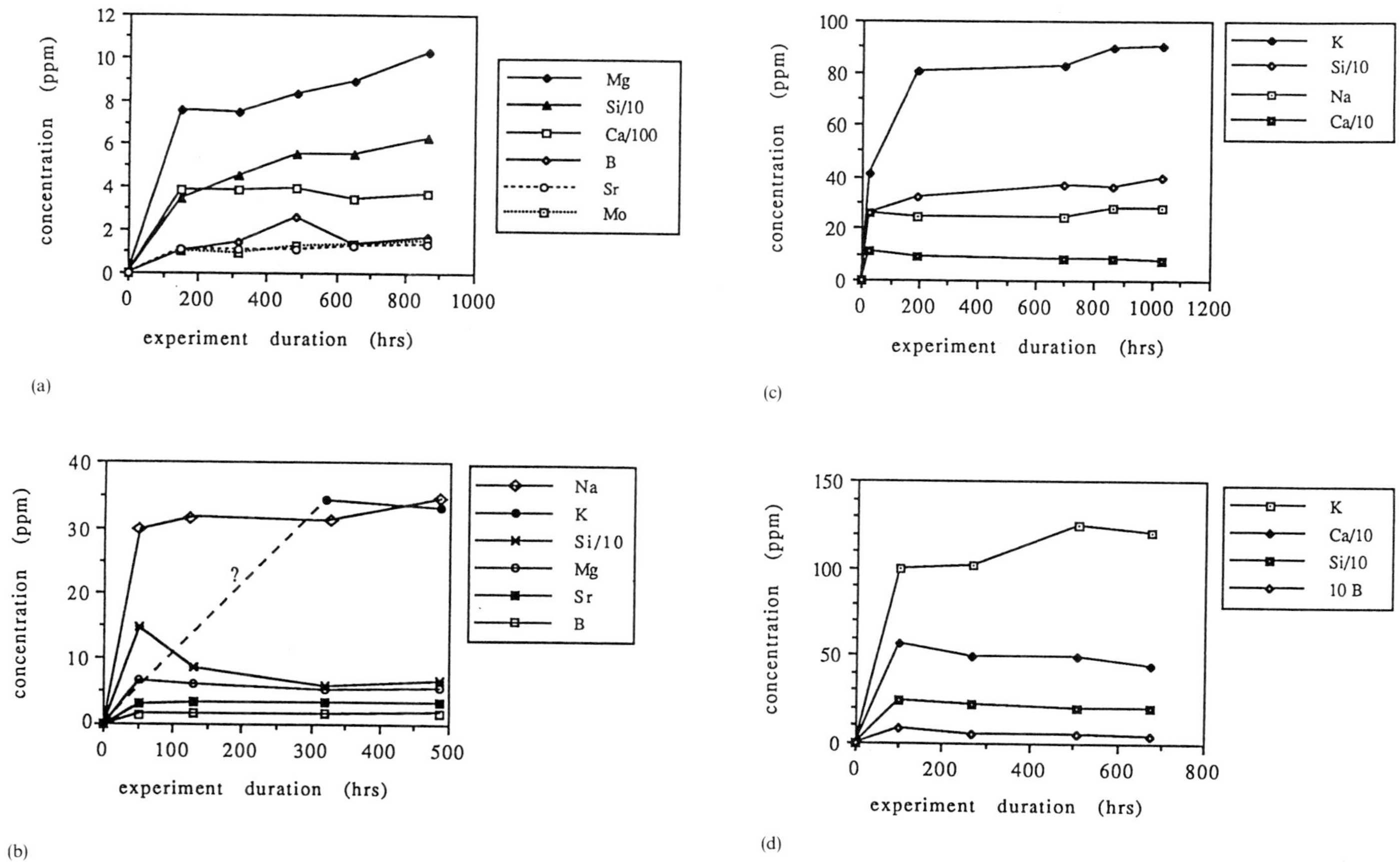

Figure 9.1 Solute yields from Kimmeridge Blackstone Band heating experiments: (a) 150°C; (b) 250°C; (c) 310°C (all with distilled water); (d) 310°C with 0·5 M NaCl brine.

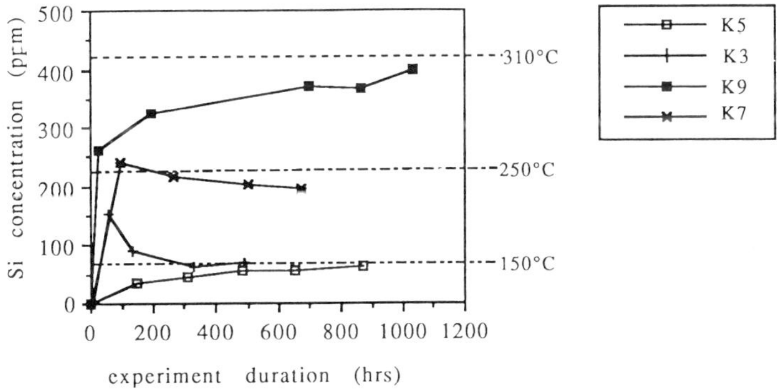

Figure 9.2 Silica solubilities for Kimmeridge Blackstone Band heating experiments as follows: K5 distilled water at 150°C; K3, distilled water at 250°C; K9, distilled water at 310°C; K7, 0.5 M NaCl brine at 310°C. Dashed lines represent silica solubilities at 500 bar at the temperatures marked for a solution saturated with respect to quartz (~60 ppm at 150°C, ~220 ppm at 250°C and ~420 ppm at 310°C; Walther and Helgeson, 1977). Note difference in silica solubilities at 310°C in distilled water and in brine (K9 and K7). Note also that all experiments are undersaturated with respect to quartz except at 150°C where saturation is approached.

show concentrations of silica in distilled water saturated with respect to quartz at 150, 250 and 310°C at 500 bar. It can be seen that fluids produced at 150°C are apparently just saturated or undersaturated with respect to quartz, while those at 250 and 310°C in both brine and distilled water are undersaturated with respect to quartz. These data are interpreted in more detail later.

9.3.1.2 MARL SLATE EXPERIMENTS

Marl Slate experiments were carried out at 310°C with brine or distilled water as the reacting fluid. Yields of solutes in each case are summarized in Figures 9.3(a), (b) and 9.4 and in Table 9.3. Fluid produced in the experiment with distilled water (M1, see Figure 9.3(a)) is dominated by silicon (~400 μg.ml^{-1}), sodium (~70 μg.ml^{-1}), potassium (~70 μg.ml^{-1}) and calcium (10–20 μg.ml^{-1}). Traces of boron were also found (3–4 μg.ml^{-1}). Fluid from the brine experiment (Figure 9.3(b)) is dominated by calcium (400–500 μg.ml^{-1}), potassium (~400 μg.ml^{-1}) and silicon (~400 μg.ml^{-1}). Also present in the solution were boron (~5 μg.ml^{-1}), lithium (3–4 μg.ml^{-1}) and strontium (3–5 μg.ml^{-1}). In both brine and distilled water experiments, solutions apparently approach or slightly exceed saturation with respect to quartz (see Figure 9.4 and detailed later discussion).

9.3.1.3 GENERAL

Aluminium was not detected in solution in any of these experiments. This may be because it was not mobilized from the mudrocks or may be due to its rapid precipitation from solution. Apart from the low-temperature release of molybdenum in Kimmeridge Blackstone Band heating experiments, no other base metals were detected in fluids produced.

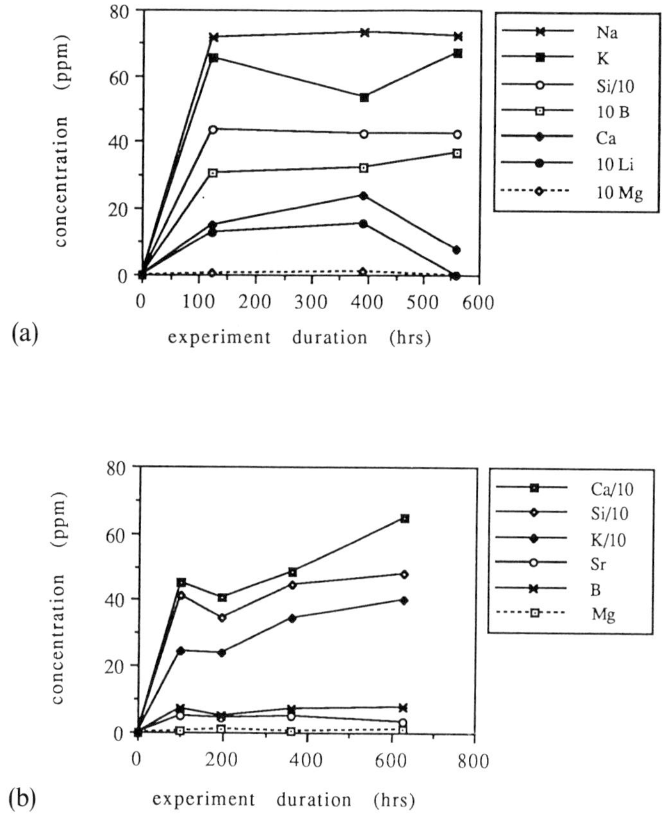

Figure 9.3 Solute yields from Marl Slate heating experiments: (a) 310°C with distilled water; (b) 310°C with 0.5 M NaCl brine.

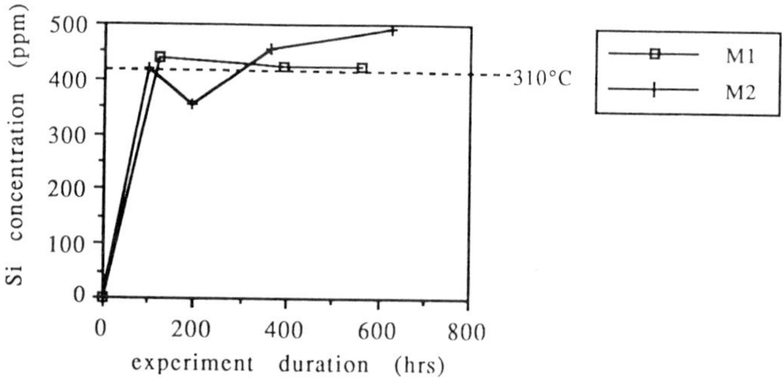

Figure 9.4 Silica solubilities for Marl Slate heating experiments as follows: M1, distilled water at 310°C; M2, 0.5 M NaCl brine at 310°C. Dashed line represents silica solubility in solution saturated with respect to quartz as discussed in caption for Figure 9.2.

The observation that molybdenum is present in only the low-temperature Kimmeridgian experiment is consistent with the known mobility of molybdenum during the weathering of the Kimmeridgian (Le Riche, 1959), and with the possibility that it is present in this rock as a non-sulphide mineral or complexed with organic matter. The Marl Slate is rich in molybdenum, a component of which may be present as molybdenite. It is probable that the lack of molybdenum in solutions derived from the Marl Slate experiments or the high-temperature Kimmeridge experiments reflects a low solubility for molybdenite under the experimental conditions. During the experiments, organic-derived molybdenum within the Kimmeridgian may be fixed as molybdenite, in a manner analogous to the formation of sphalerite from reaction between organic sulphur compounds and zinc in solution (Saxby, 1976).

9.3.2 *Fluids – organic considerations*

9.3.2.1 BITUMEN

Hydrocarbons were only collected in appreciable concentrations when reacted solids were recovered at the end of experiments; they were not detectable in samples taken for organic analysis during the course of the experiments. Bitumen yields for all experiments increased with increasing temperature, and the bitumen generated became increasingly mature with increasing temperature of experiment as expected (Table 9.4).

In order to interpret the organic geochemical data it is necessary to summarize the characteristics used to assess the maturity of organic matter (see

Table 9.4 Bitumen yield (in mg bitumen per g shale) and maturity parameters for light petroleum (saturated hydrocarbon) fraction of extracted bitumen from unheated and heated shales.

Experiment	Description	Bitumen yield	pr/n C_{17}	ph/n C_{18}	CPI C_{20}–C_{30}
KEB*	unheated	30.3	1.77	4.84	2.65
K5	150°C	7.1	0.27	0.41	1.52
K3	250°C	120.7	0.73	0.61	1.48
K4	310°C	381.4	0.30	0.16	1.16
K9	310°C	347.0	0.26	0.15	1.12
K7	310°C +brine	403.0	0.30	0.18	1.17
MS6†	unheated	7.8	1.18	1.06	1.01
M1	310°C	82.8	0.49	0.25	1.02
M2	310°C +brine	553.5	0.48	0.26	0.98

*KEB = Kimmeridge extracted bitumen, original bitumen present in the unheated Blackstone Band.
†MS6 = original bitumen present in the unheated Marl Slate.

Tissot and Welte (1984) and Mackenzie (1984) for more details). The characteristics of immature organic matter are inherited from source organisms (land plants, marine plankton and bacteria), and during diagenesis this inherited organic matter rapidly becomes unstable with respect to its surroundings and reacts to regain equilibrium. Early in the sedimentary column, the insoluble macromolecule kerogen is formed, varying in type according to the original organic matter source. Terrestrial plants contribute organic matter containing recognizable plant fragments, and the kerogen formed has high atomic O/C and low atomic H/C ratios (type III kerogen). Aquatic organisms contribute amorphous organic matter which has higher atomic H/C and lower atomic O/C ratios (type I kerogen). Type II kerogen has intermediate characteristics and is frequently a mixture of the two other types.

Kerogen composition can be represented on a Van Krevelen diagram (Figure 9.5; Tissot and Welte, 1984) which also shows how its composition changes with increasing maturity. Initial kerogen maturation results in a decrease in the atomic O/C ratio (especially for type III kerogens) with concomitant formation and loss from the kerogen of oxygen-containing moieties (e.g. carboxylic acids, phenols, ketones, etc.). With increasing thermal maturity, the H/C ratio decreases with the formation and loss from the kerogen of hydrocarbons (bitumen).

The changing nature of the bitumen generated from kerogen with increasing thermal maturity is demonstrated by the compositional ratios presented in Table 9.4. This includes data for the *n*-alkanes, which for an immature generated bitumen reflect the organic matter source. In many bitumens an odd-to-even predominance of *n*-alkanes can be seen (although an even-to-odd predominance is found in carbonate or evaporitic source rocks). This can be measured by the carbon preference index (CPI) defined by Bray and Evans (1961) as

$$\mathrm{CPI}=\frac{1}{2}\left\{\frac{C_{25}+C_{27}+C_{29}+C_{31}+C_{33}}{C_{24}+C_{26}+C_{28}+C_{30}+C_{32}}+\frac{C_{25}+C_{27}+C_{29}+C_{31}+C_{34}}{C_{26}+C_{28}+C_{30}+C_{32}+C_{34}}\right\}.$$

Immature bitumen also contains a high concentration of branched and cyclic compounds relative to the concentration of straight chain alkanes. With increasing thermal maturity, *n*-alkanes are generated with no odd or even predominance which swamp those originally present. This means that the concentration of branched and cyclic compounds relative to straight chain alkanes decreases, illustrated by the pristane (pr) to n-C_{17} and phytane (ph) to n-C_{18} ratio, and the approach to 1 of the CPI.

From the ratios given in Table 9.4 it can be seen that the initially high CPI (reflecting its immaturity) of extracted bitumen from the Kimmeridge Blackstone Band decreases with increasing temperature of reaction, as a consequence of increasing generation of *n*-alkanes from the kerogen. The decrease in the initially high branched to straight chain ratios (pr/n-C_{17} and ph/n-C_{18})

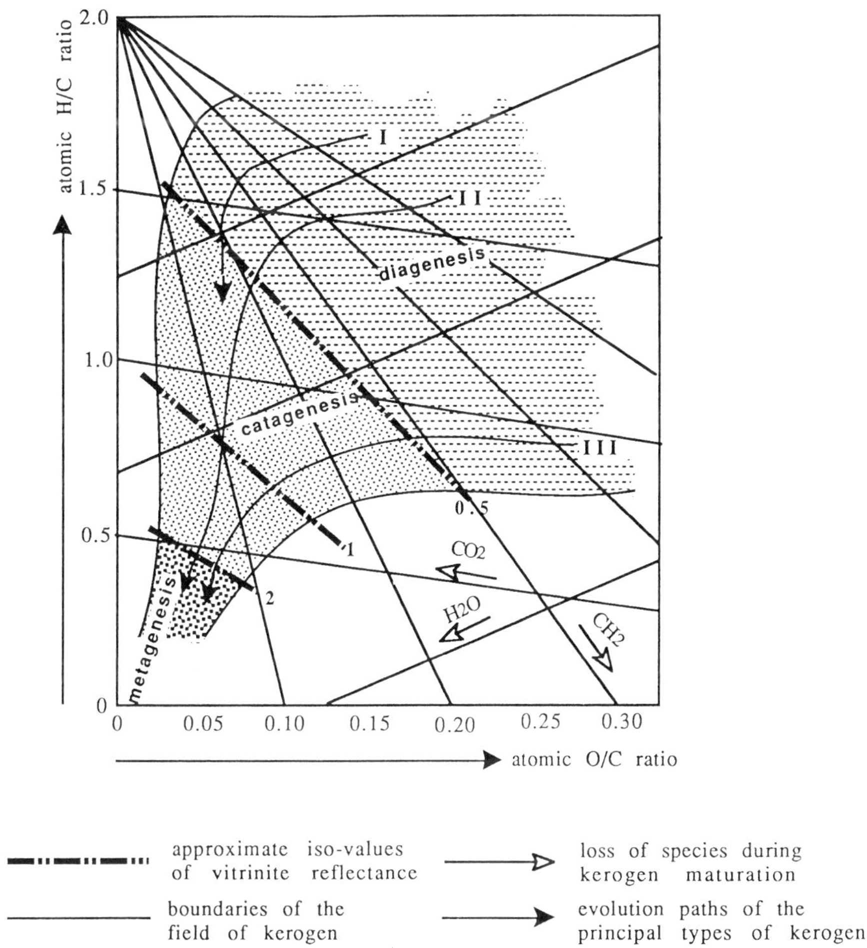

Figure 9.5 Van Krevelen diagram showing general scheme of kerogen evolution from diagenesis to metagenesis. Approximate values of vitrinite reflectance are shown for comparison. After Tissot and Welte (1984).

illustrates the increase in relative concentration of straight chain alkanes over branched and cyclic alkanes. This is also seen for the Marl Slate extracts. The initial CPI of the Marl Slate extracted bitumen is close to 1 and does not change, suggesting that organic matter present in the Marl Slate is more mature than that in the Kimmeridge Blackstone Band, but other indicators of organic maturity show that it has not passed through the oil generation window.

9.3.2.2 TOTAL SOLUBLE EXTRACTS (TSE)

As the experimental and analytical procedures used were being developed during the project it is possible that variations in compounds found in TSE of samples and quenched fluids were due in part to differences in sample handling as well as to variations due to experimental conditions. Classes of compounds discussed in the following paragraphs were present to some degree in most samples and so the data presented are composite.

The TSE of samples collected during experiments at 310°C were analysed by gas chromatography (GC) and GCMS and were found to consist of a complex mixture of organic compounds. Compound classes such as naphthalenes, benzenes, benzothiophenes, pyridines, phenols and ketones were tentatively identified on the basis of comparison with published spectra and by the use of diagnostic mass fragment ions and parent ions. Thorough study and identification of these compounds (which would involve authentic standards and elution on more than one GCMS column) have not been carried out in this exploratory study. Table 9.5 shows the main compound classes and their parent ions identified in samples.

Similarly, study of acid extracts of Kimmeridge samples from experiments at 310°C by GCMS revealed an apparent homologous series of organic acids (Figure 9.6). The series consists of straight chain, branched and cyclic acids. Compound classes were identified by the use of diagnostic fragment ions (m/z 60 representing straight chain acids; m/z 73, 74 iso-branched acids; and m/z 82, 83 cyclic acids) and parent ions (C_4 acid having a molecular weight of 88, C_5 102 etc.). No dicarboxylic acids were identified in these samples.

9.3.3 *Infrared analysis of kerogens*

Spectra were obtained from unheated and heated kerogens as described earlier (see Section 9.2.1). Absorptions at certain wavenumbers were assigned to key functional groups (1710 cm^{-1} represents carbonyl or carboxyl groups, $>C{=}O$; 2920 cm^{-1} alkyl groups, $>CH_2$ or $-CH_3$; and 3060 cm^{-1} aromatic CH groups; Robin *et al.*, 1977; Rouxhet *et al.*, 1980). Peak heights were measured and plotted on a triangular diagram to give a picture of the changing

Table 9.5 Compounds tentatively identified as present in experiments on the basis of GCMS analysis.

Compound or compound class	Parent ion, M^+
Ketones	43, 57, 71, 85
Cyclic ketones	41, 55, 69, 83
Benzene and C_1–C_4 substituted benzene	78, 91, 106, 120, 134
Phenol and C1–C_4 substituted phenol	94, 108, 122, 136, 150
Benzothiophene and C_1–C_3 substituted benzothiophene	134, 148, 162, 176
Thiophene and C_1–C_4 substituted thiophene	84, 98, 112, 126, 140
Naphthalene and C_1–C_3 substituted naphthalene	128, 142, 156, 170
Pyridine and C_1–C_2 substituted pyridine	93, 107, 121

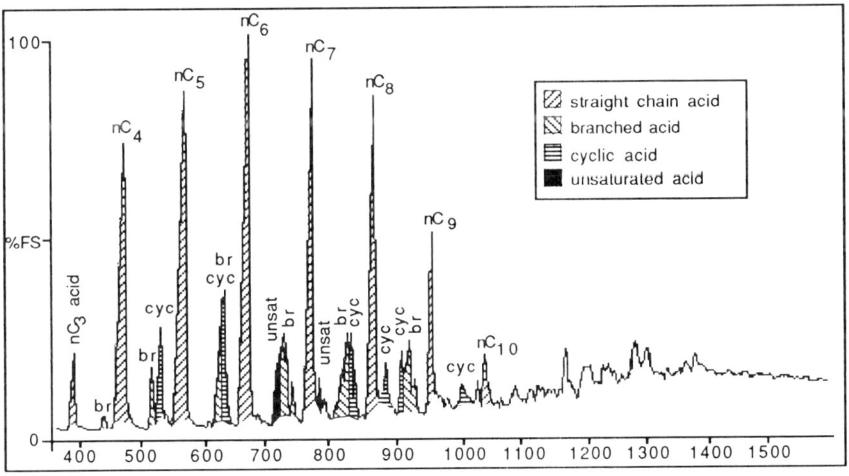

Figure 9.6 Proposed homologous series of carboxylic acids tentatively identified in several samples from Kimmeridge Blackstone Band heating experiments. Straight chain acids are denoted nC_3, nC_4 etc: branched (br), cyclic (cyc) and unsaturated (unsat) acids may also be present.

nature of the kerogen molecule with heating (Figure 9.7). As has been discussed, kerogen maturation should first result in loss of oxygen-containing functional groups and then alkyl groups, with the residual molecule becoming more aromatic in character. Thus one would initially expect to see a decrease in peak intensities at $1710\,cm^{-1}$ and then $2920\,cm^{-1}$ and an increase in peak intensity at $3060\,cm^{-1}$. Infrared analysis of unheated and heated Kimmeridge and Marl Slate kerogens from this study showed that Kimmeridge kerogen heated at 150°C and heated Marl Slate kerogens appeared to have generated hydrocarbons in preference to carbonyl- or carboxyl-containing compounds (showing a preferential decrease in the $2920\,cm^{-1}$ peak). On the other hand, Kimmeridge kerogens heated at 250 and 310°C could be seen to have generated carbonyl- and carboxyl-containing compounds (ketones or acids) in preference to hydrocarbons (a relative decrease in the peak at $1710\,cm^{-1}$). Gas chromatography mass spectrometry analysis confirms that the concentration of carboxylic acids generated by the Marl Slate during heating experiments is low. The Marl Slate is initially more mature than the Kimmeridge Blackstone Band and so may have already passed the stage of peak acid generation.

9.3.4 Mineral products

X-ray diffraction analysis of reacted Kimmeridge Blackstone Band solids reveals no detectable change in mineralogy until the shale is reacted at 310°C with 0.5 M NaCl brine, when the formation of feldspar can be clearly seen. Table 9.6 shows the minerals present before (KBB) and after heating at 310°C with NaCl brine (K7-S). Feldspar formation is accompanied by a significant loss of sodium from the reacting fluid ($\sim 3000\,\mu g.ml^{-1}$). Based on the fluid data, XRD

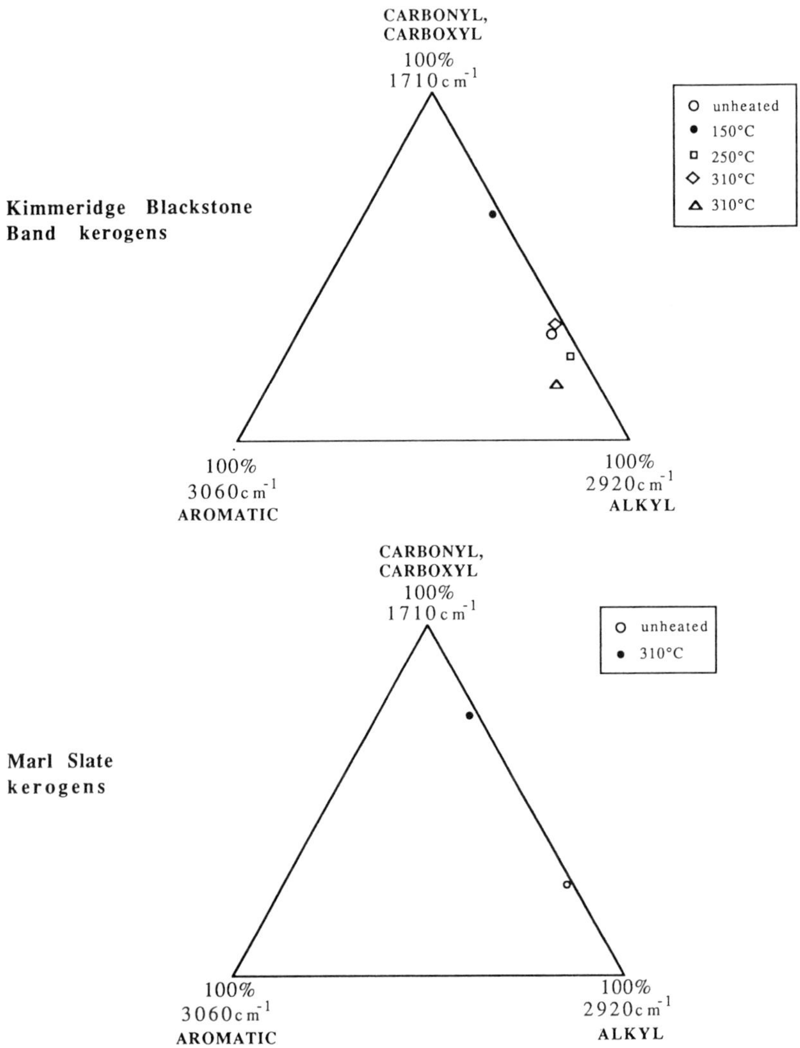

Figure 9.7 Infrared spectrometry shows changes in the nature of unheated and heated Kimmeridge and Marl Slate kerogens expressed in terms of proportions of carbonyl or carboxyl (oxygen containing) to aromatic to alkyl functional groups. For details see text.

observations and an ATEM-derived feldspar composition (see next section), the following reaction is postulated to illustrate the mineralogical changes which have taken place:

$$\underset{\text{quartz}}{4SiO_2} + \underset{\text{calcite}}{4CaCO_3} + \underset{\text{kaolinite}}{Al_4[Si_4O_{10}](OH)_8} + \underset{\text{illite}}{KAl_4[Si_7AlO_{20}](OH)_4} + 3Na^+(aq)$$

$$\rightarrow \underset{\text{feldspar}}{1.5Na_2Ca_2[Al_6Si_{10}O_{32}]} + \underbrace{K^+(aq) + Ca^{2+}(aq) + 4CO_2 + 6H_2O}_{\text{(dissolved species in fluid)}}.$$

Table 9.6 Minerals present in Kimmeridge Blackstone Band and Marl Slate experiments before and after heating on the basis of XRD analysis.

Mineral	Kimmeridge				Marl Slate					
	KBB (unheated)		K7–S (310°C + brine)		MS6 (unheated)		M1-S (310°C + dist. water)		M2-S (310°C + brine)	
	$<2\ \mu m$	$>2<8\ \mu m$	$<8\ \mu m$	$>8\ \mu m$	$<2\ \mu m$	$>2<8\ \mu m$	$<8\ \mu m$	$>8\ \mu m$	$<8\ \mu m$	$>8\ \mu m$
Quartz	–	*	*	+	*	*	*	*	*	*
Pyrite	–	*	*	+	–	*	*	*	*	*
Calcite	+	*	*	*	–	–	*	+	*	*
Dolomite	–	–	–	–	–	*	–	–	–	–
Plagioclase	–	–	*	*	–	+	*	*	*	+
Analcime	–	–	–	–	–	–	–	*	*	*
Muscovite	+	*	+	*	*	*	*	*	*	*
Kaolinite	+	*	+	*	*	*	–	–	–	–
Smectite	–	–	–	–	–	–	*	*	*	*
Apat/sid	–	–	*	*	–	+	+	*	–	–
Marcasite	–	–	*	–	–	–	–	–	–	–

Key: –, absent; +, present in trace amounts; *, present.

When Marl Slate is heated to 310°C with distilled water or brine, newly formed feldspar, smectite, calcite and analcime are all detected by XRD, with concomitant destruction of dolomite and kaolinite. Table 9.6 lists minerals present before heating (MS6), and after heating to 310°C with distilled water (M1-S) and with brine (M2-S). An equivalent equation for the Marl Slate reaction can be written, but there is greater uncertainty due to incomplete knowledge of the smectite formula:

$$\text{Quartz} + \text{dolomite} + \text{kaolinite} + \text{illite} + \text{Na}^{+}(\text{aq}) \rightarrow \\ \text{calcite} + \text{analcime} + \text{Mg-smectite}.$$

X-ray diffraction data indicate that the feldspar formed in both Kimmeridge (with brine at 310°C) and Marl Slate experiments (with brine and distilled water at 310°C) is plagioclase. Inspection of the respective $\{\bar{2}41\}$ feldspar peaks (Hutchison, 1974) suggests that plagioclase formed in the Kimmeridge experiment differs in composition from that formed in the Marl Slate experiments, which is more sodic. Even for pure feldspar samples, however, detailed chemical characterization by XRD is difficult, and consequently the technique is of limited use for the complex mixtures within the powdered rock samples of this study.

Analytical TEM was used to examine newly formed clays in reacted solids from Marl Slate experiments with both brine and distilled water at 310°C, as well as newly precipitated feldspars in reacted solids from these experiments and from the Kimmeridge Blackstone experiment at 310°C with brine.

9.3.4.1 CLAYS

Inspection of reacted solids from Marl Slate experiments revealed that newly formed and original clays are intimately admixed on a scale of tens of nanometres. Listed analyses are, therefore, unlikely to represent the composition of the isolated new precipitates because mixing contamination from 'overlapping' adjacent original clays could not be entirely avoided with confidence. This practical problem was further exacerbated due to the presence of ultra-fine contaminating salts (not always resolved optically), precipitated on sample drying. On many X-ray spectra small, though significant, Kα peaks for chlorine and sulphur were identified, and occasionally minute (<20 nm) euhedral grains were observed, though only briefly. These were very rapidly destroyed by the electron beam. Alkali salts are well known to be very unstable under most TEM operating conditions.

The above two contamination problems mean that quantitative clay compositional data could not be readily accumulated. All data presented in Table 9.7 (in the form of average and standard deviation), are therefore of qualitative value only. They are recalculated on a sodium-free basis for four reasons.

1. Sodium is widely known to be a very difficult element to determine quantitatively by TEM microanalysis. Representative experimental calibration

Table 9.7 Average and standard deviation of ATEM analyses of clays from experiments M1 and M2. Analyses given as mineral formulae calculated on the basis of 20 oxygens and 4 hydroxyls. All iron present assumed to be in the form of Fe^{3+}. Standard deviation calculated is that of a sample of a population, σ^{n-1}. OH, OCT and INT are the number of assumed hydroxyls used in the formulae, the calculated octahedral component total and the calculated interlayer total, respectively.

Species	M1		M2	
	Average	Std. dev.	Average	Std. dev.
Si	6.94	0.19	6.98	0.18
Al (TET)	1.06	0.19	1.02	0.18
Al(OCT)	2.31	0.88	2.63	0.73
Ti	–	–	0.02	0.04
Fe^{3+}	0.48	0.33	0.45	0.23
Mg	1.58	1.08	1.24	0.97
Ca	0.15	0.08	0.18	0.17
K	0.86	0.35	1.02	0.35
OCT	4.49	0.52	4.33	0.46
INT	1.01	0.35	1.12	0.34
OH	4.00	0.00	4.00	0.00

constants are very difficult to obtain because sodium behaviour under an electron beam varies considerably with operating conditions and structural coordination.

2. Careful examination of the collected data indicated that sodium concentrations were often spuriously high for the appropriate mica structural formula.
3. Attempts to analyse the salt crystals seen indicated that sodium was the principal cation present.
4. Sodium occupies highly exchangeable interlayer sites within smectite. From the perspective of this study these are of little consequence. The more permanent tetrahedral and octahedral cation occupancies are of most interest.

Recognizing the above limitations, analysed clays from M1-S and M2-S give a compositional trend as illustrated in Figure 9.8, where octahedral aluminium and interlayer potassium are plotted against octahedral magnesium for each analysis. Comparison with published illite and saponite compositions shows that the analysed clay compositions are scattered between dioctahedral and trioctahedral compositions. Clays from the unheated Marl Slate, however, were analysed and found to be dioctahedral micas/illites (see Figure 9.8(c)), and effectively indistinguishable from the dioctahedral clays analysed in the reacted samples. It appears, therefore, likely that 'overlap' type contamination is the source of the trend identified in Figure 9.8, and that the newly formed clays are saponite (trioctahedral magnesian smectite).

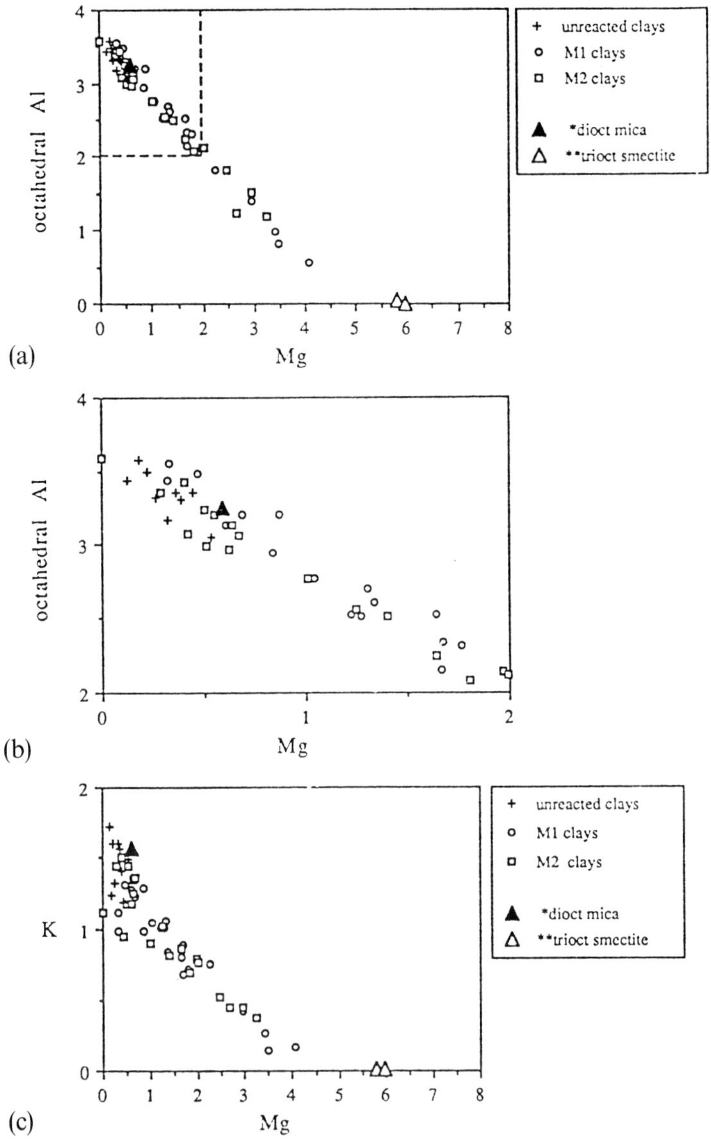

Figure 9.8 (a), (b), (c) Compositions of newly formed clay grains in experiments M1 (Marl Slate+distilled water at 310°C) and M2 (Marl Slate+0.5M NaCl brine at 310°C). (a) Shows octahedral aluminium plotted against octahedral magnesium. *dioctahedral mica (Hower and Mowatt (1966) in Newman and Brown (1987), Table 1.22, analysis 3); **trioctahedral smectites (Schmidt and Heystek (1953) and Mackenzie (1957) in Newman and Brown (1987), Table 1.17, analyses 1 and 2, respectively). (b) Shows area of graph outlined in (a) shown in expanded form with analyses of unreacted Marl Slate clays added. Key as for (a). For details of analyses see text. (c) Shows interlayer potassium plotted against octahedral magnesium and the end-member clays described for (a) are added.

9.3.4.2 FELDSPARS

Figure 9.9 illustrates the newly formed feldspar grains from Kimmeridge Blackstone and Marl Slate experiments in both SEM and TEM. In the dispersed TEM mounts newly precipitated feldspars were found to be too thick for quantitative microanalysis. Ideally, the samples should have been ion thinned; however, the poorly sorted and disparate nature of the reacted powders, coupled with time constraints, made this option impractical.

Feldspars in each dispersed mount were qualitatively analysed to permit some comparison of their compositions. Averages and standard deviation of feldspar data from Kimmeridge Blackstone (K7-S) and Marl Slate (M1-S and M2-S) reacted solids are listed in Table 9.8. Within the limits of each sample set the data are similar; silicon, aluminium and calcium concentrations are all relatively consistent. Sodium contents are less convincing, in view of the difficulties involved in its determination by ATEM in the analysis of feldspars

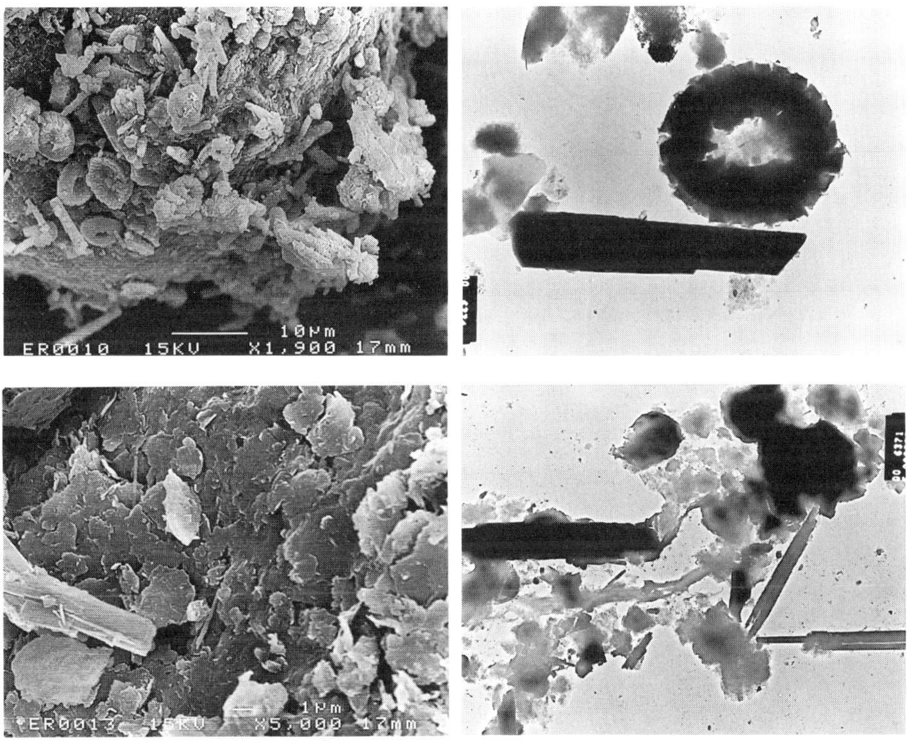

Figure 9.9 Lath- and needle-shaped newly formed feldspar grains in Kimmeridge Blackstone Band and Marl Slate heating experiments. Photos taken in SEM (a) and (c) and TEM (b) and (d). (a) Scanning electron micrograph and (b) transmission electron micrograph showing coccoliths and newly formed lath-shaped, feldspar grains (Kimmeridge Blackstone Band, 310°C + brine). (c) Scanning electron micrograph and (d) transmission electron micrograph showing newly formed feldspar laths (Marl Slate, 310°C + brine).

Table 9.8 Average and standard deviation of analyses of feldspar grains from experiments M1, M2, K7, given as mineral formulae calculated on the basis of 32 oxygens. All iron assumed to be present as Fe^{3+}. Standard deviation calculated is that of a sample of a population, σ^{n-1}.

Species	M2		M1		K7	
	Average	Std. dev.	Average	Std. dev.	Average	Std. dev.
Si	11.75	0.12	11.73	0.22	10.45	0.14
Al	4.39	0.15	4.41	0.10	5.29	0.18
Ti	0.00	0.00	0.00	0.01	–	–
Fe^{3+}	0.09	0.15	0.04	0.07	0.00	0.00
Mg	0.03	0.05	0.03	0.08	0.05	0.04
Ca	0.52	0.06	0.64	0.07	2.05	0.42
Na	2.37	0.64	2.24	1.20	2.09	0.20
K	0.07	0.15	0.10	0.07	0.03	0.04

(McGill and Hubbard, 1981). Although the data are only qualitative, the observation that Ca/Na ratios are different in the Marl Slate and Kimmeridge samples is unequivocal (M2 = 0.22, M1 = 0.29, K7 = 0.98, respectively). Analyses of newly formed feldspars from these experiments compare favourably with published data (e.g. Deer *et al.*, 1963).

9.4 Discussion

9.4.1 *New mineral formation*

Reactions of carbonates and clays have previously been noted during burial diagenesis and hydrothermal alteration. Hutcheon *et al.* (1980) put forward the reaction

$$\text{Dolomite} + \text{kaolinite} + SiO_2 + H_2O \rightarrow \text{Mg-chlorite} + \text{calcite} + CO_2,$$

in a comparative study of mineral assemblage occurrence during deep diagenesis of sedimentary sequences in British Columbia and south-western Alberta. Information on diagenesis from mineral assemblages was compared with known vitrinite reflectances from interbedded coals. It was suggested that mineral reactions such as that given above would result in large quantities of CO_2 (an inorganic source rather than from the maturation of organic matter) in sedimentary basins which might act in the form of carbonic acid to form secondary porosity. This reaction of dolomite and kaolinite to form chlorite and calcite had been put forward by Zen (1959) as occurring in devitrifying volcanic glass in recent sediments off the coasts of Peru and Chile (with quartz present in excess and the system open to CO_2 and H_2O). Similarly, Muffler and White (1969) suggested that this reaction occurs in the Salton Sea geothermal system. Formation of new minerals during artificial short-term water–rock interactions has also been widely reported (Hutcheon *et al.*, 1981; Bischoff *et al.*, 1981; Janecky and Seyfried, 1983; Rosenbauer *et al.*, 1983; Hutcheon, 1984; Eglinton, 1988; Milodowski *et al.*, 1989). Hutcheon *et al.* (1981) consider the instability of the restricted assemblage kaolinite + quartz. However, most reactions involve magnesium either as a process of magnesium fixation from seawater (e.g. Bischoff *et al.*, 1981; Janecky and Seyfried, 1983) or involving a dolomite-containing mineral assemblage (such as dolomite + kaolinite + quartz) which becomes unstable (Rosenbauer *et al.*, 1983). In either case, magnesium-bearing secondary phases are formed, such as Mg-smectite (as in this study) or a magnesium-containing mixed-layer clay; the presence of an amorphous precipitate, magnesium hydroxide sulphate hydrate (MHSH) has also been observed (Janecky and Seyfried, 1983; Milodowski *et al.*, 1989). 'Magnesium cycling' was noted by Muffler and White (1969) who pointed out that all magnesium in original, shallow clastic carbonates and clay minerals seemed to be present in Mg-chlorite at depth. In addition to magnesium-bearing phases, the reaction of dolomite, quartz and kaolinite also forms calcite. Analcime and feldspar are products of sodium fixation when brines or seawater are involved in the reactions.

The development of secondary phases may be influenced by kinetic factors or by the formation of metastable equilibrium assemblages. For example, Hutcheon *et al.* (1981) suggest that the analcime formed during steam injection of the Alberta Tar Sands is part of a metastable equilibrium assemblage. It has also been suggested that smectite formation seen in many fluid-rock reactions is metastable. As many of these reactions occur during quite short periods of time this seems likely. In studies of artificial procedures such as enhanced oil recovery, periods of the order of two years are involved (Hutcheon *et al.*, 1981; Hutcheon, 1984), whilst experimental investigations of water–rock reactions associated with geothermal power exploitation involve periods of months (Milodowski *et al.*, 1989). Relatively short-term natural geological processes have also been studied (Zen, 1959; Muffler and White, 1969). Other laboratory studies (Bischoff *et al.*, 1981; Rosenbauer *et al.*, 1983; Eglinton, 1988; this study) involve experimental fluid–rock reactions lasting from the order of days to several weeks. Of particular interest is the suggestion that smectite formation may be favoured by the sluggish kinetics of quartz precipitation, which results in solutions becoming supersaturated with respect to quartz (Thornton and Seyfried, 1985). This means that the reaction

$$\text{Smectite} + Al^{3+} + K^{+} \rightarrow \text{illite} + Si^{4+}$$

(representing in general terms the smectite–illite transformation) is inhibited or reversed, resulting in smectite formation.

9.4.2 Fluids produced

Apart from the presence of molybdenum in low-temperature Kimmeridge heating experiments, base metals in general were not detected in fluids generated from these experiments, providing no new experimental evidence that mudrocks will act as a source of metals for ore-forming solutions. This agrees with the results of related studies by Bischoff *et al.* (1981) and Rosenbauer *et al.* (1983) who heated greywacke with seawater and artificial brines to study metal mobilization. Both studies concluded that the chlorinity of seawater was insufficient to mobilize metals and that a concentrated brine was necessary. However, Thornton and Seyfried (1985) found that metals were mobilized by seawater-strength brines in experiments which involved a much lower fluid/rock ratio (3:1 as opposed to 10:1 used by Bischoff *et al.* (1981) and Rosenbauer *et al.* (1983)). Thornton and Seyfried (1985) considered that the oxidation state of the sediment was also a controlling factor in controlling metal solubility.

Fluids formed in the mudrock maturation experiments reported here are complex, containing both inorganic and organic species, including polar organic molecules, such as acids, ketones and phenols. Because of their complexity, the compositional data are not sufficient for a rigorous interpretation using solution modelling techniques. However, plots of K^+/H^+ activity ratios against silica activities for samples taken during experiments carried out at 310°C and 50 MPa (500 bar) have been produced, using activity data from EQ3NR (version 3245R124X with steam saturation database EQLIB version 3245R153X; Wolery, 1983), EQ6 (version 3245R119X) and SUPCRT92, and are shown in Figure 9.10. EQ6 recalculates fluid properties at any given temperature from a data file generated by EQ3NR from original fluid analyses. The limits of the database, however, mean that data generated for temperatures greater than 300°C will be of poor quality. Data sets generated for fluids from experiments carried out at 310°C at model temperatures of 300 and 310°C were not found to differ substantially. Data from the 300°C model were used in Figure 9.10.

Samples from experiment K7 (Kimmeridge Blackstone Band with 0.5 M NaCl) are consistently undersaturated with respect to quartz, and those from Marl Slate experiments M1 and M2 (with distilled water and 0.5 M NaCl respectively) are slightly oversaturated with respect to quartz. Despite the presence of organic speices believed to enhance silica solubility, there is no evidence for this occurring to any great extent. These plots also suggest that potassium/hydrogen ion activity ratios are consistent with the formation of feldspar as reported for the experimental product solids. It must be stressed however that modelling of this kind is limited by the lack of complete analytical data, particularly by the lack of estimation of CO_2 and consequent limited estimation of pH and $a\mathrm{H}^+$.

Fluid compositions in these experiments and in the single experiment carried out at 150°C (K5), where data points plot close to quartz saturation within the

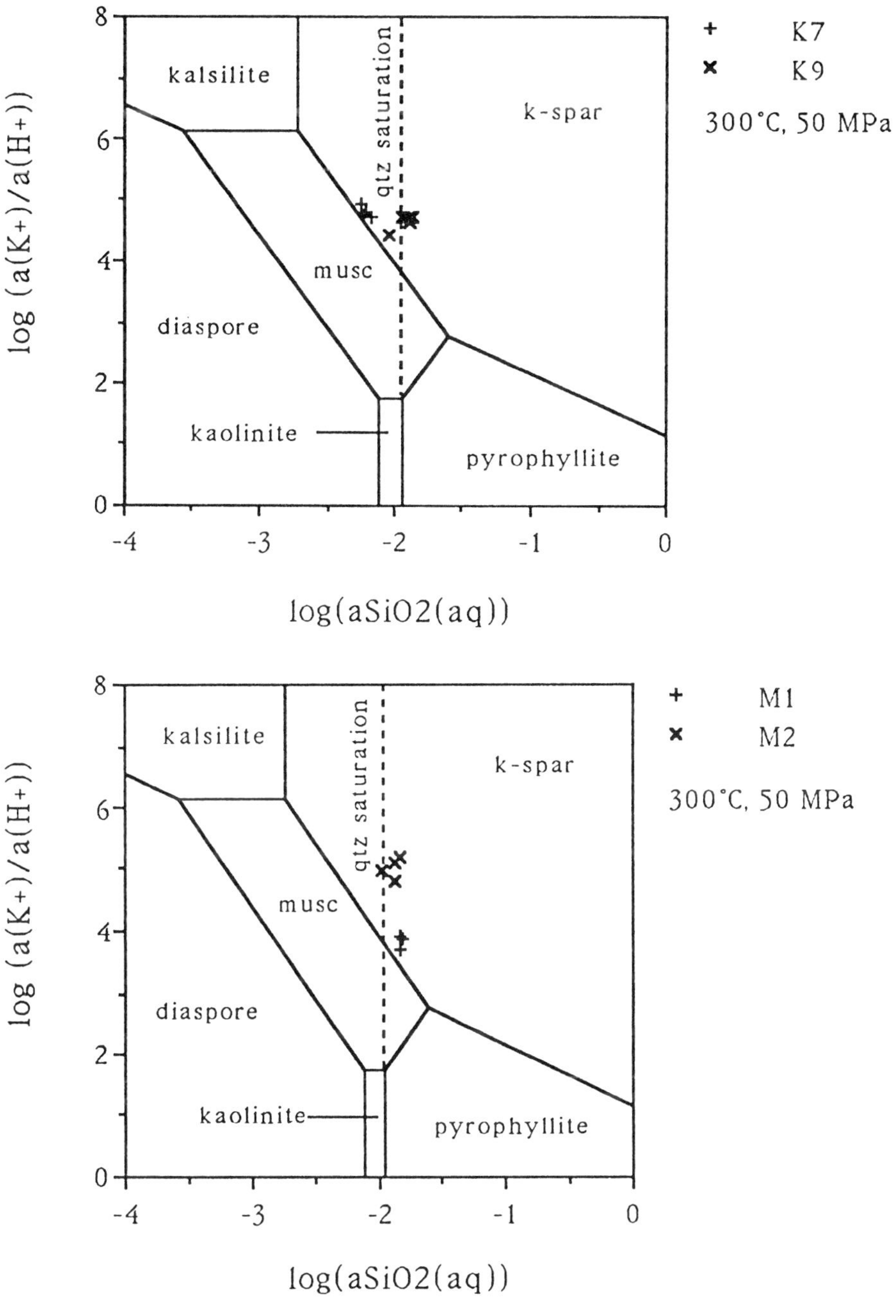

Figure 9.10 K^+/H^+ ionic activity ratio vs. silica activity diagrams for all samples taken during experiments K7, K9, M1 and M2 (details of experiments in text). Mineral phase boundaries calculated at 300°C, 50 MPa. K-spar = potassium feldspar, musc = muscovite.

muscovite stability field, seem to be controlled by mineralogical buffers. Hutcheon and Abercrombie (1990) and Hutcheon *et al.* (in press) have produced data which strongly suggest that the ability of mineral assemblages to control the pH of interstitial fluids is far greater than that of carboxylic acids or CO_2. Fluid-mineral equilibria attained during these heating experiments are the result of a closed system and so model the situation in a mudstone with no fluid flow out of the shale. In order to evaluate the potential of shales to supply solutes to a maturing basin it is necessary to model an open system where fluids produced as a result of mineral-fluid reactions within the shale migrate out of the mudstone, as is known to occur during diagenesis. This severely limits the usefulness of these experiments in this respect but suggests that the effect of fluids expelled from shales will depend on mechanisms of fluid flow and on mudstone-sandstone geometry. The experiments do suggest that if fluids are derived from a carbonate-bearing source they have the potential to precipitate carbonate cements, but the effects of changes in fluid pressure become particularly significant as a control of carbonate solubility.

9.5 Conclusions

This exploratory study has involved the use of complex materials and has yielded results which are difficult to interpret fully in the absence of further information; this limits their application. The experiment configuration was also not particularly suitable for modelling fluids expelled from mudstones. Nevertheless, the study has shown that mudrocks may act as sources of cement-forming components. The carbonate-precipitation potential of mudrock-derived pore fluids depends largely on the dissolution of their carbonate components, which will be controlled by pressure and fluid CO_2 content as well as temperature. There is no evidence that organic constituents of the pore fluids enhance transport; they may be influential in affecting pH or contributing carbonate on their decomposition. There is also no evidence from these experiments to support the possibility that mudrocks are sources rather than sinks for base metals in general; the mobility of molybdenum is, however, potentially important.

Acknowledgements

This work was carried out initially at the University of Newcastle upon Tyne and completed at the University of Manchester, with financial support from the Natural Environment Research Council (awards GR3/6141 and GST/02/157 to D.A.C.M. and D1/G1/19 to Professor C. D. Curtis). D.A.C.M. also gratefully acknowledges receipt of a Science Research Fellowship from the Nuffield Foundation. We would like to thank Dr A. P. Gize for providing assistance with organic geochemical analysis in Manchester.

References

Anderson, G. M. and Macqueen, R. W. (1982) Ore deposit models 6. Mississippi-Valley-type lead–zinc deposits. *Geoscience Canada*, **9**, 108–17.

Barth, T., Borgund, A. E. and Riis, M. (1990) Organic acids in reservoir waters—relationship with inorganic ion composition and interactions with oil and rock. *Advances in Organic Geochemistry 1989*, (eds B. Durand and F. Behar), *Organic Geochemistry*, **16**, 489–96.

Beales, F. W. and Jackson, J. A. (1966) Precipitation of lead–zinc ores in carbonate reservoirs as illustrated by Pine Point Ore Field, Canada. *Transactions of the Institution of Mining and Metallurgy*, **B75**, B278–85.

Billings, G. K., Kesler, S. E. and Jackson, S. A. (1969) Relation of zinc-rich formation waters, northern Alberta, to the Pine Point ore deposit. *Economic Geology*, **64**, 385–91.

Bischoff, J. L., Radtke, A. S. and Rosenbauer, R. J. (1981) Hydrothermal alteration of greywacke by brine and seawater: roles of alteration and chloride complexing on metal solubilisation at 200°C and 350°C. *Economic Geology*, **76**, 659–76.

Bray, E. E. and Evans, E. D. (1961) Distribution of *n*-paraffins as a clue to recognition of source beds. *Geochimica et Cosmochimica Acta*, **22**, 2–15.

Cantrell, K. J., Serkiz, S. M. and Perdue, E. M. (1990) Evaluation of acid neutralising capacity data for solutions containing natural organic acids. *Geochimica et Cosmochimica Acta*, **54**, 1247–54.

Collins, A. G. (1974) Geochemistry of liquids, gases and rocks from the Smackover Formation. *U.S. Bureau of Mines Report of Investigation 7879*, Washington, DC, 84 pp.

Crossey, L. J. (1991) Thermal degradation of aqueous oxalate species. *Geochimica et Cosmochimica Acta*, **55**, 1515–27.

Crossey, L. J., Surdam, R. C. and Lahann, R. (1986) Application of organic/inorganic diagenesis to porosity prediction. *Roles of Organic Matter in Sediment Diagenesis*, (ed. D. L. Gauthier), Society of Economic Palaeontologists and Mineralogists, Special Publication, **38**, 147–55.

Curtis, C. D. (1983a) Geochemistry of porosity enhancement and reduction in clastic sediments, in *Geological Society of London Special Publication*, No. 12, *Petroleum Geochemistry and Exploration of Europe*, (ed. J. Brooks), Blackwell Scientific Publications, Oxford, pp. 113–25.

Curtis, C. D. (1983b) Link between aluminium mobility and destruction of secondary porosity. *Bulletin of the American Association of Petroleum Geologists*, **67**, 380–3.

Deer, W. A., Howie, R. A. and Zussman, J., (1963) *Rock Forming Minerals*, vol. 4, *Framework Silicates*, Longman, London.

Dickson, F. W., Blount, C. W. and Tunell, G. (1963) Use of hydrothermal solution equipment to determine the solubility of anhydrite in water, 100–275°C, 1–1000 bars pressure. *American Journal of Science*, **261**, 61–79.

Drummond, S. E. and Palmer, D. A. (1986) Thermal decarboxylation of acetate. Part I. Boundary conditions for the role of acetate in the primary migration of gas and the transportation of metals in hydrothermal systems. *Geochimica et Cosmochimica Acta*, **50**, 825–33.

Eglinton, T. I. (1988) *An Investigation of Kerogens Using Pyrolysis Methods*, PhD Thesis, University of Newcastle upon Tyne (unpublished).

Giordano, T. H. (1985) Preliminary evaluation of organic ligands and metal–organic complexing in Mississippi-Valley-type ore solutions. *Economic Geology*, **80**, 96–106.

Hennet, R. J-C., Crerar, D. A. and Schwartz, J. (1988) Organic complexes in hydrothermal systems. *Economic Geology*, **83**, 742–64.

Hughes, C. R., Curtis, C. D., Whiteman, J. A., Heping, Sun, Whittle, C. K. and Ireland, B. J. (1990) Selected applications of analytical electron microscopy in clay mineralogy, in *CMS Workshop Lectures*, vol. 2, *Electron-optical methods in clay science*, (eds I.D.R. Mackinnon and F. A. Mumpton), Clay Minerals Society, Evergreen, Colorado, pp. 69–88.

Hutcheon, I. (1984) A review of artificial diagenesis during thermally enhanced recovery, in (eds D. A. McDonald and R. C. Surdam), *Memoirs, American Association of Petroleum Geologists*, **37**, 413–29.

Hutcheon, I. (1990) Clay-carbonate reactions in the Venture area, Scotia Shelf, Nova Scotia, Canada, in *Geochemical Society Special Publication*, No. 2, *Fluid–Mineral Interactions: A Tribute to H. P. Eugster*, (eds R. J. Spencer and I-Ming Chou), Geochemical Society, San Antonio, Texas, pp. 199–212.

Hutcheon, I. and Abercrombie H. J. (1990) Carbon dioxide in clastic rocks and silicate hydrolysis. *Geology*, **18**, 541–4.

Hutcheon, I., Abercrombie, H. J. and Shevalier, M. (1992) Metastable silicate equilibria, the dissolution of calcite and the evolution of carbon dioxide during burial diagenesis (in press).

Hutcheon, I., Oldershaw, A. and Ghent, E. (1980) Diagenesis of sandstones of the Kootenay Formation at Elk Valley (southeastern British Columbia) and Mount Allen (southwestern Alberta). *Geochimica et Cosmochimica Acta*, **44**, 1425–35.

Hutcheon, I., Oldershaw, A. and Nahnybida, C. (1981) The effects of in situ steam injection on Cold Lake Oil Sands-II. Petrophysical and petrological considerations. *Bulletin of Canadian Petroleum Geology*, **29**, 464–78. (Forms part of a combined paper by Sedimentology Research Group. *Bulletin of Canadian Petroleum Geology*, **29**, 447–78.)

Hutchison, C. S. (1974) *Laboratory Handbook of Petrographic Techniques*, Wiley Interscience, New York.

Janecky, D. R. and Seyfried, W. E. Jr. (1983) The solubility of magnesium-hydroxide-sulphate-hydrate in seawater at elevated temperatures and pressures. *American Journal of Science*, **283**, 831–60.

Kharaka, Y. F., Law, L. M., Carothers, W. W. and Goerlitz, D. F. (1986) Role of organic species dissolved in formation waters from sedimentary basins in mineral diagenesis. *Roles of Organic Matter in Sediment Diagenesis*, (ed. D. L. Gauthier), Society of Economic Palaeontologists and Mineralogists, Special Publication, **38**, 111–22.

Křibek, B. (1991) Metallogeny, structural, lithological and time controls of ore deposition in anoxic environments. *Mineralium Deposita*, **26**, 122–31.

Le Riche, H. H. (1959) The distribution of certain trace elements in the Lower Lias of southern England. *Geochimica et Cosmochimica Acta*, **16**, 101–22.

MacGowan, D. B. and Surdam, R. C. (1988) Difunctional carboxylic acid anions in oil-field waters. *Organic Geochemistry*, **12**, 245–59.

Mackenzie, A. S. (1984) Applications of biological markers in petroleum geochemistry. *Advances in Petroleum Geochemistry*, **1**, 115–214.

McGill, R. J. and Hubbard, F. H. (1981) Thin film k-value calibration for low atomic number elements using silicate mineral standards, in *Quantitative Microanalysis with High Spatial Resolution*, (eds G. W. Lorimer, M. H. Jacobs and P. Doig), *Book 277*, Metals Society, Arrowsmith, England.

Milodowski, A. E., Savage, D., Bateman, K. and Hughes, C. R. (1989) The precipitation of solids from potential heat exchange fluids for use in a HDR geothermal system in granite, *Geothermics*, **18**, 233–40.

Moncure, G. K., Lahann, R. W. and Seibert, R. M. (1984) Origin of secondary porosity and cement distribution in a sandstone/shale sequence from the Frio Formation (Oligocene), in (eds D. A.

McDonald and R. C. Surdam), *Memoirs, American Association of Petroleum Geologists*, **37**, 151–61.

Muffler, L. P. J. and White, D. E. (1969) Active metamorphism of Upper Cenozoic sediments in the Salton Sea geothermal field and the Salton Trough, southeastern California. *Geological Society of America Bulletin*, **80**, 157–82.

Newman, A. C. D. and Brown, G. (1987) The chemical constitution of clays, in *Chemistry of Clays and Clay Minerals*, (ed. A. C. D. Newman), Monograph, No. 6, Mineralogical Society, Longman, London.

Palmer, D. A. and Drummond, S. E. (1986) Thermal decarboxylation of acetate. Part I. The kinetics and mechanism of reaction in aqueous solution. *Geochimica et Cosmochimica Acta*, **50**, 813–23.

Robin, P. L., Rouxhet, P. G. and Durand, B. (1977) Caracterisation des kérogènes et de leur évolution par spectroscopie infrarouge: fonctions hydrocarbonees, (eds R. Campos and J. Goni), in *Advances in Organic Geochemistry 1975*, Empresa Nacional Adaro de Investigaciones Minerasa S.A., Madrid, pp. 693–716.

Rosenbauer, R. J., Bischoff, J. L. and Radtke, A. S. (1983) Hydrothermal alteration of graywacke and basalt by 4 molal NaCl. *Economic Geology* **78**, 1701–10.

Rouxhet, P. G., Robin, P. L. and Nicaise, G. (1980) Characterisation of kerogens and of their evolution by infrared spectroscopy, in *Kerogen*, (ed. B. Durand), Chapter 6, Editions Techniq, Paris, pp. 163–90.

Saxby, J. D. (1976) The significance of organic matter in ore genesis, in *Handbook of Strata Bound and Stratiform Ore Deposits*, (ed. K. H. Wolf), vol. 2, Elsevier, Amsterdam, pp. 111–34.

Schleusener, J. L., Barnes, H. L., Drummond, S. E. and Palmer, D. A. (1988) Activation parameters and low temperature half-lives for the decarboxylation of acetate in sedimentary basin fluids. *Geological Society of America, Abstracts with Programs*, **20**, A150.

Schleusener, J. L., Drummond, S. E., Palmer, D. A. and Barnes, H. L. (1987) Effects of common minerals on acetate decarboxylation kinetics. *Geological Society of America, Abstracts with Programs*, **19**, 832–3.

Schmidt, V. and McDonald, D. A. (1979a) Texture and recognition of secondary porosity in sandstones, in *Aspects of Diagenesis*, (eds P. A. Scholle and P. R. Schluger), Society of Economic Paleontologists and Mineralogists, Special Publication, **26**, 209–25.

Schmidt, V. and McDonald, D. A. (1979b) The role of secondary porosity in the course of sandstone diagenesis, in *Aspects of Diagenesis*, (eds P. A. Scholle and P. R. Schluger), Society of Economic Paleontologists and Mineralogists, Special Publication, **26**, 157–207.

Seward, T. M. and Kishima, N. (1987) Problems in working with hydrogen under hydrothermal conditions, in *Hydrothermal Experimental Techniques*, (eds G. C. Ulmer and H. L. Barnes), John Wiley, New York, pp. 141–56.

Seyfried, W. E. Jr., Gordon, P. C. and Dickson, F. W. (1979) A new reaction cell for hydrothermal solution equipment. *American Mineralogist*, **64**, 646–9.

Seyfried, W. E. Jr., Janecky, D. R. and Bernt, M. E. (1987) Rocking autoclaves for hydrothermal experiments II. The flexible reaction cell system, in *Hydrothermal Experimental Techniques* (eds G. C. Ulmer and H. L. Barnes), John Wiley, New York pp. 216–39.

Shock, E. L. (1988) Organic acid metastability in sedimentary basins. *Geology*, **16**, 886–90.

Shock, E. L. (1989) Corrections to "Organic acid metastability in sedimentary basins". *Geology*, **17**, 572–3.

Stoessell, R. K. and Pittman, E. D. (1990) Secondary porosity revisited: the chemistry of feldspar dissolution by carboxylic acids and anions. *Bulletin of the American Association of Petroleum Geologists*, **74**, 1795–805.

Sullivan, K. B. and McBride, E. F. (1991) Diagenesis of sandstones at shale contacts and diagenetic heterogeneity, Frio Formation, Texas. *Bulletin of the American Association of Petroleum Geologists*, **75**, 121–38.

Surdam, R. C., Crossey, L. J., Hagen, E. S. and Heasler, H. P. (1989) Organic–inorganic interactions and sandstone diagenesis. *Bulletin of the American Association of Petroleum Geologists*, **73**, 1–23.

Sverjensky, D. A. (1984) Oil field brines as ore forming solutions. *Economic Geology*, **79**, 23–7.

Tissot, B. P. and Welte, D. H. (1984) *Petroleum Formation and Occurrence*, 2nd edn, Springer-Verlag, Berlin.

Thornton, E. C. and Seyfried, W. E., Jr. (1985) Sediment–seawater interaction at 200° and 300°C, 500 bars pressure: the role of sediment composition in diagenesis and low-grade metamorphism of marine clay. *Geological Society of American Bulletin*, **96**, 1287–95.

Walther, J. V. and Helgeson, H. C. (1977) Calculation of the thermodynamic properties of aqueous silica and the solubility of quartz and its polymorphs at high pressures and temperatures. *American Journal of Science*, **277**, 1315–51.

Williams, P. F. V. and Douglas, A. G. (1979) A preliminary organic geochemical investigation of the Kimmeridgian oil shales, in *Advances in Organic Chemistry 1979*, (eds A. G. Douglas and J. R. Maxwell), Pergamon Press, Oxford, pp. 531–45.

Wolery, T. J. (1983) *EQ3NR, A Computer Program for Geochemical Aqueous Speciation–Solubility Calculations: User's Guide and Documentation, Lawrence Livermore Laboratory Publication UCRL-53414*, Lawrence Livermore National Laboratory, University of California, Livermore, California.

Yang, M. M., Crerar, D. A. and Irish, D. E. (1989) A Raman spectroscopic study of lead and zinc acetate complexes in hydrothermal solutions. *Geochimica et Cosmochimica Acta*, **53**, 319–26.

Zen, E.-An. (1959) Clay mineral–carbonate relations in sedimentary rocks. *American Journal of Science*, **257**, 29–43.

CHAPTER TEN

Weathering kinctics of montmorillonite: Investigations in batch and mixed-flow reactors

G. Furrer, M. Zysset and P. W. Schindler

10.1 Introduction

Weathering processes—mainly a consequence of the interactions of water and the atmosphere with the lithosphere—occur because the original minerals are thermodynamically unstable under the ambient conditions. The rates of chemical weathering reactions can be increased significantly by the activities of the biosphere, e.g. the formation of carbon dioxide and organic acids in soils.

The stability of minerals in aquatic environments has successfully been discussed on the basis of equilibrium concepts (Sillén 1961; 1967; Garrels and Christ, 1965; Stumm and Morgan, 1981). Nevertheless, natural aqueous solutions are often not in thermodynamic equilibrium with the mineral phases present. The extent to which the thermodynamic equilibrium is approached depends on the dissolution kinetics and the contact time.

In recent years, numerous studies on the kinetics of mineral–water interactions have been published (e.g. Wollast, 1967; Busenberg and Clemency, 1976; Holdren and Berner, 1979; Aagaard and Helgeson, 1982; Chou and Wollast 1985; Furrer and Stumm, 1986; Mast and Drever, 1987; Blum and Lasaga, 1988; Wehrli *et al.*, 1990). Many of them have emphasized the role of surface complexes in the dissolution kinetics of minerals.

10.1.1 Chemical speciation of hydrous mineral surfaces

When brought into contact with water, the surface of a mineral forms hydrated functional groups. Their reactivities determine the properties of the phase boundary between the solid surface and the electrolyte solution. According to the Surface Complexation Model (Schindler and Stumm, 1987; Dzombach and Morel, 1990), a surface hydroxyl group can become protonated or deprotonated leading to a positive or negative charge. Owing to electrostatic

Geochemistry of Clay-Pore Fluid Interactions. Edited by D. A. C. Manning, P. L. Hall and C. R. Hughes. Published in 1993 by Chapman & Hall, London. ISBN 0 412 48980 5

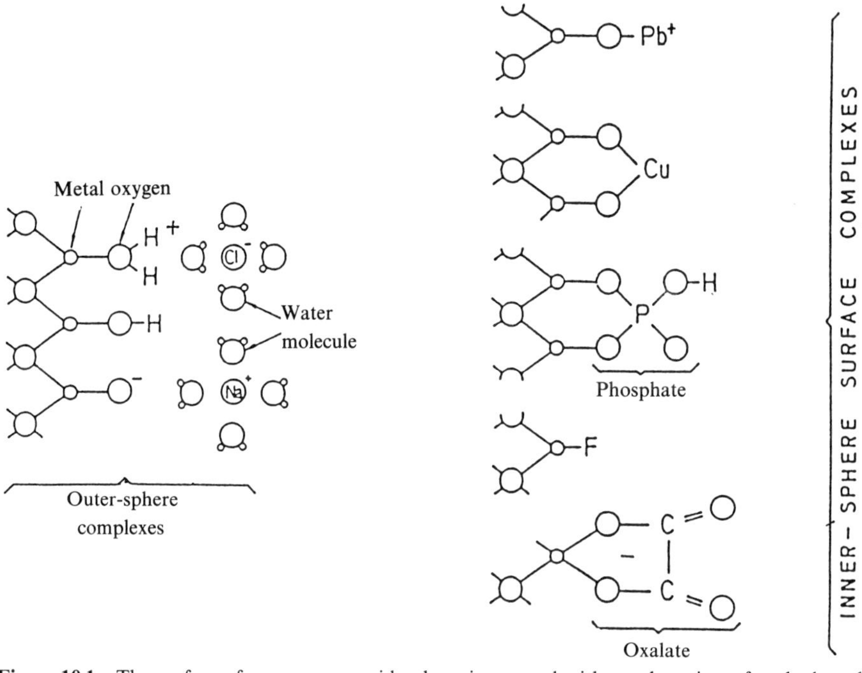

Figure 10.1 The surface of an aqueous oxide phase is covered with amphoteric surface hydroxyl groups. These can form outer-sphere complexes with hydrated ions or undergo ligand exchange reactions in order to form inner-sphere complexes.

attractions, the charged functional groups may interact with hydrated ions, resulting in the formation of outer-sphere complexes (Figure 10.1).

In recent advances in aquatic surface chemistry, the interactions of hydrous oxide surfaces with H^+, OH^-, other ions and weak acids have been interpreted as surface coordination reactions leading to the formation of inner-sphere complexes (Figure 10.1). The oxygen functional groups at a mineral surface behave potentially as ligands, towards metal ions (e.g. Pb^{2+} and Cu^{2+}), whereas anions (e.g. HPO_4^{2-}, F^- and oxalate) are capable of competing with surface hydroxyl groups for a surface metal centre.

10.1.2 Surface-controlled dissolution kinetics

Some authors have shown that rate laws for dissolution reactions can be formulated successfully in terms of the Surface Complexation Model (Pulfer *et al.*, 1984; Furrer and Stumm, 1986; Zinder *et al.*, 1986; Wieland, 1988; Blum and Lasaga, 1988; Wehrli *et al.*, 1990).

From a chemical point of view, the dissolution of an oxide mineral is characterized by a change in the coordination sphere, i.e. the original O^{2-} ligands of a metal ion in a solid oxide phase eventually become protonated or replaced by H_2O or other ligands during the detachment process. Since, in many minerals the metal–oxygen interaction is quite strong, a ligand exchange process is hampered by a high activation energy. However, if the oxide surface is covered with excess hydrogen ions, as is mostly the case in acidic solutions, these can migrate and ultimately become bound to a bridging O^{2-} ligand. The lowering of charge leads consequently to weakening of the bond and a far greater probability of detachment of the metal complex. Such a process is called 'proton-promoted dissolution'.

Furthermore, organic ligands such as oxalate may become coordinated to a metal centre located at a mineral surface (Figure 10.1). The extent of this complex formation is a function of pH, i.e. it occurs preferentially in moderately acidic media. The surface-bound oxalate with its double negative charge and the chelate structure leads to a polarization and weakening of the bonding between the metal centre and the adjacent O^{2-} bridges, resulting finally in a favoured detachment of the organic metal complex. It has been demonstrated that the rates of such a 'ligand-promoted dissolution' of δ–Al_2O_3 are linearly dependent on the surface concentration of organic ligands (Furrer and Stumm, 1986).

Thus, we can summarize that the dissolution process of most minerals consists mainly of two consecutive steps: the fast adsorption of reactants, often in thermodynamic equilibrium, is followed by the rate-limiting formation of an activated complex and its detachment from the mineral surface (surface-controlled dissolution kinetics).

Various aspects of the interaction of clay minerals with either neutral or very acidic solutions have been studied by Eeckman and Laudelout (1961), Kamil and Shainberg (1968), Kittrick (1971), and Kline and Fogler (1981). The dissolution kinetics of minerals are normally investigated using either batch or mixed flow reactors. Recent reviews of these reactors have been published by Rimstidt and Dove (1986) and Posey-Dowty *et al.* (1986). With this paper, we would like to contribute to the understanding of the dissolution kinetics of montmorillonite as related to its surface chemistry.

10.2 Materials

10.2.1 Origin of K-montmorillonite sample and preparation of stock solutions

The experiments were carried out with SWy-1 Na-montmorillonite from Crook County, Wyoming, USA. The particle fraction ($\leq 2\,\mu$m, Stokes' diameter) was obtained using the sedimentation technique described by Day (1965). X-ray analysis of this particle fraction showed an impurity of about 1% quartz. The fractionated montmorillonite was converted into the K-exchanged form as described by Galicia (1989). After the ionic strength had been adjusted to 0.1 M KCl, the stock suspensions were stored in the dark under N_2.

10.2.2 Chemical composition

The fractionated sample was analysed by X-ray fluorescence. The formula of the K-exchanged form of montmorillonite (equation (1)) was obtained from these data according to the procedure described by Van Olphen (1977):

$$K_{0.310}(Si_{3.976}Al_{0.024})(Al_{1.509}Fe_{0.205}Mg_{0.286})O_{10}(OH)_2. \tag{1}$$

The oxidation state of all Fe was assumed to be III. The formula weight of the half unit cell and the Si/Al ratio are 377.6 g.mol^{-1} and 2.59, respectively. The calculated cation exchange capacity (CEC) is 82.1 mEq per 100 g. In comparison, the experimentally determined CEC ($BaCl_2$ method; Hendershot and Duquette, 1986) of the original material is 95.2 mEq per 100 g.

10.2.3 Solubility

The Gibbs free energy of the formation of K-montmorillonite was calculated to be -5231.63 kJ.mol^{-1} using a semi-empirical method as described by Mattigod and Sposito (1978). The Gibbs free energy of the dissolution reaction

$$\begin{aligned}&\text{K-montmorillonite} + 3.904\ H_2O + 6.096\ H^+ = \\ &\quad 0.310\ K^+ + 1.533\ Al^{3+} + 0.205\ Fe^{3+} + 0.286\ Mg^{2+} + 3.976\ H_4SiO_4\end{aligned} \tag{2}$$

was obtained as -8.89 kJ.mol^{-1}. Based on this value, the equilibrium concentrations of silicon, aluminium and magnesium (Figures 10.2(a), (b)) were calculated with the aid of the computer program MICROQL (Westall, 1986). For comparison, the solubilities of kaolinite, amorphous $Al(OH)_3$ and amorphous SiO_2 are shown as well (Figure 10.2(a)). The individual data points in Figure 10.2(b) represent concentrations of silicon, aluminium and magnesium in a series of 11 montmorillonite suspensions in 0.1 M KCl after they had been shaken for 180 days. The experimental data are in good agreement with the calculated equilibrium concentrations. May *et al.* (1986) reported that the solubility equilibria of smectite suspensions cannot be determined because the composition of the solution phase appears to be controlled by solid aluminium hydroxide rather than by smectite. For our studies it is important to note that the concentrations of silicon and aluminium in the kinetic experiments were all far below saturation with respect to montmorillonite, kaolinite, SiO_2 and $Al(OH)_3$ (Figure 10.2(a)). The dissolution experiments thus were not influenced by back-reactions.

10.2.4 Solutions

Solutions were prepared from reagent grade chemicals with fresh double-distilled water and filtered through 0.1 μm filters (Sartorius).

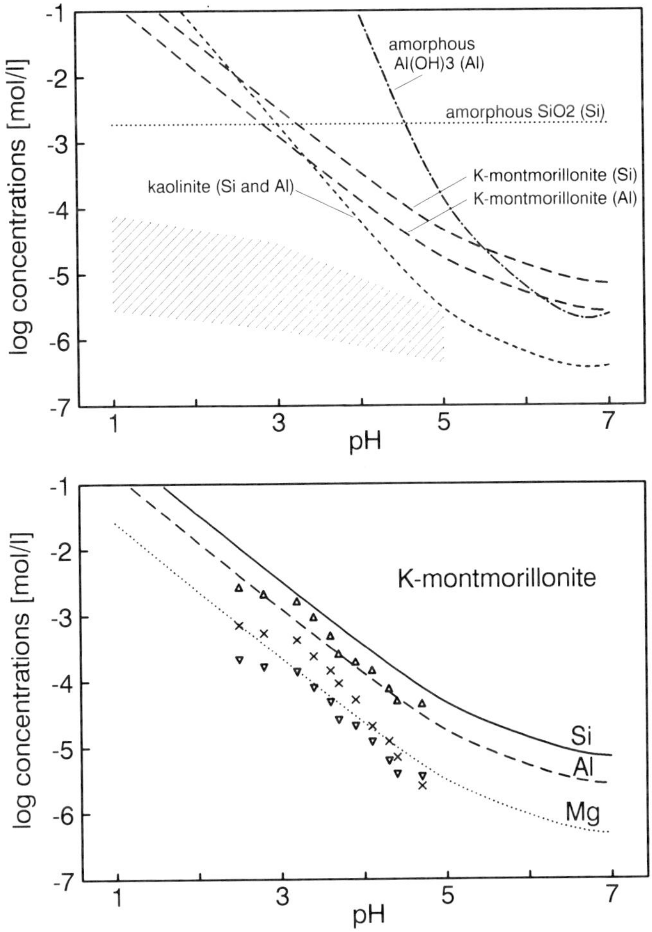

Figure 10.2 (a) Solubility of minerals at 298 K and 0.1 M KCl. The calculated concentrations of silicon and aluminium in equilibrium with K-montmorillonite, kaolinite, amorphous SiO_2, and amorphous $Al(OH)_3$ exceed by far the concentrations which were measured in the kinetic experiments (indicated by the dashed region), thus back-reactions can be neglected. The equilibrium calculations are based on the assumption of congruent dissolution. (b) Solubility of K-montmorillonite at 298 K and 0.1 M KCl. The calculated concentrations of silicon (solid line), aluminium (broken line), and magnesium (dotted line) in equilibrium with K-montmorillonite are in good agreement with the measured concentrations of silicon ($\triangle$), aluminium ($\times$), and magnesium (∇) in a series of 11 montmorillonite suspensions in 0.1 M KCl after shaking for 180 days.

10.3 Methods

10.3.1 Experiments

The dissolution kinetics of K-montmorillonite were investigated in acidic solutions of constant pH. The experiments were performed in both batch and mixed-flow reactors. The temperature was kept constant at 298 K and the ionic strength was adjusted generally to 0.1 M and in some experiments to 1 M KCl.

10.3.2 *Preconditioning*

In order to remove extra-reactive material, 30 ml portions of a stock suspension ($4.7\,g.l^{-1}$) were transferred into tubes and centrifuged for 20 min at 400–600 *g*. The supernatant solutions were carefully removed by suction and replaced by solutions used in the subsequent experiments. The centrifugation tubes were vigorously shaken by hand for 1 min and then placed on a laboratory shaker. After 1 day of shaking at 120 rpm, the suspensions were again centrifuged and the same procedure was repeated 5, 10 or 15 times. Five repetitions were found to result in initially higher dissolution rates. Therefore, the standard preconditioning included 15 pretreatment procedures.

10.3.3 *Batch reactor*

Batch experiments were carried out in polyethylene (PE) flasks. The experiments were started with portions of 200 ml suspensions. For optimum mixing conditions, the flasks were placed horizontally on the laboratory shaker operating at 120 rpm. The solution pH was periodically controlled, and if necessary adjusted to the original values by addition of small amounts of HCl. Samples of 15 ml were taken after intervals of 1–3 days, centrifuged and filtered (0.01 μm).

10.3.4 *Mixed-flow reactor*

The mixed-flow reactor was developed in our department to carry out dissolution experiments under steady-state conditions. For this purpose, the particle suspensions should be well mixed and the reaction products must be separated continuously from the particles. The cylindrical-shaped reactors are made from transparent polycarbonate material (Plexiglass) and sealed with Teflon and Viton materials. Figure 10.3(a) shows the design of the reactor. The volume of the reaction cell containing the particle suspension (1) is about 60 ml. The solution flows through a curved jet at the inlet and a polysulphone filter (with a pore diameter of less than 10 nm) at the outlet. The aqueous solutions are pumped by a peristaltic pump from PE bottles (where the solutions are protected from atmospheric CO_2) through the reactor, i.e. from (2) to (3). Up to ten mixed-flow reactors can be placed on the axis of a rotation apparatus (Figure 10.3(b)), which turns the reactors at a discrete rotation frequency by 180° forward and backward, always avoiding contact between the particles and the filters at the outlet.

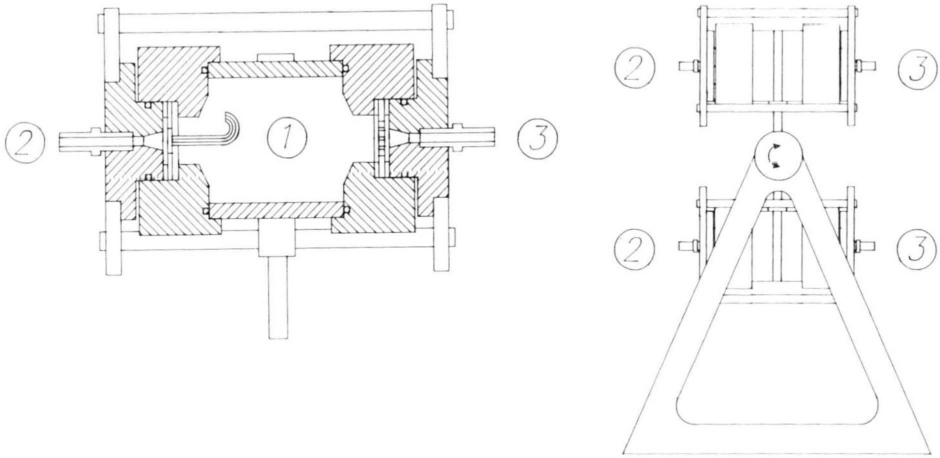

Figure 10.3 (a) Cross-section of a mixed-flow reactor. The solutions are pumped from the inlet (2) through the reaction cell containing the particle suspension (1) to the outlet (3). (b) Side view of the rotation apparatus carrying two mixed-flow reactors. The particle suspensions are mixed every few minutes by half a rotation movement forward and backward.

The chosen volumetric flow rate in our experiments was about 20 ml per day, thus the mean residence time, τ, of the solution in the reaction cell averaged 3 days. The rotation frequency was adjusted to 7 min, i.e. the reactors were turned about 600 times per τ. In order to achieve steady-state conditions more quickly, no solution was pumped through for the first τ. The sampling period varied between 3 and 10 days and steady-state conditions were reached after several weeks (see Section 10.4, Figure 10.4(b)). The samples were weighed and stored in a refrigerator until they were analysed.

10.3.5 Analyses

For the pH measurements, a combined glass electrode was used, which was calibrated with the aid of a set of solutions of known hydrogen ion concentrations. The concentrations of H_4SiO_4 and total aluminium were measured in a spectrophotometer (Perkin Elmer Lambda 3) with the molybdenum blue method (Koch and Koch, 1964) and the catechol violet method (Dougan and Wilson, 1974), respectively.

10.3.6 Titration of the montmorillonite surface

The extent of surface protonation as a function of solution pH was mesured by acid–base titrations of the stock suspension (4.7 g.l^{-1}) where the ionic strength was adjusted to 0.1 M and 1.0 M KCl, respectively. The initial pH of around 7 was lowered by 30 consecutive additions of 0.25 ml of 1.75 mM HCl. The time elapsed between two consecutive additions was 30 min except when the

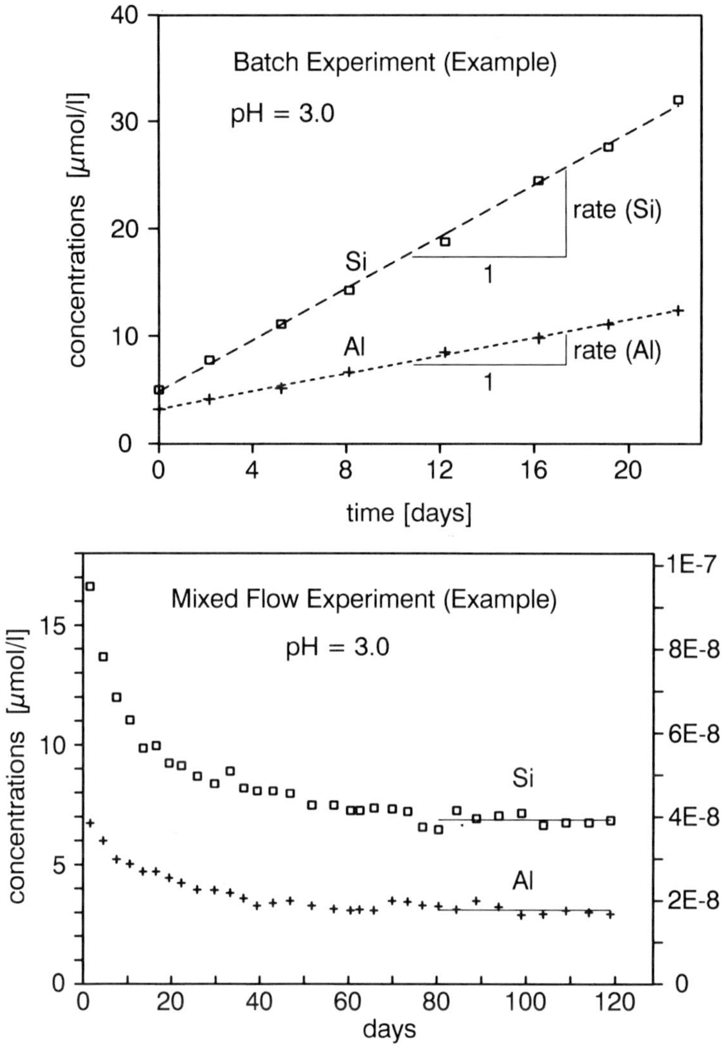

Figure 10.4 (a) After sufficient pretreatment, time-invariant dissolution rates in batch reactors could be observed. The experiments were carried out at pH 3.0 with 0.7 g.l^{-1} montmorillonite. (b) Owing to aggregation occurring in the mixed-flow reactor at the beginning of the experiments, steady-state conditions were reached only after several days or weeks. The particle concentration was 2.2 g.l^{-1}.

potential became stable within a shorter time. For every titration step, the concentration of adsorbed protons at the mineral surface, C_H^s, was calculated from

$$C_H^s = \frac{[H^+(tot)] - [H^+]}{[montmorillonite]} \tag{3}$$

where $[H^+(tot)]$ and $[H^+]$ are the total and free concentration ($mol.l^{-1}$) of H^+ and [montmorillonite] signifies the particle concentration ($g.l^{-1}$). Titrations of the stock suspensions were carried out between pH 7 and 2.8. At pH values below 2.8, the evaluation of reliable C_H^s values is impaired by the fact that $[H^+(tot)]-[H^+]$ (equation (3)) becomes a small difference of two large quantities.

10.3.7 Calculation of dissolution rates

The rates of the proton-promoted dissolution, R_H, are based on the release rate of silica (R_H(Si)) or aluminium (R_H(Al)). If dissolution occurs congruently, R_H(Si) and R_H(Al) are related by the stoichiometry of silicon and aluminium in the mineral:

$$R_H = R_H(\mathrm{Si}) = 2.59\ R_H(\mathrm{Al}). \tag{4}$$

Where appropriate, the abbreviation R_H is used instead of R_H(Si) or R_H(Al).

Figure 10.4(a) shows an example of a batch experiment. Both silicon and aluminium concentrations show a linear increase with time. The slopes $[mol.(l.h)^{-1}]$ are divided by the particle concentration ($g.l^{-1}$) in order to obtain the rates of the proton-promoted dissolution, R_H, in $[mol.(g.h)^{-1}]$. In some cases insufficient preconditioning gave rise to initially enhanced dissolution rates. Deviations from linearity were also observed at pH values below 2, where after several days a sudden increase of the dissolution rates was observed. In both cases, the evaluation of the dissolution rate, R_H, was restricted to the linear region.

The dissolution rates, R_H, of the mixed-flow experiments (Figure 10.4(b)) were calculated as follows: the average concentrations ($mol.l^{-1}$) of nine samples between 80 and 118 days were divided by the mean residence time (h) of the solution in the flow reactor and by the particle concentration ($g.l^{-1}$). The decay of the initial concentrations to the steady-state values (relaxation) might be primarily caused by physical aggregation processes (see Section 10.4.1) rather than by simple dilution: the calculated response curve for an inert tracer (see e.g. Levenspiel, 1972) indicates that initial deviation from steady state is depleted to less than 5% after 3τ (mean residence time), which equals 9 days in our experiments. Since the initial deviation at pH 3.0 (Figure 10.4(b)) is decayed after 9 days to only 50%, we may conclude that other processes than simple dilution are causing the observed relaxation behaviour.

Several authors prefer ($mol.(m^2.h)^{-1}$) as units of mineral dissolution rates. However, this is not suitable for montmorillonite, since it is not possible to measure exactly the extent of the edge surface (where dissolution reactions are assumed to occur).

10.4 Results

10.4.1 Effect of solution pH

In Figure 10.5 the dissolution rates of silicon and aluminium are plotted against the solution pH for both batch and mixed-flow experiments. Increasing dissolution rates were measured with decreasing pH. In contrast to δ-Al_2O_3 (Figure 10.5; Furrer and Stumm, 1986) and kaolinite (Wieland, 1988) no maximum of the dissolution rate was found. At pH 4 the dissolution rates in batch and mixed-flow reactors are very similar. At lower pH values the rates in the batch reactor exceed those in the mixed-flow reactor. This difference increases with decreasing pH and reaches a factor of approximately 3 at pH 1.5 and 1.0 for silicon and aluminium, respectively. It was noted that the suspended clay particles in the mixed-flow reactors were not homogeneously distributed. They rather formed a gel-like aggregate, the formation of which took several days or even weeks. The occupied volume was observed to decrease with decreasing pH.

10.4.2 Si/Al ratio

The values of the Si/Al ratios are shown in Figure 10.6. Most of the observed Si/Al ratios are close to congruency. Nevertheless it seems that the dissolution

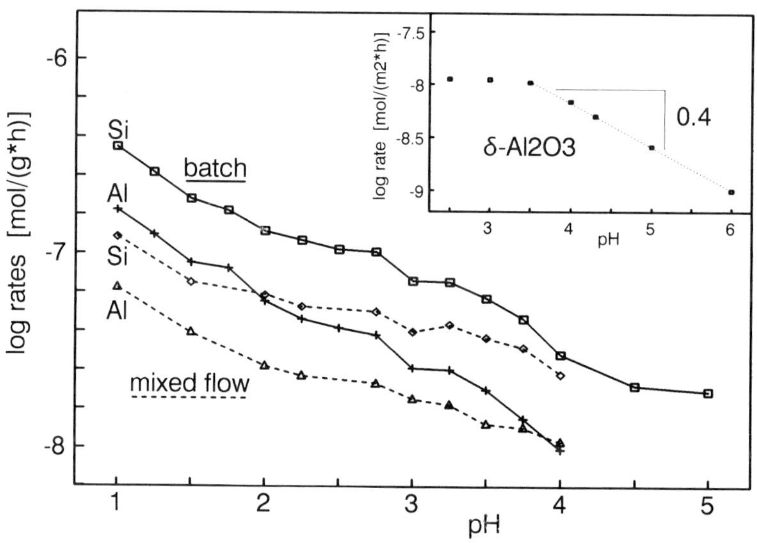

Figure 10.5 Comparison of dissolution rates. Silicon and aluminium dissolution rates, $R_H(Si)$ and $R_H(Al)$, in batch (solid lines) and mixed-flow reactors (dashed lines) are both increasing with free hydrogen ion concentration throughout the investigated pH range. The insert shows the dependence of the rate of proton-promoted dissolution, R_H, of δ-Al_2O_3 on the solution pH. The slope of the log rate/pH plot is 0.4 (Furrer and Stumm, 1986).

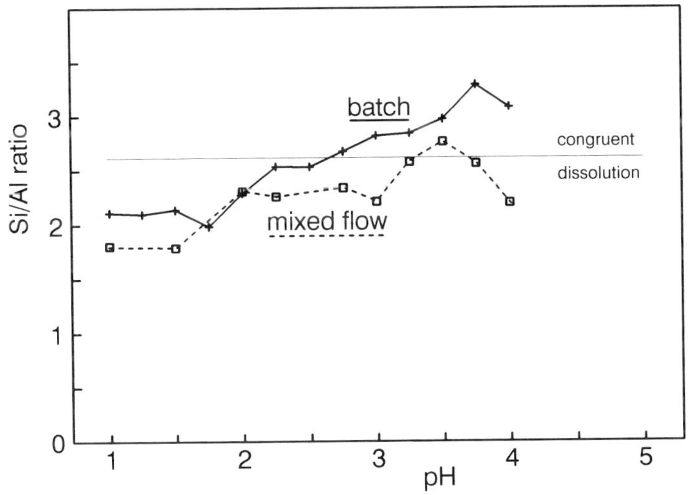

Figure 10.6 Comparison of Si/Al ratios. The stoichiometry of congruent dissolution is indicated by a straight line. The Si/Al ratios in batch (solid line) and mixed-flow experiments (dashed line) depend on the solution pH; however, they do not deviate very far from congruency. The uncertainty in the Si/Al ratios is about ± 0.2.

of silicon is favoured at high pH (batch), whereas aluminium is preferentially dissolved at low pH values (batch and mixed flow).

10.4.3 Particle concentration and ionic strength

The influence of particle concentration was investigated in the batch reactors. Two series of experiments were carried out at pH 2.5, the first in 0.1 M KCl, the second in 1.0 M KCl. The concentrations of montmorillonite in both series were 0.7, 1.4, 2.1, 2.8, and 3.5 $g.l^{-1}$ (Figure 10.7). Significantly larger rates were measured in the experiments with the higher ionic strength. The influence of the montmorillonite concentration on $R_H(Si)$ and $R_H(Al)$ was found to be of minor importance.

10.4.4 Inhibition of dissolution rates by Al(III)

The influence of dissolved Al(III) on the dissolution kinetics was studied in mixed-flow reactors by addition of $AlCl_3$ to the inflowing solutions. In the course of ten mixed-flow experiments at pH values ranging from 4 to 1, the aluminium concentrations were increased in three steps: from 0 to 10 $\mu mol.l^{-1}$ after 175 days, from 10 to 32 $\mu mol.l^{-1}$ after 227 days and from 32 to 100 $\mu mol.l^{-1}$ after 263 days. In the experiments at pH 3 and lower, $R_H(Si)$ was not affected at all by added $AlCl_3$, whereas at pH values 3.25, 3.50, 3.75 and 4.0 the inhibition became gradually more significant, reaching almost completion at pH 4.0 (Figure 10.8). The dissolution rate at pH 4 is lowered progressively with increasing aluminium concentration.

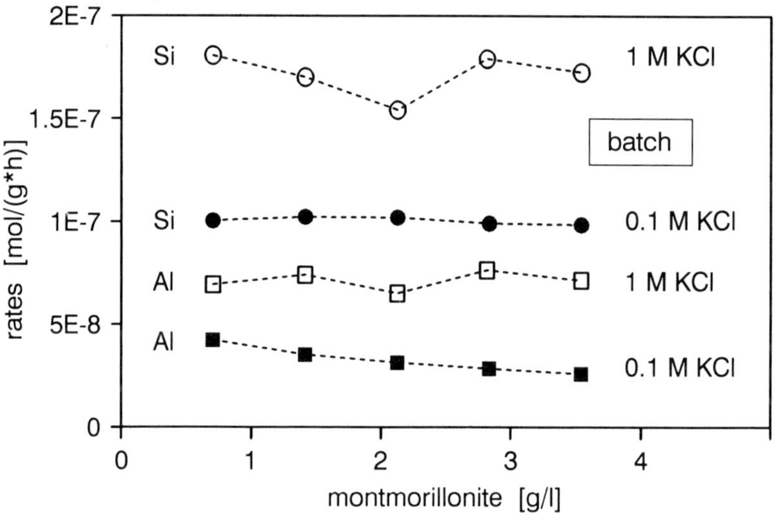

Figure 10.7 The influence of particle and KCl concentrations at pH 2.5 in batch experiments. While the effect of the KCl concentration on the dissolution rates, $R_H(Si)$ and $R_H(Al)$, is significant, the influence of the particle concentration is rather minor.

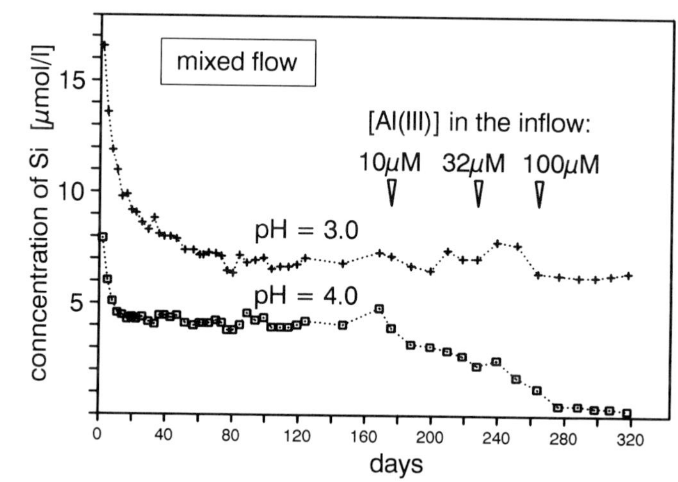

Figure 10.8 Inhibition of dissolution rates by Al(III) in mixed-flow experiments. At pH 4, Al(III) can reduce the dissolution rate, $R_H(Si)$, to almost zero, while it has virtually no effect at pH 3. The experiments were carried out in mixed-flow reactors containing 2.2 $g.l^{-1}$ montmorillonite.

10.4.5 Titration of montmorillonite

The results from the titration of K-montmorillonite in 0.1 M KCl are included in Figure 10.9. No significant differences were observed for the titration of K-montmorillonite in 1.0 M KCl.

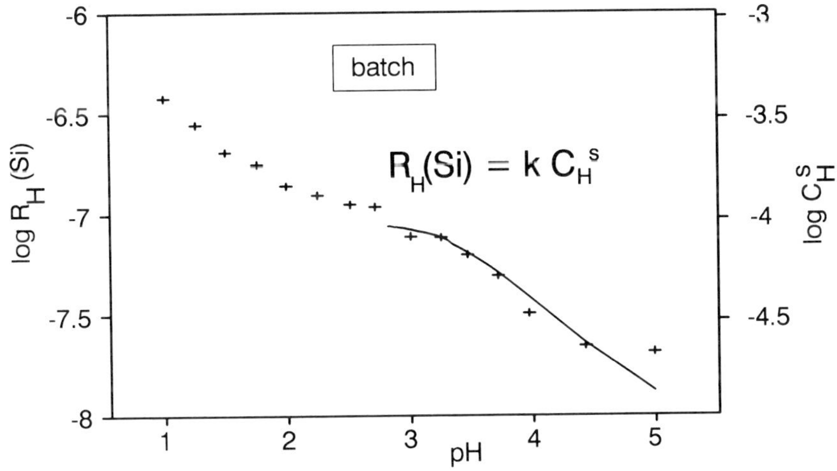

Figure 10.9 Dissolution rates and surface protonation in batch experiments. Between pH 4.5 and 3.0, the rate of proton-promoted dissolution, $R_H(Si)$, is directly proportional to the surface concentration of protons, C_H^s. The individual data points show the values of $\log R_H(Si)$ in $mol.(g.h)^{-1}$ and the solid line represents $\log C_H^s$ in mol adsorbed H^+ per g montmorillonite.

10.5 Discussion

10.5.1 Batch reactors versus mixed-flow reactors

Both batch and mixed-flow reactors were used in this study because both of them are appropriate to investigate certain phenomena occurring in soils such as particle aggregation, flow-through conditions and re-adsorption processes.

Batch reactors are suitable for the study of rather slow reactions because the reaction products are accumulated in the solution (detection limit). As discussed below, edge-to-face aggregation of clay particles is favoured especially in acidic solutions; this process can be minimized if the batch reactors are vigorously shaken. Furthermore batch reactors allow a sensitive monitoring of solution–surface interactions (e.g. re-adsorption of reaction products), in contrast to mixed-flow reactors where the solutions are displaced continuously by the inflow.

The main advantage of mixed-flow reactors is that steady-state conditions can be realized without buffers or chemostatic equipment, i.e. at steady state the chemical composition of the solutions are time invariant and determined by the flux balances:

$$\text{flux}_{\text{inflow}} + \text{flux}_{\text{reactions}} = \text{flux}_{\text{outflow}}, \tag{5}$$

where the fluxes in the outflow ($mol.s^{-1}$) are equal to the flow velocity ($l.s^{-1}$) multiplied by the concentrations ($mol.l^{-1}$) in the mixed-flow reactor. Based on continuous removal of the reaction products, the experimental window can be expanded towards the region of lower solubility (higher pH values, see Figure 10.2) without the risk of reaching saturation conditions. Furthermore, the reservoirs with the inflowing solutions are easily disconnected from the reactors and exchanged (i.e. the clay minerals can be used several times for different experiments) and the solution samples are filtered continuously at the outflow of the reactors (Figure 10.3(a)). In addition, a serial set-up of several mixed-flow reactors may permit investigation of specific problems.

10.5.2 Dissolution rates

Figure 10.5 shows that the dissolution rates are progressively larger in acidic solutions. From other studies we have ample evidence that the rates of the proton-promoted dissolution reactions of oxide minerals and feldspars are directly dependent on the surface concentration of protons (Furrer and Stumm, 1986; Wieland, 1988; Blum and Lasaga, 1988) rather than on the concentration of protons in solution. In the case of δ-Al_2O_3 (insert in Figure 10.5; Furrer and Stumm, 1986), the reaction rate is pH independent at values below 3.5, which suggests that the oxide surface is saturated with protons.

Based on the observed Si/Al ratio, which is relatively close to congruency, we assume that the dissolution of montmorillonite occurs at edge surfaces rather than at plane surfaces. The edge surfaces of montmorillonite particles contain functional groups similar to these in oxide minerals. One would expect, therefore, the same type of rate–pH relationship. In fact, at pH values between 4.5 and 3.0, the dissolution rates of montmorillonite are linearly correlated to the surface protonation which was determined from separate titration experiments (Figure 10.9). Thus, for pH values between 4.5 and 3.0, the rate law can be written as

$$R_{\mathrm{H}}(\mathrm{Si}) = k_{\mathrm{H}}\ \mathrm{C}_{\mathrm{H}}^{\mathrm{s}} \tag{6}$$

where $R_{\mathrm{H}}(\mathrm{Si})$ signifies the rate of proton-promoted dissolution (calculated from dissolved Si), k_{H} is the corresponding rate constant ($9.4 \times 10^{-4}\ \mathrm{h}^{-1}$) and $C_{\mathrm{H}}^{\mathrm{s}}$ indicates the surface concentration of protons. In other words, the dissolution of montmorillonite is a first-order reaction with respect to the surface concentrations of hydrogen ions. The same rate law was found for kaolinite (Wieland, 1988) and for albite and olivine (Blum and Lasaga, 1988). If the rate law of the proton-promoted dissolution (equation (6)) is expressed in terms of the free hydrogen ion concentration in solution, a rate equation with a fractional order, n, of $[\mathrm{H}^+]$ would result:

$$R_{\mathrm{H}} = k'_{\mathrm{H}}\ [\mathrm{H}^+]^n \tag{7}$$

where n is typically smaller than 1, indicating a non-linear dependence of the extent of surface protonation on $[\mathrm{H}^+]$ in solution. While n was reported to be

0.4 in the case of δ-Al_2O_3 (insert Figure 10.5), we found values for n of about 0.35 for batch experiments and 0.22 for mixed-flow experiments (Figure 10.5). For soils we may speculate that—due to extended particle aggregation—the pH dependence might be even less distinct than was found for the mixed-flow experiments.

The rate-limiting step in the dissolution reaction of montmorillonite involves the formation of a specific surface complex. This specific surface complex is probably located at an edge surface aluminium centre and contains one excess proton. However, at pH values below 3, where the surface protonation cannot be measured for practical reasons (see Section 10.3), the dissolution rate increases further without reaching a maximum value. In contrast to oxide minerals, K-montmorillonite particles are composed of negatively charged layers. These are held together by potassium ions, which may be exchanged for protons at low pH values (Vanselov equilibrium) leading to an increased concentration of H^+ in the near vicinity of dissolution active aluminium sites. This means that the protonation of montmorillonite is not limited by the surface density of aluminium centres as is the case for aluminium oxide. Although we propose that the formation of a protonated specific surface complex is inevitable for the dissolution of montmorillonite, we have no direct method to determine its particular structure.

10.5.3 Comparison of dissolution rates in batch versus mixed-flow reactors

The differences of dissolution rates in batch versus mixed-flow experiments (Figure 10.5) probably result from the two dissimilar mixing methods, which allow different degrees of particle aggregation. The initial dissolution rates in the mixed-flow reactors were about the same as the rates in the batch reactors. However, Figure 10.4(b) shows that the initial rates in the mixed-flow experiments decreased significantly in the first days, reaching steady-state values only after several days or weeks (depending on solution pH). This is possibly a consequence of the observed slow formation of the gel-like aggregates (see Section 10.4.1). From these parallel phenomena we conclude that the aggregation processes, which preferentially take place in the mixed-flow reactors, continue to progress during several days or weeks.

The aggregation of montmorillonite particles is favoured especially in acidic solutions: whereas the faces of this clay mineral are permanently negatively charged, the charge on the edges is variable and can be positive (in acidic solutions) or negative (in alkaline media). This means that increasing edge-to-face aggregation (Van Olphen, 1977) is expected with decreasing pH down to a pH value where the edge surface is saturated with protons. In fact, at low pH values, the dissolution rates in the mixed-flow reactors are lower compared with those in the batch reactors, indicating extended edge-to-face aggregation.

Although we present some evidence for surface-controlled dissolution reactions in the batch reactors, it is not possible to give a final interpretation on the mechanism of the reduced dissolution rate in the mixed-flow reactors. The interaction between the particles in the edge-to-face aggregation would be at best an outer-sphere type, i.e. no surface area would be blocked by chemical reactions. Thus, the concept that the dissolution becomes controlled by the diffusion of the reaction products through the gel-like aggregates cannot be excluded.

10.5.4 Si/Al ratio

In mixed-flow reactors at steady state the dissolution should occur congruently. However as shown in Figure 10.6, the Si/Al ratio diverges from congruency in both reactors, although we should emphasize the relatively large uncertainties in the experimental data (± 0.2). The apparent tendency of silicon to dissolve faster than aluminium at pH values above 2.5 in the batch reactor may actually be caused by the re-adsorption of dissolved aluminium. Below pH 2.5, aluminium might dissolve selectively from the edge surface leading to a lowered Si/Al ratio. From the amount of totally dissolved silicon and aluminium, we estimated that in the batch experiments only a fraction of one edge surface layer was depleted in aluminium.

As reported by Wieland (1988), kaolinite dissolves far from congruently ($I = 0.1$ M $NaNO_3$). Especially at pH values above 3, the Si/Al ratio exceeds congruency by a factor of up to 8. Compared with kaolinite we may conclude, despite the observed variation in Si/Al ratio, that montmorillonite dissolution occurs almost congruently.

10.5.5 Influence of particle concentration

The variation of the particle concentration, investigated in batch experiments at pH 2.5, did not show a distinctive impact on the dissolution rate (Figure 10.7). The successively decreasing aluminium dissolution rate with increasing particle concentration in 0.1 M KCl media is not understood so far and is subject to further investigations. However, this effect is undoubtedly minor compared with that of other parameters (pH, electrolyte concentration, dissolved aluminium).

10.5.6 Influence of electrolyte concentration

The dissolution rates at pH 2.5 are almost doubled when the KCl concentration is increased from 0.1 to 1.0 M (Figure 10.7). The same effect has been observed even more distinctly by Charlet *et al.* (1992) in the case of δ-Al_2O_3 on comparing the dissolution rates at pH 3.0 in 0.1 M and 0.01 M $NaNO_3$. However, the opposite effect for the dissolution kinetics of labradorite feldspar at pH 2 was reported by Sjöberg (1989). From the postulated mechanism for the proton-promoted dissolution one would expect that an increased dissolution rate must be based on a larger surface concentration of protons. Since no effect of electrolyte concentration on the surface proton concentration was observed in the titration experiments, the question arises whether a high ionic strength may cause, indirectly, an improved mobility of protons (due to modification of the electric double layer), which probably would result in higher dissolution rates. In order to obtain a better understanding of this effect, the influence of electrolyte concentration is subject to further investigations in our laboratories.

10.5.7 Inhibition of dissolution by aluminium

The strong inhibition effect observed in the mixed-flow experiments, mainly at pH 4, is certainly caused by the adsorption of Al(III) at the montmorillonite surface. Similar inhibition effects were reported by Chou and Wollast (1985) and by Wehrli *et al.* (1990): the dissolution rates of albite at pH 3 were reduced by about 90% when approximately 1 $mmol.l^{-1}$ Al(III) was added to the reaction medium, and the dissolution rate of δ–Al_2O_3 was reduced by about one third when as little as 20 $\mu mol.l^{-1}$ VO^{2+} was added to the solution. Wehrli *et al.* (1990) estimated from these results that the fraction of active to total sites is typically only 0.1% or less. In the case of montmorillonite the observed inhibition effect might be caused not only by the adsorption of Al(III) at dissolution active sites but also by additional aggregation, which itself could be enhanced by high aluminium concentrations.

10.6 Conclusions

The dissolution kinetics of montmorillonite and most other minerals are controlled by specific surface complexes. It is evident that the dissolution of montmorillonite occurs at edge surfaces since the Si/Al ratio remains relatively close to congruency.

The strongest effect on the dissolution kinetics was observed from the solution pH. Increasing the ionic strength from 0.1 M to 1 M KCl leads to a doubling of the dissolution rates, whereas the particle concentration had no such remarkable influence on the rates. Furthermore the adsorption of aluminium, favoured at pH values above 3, can lead to an almost complete inhibition of the dissolution process.

The reported pH effect is counterbalanced partially if the clay particles are allowed to aggregate. The observed aggregation in the mixed-flow reactors may lead to diffusion-controlled dissolution, where transport of the reaction products through the gel-like aggregates becomes rate limiting.

Both inhibition processes, adsorption of aluminium at the particle surfaces and edge-to-face aggregation are probably common phenomena in soils.

The mixed-flow reactors, which were developed for this study, have been successfully used to investigate the kinetics of heterogeneous reactions under steady-state conditions.

Acknowledgements

Bernhard Trusch is gratefully acknowledged for his assistance in the performance of the mixed-flow experiments. The authors extend special thanks to Ulrich Kindler and Peter Schuetz who manufactured the flow reactors and the rotation apparatus. Markus Stadler provided the equipment for the titration experiments. Gabriela Frei contributed to the measurements of the CEC. Rudolf Giovanoli, Emma Ettinger and Beatrice Wild are gratefully acknowledged for the X-ray analysis. We thank two anonymous reviewers for several helpful suggestions.

Our research on mobilization, speciation and immobilization of aluminium in soil is supported by the Swiss National Science Foundation (Project 2–5636).

References

Aagaard, P. and Helgeson, H. C. (1982) Thermodynamic and kinetic constraints on reaction rates among minerals and aqueous solutions. *American Journal of Science*, **282**, 237–85.

Blum, A. E. and Lasaga, A. C. (1988) Role of surface speciation in the dissolution of minerals. *Nature*, **331**, 431–3.

Busenberg, E. and Clemency, C. V. (1976) The dissolution kinetics of feldspars at 25°C and 1 atm CO_2 partial pressure. *Geochimica et Cosmochimica Acta*, **40**, 41–9.

Charlet, L., Dise, N. and Stumm, W. (1992) Sulfate adsorption on a variable charge soil and on aluminium oxide. *Agriculture Ecosystems and the Environment* (accepted).

Chou, L. and Wollast, R. (1985) Steady-state kinetics and dissolution mechanisms of albite. *American Journal of Science*, **285**, 963–93.

Day, P. R. (1965) Particle fractionation and particle-size analyses, in *Methods of Soil Analyses*, (ed. C. A. Black, American Society of Agronomy, Madison, Wisconsin, pp. 545–67.

Dougan, W. K. and Wilson, A. L. (1974) The adsorptiometric determination of aluminium in water. A comparison of some chromogenic reagents and the development of an improved method. *Analyst*, **99**, 413–30.

Dzombach, D. A. and Morel, F. M. M. (1990) *Surface Complexation Modeling*, Wiley-Interscience, New York.

Eeckman, J. P. and Laudelout, H. (1961) Chemical stability of hydrogen-montmorillonite suspensions. *Kolloid-Zeitschrift*, **178**, 99–107.

Furrer, G. and Stumm, W. (1986) The coordination chemistry of weathering: I. Dissolution kinetics of δ-Al_2O_3 and BeO. *Geochimica et Cosmochimica Acta*, **50**, 1847–60.

Galicia, H. (1989) *Adsorption of Cadmium and Copper by Forest Humic Acids and (K, H)-Montmorillonite*, PhD Thesis, University of Bern, Switzerland.

Garrels, R. M. and Christ, C. L. (1965) *Solutions, Minerals and Equilibria*, Harper and Row, New York.

Hendershot, W. H. and Duquette, M. (1986) A simple barium chloride method for determining cation exchange capacity and exchangeable cations. *Soil Science Society of America Journal*, **50**, 605–8.

Holdren, G. R. and Berner, R. A. (1979) Mechanisms of feldspar weathering. I. Experimental studies. *Geochimica et Cosmochimica Acta*, **43**, 1161–71.

Kamil, J. and Shainberg, I. (1968) Hydrolysis of sodium montmorillonite in sodium chloride suspensions. *Soil Science*, **106**, 193–9.

Kittrick, J. A. (1971) Stability of montmorillonites: I. Belle Fourche and Clay Spur montmorillonites. *Soil Science Society of America, Proceedings*, **35**, 140–5.

Kline, W. E. and Fogler, H. S. (1981) Dissolution kinetics: Catalysis by strong acids. *Journal of Colloid and Interface Science*, **82**, 93–102.

Koch, O. G. and Koch, G. A. (1964) *Handbuch der Spurenanalyse*, Springer Verlag, Berlin.

Levenspiel, O. (1972) *Chemical Reaction Engineering*, Wiley International, New York.

Mast, M. A. and Drever, J. I. (1987) The effects of oxalate on the dissolution rates of oligoclase and tremolite. *Geochimica et Cosmochimica Acta*, **51**, 2559–68.

Mattigod, V. and Sposito, G. (1978) Improved method for estimating the standard free energies of formation ($\Delta G^\circ_{f,298.15}$) of smectites. *Geochimica et Cosmochimica Acta*, **42**, 1753–62.

May, H. M., Kinniburgh, D. G., Helmke, P. A. and Jackson, M. L. (1986) Aqueous dissolution, solubilities and thermodynamic stabilities of common aluminosilicate clay minerals: Kaolinite and smectites. *Geochimica et Cosmochimica Acta*, **50**, 1667–77.

Posey-Dowty, J., Crerar, D., Helmann, R. and Chang, C. D. (1986) Kinetics of mineral–water reactions: theory, design and application of circulating hydrothermal equipment. *American Mineralogist*, **71**, 85–94.

Pulfer, K., Schindler, P. W., Westall, J. C. and Grauer, R. (1984) Kinetics and mechanisms of dissolution of bayerite (γ-$Al(OH)_3$) in HNO_3–HF solutions at 298.2 K. *Journal of Colloid and Interface Science*, **101**, 554–64.

Rimstidt, J. D. and Dove, P. M. (1986) Mineral/solution reaction rates in a mixed flow reactor: Wollastonite hydrolysis. *Geochimica et Cosmochimica Acta*, **50**, 2509–16.

Schindler, P. W. and Stumm, W. (1987) The surface chemistry of oxides, hydroxides and oxide minerals, in *Aquatic Surface Chemistry*, (ed. W. Stumm), Wiley-Interscience, New York, pp. 83–110.

Sillén, L. G. (1961) The physical chemistry of sea water, in *Oceanography* (ed. M. Sears), American Association for Advancement of Science, Washington, DC, pp. 549–81.

Sillén, L. G. (1967) Gibbs phase rule and marine sediments, in *Advances in Chemistry Series*, No. 67, *Equilibrium Concepts in Natural Water Systems*, American Chemical Society, Washington, DC, pp. 57–69.

Sjöberg, L. (1989) Kinetics and non-stoichiometry of labradorite dissolution, in *Water–Rock Interaction, WRI-6*, (ed. D. L. Miles), Balkema, Rotterdam, pp. 639–42.

Stumm, W. and Morgan, J. J. (1981) *Aquatic Chemistry*, 2nd edn, Wiley-Interscience, New York.

Van Olphen, H. (1977) *An Introduction to Clay Colloid Chemistry*, 2nd edn, Wiley Interscience, New York.

Wehrli, B., Wieland, E. and Furrer, G. (1990) Chemical mechanisms in the dissolution kinetics of minerals, the aspect of active sites. *Aquatic Sciences*, **52**, 3–31.

Westall, J. C. (1986) *MICROQL Version 2 for PCs, Report 86–02*, Oregon State University, Corvallis.

Wieland, E. (1988) *Die Verwitterung Schwerlöslicher Mineralien—ein Koordinationschemischer Ansatz zur Auflösungskinetik*, Dissertation ETH No. 8532, Eidgenössische Technische Hochschule, Zürich.

Wollast, R. (1967) Kinetics on the alteration of K-feldspars in buffered solutions at low temperature. *Geochimica et Cosmochimica Acta*, **31**, 635–48.

Zinder, B., Furrer, G. and Stumm, W. (1986) The coordination chemistry of weathering: II. Dissolution of Fe(III) oxides. *Geochimica et Cosmochimica Acta*, **50**, 1861–9.

PART FOUR
Applications in Civil and Petroleum Engineering

CHAPTER ELEVEN

Mechanisms of overpressuring: an overview

P. L. Hall

11.1 Introduction

During the evolution of sedimentary basins, gravitational and tectonic forces cause compaction and deformation to occur. Simultaneously, pore fluids flow from regions of higher chemical potential to regions of lower potential under the influence of hydraulic, osmotic and capillary forces. Fluid pressures are also modified by thermal effects and diagenesis.

When the fluid migration processes necessary to restore a state of equilibrium are hindered, whether by low permeabilities, or by the structure of traps and fault blocks, accumulations of brine or hydrocarbons can exist whose pressures may be higher or lower than hydrostatic. The hydrostatic pressure at any depth z below the surface for a single fluid of constant density ρ is given by the equation $P = \rho g z$, where g is the gravitational acceleration. If there are zones containing fluids of different density (e.g. gas, oil and water) or if the density of a single fluid phase varies significantly with depth, a more general expression for the hydrostatic pressure will be required. Nevertheless, provided the location and densities of the fluid zones are known, the hydrostatic pressure at any given depth may be calculated. It is, of course, the pressure which would be expected to exist within the pores of a fluid-saturated, permeable soil or rock with good drainage pathways to the surface.

The terms 'overpressuring' and 'underpressuring' refer to pore fluid pressures greater or smaller than hydrostatic, respectively. Such pressures are often collectively referred to as 'abnormal formation pressures'. This may be regarded as something of a misnomer, since they are commonly encountered world-wide in deep boreholes, especially those drilled for hydrocarbon exploration and production. The persistence of abnormal formation pressures over geological timescales implies the existence of either impermeable seals, or thick strata of finite but low permeability. These act as barriers which prevent or hinder the equilibration of abnormal formation pressures with neighbouring zones of hydrostatic pressure.

Geochemistry of Clay-Pore Fluid Interactions. Edited by D. A. C. Manning, P. L. and C. R. Hughes. Published in 1993 by Chapman & Hall, London. ISBN 0 412 48980 5

Underpressuring is rather less common than overpressuring. It is often associated with groundwater removal or hydrocarbon reservoir depletion, although in some instances it may be associated with uplift and/or erosion and cooling of impermeable strata (Russell, 1972; Fertl, 1976; Dickey and Cox, 1977).

Overpressuring is of major importance to the oil industry, both for economic and safety reasons. Specifically, accurate monitoring and control of elevated fluid pressures is crucial for optimizing the efficiency of hydrocarbon drilling operations, and also for avoiding various potential pressure-related problems. The latter include influxes of brine or gas, which if uncontrolled can result in a spectacular but catastrophic 'blow-out'. Other associated problems include various manifestations of borehole instability such as shale sloughing and spalling, borehole collapse, sticking of the drillpipe, or inadvertent fracturing of the borehole surface. In all cases, problem avoidance or minimization requires the establishment and maintenance of a small positive pressure differential between the drilling fluid and the formation fluids. Fertl (1976), Whittaker (1985) and Mouchet and Mitchell (1989) give general descriptions of various methods for detecting and quantifying elevated fluid pressures in rocks of low permeability.

In overpressured zones, pore pressures may range from values slightly above hydrostatic up to pressures equivalent to the weight of the overlying rock column. The latter is referred to as the overburden or lithostatic pressure.* Since sedimentary rocks have densities which are typically 2–3 times that of water, pore pressures of more than twice hydrostatic values may be encountered. In order to control such pressures, drilling fluids must be 'weighted' by the addition of high-density solids such as baryte ($BaSO_4$) or haematite. Pore pressures higher than the overburden pressure cannot be maintained, and are relieved by natural hydraulic fracturing of the rock. This occurs because rocks have comparatively small tensile strengths (Hubbert and Willis, 1957, Jaeger and Cook, 1979). When the pore pressure is close to the overburden pressure, a significant reduction occurs in the friction between adjacent rock masses, which facilitates extensive thrust or growth fault development (Hubbert and Rubey, 1959; Bruce, 1973).

While the major hazards and economic benefits associated with overpressures are encountered in permeable strata such as sandstone or carbonate reservoirs, nevertheless the variation of fluid pressure with depth in any given basin depends on its detailed stratigraphy, especially the patterns of faulting and the variation of permeability with depth. In particular, the evolution of fine-grained argillaceous sediments during compaction and burial diagenesis is of key importance to the generation and retention of overpressures. The release of overpressured brines from argillaceous rocks has been proposed as a

*More correctly, the overburden or lithostatic stress, since the state of stress in sedimentary rocks is usually anisotropic, sometimes to a considerable extent.

possible driving force assisting the migration of hydrocarbons from organic-rich and impermeable source rocks to reservoir rocks (Powers, 1967; Burst, 1969). Elevated fluid pressures also preserve porosity at large depths in both shales and reservoir rocks. Overpressuring phenomena are, therefore, quite intimately associated with the creation of economically viable hydrocarbon reservoirs in the deep subsurface.

This article reviews the principal mechanisms for the generation of overpressures in sedimentary basins. It attempts to establish that, despite the existence of numerous pressuring mechanisms – often more than one may operate simultaneously – nevertheless a relatively small number of types of overpressured environment may be identified (see Table 11.1). These environments are defined by their broad geological setting (e.g. lithology, age, burial history, and the presence or absence of discrete seals). Within each environment, it may be possible, given enough information, to infer the major mechanisms contributing to the observed fluid pressures.

This review focusses essentially on those mechanisms associated with the evolution of stress, temperature and chemical (diagenetic) alteration processes during the development of the sedimentary basin. I shall not consider any examples of abnormal fluid pressure originating in engineered environments (e.g. osmotic pressures generated by contact between drilling fluids or cement filtrates and exposed rocks). Similarly, abnormal formation pressures associated with specific structural features (e.g. artesian systems, dipping sealed lenticular systems or capillary pressures associated with oil/water contacts) will not be discussed here. A number of examples of this type are discussed by Fertl (1976). An earlier general review of abnormal formation pressures has been published by Gretener and Feng (1985). A number of the classic papers on the subject published in the Bulletin of the American Association of Petroleum Geologists have been collected together as a reprint volume edited by Vockroth (1974).

In this article I follow the practice of the petroleum engineering literature insofar as that all clay-rich sedimentary rocks which have not yet undergone metamorphism will be referred to as shales.* For detailed discussions of the sedimentology, nomenclature, classification and typical compositions of argillaceous sediments and shales, see Shaw and Weaver (1965), Underwood (1967), Morgernstern and Eigenbrod (1974) and Potter *et al.* (1980). See also Dean *et al.* (1985) for a classification scheme of relatively organic-rich deep-sea sediments. Comprehensive accounts of shale compaction and fluid migration in sedimentary basins are given in the books by Rieke and Chilingarian (1974), Magara (1978) and Chapman (1981). Neuzil (1986) has extensively reviewed a variety of hydraulic and non-hydraulic flow processes occurring in low-permeability systems.

*As opposed to the usual geological term 'mudrocks', which may be subdivided into shales and mudstones according to whether or not they are fissile, with further grain-size-related subdivisions (e.g. into claystones, clay shales, siltstones, silty shales, etc.).

A further terminology adopted from the petroleum engineering literature will be defined here. The term 'pore pressure gradient' refers to the total fluid pressure divided by the total vertical depth below the piezometric surface, expressed in $kPa.m^{-1}$, or in the American literature usually as $psi.ft^{-1}$. Hydrostatic pressure gradients for water-bearing rocks are approximately $10.5\,kPa.m^{-1}$ or $0.465\,psi.ft^{-1}$, varying somewhat with the salinity of the pore-filling brine. If the pore pressure is everywhere hydrostatic, then the 'local' pore pressure gradient at any depth will not vary from the overall or cumulative pore pressure gradient. However, in wells containing abnormal formation pressures, this will not generally be true. In the latter case, it is the total pressure divided by depth, and not the local rate of change of pressure with depth which is referred to. The former quantity determines the minimum drilling fluid density required to maintain a borehole pressure which is equal to, or greater than, the pressure of the pore fluids at all depths within the open-hole section of the well.

The major mechanisms of generation of excess fluid pressures which will be discussed here may be conveniently subdivided into (1) mechanical and (2) chemical/diagenetic. In the first case, the pressures arise from gravitational loading due to sedimentation and burial, and/or lateral loading due to tectonic or other factors. Chemical and diagenetic mechanisms include:

1. Smectite-to-illite diagenesis.
2. Biogenic methane generation at relatively shallow depths.
3. Maturation of kerogen producing oil and gas at rather greater depths and temperatures.
4. Osmotic pressures.
5. In evaporite environments (e.g. the Permian Zechstein) the dehydration of gypsum to anhydrite.

In addition to the above mechanisms thermal contributions to elevated pore pressures can arise in sealed or isolated rock masses, since their pore fluids have significantly greater coefficients of thermal expansion than the minerals which constitute the rock framework. Where the fluids are confined, this potential relative thermal expansion is restricted, and elevated pressures will be generated.

A variety of large-scale structural or stratigraphic features such as the pattern of growth faulting and consequent isolation of sands in sand/shale sequences will determine which permeable zones can contain elevated fluid pressures. A number of these are illustrated in Figures 11.1 and 11.2.

The most widely studied area, due to the large number of wells which have been drilled, is the United States Gulf Coast. Here, the onset of overpressuring generally lies within the oil maturation window, but is also generally associated with the top of massive deltaic shale formations with relatively few interbedded sands (see Figure 11.3). For further details, see Dickinson (1953), Dickey *et al.* (1968), Harkins and Baugher (1969), Bruce (1973, 1984) and Galloway *et al.* (1982).

(a)

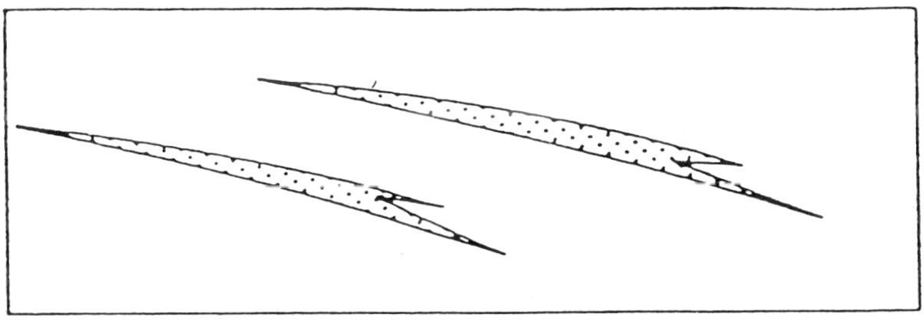

Small reservoir sealed by Pinchout

(b)

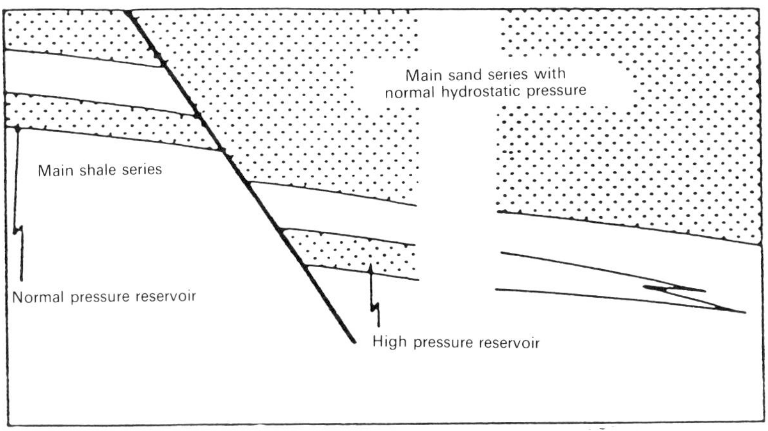

Large reservoir sealed updip by faulting down against thick shale series, sealed downdip by regional facies change

(c)

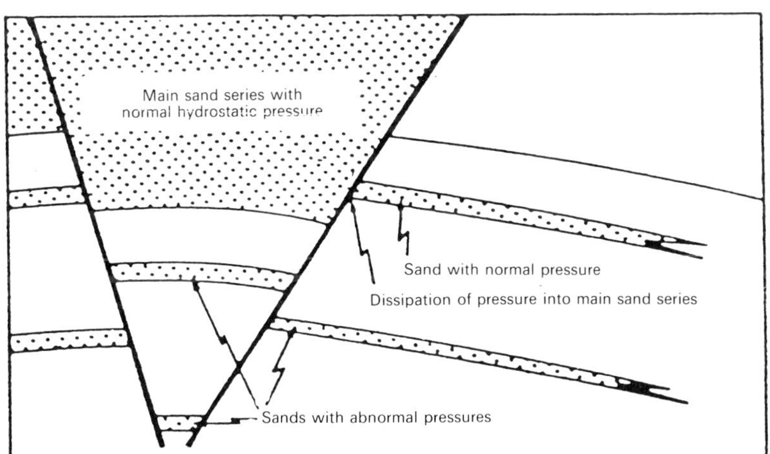

Relative position of fault seals in upthrown and down-thrown blocks

Figure 11.1 Types of reservoir seal associated with the preservation of overpressures (Dickinson, 1953).

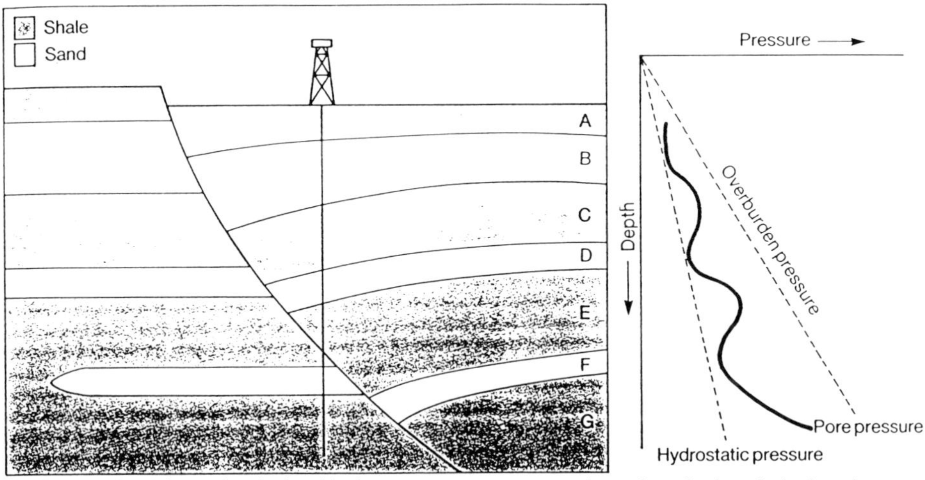

Figure 11.2 Schematic relationship between pore pressure and stratigraphy in a faulted environment (Hall *et al.*, 1988).

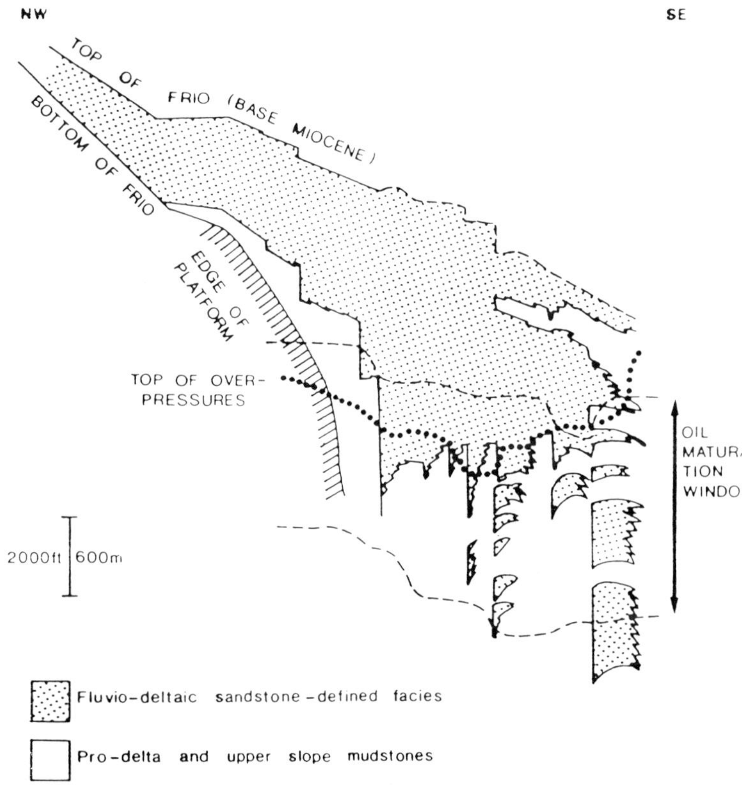

Figure 11.3 Schematic section through the Frio Formation of the western US Gulf Coast, showing the relationship between lithology, location of the oil maturation window and the depth of onset of abnormal pressures (Galloway *et al.*, 1982).

The next section contains a brief summary of some key experimental and theoretical aspects of the mechanical behaviour of compacting clays. This is followed by a more detailed discussion of mechanical origins of overpressuring. Most emphasis is placed on the relatively common phenomenon of undercompaction (compaction disequilibrium) which is encountered frequently in thick, rapidly buried shales with few interbedded permeable formations. Some cases of elevated pressure attributable to lateral compression are also discussed. Subsequent sections discuss the possible role of osmotic pressures due to ionic concentration gradients in overpressure generation, as well as the influence of a variety of organic and inorganic diagenetic processes. These include hydrocarbon generation, and dehydration reactions such as the smectite-to-illite conversion and the gypsum-to-anhydrite reaction. The last section discusses the relative importance of the various possible mechanisms of overpressuring in different geological settings, specifically sealed and non-sealed (compacting) environments.

11.2 Mechanics of clay compaction: experimental and theoretical aspects

The behaviour of clays or clay soils when subjected to elevated pressures in the laboratory has been studied extensively in the field of soil mechanics (Rieke and Chilingarian, 1974; Atkinson and Bransby, 1978). A brief outline is given here of some basic mechanical principles which will be drawn upon in subsequent sections. The influence of chemical factors (e.g. electrolyte concentrations) on the compaction state of clays will be discussed later.

When stresses are applied to a fully fluid-saturated, porous soil or rock sample, the load will be partitioned between the solid framework of the material and the pore fluids within the framework. This results in the development of grain-to-grain stresses, which increase on the sample compacts and expels pore fluid. Elevated pore fluid pressures, which tend to oppose compaction and reduce the grain-to-grain stresses. The initial pore pressure induced by the applied stresses may have any value between ambient and the mean total applied stress, depending on the load-bearing capacity of the sample. If pore fluid is allowed to drain from the sample, the pore pressure will be gradually reduced until it is in equilibrium with its external environment. During this process the sample compacts and the intergranular contacts increase in number and/or area, and are hence able to bear an increasing fraction of the total applied stresses. When the excess pore pressure is completely dissipated, no further compaction or fluid flow occurs. The sample is then said to have reached a state of 'compaction equilibrium' where its internal structure is able to support the external stresses applied to it.

Suppose that a water-saturated clay or soil is subjected to principal stresses σ_i, where $i=1$, 2 or 3 labels the three orthogonal principal axes of the stress field. The sample is assumed to have an internal pore fluid pressure u. Here compressive stresses are assigned a positive sign.

The 'effective stress principle' (Terzaghi and Peck, 1967) states that the deformation or degree of compaction which the specimen will attain at equilibrium will be a function of the reduced or effective stresses

$$\sigma_i' = \sigma_i - u. \quad (1)$$

This principle is empirical but based on extensive experimental evidence. It implies that the stresses which drive the compaction and deformation of clays and soils are not the total applied stresses, but the differences between these stresses and the internal pore pressure. The empirical Terzaghi effective stress principle should not be confused with a similar result which arises in the linear, reversible theory of elastic deformation of saturated porous media, usually referred to as poro-elasticity (Biot, 1940; Nur and Byerlee, 1971; Rice and Cleary, 1976):

$$\sigma_i' = \sigma_i - \alpha u, \quad (2)$$

where α, known as the Biot constant, is given by

$$\alpha = 1 - M_d/M_g, \quad (3)$$

M_d and M_g being the bulk moduli of elasticity of the drained soil or rock framework and of the constituent grains or particles, respectively, For a relatively soft soil- or clay-like material $M_g \gg M_d$, so that $\alpha = 1$ to a good approximation. This is not true, however, for stiffer or grain-cemented rocks. Nevertheless, it should be stressed that soils and sediments show non-elastic behaviour in which both pore pressures (in undrained tests) and void ratios may be influenced by both mean and deviatoric stresses (Skempton, 1954; Atkinson and Bransby, 1978).

Strength and brittle failure properties of porous, fluid-saturated rocks are also largely controlled by effective rather than total stresses (Jaeger and Cook, 1979). In particular, if $u > \sigma_m + T$, where σ_m is the minimum principal stress and T is the tensile strength of the rock, spontaneous tensile hydraulic fracturing will occur in a plane perpendicular to the minimum stress. If the state of stress on the rock is isotropic ($\sigma_1 = \sigma_2 = \sigma_3 = \sigma$), the maximum possible value of the pore pressure is approximately equal to σ since T is relatively small.

In a soil or rock mechanics failure test, it is usual to apply an isotropic confining stress σ_3, and then increase one of the principal stresses (e.g. the vertical stress) to a succession of higher values σ_1 until the sample fails. In such a test, the state of the material is defined by the mean effective stress p', the deviatoric stress q' and the specific volume v, where

$$\begin{aligned} p' &= (\sigma_1' + 2\sigma_3')/3 \\ &= (\sigma_1 + 2\sigma_3)/3 - u \end{aligned} \quad (4)$$

and

$$q' = \sigma_1' - \sigma_3'$$
$$= \sigma_1 - \sigma_3. \tag{5}$$

The specific volume v is defined as the ratio of the total volume of the sample to the volume of solid material alone. It is related to the void ratio e (the ratio of fluid to solid volumes) and the porosity ϕ (the ratio of fluid to total volumes) by the expression

$$v = 1 + e = \frac{1}{1-\phi}. \tag{6}$$

Figure 11.4 illustrates data from drained compaction of a kaolinite clay, the arrows indicating the sequence of stresses applied. These data are from a one-dimensional consolidation test, in which the sample is laterally constrained,

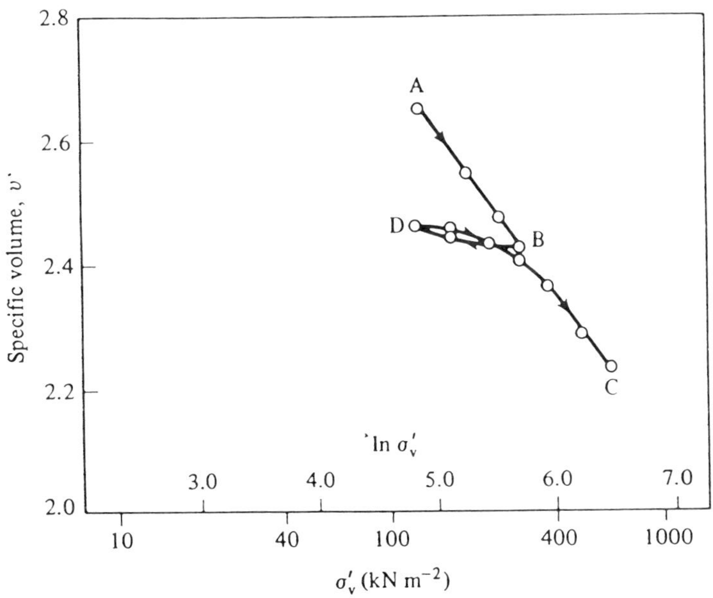

Figure 11.4 One-dimensional consolidation behaviour of a kaolin clay, showing the variation of specific volume v with the vertical effective stress σ_v'. True compaction only occurs on the paths AB and CD where the previous maximum effective stress is exceeded. Unloading below the maximum stress does not reverse the compaction, but generates a smaller 'elastic rebound' (BD) which is essentially reversible up to the previous maximum stress (DB). A similar behaviour might be expected for relatively non-lithified argillaceous sediments subjected to variations of effective stress arising from changes of sedimentary or tectonic loading or changes of pore pressure (Atkinson and Bransby, 1978).

and the lateral stress will be determined by the internal friction angle of the material (Atkinson and Bransby, 1978; Jaeger and Cook, 1979). For a soil-like material the stress field will be approximately isotropic. It can be seen that the initial compaction from A to B follows a steeper path than the unloading from B to D. The unloading path is essentially re-traversed until the maximum previous stress (corresponding to the point B) is attained, after which the specimen follows a second steeper path essentially parallel to AB. The line AC corresponds to the normal consolidation line. If at any point along this line the vertical effective stress is reduced, a shallower 'swelling' line is followed, and the sample is said to be over-consolidated. The reversible swelling and re-compaction lines are assumed to correspond approximately to elastic deformation. The irreversibility of the normal consolidation line indicates that compaction is a non-elastic and largely non-recoverable phenomenon. A relatively non-lithified argillaceous sediment which experiences sequential increases and decreases in mean effective stress, for whatever reason (e.g. burial followed by uplift or erosion, changes in tectonic loading, or changes in internal pore pressure) is likely to respond in this manner.

Normal consolidation and swelling lines are generally represented by an equation of the form

$$v = N - \lambda \ln p', \tag{7}$$

where N and λ are material constants. This implies that the porosity of a normally consolidated (normally compacted) sediment is a function of the mean effective stress p'.

It can also be shown that the characteristic time required for consolidation is given by

$$t \sim \frac{H^2 \eta \beta}{K}, \tag{8}$$

where H is the thickness of the consolidating layer, η is the viscosity of the fluid and K is the intrinsic permeability, and

$$\beta = \frac{\lambda}{v p'} \tag{9}$$

is the (non-elastic) compressibility of the compacting clay (Atkinson and Bransby, 1978). The intrinsic permeability is defined by D'Arcy's law (Chapman, 1981):

$$Q = \frac{KA}{\eta} \frac{\Delta P}{l}, \tag{10}$$

where Q is the volume of fluid per second flowing past a cross-sectional area A with a pressure difference ΔP acting over a distance l. K has the dimensions of an area, and is expressed in darcies, 1 darcy being $10^{-12}\,m^2$. It is a function only of the geometry and pore sizes of the porous medium. Some workers prefer to use the hydraulic conductivity, or coefficient of permeability κ. The hydraulic conductivity and intrinsic permeability are related by the equation

$$\kappa = \frac{K\rho g}{\eta}, \tag{11}$$

where ρ is the density of the fluid and g is the gravitational acceleration (Chapman, 1981). The hydraulic conductivity depends not only on the structure of the porous medium, but also on the nature of the fluid and the temperature (through the viscosity η).

The failure of a sediment subjected to large deviatoric or shear stresses is defined by a line known as the 'critical state line' (Atkinson and Bransby, 1978) (see Figure 11.5). The projection of the critical state line on to the v, p' plane lies parallel to and below the normal consolidation line. The behaviour of sediments subjected to increasing shear stresses will depend on whether they

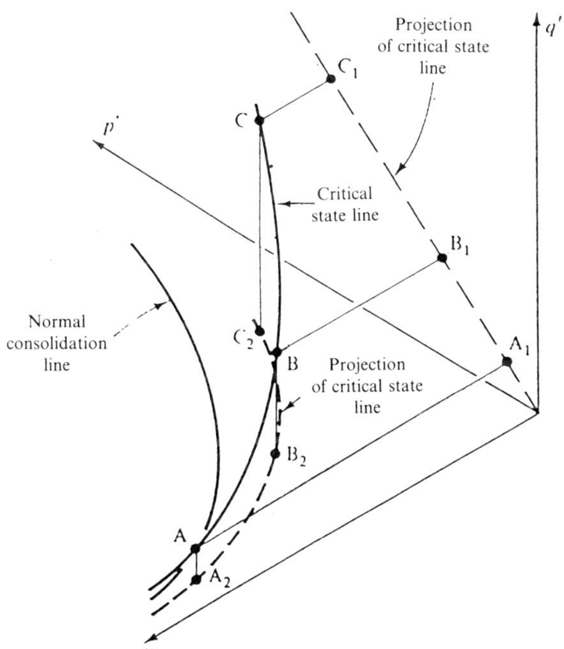

Figure 11.5 Relationship between normal consolidation and critical state lines (Atkinson and Bransby, 1978).

are normally consolidated or over-consolidated. A normally consolidated material will compact further under the action of shear stresses until the critical state line is reached. An over-consolidated material will either compact or dilate as its stress state changes from a point on the swelling line to a point on the critical state line. Whether compaction or dilation occurs depends on the degree of overconsolidation which existed prior to shearing (see Figure 11.6).

11.3 Under-compaction and other mechanical mechanisms

11.3.1 Influence of gravitational or sedimentary loading on fluid pressure

With increasing depth of burial and hence gravitational loading, argillaceous sediments expel pore fluid and become compacted during their evolution from muds via shales to metamorphic rocks (Weaver and Beck, 1971; Weaver, 1989). Compaction trends received early study in oil-containing sedimentary environments in the USA and elsewhere (Athy, 1930; Hedberg, 1936). The trend of porosity with depth estimated in ten different sedimentary basins is illustrated in Figure 11.7. In addition to total porosity, mean pore diameters also decrease markedly with depth (Borst, 1982). Recent studies of shaly rocks using small-angle scattering techniques have shown that the pore surfaces in these materials exhibit self-similar (fractal) roughness over a fairly wide range of length scales down to at least a few nanometres (see Chapter 1).

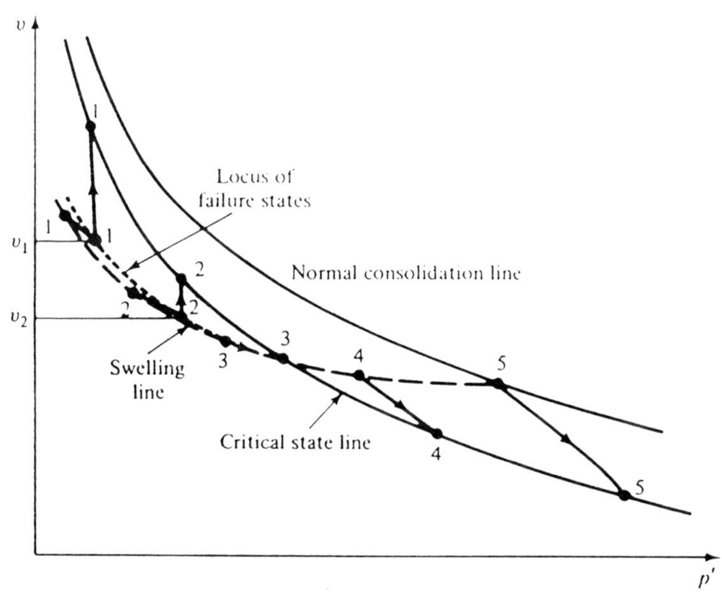

Figure 11.6 Failure pathways for normally consolidated and over-consolidated clays (Atkinson and Bransby, 1978).

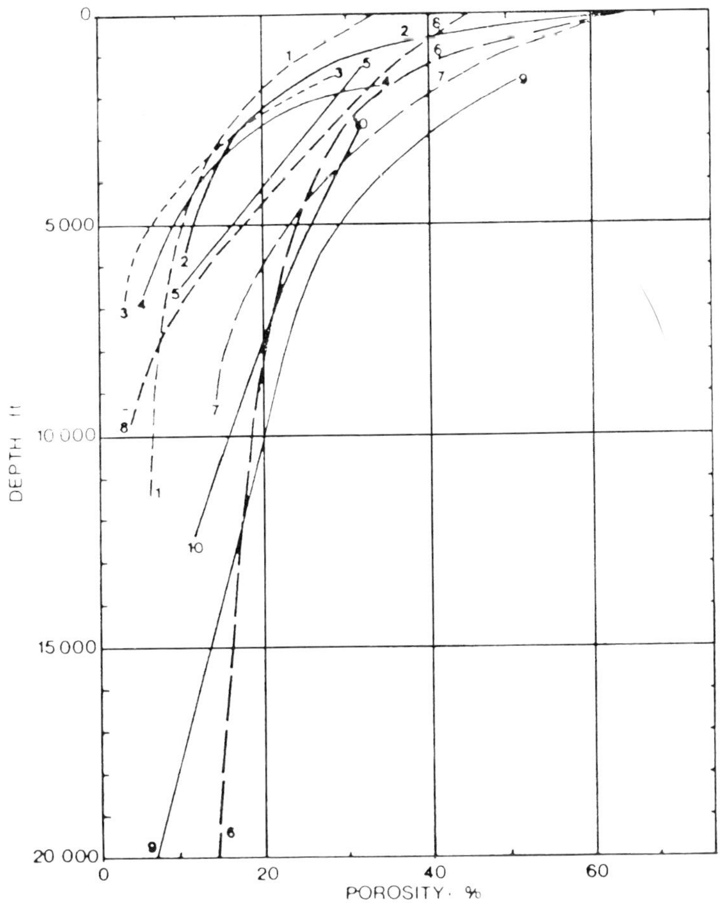

Figure 11.7 Dependence of porosity on depth for shales in various sedimentary basins (Rieke and Chilingarian, 1974). Sources of the data are: 1, Proshlyakhov (1960); 2, Meade (1966); 3, Athy (1930); 4, Hosoi (1963); 5, Hedberg (1936); 6, Dickinson (1953); 7, Magara (1968); 8, Weller (1959); 9, Ham (1966); 10, Foster and Whalen (1966).

Magara (1978) proposed an empirical relationship between porosity and depth z for a normally compacted shale of the form

$$\phi = \phi_0 \exp(-cz) \tag{12}$$

in which the parameters ϕ_0 and c may depend on a variety of factors such as mineralogy, formation age, position within the basin and the presence or absence of tectonic stresses. From Athy's (1930) data set, $c = 1.42 \times 10^{-3}\ \mathrm{m}^{-1}$. In an analysis of North Sea subsidence, Thorne and Watts (1989) have proposed a different form of normal compaction trend in which the void ratio decreases linearly with the logarithm of the overburden pressure. It should be noted that this is identical to the empirical relationships obtained from soil mechanics data. Because of the irreversible nature of the compaction process, trends of this kind are only applicable to normally consolidated sediments (currently at their maximum mean effective stress or depth of burial). They are not necessarily applicable to over-consolidated shales (during periods of uplift or erosion), and/or shales in which the porosity has been modified by diagenetic processes.

The decrease in both porosity and pore sizes with depth of burial implies a relatively marked decrease with depth of the vertical permeability of compacting sediments. Where permeabilities are sufficiently low, and the thickness of the argillaceous sediment is sufficiently great, the compaction equilibrium may not have had sufficient time to become established. This is most likely to occur in relatively young (e.g. Tertiary or Quaternary) shale formations in zones of relatively rapid sedimentation. Such shales are said to be under-compacted, and have higher porosities and pore pressures, and correspondingly lower mean effective stresses p', than would a normally compacted shale under the same total stresses at the same depth. With reference to Figure 11.4, normally compacted and under-compacted shales, at a given depth, correspond to different points of the normal consolidation line AC. The greater the excess porosity in comparison with the normal value for the depth in question, the greater the excess pore pressure above hydrostatic. Under-compacted shales have abnormally low densities, and well-logs indicate that they exhibit abnormally high electrical conductivities and slow seismic and compressional sonic velocities, consistent with their enhanced fluid contents (Fertl, 1976; Magara, 1978). Figure 11.8 illustrates schematically the nature of the trends of porosity and mean effective stress with depth associated with undercompacted zones.

One-dimensional models for under-compaction overpressuring, assuming vertical fluid flow in the compacting shales, have been proposed by a number of workers (Gibson *et al.*, 1967; Bredehoeft and Hanshaw, 1968; Smith, 1971;

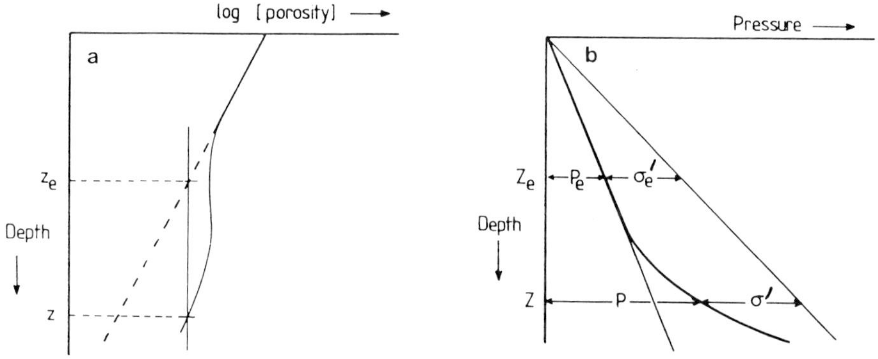

Figure 11.8 Relationship between (a) porosity and depth, and (b) pore pressure u and mean effective stress σ' with depth in zones exhibiting under-compaction overpressuring. An overpressured shale at depth z has the same effective stress as a normally compacted shale at the 'equivalence depth' z_e.

Sharp, 1976; Sharp and Domenico, 1976; Bishop, 1979; Walder and Nur, 1984; Keith and Rimstidt, 1985). These models differ in some detail, but all essentially invoke the conservation of mass of fluid and solids, and assume that the flux of fluid relative to the compacting solids is vertical and obeys D'Arcy's law. These assumptions result in non-linear diffusion-like equations which may be expressed either in terms of porosity or pore pressure. In particular, Walder and Nur (1984) have shown that the magnitude of excess pore pressures generated is proportional to a dimensionless group F given by

$$F = \frac{\eta H \dot{\phi}}{\Delta \rho g K}, \tag{13}$$

where H is the vertical thickness of the sediment, $\dot{\phi}$ is the rate of reduction with time of the porosity ϕ which is required in order to maintain the normal compaction trend, $\Delta\rho$ is the difference in density between rock grains and fluid and g is the gravitational acceleration. The quantity F increases with the rate of burial of the sediment. The timescale of pressure relaxation is given by

$$t = \frac{H^2 \eta \phi (\beta_f + \beta_\phi)}{K}, \tag{14}$$

where β_f and β_ϕ are, respectively, the compressibility of the fluid and the (non-elastic) compressibility of the pore space.

For compacting sediments, the values of the latter are typically very much larger than short timescale elastic compressibilities as measured in rock mechanics tests, and are also largely irreversible. A limited analogy may be made between compacting sediments and visco-elastic materials described by spring and dashpot models. A different name, such as compactability, should be used to distinguish compactional deformation from short timescale, reversible elastic deformation. The compactability of a typical Tertiary shale may be estimated from its compaction trend, and is typically two orders of magnitude larger than the instantaneous elastic compressibility of the same rock. Thus, the relationship between changes of rock volume and changes of pore pressure may differ considerably between shales on their initial burial pathway, and older, uplifted and/or cemented lithologies.

Equation (14) is essentially identical to equation (8). The condition for the existence of under-compacted conditions is essentially a combination of large thickness and low vertical permeability such that the time t exceeds the geological age of the sediment. Highly compacted shales buried to depths of 2 km or more may have permeabilities of the order of nano-darcies ($\sim 10^{-21}$ m^2) and non-elastic compressibilities of typically 10^{-3} bar^{-1}. Using these values, one may estimate that the characteristic consolidation times might not yet have been reached for shale formations of Tertiary or Quaternary age in excess of typically one or two hundreds of metres in thickness.

Two-dimensional and even three-dimensional models of compaction and fluid migration have been developed which take account of various other factors such as convective heat flow and diagenetic volume changes. In some instances these have been used to estimate the evolution of pore pressures as a function of time and position within specific basins (e.g. Ungerer *et al.*, 1984; Bethke *et al.*, 1988). While models of this kind may reproduce the gross features of pressure evolution, their quantitative value is limited by the uncertainty of some of the functional relationships involved, specifically the relationship between shale permeability and porosity, and how it is influenced by clay type, microfabric, anisotropy, the extent of natural fracturing and faulting and possibly other factors. For these reasons they are probably best regarded as heuristic rather than accurately predictive models. A more complex one-dimensional treatment of compaction has been proposed by McKenzie (1987) which predicts solitary waves of porosity and pore pressure which propagate upwards.

There is no general model for permeability/porosity relationships in clays or shales. A relationship of the form $k \sim \phi^n$ has been found for compacted montmorillonite (Denis *et al.*, 1991) in which the exponent depends upon the exchange cation type. An exponent of 5 has been used by Jacquin and Poulet (1970) in modelling basin subsidence. The Kozeny–Carman model for flow in granular materials predicts a power-law relationship with an exponent of 3 (Chapman, 1981). However, other data sets suggest a relationship of the form $\log k \sim \phi$ (Rieke and Chilingarian, 1974; Magara, 1978).

When thick overpressured shale sequences alternate with more permeable sands, then lateral flows of pore fluid through the sands may contribute significantly to the overall compactional dewatering (Magara, 1978; Walder and Nur, 1984; Bredehoeft *et al.*, 1988). Rapid burial of an alternating sand/shale sequence would initially give rise to excess fluid pressures everywhere, but these are able to bleed off much more rapidly in the sands because of their higher permeabilities. Thus, in general in overpressured environments pore pressures will be higher in thick impermeable shales than in bounding sand layers, reaching a maximum near the centre of the shale formation, and decreasing both upwards and downwards into the sands, unless the latter are isolated by faults or pinchouts. The way in which porosity and fluid pressure vary with depth in an alternating sand/shale sequence is illustrated in Figure 11.9.

11.3.2 Influence of lateral compression on fluid pressure

Normally, in a basin where no lateral compressive effects occur, the horizontal stresses at any depth will be less than, or at most equal to, the vertical stresses due to the rock overburden. However, wherever the pattern of faulting or folding indicates large lateral compressive stresses, high pore pressures can be generated, which arise both from the increase in mean confining stress, and from

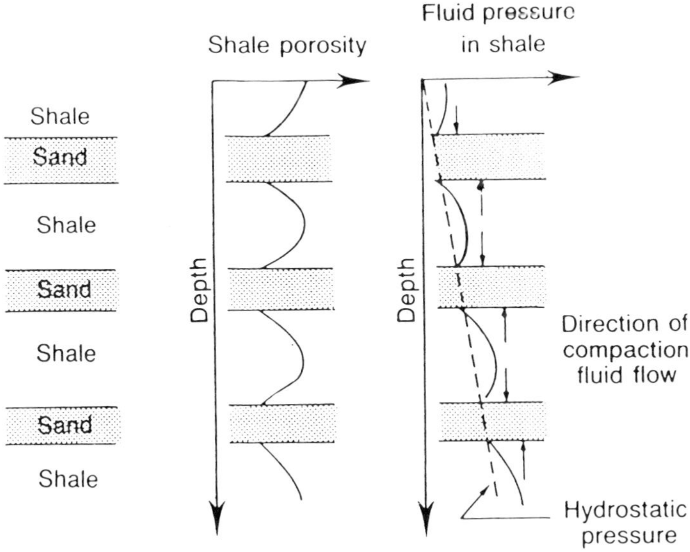

Figure 11.9 Schematic illustration of the likely variation of porosity and pore pressure with depth in an alternating shale/sand sequence (Magara, 1974).

the compactant effect of deviatoric stresses on normally consolidated sediments (Atkinson and Bransby, 1978; see Figure 11.6). Applications of soil mechanics concepts, including critical state theory, to sediment deformation and compaction have been made by Jones and Addis (1986).

Elevated pressures due to tectonic loading can occur at convergent plate margins where thick accretionary wedges of deforming sediment are piled up against the overthrusting plate while the sediments resting on the subducting plate may be relatively undeformed (Davis *et al.*, 1983; see Figure 11.10). Hottman *et al.* (1979) reported elevated pore pressures attributed to active plate subduction in the Gulf of Alaska region. Qualitatively similar effects may occur in the vicinity of salt domes (Harkins and Baugher, 1969; see Figure 11.11).

Where the necessary dewatering has had sufficient time to occur, the deformed sediments in accretionary wedges exhibit lower porosity and higher cohesive strength than adjacent undeformed sediments (Bray and Karig, 1985; Zhao *et al.*, 1986).

11.3.3 Influence of temperature on fluid pressure

Some models have included the effect of temperature as well as compaction, and a considerable debate occurred in the petroleum geology literature concerning the relative importance of mechanical and thermal contributions to pore pressures, especially in the Gulf Coast of the USA (Barker, 1972; Bradley, 1975; Magara, 1975b; Dickey, 1976; Chapman, 1980; Daines, 1982; Sharp, 1983). The

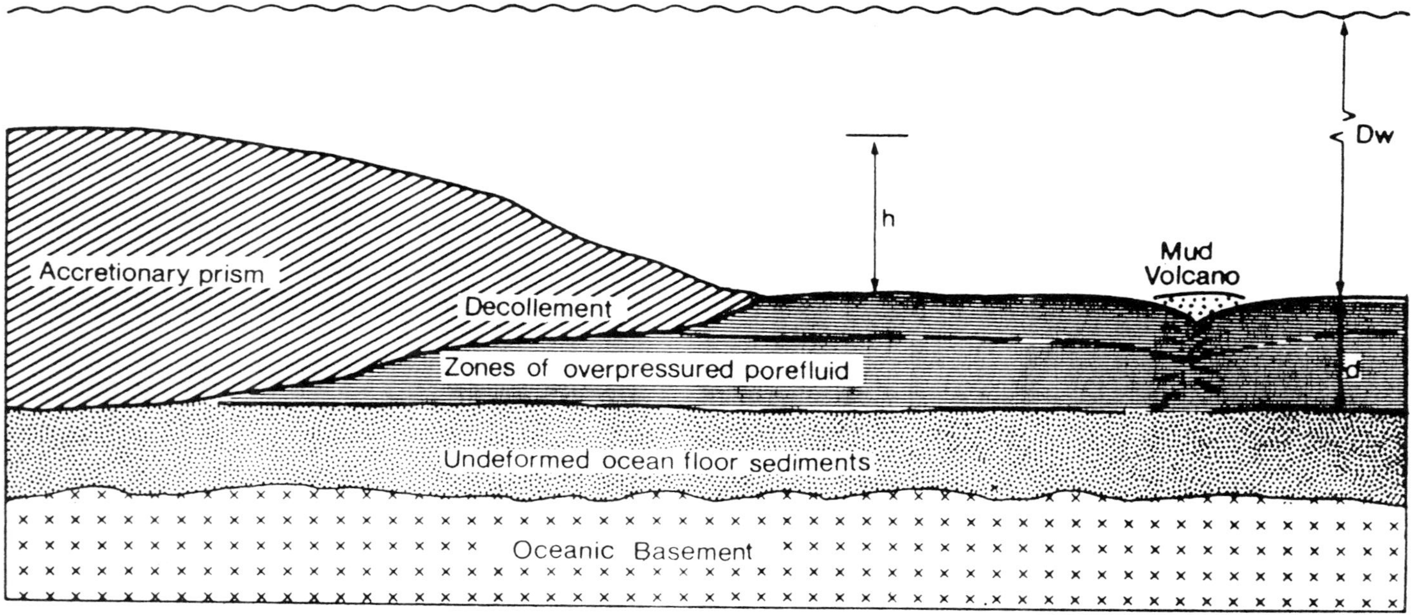

Figure 11.10 Overpressuring phenomena associated with tectonic motion of an accretionary wedge (Davis *et al.*, 1983).

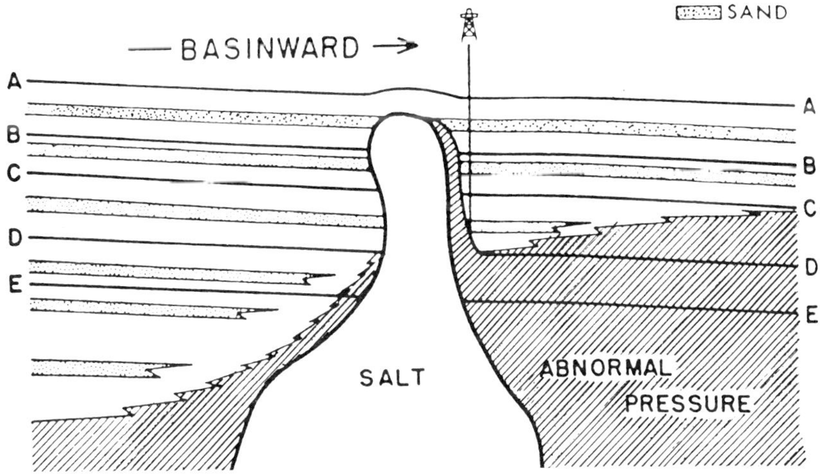

Figure 11.11 Overpressuring phenomena associated with halokinesis (Harkins and Baugher, 1969).

most prevalent view is that the relatively small volume changes arising from differential thermal expansion of pore fluids relative to rock grains are probably of secondary importance except where permeabilities are extremely low. An analysis of the factors which can influence the relative contributions of stress and temperature to pore pressures in a porous solid shows that the crucial controlling parameter is the compliance of the rock, particularly the difference between elastic compressibility and the long timescale sediment compactability as defined above. In non-sealed sediments of relatively high compactability, thermal effects are virtually negligible. In contrast, in sealed compartments which have relatively low compressibilities over the timescale of pore pressure generation, the effect of relative thermal expansion of fluids may be of major importance. In an environment where temperature is the dominant control on fluid pressures, the relationship between temperature and pore pressure will follow lines of constant density for the pore fluid (Barker, 1972; see Figure 11.12). Examples of such environments are discussed in the last section of this chapter. See also the paper by Hunt (1990).

11.3.4 Influence of pore fluid chemistry and osmosis on fluid pressure

The previous sections have dealt solely with mechanical and thermal influences on pore pressures in compacting argillaceous sediments, neglecting any possible chemical effects. However, clay minerals have both fixed and pH-dependent electrical charges which give rise to cation exchange phenomena and a variety of associated colloidal and electrochemical properties (Swartzen-Allen and Matijevic, 1974; Van Olphen, 1977). Both laboratory compaction experiments on clays and field studies of sediments have shown that the relationship between water content and compaction pressure or depth depends upon the

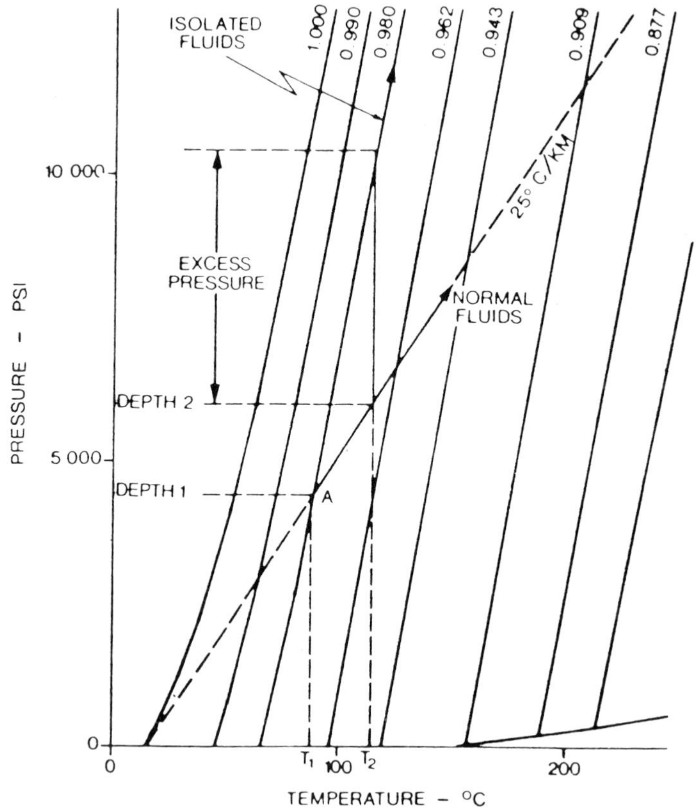

Figure 11.12 Pressure–temperature–density relations for water, superimposed on a trajectory corresponding to a geothermal gradient of 25°C per km, showing potential thermal pressure development in sealed fluid compartments (after Barker, 1972).

clay type, the nature of the exchange cation, brine concentration and pH (Chilingar and Knight, 1960; Meade, 1966; Olson and Mesri, 1970; Mesri and Olson, 1971; Rieke and Chilingarian, 1974). Other factors such as the nature of the depositional environment and the concentration of organic matter present may influence the fabric of clay sediments (O'Brien, 1970; O'Brien *et al.*, 1980; Bennett *et al.*, 1981; Moon and Hurst, 1984).

Figure 11.13 summarizes the effect of exchange cation type and electrolyte concentration on the equilibrium void ratios obtained in one-dimensional consolidation and re-swelling tests on montmorillonite. Both exchange cation forms show a large irreversibility between compaction and re-swelling, similar to the data for the kaolin clay illustrated in Figure 11.4. The compaction state of the sodium-exchanged montmorillonite is much more sensitive to the electrolyte concentration than the calcium-exchanged form, which may be attributed to the strong double-layer interactions and gel structure in the initial suspensions (Olson and Mesri, 1970; Mesri and Olson, 1971).

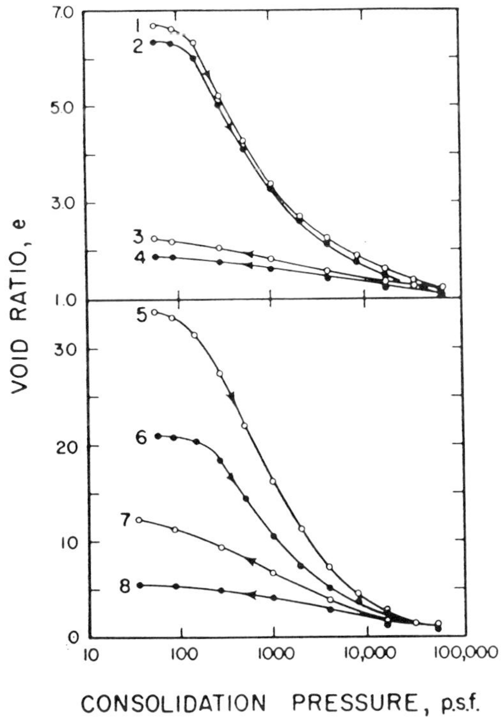

Figure 11.13 One-dimensional consolidation curves for sodium- and calcium-exchanged smectite in the presence of different concentrations of the common cation electrolyte. Directions of arrows indicate compression or re-swelling. 1/3: 0.001 N calcium. 2/4: 1.0 N calcium. 5/7: 0.0005 N sodium. 6/8: 0.1 N sodium (Olson and Mesri, 1970.) Note the large irreversibility between compaction and re-swelling (cf. Figure 11.4).

A number of workers have demonstrated that the nature of electrolyte flow through (and expulsion from) compacted clays indicates that these systems exhibit the properties of semi-permeable, cation-exchange and anion-excluding membranes (McKelvey and Milne, 1962; Von Engelhardt and Gaida, 1963; Young and Low, 1965; Mitchell *et al.*, 1973; Rosenbaum, 1976). The transport of fluid and ions through compacting argillaceous sediments depends on chemical potential gradients rather than simple hydraulic pressure gradients. At equilibrium, the relationship between the ionic concentrations in compacted clays or shales and those found in adjacent reservoirs or aquifers may be modelled in terms of electrical double-layer or Donnan theories (Helfferich, 1962; Van Olphen, 1977).

For water, the condition for equilibrium of chemical potential between a shale and an external fluid reservoir at the same depth and temperature is

$$P_{\mathrm{sh}} = P_{\mathrm{res}} + \frac{RT}{V} \ln\left(\frac{a_{\mathrm{res}}}{a_{\mathrm{sh}}}\right), \qquad (15)$$

where P_{sh}, P_{res}, a_{sh} and a_{res} are the pressures and water activities in the shale and the reservoir, respectively, R is the gas constant, T is the absolute temperature, and V is the molar volume of water (Chenevert, 1970). A difference in activity or brine concentration between the shale and the reservoir can be balanced by a difference in fluid pressure, with no flow taking place. However, if the situation is disturbed, e.g. by drilling into a shale formation using a fluid whose ionic composition differs from the reservoir, the osmotic pressure will be changed and fluid will flow into or out of the shale. This swelling or dewatering of shales in contact with aqueous drilling fluids can contribute to swelling, erosion and related borehole stability problems (see e.g. Chenevert, 1970).

Since the water activity of a compacted smectite/electrolyte system decreases with increasing degree of compaction, as free water is progressively expelled and double-layer and specific cation hydration effects become dominant (Denis, 1991), it follows that the equilibrium pore pressure in highly compacted smectite-containing shales must eventually develop an osmotic component and should not necessarily revert to a hydrostatic value with time. Marine and Fritz (1981) have given a convincing example from the Dunbarton Basin in South Carolina, USA in which elevated fluid pressures appear to arise from osmotic flow downwards through a clay-rich sediment from a less saline to a more saline aquifer. Hanshaw and Zen (1965) proposed that osmotic pressures may be involved in thrust faulting. Jones (1969) suggested that osmotic pressures, together with diagenesis, may contribute to overpressuring in the Northern Gulf of Mexico basin. For an ideal membrane, the maximum osmotic pressure developed can be predicted by equation (15). However, clays are frequently not ideal membranes. Fritz (1986) has reviewed the factors controlling the degree of ideality of clay membranes.

Brine salinities in the subsurface normally increase with depth (Dickey, 1966). However, Schmidt (1973) showed that the salinities of the waters in both overpressured reservoirs and adjacent shales in the US Gulf Coast were significantly lower than those of the waters present in normally pressured reservoirs and adjacent shales. This does not necessarily imply the occurrence of osmotic phenomena, however, since similar effects would be expected when large volumes of fresh water are expelled from shales following the diagenetic loss of interlamellar water from smectites. More recent studies by Morton and Land (1987) support this conclusion, but also show that subsurface brines in the US Gulf Coast area show a complex pattern of zonal compositional variations, which is strongly correlated with both proximity to salt domes and faulting patterns or other factors promoting salt diffusion.

Other workers (Graf *et al.*, 1966) have proposed reverse osmosis (hyperfiltration) as a mechanism for the observed brine concentrations in sediments. Overton and Zanier (1971) proposed a reverse osmosis mechanism, together with gas generation and cementation, to explain many observed features of overpressured zones in numerous different locations. Overall, however, the evidence for osmosis or reverse osmosis as a widespread overpressuring mechanism remains questionable.

11.4 Chemical and diagenetic mechanisms

11.4.1 Diagenesis of argillaceous sediments

Diagenetic reactions occurring within a sedimentary basin can contribute to the generation and retention of excess fluid pressures in various possible ways:

1. If the reaction is accompanied by a positive potential volume increase, or a decrease in available porosity, an increase in fluid pressure is likely to occur.
2. The conversion of bound water to free water during a mineral dehydration reaction would be expected to disturb the normal compaction trend, in particular providing excess fluid pressure and a renewed potential for compaction even in the absence of any volume change.
3. Any accompanying decrease in rock permeability due to cementation or any change of microstructural fabric (e.g. following clay diagenesis) can extend the timescale of excess pressure retention.

The diagenesis of argillaceous sediments is a complex subject, and only a brief outline of the major processes which may influence fluid pressures will be given here. For more details see Berner (1971), Curtis (1977; 1978; 1983; 1985) Kisch (1983), Parker and Sellwood (1983), Singer and Müller (1983), Surdam and Crossey (1985), Burley *et al.* (1985), Weaver (1989) and Surdam *et al.* (1989).

Essentially, five major zones of diagenesis are recognized, representing increasing depth below the sediment/water interface, increasing temperature and decreasing porosity. These are illustrated in Figure 11.14. At rather higher temperatures ($T > 200°C$) sediments enter the zone of metamorphism, and recrystallize to form slate, schist or gneiss, with further gas generation.*

Zone I, which is limited to essentially the top centimetre or so of sediment, is a zone in which organic matter is destroyed by bacterial oxidation. Zone II, which typically occupies the next 10 m of the sedimentary column, is characterized by the action of sulphate-reducing bacteria, which convert iron oxides to pyrite and organic matter to carbonate ions which may initiate the precipitation of iron-poor carbonates. Zone III, extending down to approximately 1 km, is characterized by fermentation reactions, as well as further iron reduction resulting in precipitation of iron-rich carbonates such as siderite and also the generation of considerable quantities of methane.

*Some workers distinguish the low-temperature and thermal diagenetic processes as 'diagenesis' and 'catagenesis', respectively. Here, all sub-metamorphic processes are termed diagenetic.

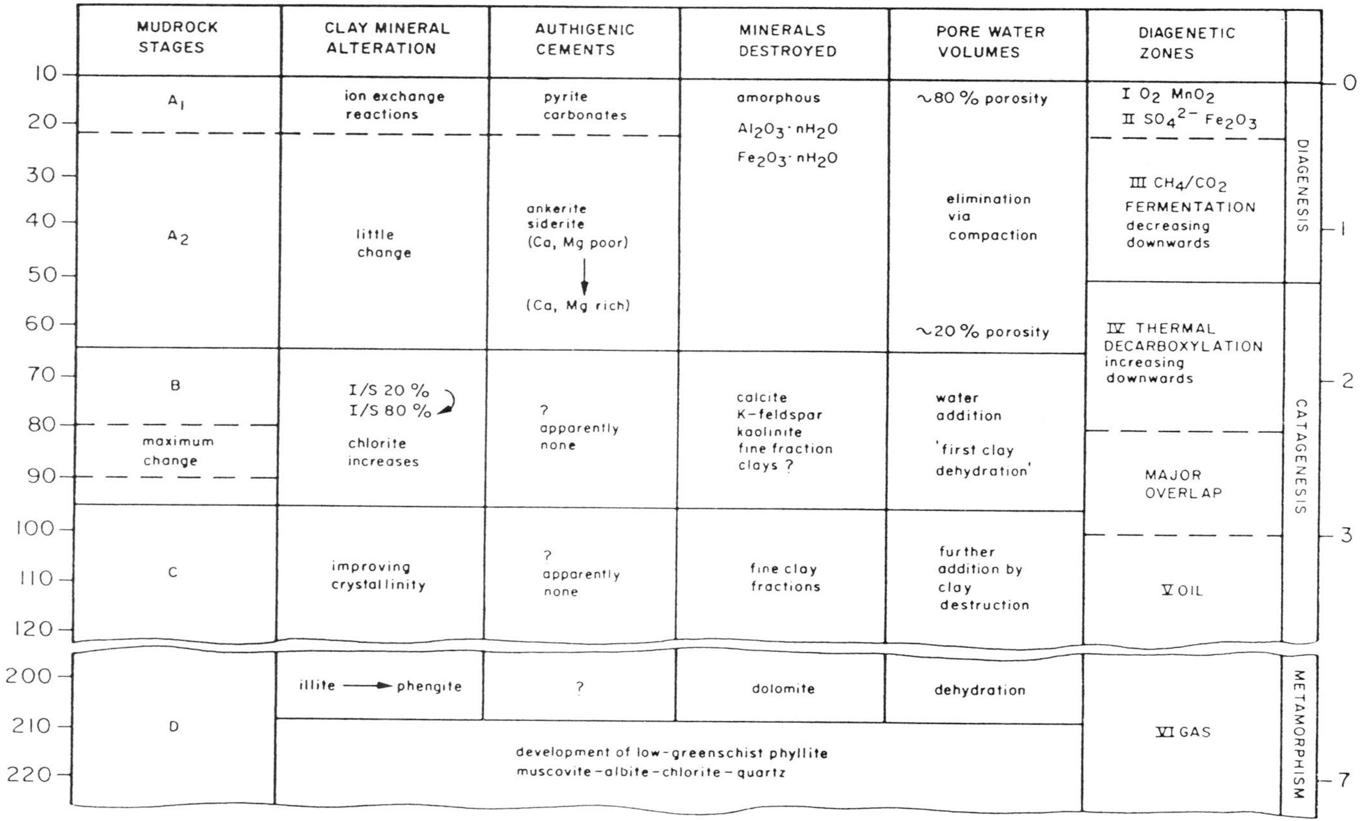

Figure 11.14 Zones and principal reactions in shale diagenesis (Curtis, 1983).

At the base of Zone III lie two much thicker, overlapping zones (Zones IV and V) characterized by thermal maturation of organic matter. Zone IV is characterized by the decarboxylation of kerogen, producing further carbonate mineralization, and also the generation of short-chain aliphatic acids such as acetic acid. Zone V is the major region of generation of liquid and gaseous hydrocarbons, although this occurs also to some extent in Zone IV. Overall, the zone of thermal hydrocarbon generation spans a temperature range from 65 to 150°C (Curtis, 1983). Figure 11.15 illustrates the extent of the evolution of carboxylic acids as a function of temperature across the thermal maturation zone. The diagenesis reactions have largely been characterized in sandstones, although some (e.g. K-feldspar dissolution, albite formation, and various clay diagenetic reactions) are also observed in shales (Hower *et al.*, 1976).

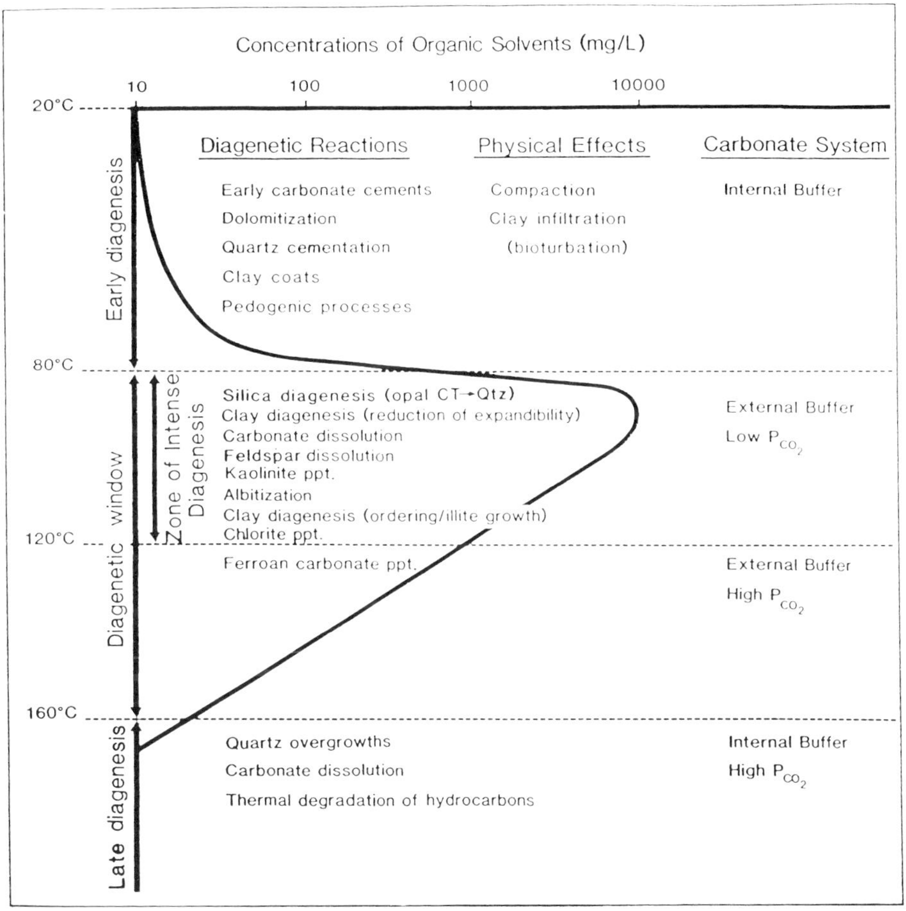

Figure 11.15 Evolution of carboxylic acid concentration with temperature in oilfield waters (from Surdam *et al.*, 1989).

In addition to organic matter diagenesis, and the precipitation of pyrite and carbonates, a complex series of diagenetic transformations of silica takes place (Berner, 1971; Singer and Müller, 1983), in which amorphous silica consisting of biogenic opal A (from the skeletal remains of planktonic organisms) and also volcanic glass are converted via solution and recrystallization processes into opal CT (related to the cristobalite and trydimite polymorphs of quartz) and eventually to quartz. These processes take place mainly in Zones III and IV, and result in a reduction by more than an order of magnitude of the equilibrium (saturation) concentration of silicic acid (H_4SiO_4) in solution.

The simultaneous dissolution of potassium feldspars increases the activity of K^+ in solution. Together, these changes in brine composition provide a thermodynamic drive for clay mineral diagenesis, most importantly the conversion of smectite via mixed-layer illite–smectite to illite. This process releases large quantities of bound interlayer water into the pore spaces of shales under deep burial conditions. Thermodynamic stability fields for various aluminosilicate phases at ambient pressure and temperature as a function of silicic acid, K^+ and H^+ concentrations are illustrated in Figure 11.16 (from Aagaard and Helgeson, 1983).

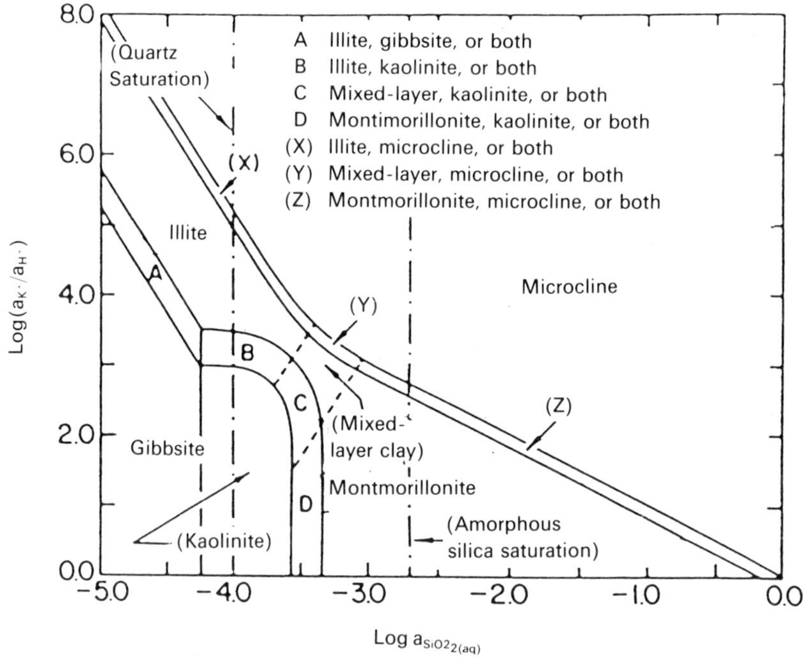

Figure 11.16 Activity diagram for clay mineral transformations (Aagaard and Helgeson, 1983).

The diageneses of organic and inorganic phases interact in a number of ways, in addition to the formation of pyrite and carbonates discussed above. The increased acidity in the zone of kerogen decarboxylation drives the dissolution of potassium feldspars, providing a source of potassium. In addition, low pH values promote clay diagenesis by mobilizing aluminium in the form of organo-aluminium complexes (Curtis, 1985; Surdam and Crossey, 1985; Surdam *et al.*, 1989). Conversely, Brönsted acidity developed during the smectite-to-illite reaction catalyses the thermally driven hydrocarbon formation reactions (Johns and Shimoyama, 1972; Johns and McKallip, 1989).

Hydrocarbon generation from kerogen, as well as water expelled during the smectite-to-illite reaction, have both been proposed as causes of overpressuring (Rieke and Chilingarian, 1974; Fertl, 1976; Magara, 1978). Some examples of hydrocarbon-related overpressuring are discussed in the next section. After that, the smectite-to-illite transformation and its influence on fluid pressures is considered in detail.

11.4.2 Biogenic and thermally generated hydrocarbons

Hedberg (1974) considered that both biogenic methane (in Zone III) and thermally generated gas (in Zones IV and V) contribute significantly to overpressuring. Trapped biogenic gas may contribute to under-compaction, and phenomena associated with density reversals (e.g. diapirism and mud volcanoes). Barker (1987) has recently argued that bacterial gas may give rise to overpressures or underpressures, depending on the nature of compaction and isolation of the gas-bearing strata, and the amount of gas generated in relation to its water solubility.

According to Hunt (1979) the thermal cracking of kerogen to oil and gas at peak oil generation corresponds to a potential volume increase of *c.* 25%. Milliken *et al.* (1981) point out that the onset of elevated pore pressures frequently coincides with the maximum of thermal hydrocarbon generation in Zones IV and V. It has been suggested that primary hydrocarbon migration from source rocks to reservoir rocks is driven by natural tensile fracturing of shales when such pressures become large enough (Du Rouchet, 1981).

Goff (1983) made a detailed study of the maturation of Upper Jurassic Kimmeridgian source rocks in the northern North Sea. His data show that the generation of light hydrocarbons has significantly enhanced overpressuring originally associated with under-compaction. Some evidence of highly lithified or cemented cap rocks above the overpressured zones was reported.

Stuart and Kozik (1977) have attributed pressures close to the overburden pressure encountered in the Jurassic Smackover gas reservoirs to the trapping of hydrocarbons within a sealed cell which is subsequently buried and heated. From a knowledge of the sealing depths, the amount of pressure generated could be inferred from the equation for real gases combining Boyle's and

Charles' laws. The porosity in these formations is essentially secondary and cannot be predicted from compaction trend analysis (Parker, 1974).

Another example of overpressuring directly associated with hydrocarbon maturation is the Antelope Field in the Williston Basin of North Dakota, where pressures up to 20 MPa above hydrostatic values may be attributed to thermal maturation of the organic-rich Bakken Formation of Mississippian age (Hunt, 1979; Meissner, 1984). Overpressuring is encountered only when this formation is at, or below, the depth of onset of thermal maturation (see Figure 11.17). Spencer (1987) documents the existence of overpressures in low-permeability (tight) reservoirs having pore pressures ranging up to lithostatic values. These are located in formations ranging from Devonian to Tertiary age in the deeper parts of many basins in the Rocky Mountain region of the USA. These pressures are attributed to the hydrocarbon maturation process.

11.4.3 Phyllosilicate diagenesis: the smectite-to-illite transition in shales

This reaction has been documented in many different parts of the world, in diagenetic environments such as the US Gulf Coast (Hower *et al.*, 1976; Boles

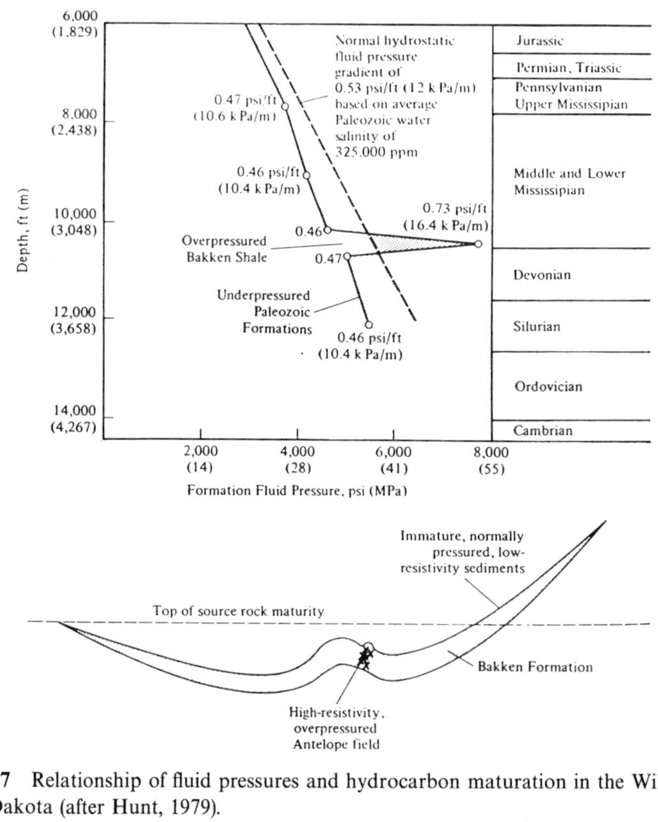

Figure 11.17 Relationship of fluid pressures and hydrocarbon maturation in the Williston Basin of North Dakota (after Hunt, 1979).

and Franks, 1979; Ahn and Peacor, 1986; Freed and Peacor, 1989); continental North America (Powell *et al.*, 1978; Eslinger and Sellars, 1981); the North Sea Basin (Pearson *et al.*, 1982; Dypvik, 1983; Glasmann *et al.*, 1989); the Rhinegraben (Heling, 1974; 1978) and Cameroun (Dunoyer de Segonzac, 1970). It also occurs in zones of contact metamorphism (e.g. Nadeau and Reynolds, 1981) and hydrothermal environments such as the Salton Sea, California (Muffler and White, 1969; Yau *et al.*, 1987) and Shinzan, Japan (Inoue *et al.*, 1987). X-ray diffraction data indicate the fine fractions of these clays to be predominantly a mixed-layered illite–smectite, with a progressive decrease in the percentage of expandable (smectite-like) interlayers with increasing depth or temperature, together with a change from random to ordered interlayer stacking. A typical variation of expandability with depth for US Gulf Coast shales is illustrated in Figure 11.18.

Chemical analyses showed a progressive increase in potassium and aluminium concentration, and a decrease in silicon as the degree of expandability of the

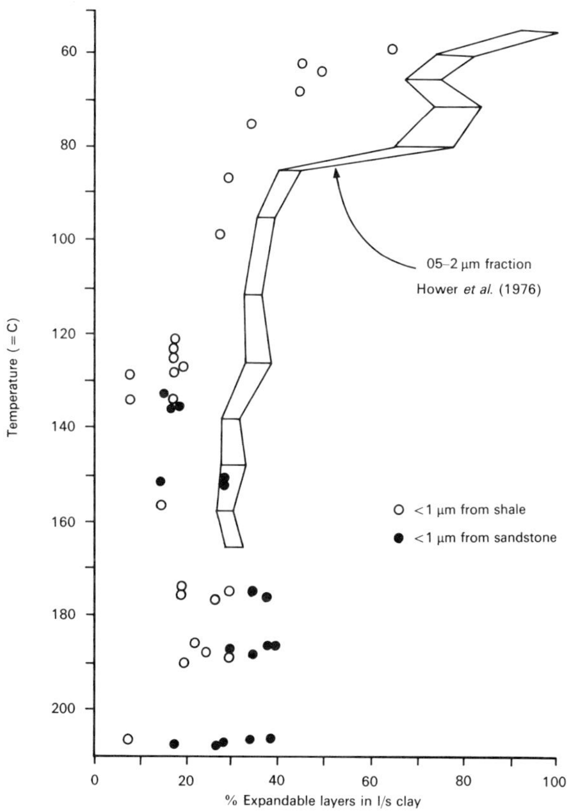

Figure 11.18 Variation with depth of the expandability of illite–smectite clays from wells in the US Gulf Coast region (Boles and Franks, 1979).

fine illite–smectite clays decreased from approximately 80% to 20% with increasing depth (Hower *et al.*, 1976). Initially, a layer-by-layer transformation mechanism was proposed, corresponding to an essentially stoichiometric reaction of the form

$$\text{smectite} + Al^{3+} + K^{+} \rightarrow \text{illite} + Si^{4+}$$

(Weaver and Beck, 1971; Hower *et al.*, 1976). However, Boles and Franks (1979) suggested that the source of aluminium was not external but involved destruction of approximately one-third of the original smectite layers, releasing a proportionately larger quantity of silica into solution:

$$\text{smectite} + K^{+} \rightarrow \text{illite} + \text{chlorite} + Si^{4+}.$$

In the macroscopic pores of sandstones, filamentous authigenic illite can form without any smectite precursor (e.g. McHardy *et al.* 1982). Discrepancies between XRD and HRTEM data for clays from sandstones and bentonites led Nadeau *et al.* (1984; 1985) to propose the concept that mixed-layer illite–smectites are comprised of very thin aggregates of 'fundamental particles' whose interlayers are expandable. They further argued that the XRD data from the smectite-to-illite transition could readily be explained in terms of dissolution of smectite with concomitant crystallization and growth of thin illite particles. Magic-angle spinning NMR studies by Altaner *et al.* (1988) show unequivocally that the external surfaces of illite particles have an essentially smectite-like layer charge associated with a lower level of aluminium for silicon substitution, presumably balanced by exchangeable rather than fixed cations, accounting for the measured cation exchange capacities of illites.

Recent work on the smectite-to-illite transition has utilized a variety of electron optical techniques such as SEM (Keller *et al.*, 1986), HRTEM and ATEM (Ahn and Peacor, 1986; 1989; Inoue *et al.*, 1987; Yau *et al.*, 1987). Recent improvements in the interpretation of lattice fringe image patterns (Veblen *et al.*, 1990), and the recognition that individual 10 Å layers may have asymmetrical layer charge distributions (i.e. both a high charge (illitic) and lower charge (smectitic) tetrahedral layer (Guven, 1991)) effectively reconciles previous apparent discrepancies between XRD and HRTEM data. These studies, together with hydrothermal laboratory syntheses, lead to an interpretation of the smectite–illite transformation as a complex series of chemical and morphological transformations, involving at least partial dissolution, solute transport and recrystallization, and a progressive coarsening (Ostwald ripening) of fundamental particles or MacEwan crystallites (see Figure 11.19, and also Eberl *et al.*, 1978; 1990; Whitney, 1990).*

*It should be noted here that Small and Manning (Chapter 8) have observed the neo-formation of both fibrous and platy illite crystals on sandstone substrates under different solution conditions in laboratory experimental simulations of diagenesis.

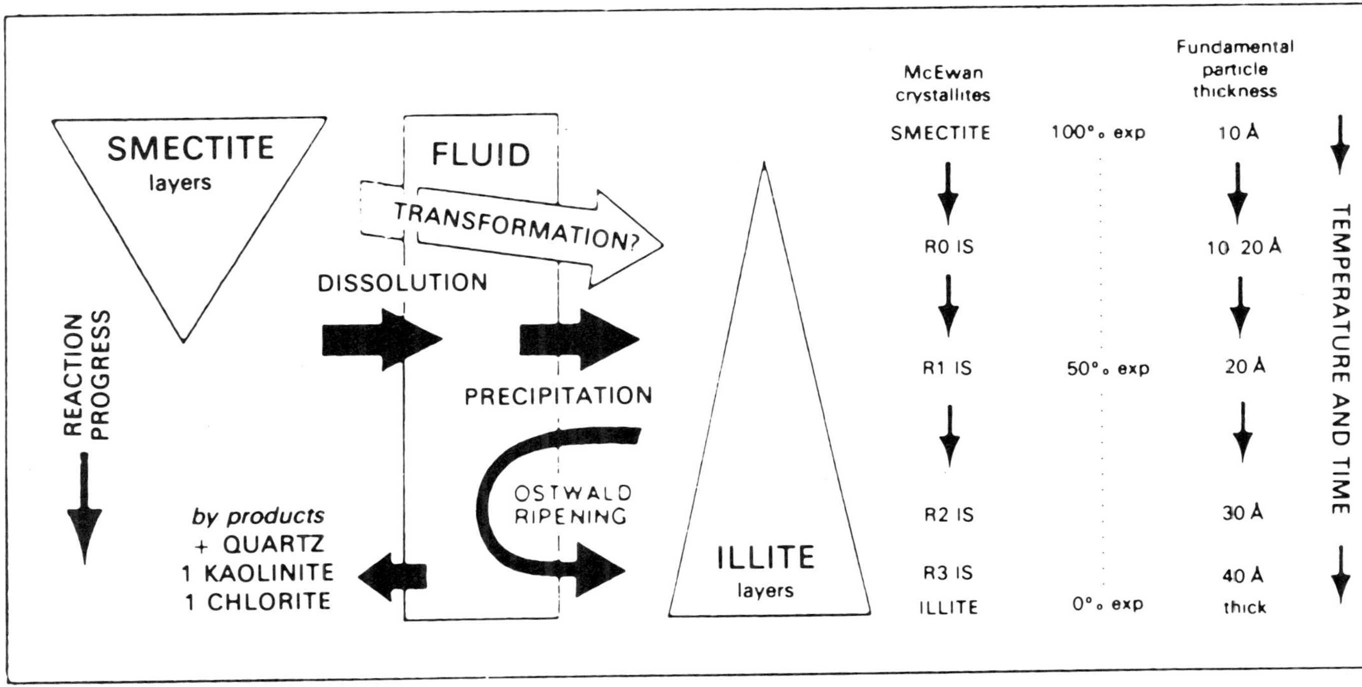

Figure 11.19 Schematic illustration of the smectite-to-illite reaction sequence (Whitney, 1990).

The evidence from many studies world-wide indicates that the smectite–illite conversion occurs over a wide range of temperatures and timescales (Singer and Müller, 1983; Weaver, 1989). This is characteristic of a kinetically hindered reaction, like silica diagenesis, in which the cumulative effect of time and temperature controls the degree of reaction (Putnis and McConnell, 1980; Siever, 1983). Other factors such as fabric and permeability may also influence the reaction kinetics by hindering transport of reactant and product solution species.

A typical temperature–time transformation diagram for a kinetically hindered reaction is illustrated in Figure 11.20(a). The upper parts of the curves, close to the thermodynamic equilibrium temperature of the reaction T_c, are controlled by the rate of nucleation. The activation energy ΔG^* for the nucleation phase depends on the degree of supersaturation of the solution. ΔG^* becomes infinite at $T = T_c$ when the degree of supersaturation is zero. At lower temperatures, farther from equilibrium, the reaction is controlled kinetically by

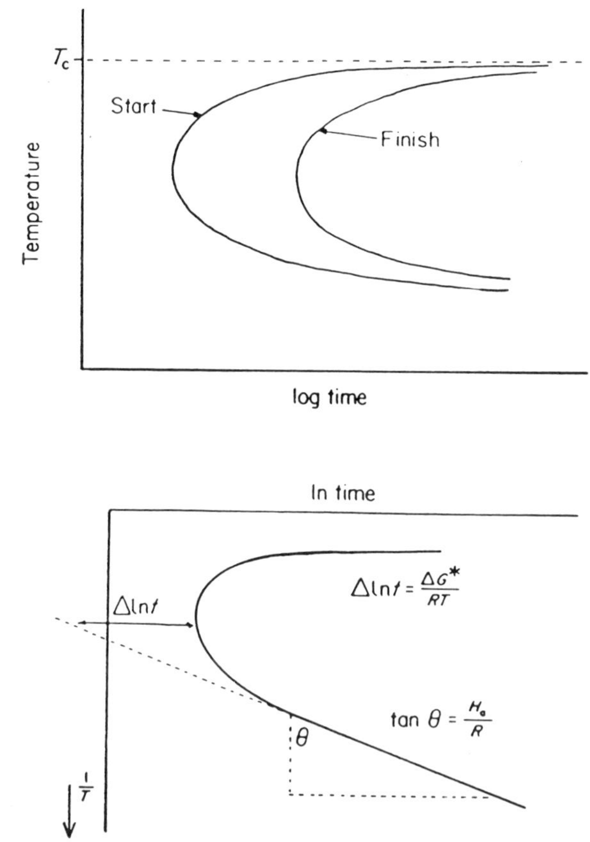

Figure 11.20 Time–temperature transformation diagram for a kinetically hindered reaction (Putnis and McConnell, 1980).

the rate of thermally activated processes such as solid-state diffusion or solution transport. In this region, a plot of the logarithm of time against $1/T$ is essentially an Arrhenius plot whose slope is $-H_a/R$, where H_a is the activation enthalpy of the diffusion process (see Figure 11.20(b)). When the temperature is well below T_c, the extent of completion of the reaction is, therefore, determined by the intersection of the time–temperature trajectory of the formation with the curves for the reaction. One can, therefore, predict that the illitization reaction should be more rapid, i.e. should be observed in younger sediments, but should occur at higher temperatures, in zones of more rapid heating, e.g. in contact metamorphic and hydrothermal environments compared with diagenetic environments. Comparing the data of Hower *et al.* (1976) with the studies of Muffler and White (1969) in the Salton Sea area, this appears to be the case. Also Heling (1974; 1978) indicated that a time/temperature coupling appeared to control the rate of illitization in different parts of the German Rhinegraben.

In considering the timing and number of possible stages involved in the loss of interlamellar water from smectites in sedimentary basins, it is necessary to consider the possible occurrence of a simple and reversible dehydration of smectite, in addition to the well-documented diagenetic illitization reaction discussed up to now. Investigations of the conditions under which smectites dehydrate at elevated pressures and temperatures in the laboratory (Koster van Groos and Guggenheim, 1984; Colten, 1986; Hall *et al.*, 1986) suggest that the last interlamellar bilayer of water is stable to temperatures rather higher than those encountered under diagenetic conditions. This implies that, at least in diagenetic environments, the later stages of water loss from smectite are controlled by the kinetics of the thermally-activated illitization reaction rather than by the hydration/dehydration thermodynamics of the water–smectite system. Colten-Bradley (1987) has suggested the partial intervention of a thermodynamically controlled dewatering process in buried sediments, but the calculations made rest on data and assumptions which may be questioned.

Recent Monte Carlo simulations of the interlayer water structure in magnesium and sodium montmorillonite (Chapter 2) have determined that the interlayer water in both cases has a significantly higher density than bulk water, which supports the likelihood of an overall positive volume change on illitization. Corroborative evidence comes from neutron diffraction studies of the montmorillonite water system (Hawkins and Egelstaff, 1980). It has also been observed that elevated pressures tend to suppress illite formation in laboratory hydrothermal experiments (Eberl *et al.*, 1978) and in some high-pressure wells (e.g. Freed, 1980). These results contradict the findings of early studies at ambient pressure using techniques of low accuracy like capillary rise methods (e.g. Anderson and Low, 1958). The overall volume change accompanying the complex smectite-to-illite reaction series is, however, not currently known.

Despite the original suggestions that the water released during illitization may partially drive hydrocarbon migration (Powers, 1967; Burst, 1969), and a considerable amount of other circumstantial evidence, particularly from the US Gulf of Mexico region, assessing the importance of the transformation in relation to other overpressuring mechanisms such as under-compaction remains difficult. Magara (1975a) considered smectite dewatering to be of secondary importance in comparison with under-compaction for overpressure generation, while conceding the significance of illitization in providing a hydraulic driving force for the migration process, given sufficient drainage. Conversely, Bruce (1984) claimed that there was a close relationship between smectite dehydration and high fluid pressures in both the US Gulf Coast and the Niger delta, identifying it as a major overpressuring mechanism in both regions. The flow of diagenetically released pore fluids was a significant control for both the development of growth fault structures and the accumulation of hydrocarbons.

Plumley (1980) pointed out that the porosities of overpressured shales in the US Gulf Coast are too low for the mechanism of pressure generation to be purely under-compaction. He makes the important point that if under-compaction is the only mechanism operating to raise pore pressures, and if the shale is currently at its maximum depth of burial, the present mean effective stress is the maximum value experienced by the shale in its burial history. In contrast, if after burial and partial or complete compaction the pore pressure is subsequently 'pumped up' by another process (thermal, osmotic or diagenetic) the present mean effective stress will be lower than the historical maximum. The shale will, therefore, be 'over-consolidated' in soil mechanics terms, as defined earlier. Its porosity at any given effective stress will correspond to some point on the elastic rebound or 'swelling' line of Figure 11.4, and will be lower than for a shale of the same composition and effective stress which has arrived at its present state entirely as a result of compaction disequilibrium. One would also expect the change in porosity for a given change in effective stress or pore pressure to be rather smaller for the over-consolidated shale.

Berg and Habeck (1982) investigated the causes of overpressure in a well in South Texas, and identified two different zones of elevated pressure, the deeper of which was attributed to a combination of smectite dehydration and hydrocarbon maturation. Freed and Peacor (1989) show examples where the onset of elevated pore pressures coincides with the onset of illitization. Observing no evidence for cementation, they conclude that the change in shale fabric tends to reduce the permeability. However, such a reduction in permeability is not required if the release of water during illitization raises the pore pressure sufficiently rapidly as a function of depth or time.

11.4.4 *The gypsum-to-anhydrite reaction and overpressuring in evaporite lithologies*

Heard and Rubey (1966) and Hanshaw and Bredehoeft (1968) considered the implications of the gypsum–anhydrite reaction in overpressure development and thrust faulting. Unlike the smectite-to-illite reaction series, this reaction appears to be thermodynamically rather than kinetically controlled (MacDonald, 1953).

Figure 11.21 illustrates schematically the pressure/temperature phase diagram of a mineral which can exist in either hydrated or dehydrated states in the presence of water in liquid or vapour forms. The significant line for sedimentary rocks is the line DE, which represents the line of coexistence at equilibrium of hydrated and dehydrated forms of the mineral in the presence of water. According to the Clapeyron–Clausius equation, the slope of this line is expressed by the relationship

$$\frac{\mathrm{d}P}{\mathrm{d}T} = \frac{\Delta S}{\Delta V} \tag{16}$$

at constant fluid composition and pressure, where ΔS and ΔV are the entropy and molar volume changes occurring during the reaction. The positive slope of the line DE as drawn corresponds to the case where the entropy and volume changes during the reaction are of the same sign. The position of the line DE may be influenced by both the pressure and water activity of the fluid phase, both of which will modify its free energy.

The thermodynamic data for the dehydration of gypsum ($CaSO_4.2H_2O$) to anhydrite ($CaSO_4$) are well established (MacDonald, 1953), the reaction occurring at 40°C at ambient pressure. The coexistence line has a slope of *c.* 85 bar per degree, assuming that the fluid released is at the lithostatic pressure. If the fluid pressure is lower than this (e.g. hydrostatic), then the

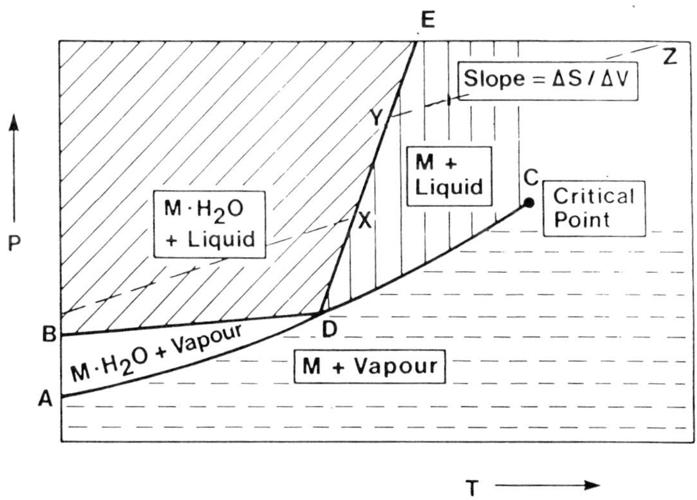

Figure 11.21 Pressure/temperature phase diagram for a dehydration reaction.

Gibbs free energy of the fluid will be reduced by an amount $V\Delta P$, where V is the molar volume of water and $\Delta P = P_{\text{lith}} - P_{\text{hyd}}$. This causes the equilibrium temperature to be reduced somewhat at any given pressure. For typical geothermal temperature and lithostatic pressure gradients, the reaction is expected to occur within approximately the first kilometre of burial (see Figure 11.22). Both the entropy and volume changes on dehydration are positive; the latter is *c.* 7 cm^3 per mol of anhydrite, or a fractional potential volume increase of *c.* 10% compared with the starting volume of gypsum. Calculations show that in a closed system, the reaction can potentially generate a fluid pressure significantly in excess of the overburden pressure at 1 km.

The general stratigraphy of the Permian Zechstein evaporite formations is illustrated in Figure 11.23. In this region, elevated fluid pressures may be attributed to gypsum dehydration, and are associated with extensive fracture systems in the anhydrites and dolomites. Thomeer and Bottema (1961) reported elevated pore pressures and borehole stability problems in gas wells in the Permian of the north-west German plain. Drilling fluid specific gravities in excess of 2.0 were required, indicating pore pressures close to lithostatic values. Similar occurrences have been reported in the Permian Zechstein of the southern North Sea Basin, in wells targeting the gas-bearing Rotliegendes sandstones. The reaction has also been implicated as an overpressuring mechanism in the Buckner Formation of Mississippi, USA (Mouchet and Mitchell, 1989).

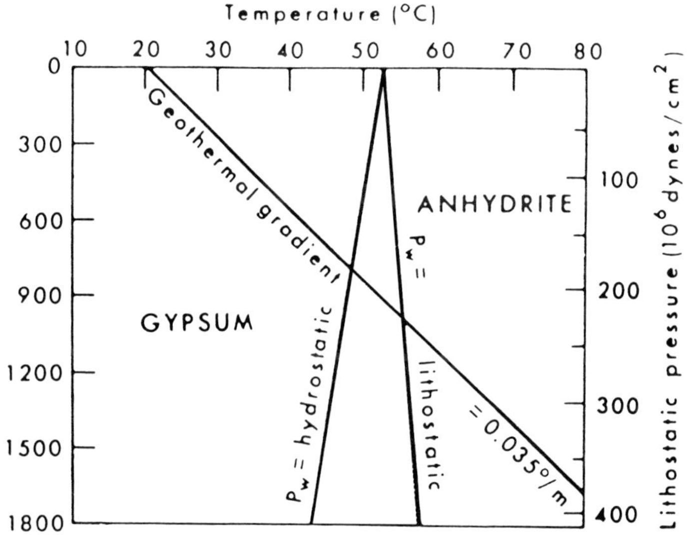

Figure 11.22 Stability fields of gypsum and anhydrite for a water activity of 0.93 (from Hanshaw and Bredehoeft, 1968).

GROUP		FORMATION	LITHOLOGY	DESCRIPTION
BACTON TRIASSIC		BUNTER		Siltstone, fine sandstone
ZECHSTEIN PERMIAN	Z 4	GRENZANHYDRIT		Anhydrite
		ALLER HALITE		Halite, locally potassium salts
		PEGMATIT-ANHYDRIT		Anhydrite
		ROTER SALZTON		Claystone, saliferous
	Z 3	LEINE HALITE		Halite, with carnalite
		HAUPT-ANHYDRIT		Anhydrite
		PLATTEN-DOLOMIT		Dolomite, carbonaceous, pyritic
		GRAUER SALZTON		Claystone, anhydritic
	Z 2	DECKANHYDRIT		Anhydrite
		STRASSFRUT-HALITE		Halite, occasionally K Mg salts
		BASALANHYDRIT		Anhydrite, commonly dolomitic
		HAUPTDOLOMIT		Dolomite
	Z 1	WERRAANHYDRIT		Anhydrite, layers of dolomite
		ZECHSTEINKALK		Dolomite, calcitic
		KUPFERSCHIEFER		Shale, dolomitic
ROTLIGEN-DES				Sandstone, thin layers of claystone

Figure 11.23 Stratigraphy of the Permian Zechstein evaporite formations.

11.5 Relationship between overpressures in sealed and non-sealed environments

Up to now, the emphasis of the discussion has focused on the various mechanisms by which overpressures may be generated. Attention is now turned to the variable nature or setting of overpressured environments, and the means by which pressures can be maintained over geological timescales.

There is considerable evidence from well-logs that diagenetic processes can create hard, impermeable seals or cap rocks which may effectively isolate sedimentary basins vertically into two or more 'compartments' (Hunt, 1990). Such cap rocks are characterized by anomalously high densities, electrical resistivities, or compressional sonic velocities (Fertl and Timko, 1970; Fertl, 1976). An example from the US Gulf Coast is illustrated in Figure 11.24. The processes of generation of fluid pressures in each compartment (by whatever

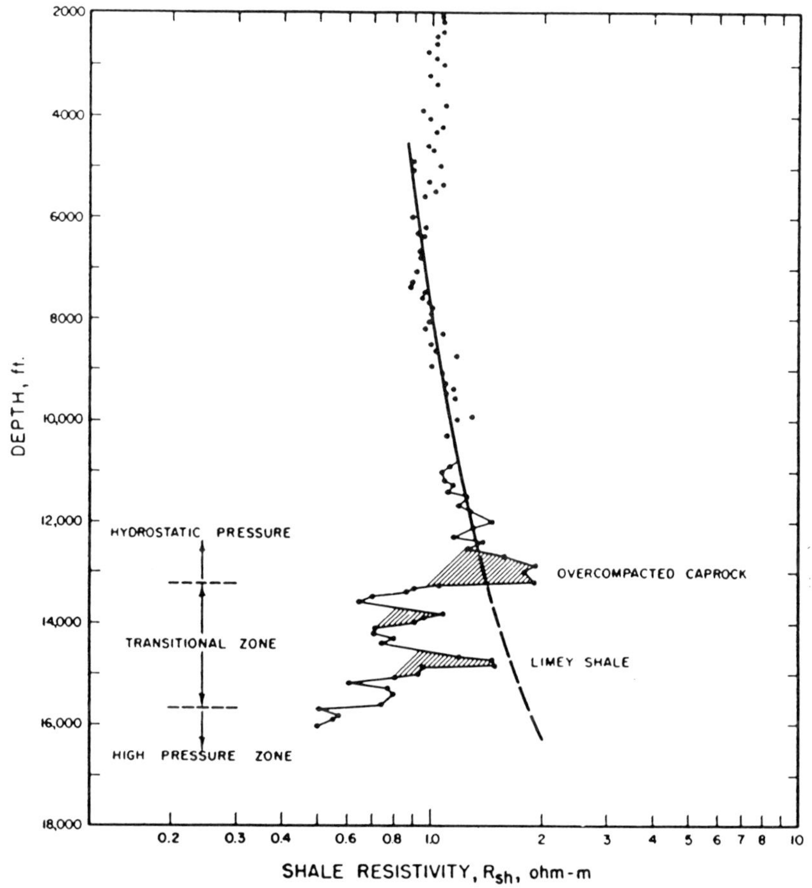

Figure 11.24 Evidence of indurated 'cap rocks' in a well in the Gulf Coast region of the USA (from Fertl and Timko, 1970).

mechanism or combination of mechanisms) become independent of one another. Vertically sealed compartments may contain permeable zones in which the 'local' fluid pressure gradient will be hydrostatic, with a discontinuous jump in hydrostatic head across the seal. This differs from the situation in compacting argillaceous sediments, in which the vertical permeability, while low, is sufficient to ensure that the fluid pressure gradient, whether hydrostatic or not, is a continuous function of depth with no abrupt jumps. Gretener and Feng (1985) use the terms 'aquifers', 'aquitards' and 'aquicludes' to describe permeable zones, compacting low-permeability systems, and impermeable seals, respectively. Thus, overpressuring may be either a long timescale relaxation process (within an aquitard or aquitard/aquifer system) or due to complete fluid isolation in an aquifer/aquiclude system. I propose the terminology Type 1 and Type 2 for these cases, respectively.

Magara (1981) proposed that the low pH environments associated with decarboxylation of kerogen in Zone IV may promote the dissolution of carbonates and feldspars by brines expelled vertically and horizontally during compaction. These dissolved species precipitate calcite and/or silica at shallower depths from brines of lower temperature and higher pH. Katz and Ibrahim (1971) considered that such cap rocks, depending on their thickness, could withstand fluid pressure differentials of up to a few tens of megapascals, such that sudden jumps of pore pressure by amounts up to this magnitude may be experienced on penetrating a cap rock.

Carstens and Dypvik (1981) found evidence for a sealing cap due to extensive carbonate mineralization in Upper Jurassic shales of the North Sea, dating from the major break in sedimentation between the Jurassic and the Cretaceous. Elevated fluid pressures were thought to have been generated subsequent to burial under the Cretaceous and Tertiary overburden. The mudrock sequences show no porosity anomalies or evidence of present under-compaction (see Figure 11.25).

Carstens and Dypvik did not positively identify the mechanism(s) of overpressuring, although they considered the effect of temperature to be important. Chiarelli and Duffaud (1980) also give evidence for cemented cap rock in the North Sea in formations dating from late Jurassic to early Cretaceous age. The highest reported pore pressures in the Viking Graben appear to be associated with zones of maximum subsidence, i.e. highest temperature and degree of hydrocarbon maturation. There are a considerable number of lateral fluid pressure discontinuities associated with fault-block structures.

Hunt (1990) has proposed that seals dividing sedimentary basins into two or more hydrogeological zones are essentially three-dimensional, as proposed earlier by Bradley (1975). Provided that the vertical permeability within each zone is sufficiently high, the 'local' pore pressure gradient will be the hydrostatic value for the fluid concerned, with a step change or discontinuity at the seal. The sealed fluid compartments essentially maintain a constant volume as

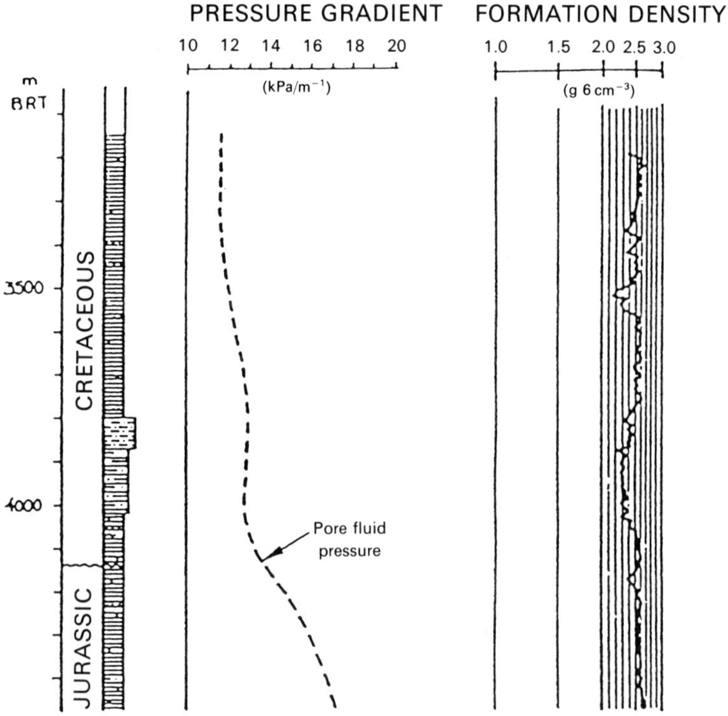

Figure 11.25 Pore pressure gradients and formation densities in part of the Cretaceous and Jurassic of the North Sea (Carstens and Dypvik, 1981).

they are buried, and the magnitude of the pressures generated within them will depend on the potential volume changes associated with diagenetic reactions (especially hydrocarbon maturation) and relative thermal expansion of fluids. An actual example is illustrated in Figure 11.26. In this and other similar examples cited by Hunt (1990) the fluid pressures were measured directly by formation testing.

Where steady generation of pressures occurs in a sealed compartment, the pore pressure beneath the seal rises until a value close to the lithostatic (geostatic) pressure is reached, at which point the top seal will fracture, and the pore fluids are able to escape vertically upwards, reducing the pressure. The seal may later be re-established, after which the generative mechanisms will cause the fluid pressures beneath it to build up again, and the entire process will repeat itself. Repetitions of fracturing and re-sealing provides an episodic dewatering process, exactly as proposed by Cathles and Smith (1983) to explain the nature of the observed Mississippi-Valley-type lead–zinc deposits. Similar concepts have also been enunciated by Gold (1987) in the book outlining his

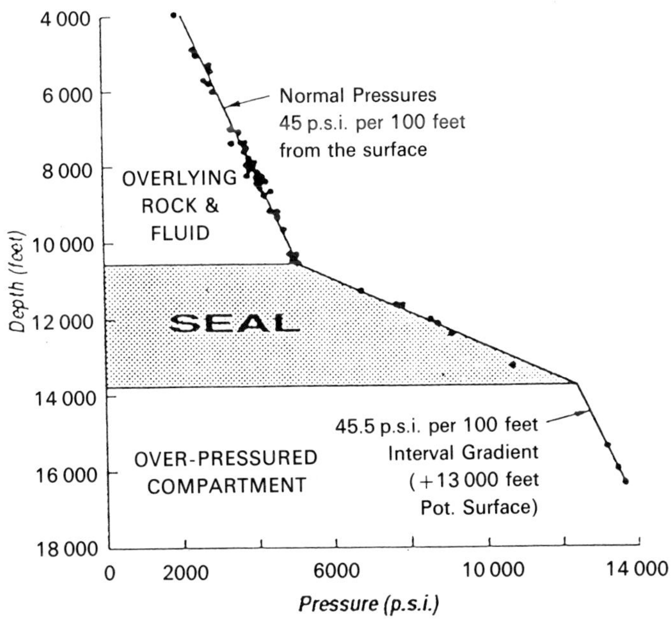

Figure 11.26 Pore pressure against depth in the Cook Inlet fields, Alaska (after Hunt, 1990).

controversial theory of gas generation from deep within the Earth. Natural hydraulic fractures in many parts of the world, e.g. in North Sea chalks (Watts, 1983) may be attributed to this process.

11.6 Summary and conclusions

From the above discussion, it is clear that two distinct types of overpressuring environment may be identified. These are distinguished by the presence or absence of a well-defined cap rock or impermeable sealing layer. Environments without such a sealing layer, termed Type 1 overpressure environments, or aquitard/aquifer systems (Gretener and Feng, 1985) are associated with young (typically Tertiary or Quaternary) thick and/or rapidly buried argillaceous sediments, consisting of shales alternating with permeable zones (sandstones or carbonates). Both the long timescale compactability of such sediments, and the distance scale over which pore pressure gradients change, are relatively large. Changes of pore pressure regime are associated with significant and readily detectable changes in shale porosity or density. One may subdivide this environment into two sub-classes:

1. Type 1A. A basin which is currently at its maximum historical mean effective stress, whether due to sedimentary loading, tectonic factors or a combination of both. The overpressures, whether directly measurable in a contained aquifer, or inferred within the shale from well-log or drilling data

(Whittaker, 1985: Mouchet and Mitchell, 1989) are entirely due to incomplete compaction.
2. Type 1B. This differs from Type 1A in that the pore pressures arising from compaction disequilibrium have been increased by diagenesis or other mechanisms subsequent to burial, so that the present mean effective stress in both aquitard and aquifer is lower than its historical maximum value. Although perhaps complicated by the overprint of diagenesis, in this case the relationship between effective stress (hence pore pressure) and porosity will differ from that which applies in the Type 1A case, because of the mechanical irreversibility of compaction, as discussed in section 11.2.

In contrast, the cemented or lithified seals responsible for Type 2 environments generate a pressured environment of low compressibility and essentially constant volume. This implies that any expansion of fluids relative to rock grains when the temperature increases due to burial can make a significant contribution to fluid pressures and rock stresses in this type of environment, while they may be minor or insignificant in Type 1 environments.

Again, we may perhaps postulate two sub-types of this environment:

1. Type 2A. In this case the overpressures were essentially generated prior to the formation of the seal, by compactional dewatering of adjacent shales, perhaps subsequently modified by thermal expansion of fluids relative to rock grains.
2. Type 2B. In this case, the fluid pressures have mainly been generated following the formation of the seal. Here any diagenetic reaction associated

Table 11.1 Probable examples of proposed classes of overpressured environments.

Type	Location	Lithology	Mechanism(s)	References
1A	N. Sea (Tertiary)	Sh, Sa, Ca	U	1
1B	Gulf of Mexico (<10 000 ft ?)	Sh, Sa	U, S?	2, 3, 4
1B	Niger Delta	Sh, Sa	U, S	4
2A	Gulf of Mexico (>10 000 ft ?)	Sh, Sa	U? S? H? T	5, 6
2B	N. Sea (Jurassic)	Sh, Sa, Ca	H?, T	7, 8
2B	Williston, N. Dakota	Sh, Sa	H	9
2B	Rocky Mountain area	Sh, Sa	H	10
2B	Zechstein (Germany)	E	G, T?	11
3	Dunbarton, S. Carolina	Sh, Sa	O	12

Mechanisms: U = under-compaction; S = smectite diagenesis; H = hydrocarbon maturation; T = thermal; G = gypsum–anhydrite; O = osmosis.
Lithologies: Sh = shale; Sa = sandstone; Ca = carbonates; E = evaporites.
References: 1 = Herring (1973); 2 = Plumley (1980); 3 = Berg and Habeck (1982); 4 = Bruce (1984); 5 = Fertl and Timko (1970); 6 = Fertl (1976); 7 = Chiarelli and Duffaud (1980); 8 = Carstens and Dypvik (1981); 9 = Hunt (1979); 10 = Thomeer and Bottema (1961); 11 = Spencer (1987); 12 = Marine and Fritz (1981).

with an increase in molar volume (e.g. illitization, gypsum dehydration, hydrocarbon maturation, precipitation of natural cements) may contribute significantly to the observed pressures. Relative thermal expansion effects may also be significant or even dominant. Such pressures may build up essentially to lithostatic values, and may be expected to be relieved by episodic fracturing events, releasing occasional pulses of overpressured fluids.

Type 2 environments provide the greatest difficulties and challenges for prediction, monitoring, and drilling engineering operations.

A further type of overpressured environment which does not fit neatly into either category discussed above is the 'osmotic cell' type as described by Marine and Fritz (1981), which might be proposed as a Type 3 environment.

Table 11.1 attempts a tentative classification of specific basins in terms of the scheme proposed above, including the probable or possible mechanism(s) operating in each case. It is hoped that this attempted classification scheme will form a basis to be confirmed or refuted, and certainly to be extended, by future workers.

Acknowledgements

I am grateful to David Manning for inviting me to give the talk from which this chapter has evolved, and also to many present and former colleagues (engineers, earth scientists, mathematicians, physicists and chemists) who have contributed to my continuing education in this broad, multidisciplinary and complex subject. Finally, I thank Suzanne D. Weedman of Pennsylvania State University who reviewed an earlier draft of this chapter.

References

Aagaard, P. and Helgeson, H. C. (1983) Activity/composition relationships among silicates and aqueous solutions: II. Chemical and thermodynamic consequences of ideal mixing of atoms on homological sites in montmorillonites, illites and mixed-layer clays. *Clays and Clay Minerals*, **31**, 207–17.

Ahn, J. H. and Peacor, D. R. (1986) Transmission and analytical microscopy of the smectite-to-illite transition. *Clays and Clay Minerals*, **34**, 180–6.

Ahn, J. H. and Peacor, D. R. (1989) Illite/smectite from Gulf Coast shales: a reappraisal of transmission electron microscope images. *Clays and Clay Minerals*, **37**, 542–6.

Altaner, S. P., Weiss, C. A., Jr. and Kirkpatrick, R. J. (1988) Evidence from ^{29}Si NMR for the study of mixed-layer illite/smectite in clay minerals. *Nature*, **331**, 699–702.

Anderson, D. M. and Low, P. F. (1958) The density of water adsorbed by lithium-, sodium- and potassium bentonite. *Soil Science Society of America, Proceedings*, **22**, 99–103.

Athy, L. F. (1930) Density, porosity and compaction of sedimentary rocks. *American Association of Petroleum Geologists Bulletin*, **14**, 1–24.

Atkinson, J. H. and Bransby, P. L. (1978) *The Mechanics of Soils*, McGraw-Hill, New York.

Barker, C. (1972) Aquathermal pressuring-role of temperature in development of abnormal pressure zones. *American Association of Petroleum Geologists Bulletin*, **56**, 2068–71.

Barker, C. (1987) Development of abnormal and subnormal pressures in reservoirs containing bacterially generated gas. *American Association of Petroleum Geologists Bulletin*, **71**, 1404–13.

Bennett, R. H., Bryant, W. R. and Keller, G. H. (1981) Clay fabric of selected submarine sediments—fundamental properties and models. *Journal of Sedimentary Petrology*, **51**, 217–32.

Berg, R. R. and Habeck, M. F. (1982) Abnormal pressures in the Lower Vicksburg, McAllen Ranch Field, South Texas. *Transactions of the Gulf Coast Association of Geological Societies*, **32**, 247–53.

Berner, R. A. (1971) *Principles of Chemical Sedimentology*, McGraw-Hill, New York.

Bethke, C. M., Harrison, W. J., Upson, C. and Altaner, S. P. (1988) Supercomputer analysis of sedimentary basins. *Science*, **239**, 261–7.

Biot, M. A. (1940) General theory of three-dimensional consolidation. *Journal of Applied Physics*, **12**, 155–64.

Bishop, R. S. (1979) Calculated compaction states of thick, abnormally pressured shales. *American Association of Petroleum Geologists Bulletin*, **63**, 918–33.

Boles, J. R. and Franks, S. G. (1979) Clay diagenesis in Wilcox sandstones of southwest Texas: implications of smectite diagenesis on sandstone cementation. *Journal of Sedimentary Petrology*, **49**, 55–70.

Borst, R. L. (1982) Some effects of compaction and geological time on the pore parameters of argillaceous rocks. *Sedimentology*, **29**, 291–8.

Bradley, J. S. (1975) Abnormal formation pressure. *American Association of Petroleum Geologists Bulletin*, **59**, 957–73.

Bray, C. J. and Karig, D. E. (1985) Porosity of sediments in accretionary prisms and some implications for dewatering processes. *Journal of Geophysical Research B*, **90**, 768–78.

Bredehoeft, J. D. and Hanshaw, B. B. (1968) On the maintenance of abnormal fluid pressures. I. Thick sedimentary sequences. *Geological Society of America Bulletin*, **79**, 1097–106.

Bredehoeft, J. D., Djervanshir, R. D. and Belitz, K. R. (1988) Lateral fluid flow in a compacting sand–shale sequence: South Caspian Basin. *American Association of Petroleum Geologists Bulletin*, **72**, 416–24.

Bruce, C. H. (1973) Pressured shales and related sediment deformation: mechanisms for development of regional contemporaneous faults. *American Association of Petroleum Geologists Bulletin*, **57**, 878–86.

Bruce, C. H. (1984) Smectite dehydration – its relation to structural development and hydrocarbon accumulation in Northern Gulf of Mexico basin. *American Association of Petroleum Geologists Bulletin*, **68**, 673–83.

Burley, S. D., Kantorowicz, J. D. and Waugh, B. (1985) Clastic diagenesis, in *Geological Society of London Special Publication*, No. 18, *Sedimentology: Recent Developments and Applied Aspects*, (eds P. J. Brenchley and B. P. J. Williams), Blackwell Scientific Publications, Oxford, pp. 189–226.

Burst, J. F. (1969) Diagenesis of Gulf Coast clayey sediments and its possible relation to petroleum migration. *American Association of Petroleum Geologists Bulletin*, **53**, 73–93.

Carstens, H. and Dypvik, H. (1981) Abnormal formation pressure and shale porosity. *American Association of Petroleum Geologists Bulletin*, **65**, 344–50.

Cathles, L. M. and Smith, J. E. (1983) Thermal constraints on the formation of Mississippi Valley-type lead–zinc deposits and their implications for episodic basin dewatering and deposit genesis. *Economic Geology*, **78**, 983–1002.

Chapman, R. E. (1980) Mechanical versus thermal causes of abnormally high pore pressures in shales. *American Association of Petroleum Geologists Bulletin*, **64**, 2179–83.

Chapman, R. E. (1981) *Geology and Water*, Martinus Nijhoff, The Hague.

Chenevert, M. (1970) Shale alteration by water adsorption. *Journal of Petroleum Technology*, **22**, 1141–9.

Chiarelli, A. and Duffaud, F. (1980) Pressure origin and distribution of Viking Basin (United Kingdom–Norway). *American Association of Petroleum Geologists Bulletin*, **64**, 1245–66.

Chilingar, G. V. and Knight, L. (1960) Relationship between pressure and moisture content of kaolinite, illite and montmorillonite clays. *American Association of Petroleum Geologists Bulletin*, **44**, 101–6.

Colten, V. A. (1986) Hydration states of smectite in NaCl brines at elevated pressures and temperatures. *Clays and Clay Minerals*, **34**, 385–9.

Colten-Bradley, V. A. (1987) Role of pressure in smectite dehydration: effects on geopressure and smectite-to-illite transformation. *American Association of Petroleum Geologists Bulletin*, **71**, 1414–27.

Curtis, C. D. (1977) Sedimentary geochemistry: environments and processes dominated by the involvement of an aqueous phase. *Philosophical Transactions of the Royal Society of London, Series A*, **286**, 353–72.

Curtis, C. D. (1978) Possible links between sandstone diagenesis and depth-related geochemical reactions occurring in enclosing mudstones. *Journal of the Geological Society of London*, **135**, 107–17.

Curtis, C. D. (1983) Geochemistry of porosity enhancement and reduction in clastic sediments, in *Geological Society of London Special Publication*, No. 12, *Petroleum Geochemistry and Exploration of Europe* (ed. J. Brooks), Blackwell Scientific Publications, Oxford, pp. 113–25.

Curtis, C. D. (1985) Clay mineral precipitation and transformation during burial diagenesis. *Philosophical Transactions of the Royal Society of London, Series A*, **315**, 91–105.

Daines, S. R. (1982) Aquathermal pressuring and geopressure evaluation. *American Association of Petroleum Geologists Bulletin*, **66**, 931–9.

Davis, D. M., Suppe, J. and Dahlen, F. A. (1983) Mechanics of fold-and-thrust belts and accretionary wedges. *Journal of Geophysical Research B*, **88**, 1153–72.

Dean, W. E., Leinen, M. and Stow, D. A. V. (1985) Classification of deep-sea, fine-grained sediments. *Journal of Sedimentary Petrology*, **55**, 250–6.

Denis, J. H. (1991) Compaction and swelling of Ca-smectite in water and in $CaCl_2$ solutions: water activity measurements and matrix resistance to compaction. *Clays and Clay Minerals*, **39**, 35–42.

Denis, J. H., Keall, M. J., Hall, P. L. and Meeten, G. H. (1991) Influence of potassium concentration on the swelling and compaction of mixed (Na,K) ion-exchanged montmorillonite. *Clay Minerals*, **26**, 255–68.

Dickey, P. A. (1966) Patterns of chemical composition in deep subsurface waters. *American Association of Petroleum Geologists Bulletin*, **50**, 2472–8.

Dickey, P. A. (1976) Abnormal formation pressure: discussion. *American Association of Petroleum Geologists Bulletin*, **60**, 1124–8.

Dickey, P. A. and Cox, W. C. (1977) Oil and gas in reservoirs with subnormal pressure. *American Association of Petroleum Geologists Bulletin*, **61**, 2134–42.

Dickey, P. A., Shiram, C. R. and Paine, W. R. (1968) Abnormal pressures in deep wells of Southwestern Louisiana. *Science*, **160**, 609–15.

Dickinson, G. (1953) Geological aspects of abnormal reservoir pressures in Gulf Coast Louisiana. *American Association of Petroleum Geologists Bulletin*, **37**, 410–32.

Dunoyer de Segonzac, G. (1970) The transformation of clay minerals during diagenesis and low-grade metamorphism: a review. *Sedimentology*, **15**, 281–346.

Du Rouchet, J. (1981) Stress fields: a key to oil migration. *American Association of Petroleum Geologists Bulletin*, **65**, 74–85.

Dypvik, H. (1983) Clay mineral transformations in Tertiary and Mezozoic sediments from North Sea. *American Association of Petroleum Geologists Bulletin*, **67**, 160–5.

Eberl, D. D., Whitney, G. and Khoury, H. (1978) Hydrothermal reactivity of smectite. *American Mineralogist*, **63**, 401–9.

Eberl, D. D., Środon, J., Kralik, M., Taylor, B. E. and Peterman, B. E. (1990) Ostwald ripening of clays and metamorphic minerals. *Science*, **248**, 474–7.

Eslinger, E. and Sellars, B. (1981) Evidence for the formation of illite from smectite during burial metamorphism in the Belt Supergroup, Clark Fork, Idaho. *Journal of Sedimentary Petrology*, **51**, 203–16.

Fertl, W. H. (1976) *Developments in Petroleum Science*, vol. 2, *Abnormal Formation Pressures*, Elsevier, Amsterdam.

Fertl, W. H. and Timko, D. J. (1970) Occurrence of cemented roof rock and geopressure caprock and its implications in petroleum geology and geohydrology, in *45th Annual Fall Meeting of the Society of Petroleum Engineers*, Houston, Texas, Paper No. SPE 3085, 5pp.

Foster, J. B. and Whalen, H. E. (1966) Estimation of formation pressures from electrical surveys: offshore Louisiana. *Journal of Petroleum Technology*, **18**, 165–71.

Freed, R. (1980) Shale mineralogy in the No.1 Pleasant Bayou test well, in *Proceedings of the Fourth Geopressured–Geothermal Energy Conference, October 29–31, 1979* (ed. M. H. Dorfman and W. L. Fisher), vol. 1, University of Texas, Austin, pp. 153–65.

Freed, R. and Peacor, D. R. (1989) Geopressured shale and sealing effect of smectite to illite transition. *American Association of Petroleum Geologists Bulletin*, **73**, 1223–32.

Fritz, S. (1986) Ideality of clay membranes in osmotic processes: a review. *Clays and Clay Minerals*, **34**, 214–23.

Galloway, W. E., Hobday, D. K. and Magara, K. (1982) Frio Formation of Texas Gulf Coastal Plain: depositional systems, structural framework and hydrocarbon distribution. *American Association of Petroleum Geologists Bulletin*, **66**, 649–88.

Gibson, R. E., England, G. L. and Hussey, M. J. L. (1967) The theory of one-dimensional consolidation of saturated clays. *Géotechnique*, **17**, 261–73.

Glasmann, J. R., Larter, S., Briedis, N. A. and Lundegard, P. D. (1989) Shale diagenesis in the Bergen High area, North Sea. *Clays and Clay Minerals*, **37**, 97–112.

Goff, J. C. (1983) Hydrocarbon generation and migration from Jurassic source rocks in the E. Shetland Basin and Viking Graben of the northern North Sea. *Journal of the Geological Society of London*, **140**, 445–74.

Gold, T. (1987) *Power From The Earth*, J. M. Dent and Sons, London.

Graf, D. L., Meents, W. F., Friedman, T. and Shimp, N. F. (1966) The origin of saline formation waters. III. Calcium chloride waters. *Illinois State Geological Survey Circular*, **397**, 60 pp.

Gretener, P. and Feng, Z.-M. (1985) Three decades of geopressure—insights and enigmas. *Bulletin der Vereinigung Schweizerischen Petroleum-Geologen und-Ingenieure*, **51**, 1–34.

Guven, N. (1991) On a definition of illite/smectite mixed layer. *Clays and Clay Minerals*, **39**, 661–2.

Hall, P. L., Astill, D. M. and McConnell, J. D. C. (1986) Thermodynamic and structural aspects of the dehydration of smectites in sedimentary rocks. *Clay Minerals*, **21**, 633–48.

Hall, P. L., Rasmus, J. C. and Aldred, W. (1988) Shales and overpressuring: a current perspective. *Drilling and Pumping Journal (Schlumberger)*, **5**, 32–40.

Ham, H. H. (1966) New charts help estimate formation pressures. *Oil and Gas Journal*, **64**, 58–63.

Hanshaw, B. B. and Bredehoeft, J. D. (1968) On the maintenance of anomalous fluid pressures. II. Source layer at depth. *Geological Society of America Bulletin*, **79**, 1107–22.

Hanshaw, B. B. and Zen, E. A. (1965) Osmotic equilibrium and overthrust faulting. *Geological Society of America Bulletin*, **76**, 1379–86.

Harkins, K. L. and Baugher, J. W., III (1969) Geological significance of abnormal formation pressures. *Journal of Petroleum Technology*, **21**, 961–6.

Hawkins, R. H. and Egelstaff, P. A. (1980) Interfacial water structure in montmorillonite from neutron diffraction experiments. *Clays and Clay Minerals*, **28**, 19–28.

Heard, H. C. and Rubey, W. W. (1966) Tectonic implications of gypsum dehydration. *Geological Society of America Bulletin*, **77**, 741–60.

Hedberg, H. D. (1936) Gravitational compaction of clays and shale. *American Journal of Science*, **31**, 241–87.

Hedberg, H. D. (1974) Relation of methane generation to undercompacted shales, shale diapirs and mud volcanoes. *American Association of Petroleum Geologists Bulletin*, **58**, 661–73.

Helfferich, F. (1962) *Ion Exchange*, McGraw-Hill, New York.

Heling, D. (1974) Diagenetic alteration of smectite in argillaceous sediments of the Rhinegraben. *Sedimentology*, **21**, 463–72.

Heling, D. (1978) Diagenesis of illite in argillaceous sediments of the Rhinegraben. *Clay Minerals*, **13**, 211–20.

Herring, E. (1973) Estimating abnormal pressures from log data in the North Sea, in *Society of Petroleum Engineers of AIME*, Dallas, Texas, Paper No. SPE 4301.

Hosoi, H. (1963) First migration of petroleum in Akita and Yamagata Prefectures. *Japanese Association of Mineralogy, Petrology, and Economic Geology Journal*, **49**, 43–55, 101–14.

Hottman, C. E., Smith, J. H. and Purcell, W. R. (1969) Relationship among earth stresses, pore pressure and drilling problems offshore Gulf of Alaska. *Journal of Petroleum Technology*, **21**, 1477–84.

Hower, J., Eslinger, E. V., Hower, M. E. and Perry, E. A. (1976) Mechanism of burial metamorphism of argillaceous sediment: I. Mineralogical and chemical evidence. *Geological Society of America Bulletin*, **87**, 725–37.

Hubbert, M. K. and Rubey, W. W. (1959) Role of fluid pressure in mechanics of overthrust faulting. I. *Geological Society of America Bulletin*, **70**, 115–66.

Hubbert, M. K. and Willis, D. B. (1957) Mechanics of hydraulic fracture. *AIME Transactions*, **210**, 153–68.

Hunt, J. M. (1979) *Petroleum Geochemistry and Geology*, W. H. Freeman & Co., San Francisco.

Hunt, J. M. (1990) Generation and migration of petroleum from abnormally pressured fluid compartments. *American Association of Petroleum Geologists Bulletin*, **74**, 1–12.

Inoue, A., Kohyama, N., Kitagawa, R. and Watanabe, T. (1987) Chemical and morphological evidence for the conversion of smectite to illite. *Clays and Clay Minerals*, **35**, 111–20.

Jacquin, C. and Poulet, M. J. (1970) Study of the hydrodynamic pattern in a sedimentary basin subject to subsidence. *Society of Petroleum Engineers of AIME*, Dallas, Texas, Paper, No. SPE 2988.

Jaeger, J. C. and Cook, N. G. W. (1979) *Fundamentals of Rock Mechanics*, 3rd edn, Chapman and Hall, London.

Johns, W. D. and McKallip, T. E. (1989) Burial diagenesis and specific catalytic activity of illite–smectite clays from the Vienna Basin, Austria. *American Association of Petroleum Geologists Bulletin*, **73**, 472–82.

Johns, W. D. and Shimoyama, A. (1972) Clay minerals and petroleum forming reactions during burial and diagenesis. *American Association of Petroleum Geologists Bulletin*, **56**, 2160–7.

Jones, M. E. and Addis, M. A. (1986) The application of stress path and critical state analysis to sedimentary deformation. *Journal of Structural Geology*, **8**, 575–80.

Jones, P. H. (1969) Hydrodynamics of geopressures in the Northern Gulf of Mexico basin. *Journal of Petroleum Technology*, **21**, 803–10.

Katz, D. L. and Ibrahim, M. A. (1971) Threshold displacement pressure considerations for caprocks of abnormally pressured reservoirs. *Society of Petroleum Engineers of AIME, Papers*, No. SPE 3222.

Keith, L. A. and Rimstidt, J. D. (1985) A numerical compaction model of overpressuring in shales. *Mathematical Geology*, **17**, 115–35.

Keller, W. D., Reynolds, R. C. and Inoue, A. (1986) Morphology of clay minerals in the smectite-to-illite conversion series by scanning electron microscopy. *Clays and Clay Minerals*, **34**, 187–97.

Kisch, H. J. (1983) Mineralogy and petrology of burial diagenesis (burial metamorphism) and incipient metamorphism in clastic rocks, in *Developments in Sedimentology*, vol. 25B, *Diagenesis of Sediments and Sedimentary Rocks, 2*, Chapter 5, (eds G. Larsen and G. V. Chilingar), Elsevier, Amsterdam, pp. 289–493.

Koster van Groos, A. F. and Guggenheim, S. (1984) The effect of pressure on the dehydration reaction of interlayer water in Na-montmorillonite. *American Mineralogist*, **69**, 872–9.

MacDonald, G. J. F. (1953) Anhydrite–gypsum equilibrium relations. *American Journal of Science*, **251**, 884–98.

Magara, K. (1968) Compaction and migration of fluids in Miocene mudstone, Nagaoka Plain, Japan. *American Association of Petroleum Geologists Bulletin*, **52**, 2466–501.

Magara, K. (1974) Compaction, ion filtration and osmosis in shale and their significance in primary migration. *American Association of Petroleum Geologists Bulletin*, **58**, 283–90.

Magara, K. (1975a) Reevaluation of montmorillonite dehydration as cause of abnormal pressure and hydrocarbon migration. *American Association of Petroleum Geologists Bulletin*, **59**, 292–302.

Magara, K. (1975b) Importance of aquathermal pressuring effect in Gulf Coast. *American Association of Petroleum Geologists Bulletin*, **59**, 2037–45.

Magara, K. (1978) *Developments in Petroleum Science*, vol. 9, *Compaction and Fluid Migration*, Elsevier, Amsterdam.

Magara, K. (1981) Fluid dynamics for cap-rock formation in Gulf Coast. *American Association of Petroleum Geologists Bulletin*, **65**, 1334–43.

Marine, I. W. and Fritz, S. J. (1981) Osmotic model to explain anomalous hydrostatic heads. *Water Resources Research*, **17**, 78–82.

McHardy, W. J., Wilson, M. J. and Tait, J. M. (1982) Electron microscope and X-ray diffraction studies of filamentous illite clay from sandstones of the Magnus field. *Clay Minerals*, **17**, 23–40.

McKelvey, J. G. and Milne, I. H. (1962) Flow of salt solutions through compacted clay. *Proceedings of National Conference, Clays and Clay Minerals*, **9**, 248–59.

McKenzie, D. P. (1987) The compaction of igneous and sedimentary rock. *Journal of the Geological Society of London*, **144**, 299–307.

Meade, R. H. (1966) Factors influencing the early stages of the compaction of clays and sands – review. *Journal of Sedimentary Petrology*, **36**, 1085–1101.

Meissner, F. F. (1984) Petroleum geology of the Bakken formation. Williston Basin, North Dakota and Montana. *Memoirs, American Association of Petroleum Geologists*, **35**, 159–80.

Mesri, G. and Olson, R. E. (1971) Consolidation characteristics of montmorillonite. *Géotechnique*, **21**, 341–52.

Milliken, K. L., Land, L. S. and Loucks, R. G. (1981) History of burial diagenesis determined from isotope geochemistry, Frio formation, Brazoria County, Texas. *American Association of Petroleum Geologists Bulletin*, **65**, 1397–413.

Mitchell, J. K., Greenberg, J. A. and Witherspoon, P. A. (1973) Chemico-osmotic effects in fine-grained soils. *American Society of Civil Engineers, Journal of the Soil Mechanics and Foundation Division*, **99**, 307–22.

Moon, C. F. and Hurst, C. W. (1984) Fabric of mud and shales; an overview, in *Fine Grained Sediments: Deep Water Processes and Facies*, (eds D. A. V. Stow and D. J. W. Piper), Geological Society, Blackwell Scientific Publications, Oxford.

Morgernstern, N. R. and Eigenbrod, K. D. (1974) Classification of argillaceous soils and rocks. *American Society of Civil Engineering, Journal of Geotechnical Engineering Division*, **100**, 1137–56.

Morton, R. A. and Land, L. S. (1987) Regional variations in formation water chemistry, Frio Formation (Oligocene), Texas Gulf Coast. *American Association of Petroleum Geologists Bulletin*, **71**, 191–206.

Mouchet, J. P. and Mitchell, A. (1989) Abnormal pressures while drilling. *Manuels Techniques Elf Aquitaine*, No. 2, 264 pp.

Muffler, L. J. P. and White, D. E. (1969) Active metamorphism of Upper Cenozoic sediments in the Salton Sea geothermal field, and the Salton Trough, Southeastern California. *Geological Society of America Bulletin*, **80**, 157–82.

Nadeau, P. H. and Reynolds, R. C., Jr. (1981) Burial and contact metamorphism in the Mancos shale. *Clays and Clay Minerals*, **29**, 249–59.

Nadeau, P. H., Wilson, M. J., McHardy, W. J. and Tait, M. J. (1984) Interstratified clays as fundamental particles. *Science*, **225**, 923–5.

Nadeau, P. H., Wilson, M. J., McHardy, W. J. and Tait, M. J. (1985) The conversion of smectite to illite clays during diagenesis: evidence from some illitic clays from bentonites and sandstones. *Mineralogical Magazine*, **49**, 393–400.

Neuzil, C. E. (1986) Groundwater flow in low-permeability environments. *Water Resources Research*, **22**, 1163–95.

Nur, A. and Byerlee, J. D. (1971) An exact effective stress law for elastic deformation of rock with fluids. *Journal of Geophysical Research*, **76**, 6414–19.

O'Brien, N. R. (1970) The fabric of shales: an electron microscope study. *Sedimentology*, **15**, 229–46.

O'Brien, N. R., Nakazawa, K. and Tokumashi, S. (1980) Use of clay fabric to distinguish turbiditic and hemipelagic siltstones and silts. *Sedimentology*, **27**, 47–61.

Olson, R. E. and Mesri, G. (1970) Mechanisms controlling compressibility of clays. *American Society of Civil Engineers, Journal of the Soil Mechanics and Foundation Division*, **96**, 1863–78.

Overton, H. L. and Zanier, A. M. (1971) An osmotic model for gas and overpressured formations. *Society of Petroleum Engineers of AIME*, Dallas, Texas, Paper No. SPE 3221.

Parker, A. and Sellwood, B. W. (1983) *Sediment Diagenesis*, Reidel, Dordrecht.

Parker, C. A. (1974) Geopressures and secondary porosity in the deep Jurassic of Mississippi. *Transactions of the Gulf Coast Association of Geological Societies*, **24**, 69–80.

Pearson, M. J., Watkins, D. and Small, J. S. (1982) Clay diagenesis and organic maturation in Northern North Sea sediments, in *Proceedings of the International Clay Conference*, Bologna and Pavia, 1981, *Developments in Sedimentology*, vol. 35, Elsevier, Amsterdam, pp. 665–75.

Plumley, W. J. (1980) Abnormally high fluid pressure: survey of some basic principles. *American Association of Petroleum Geologists Bulletin*, **64**, 414–30.

Potter, P. E., Maynard, J. B. and Pryor, W. A. (1980) *Sedimentology of Shale*, Springer-Verlag, New York.

Powell, T. G., Foscolos, A. E., Gunther, P. R. and Snowdon, L. R. (1978) Diagenesis of organic matter and fine clay minerals: a comparative study. *Geochimica et Cosmochimica Acta*, **42**, 1181–97.

Powers, M. C. (1967) Fluid release mechanisms in compacting marine mudrocks and their importance in oil exploration. *American Association of Petroleum Geologists Bulletin*, **51**, 1240–54.

Proshlyakov, B. K. (1960) Reservoir properties of rocks as a function of their depth and lithology. *Geologiya Nefti i Gaza*, **1960(12)**, 24–9.

Putnis, A. and McConnell, J. D. C. (1980) *Principles of Mineral Behaviour*, Elsevier, New York.

Rice, J. R. and Cleary, M. P. (1976) Some basic stress diffusion solutions for fluid-saturated elastic porous media with compressible constituents. *Reviews of Geophysics and Space Physics*, **14**, 227–41.

Rieke, H. H., III and Chilingarian, G. V. (1974) *Developments in Sedimentology*, vol. 16, *Compaction of Argillaceous Sediments*, Elsevier, Amsterdam.

Rosenbaum, M. S. (1976) Effect of compaction on the pore fluid chemistry of montmorillonite. *Clays and Clay Minerals*, **24**, 118–21.

Russell, W. L. (1972) Pressure–depth relations in Appalachian region. *American Association of Petroleum Geologists Bulletin*, **56**, 528–36.

Schmidt, G. W. (1973) Interstitial water compositions and geochemistry of deep Gulf Coast shales and sandstones. *American Association of Petroleum Geologists Bulletin*, **57**, 321–37.

Sharp, J. M., Jr. (1976) Momentum and energy balance equations for compacting sediments. *Mathematical Geology*, **8**, 305–22.

Sharp, J. M., Jr. (1983) Permeability controls on aquathermal pressuring. *American Association of Petroleum Geologists Bulletin*, **67**, 2057–61.

Sharp, J. M., Jr. and Domenico, P. A. (1976) Energy transport in thick sequences of compacting sediment. *Geological Society of America Bulletin*, **87**, 390–400.

Shaw, D. B. and Weaver, C. E. (1965) The mineralogical composition of shales. *Journal of Sedimentary Petrology*, **35**, 213–22.

Siever, R. (1983) Burial history and diagenetic reaction kinetics. *American Association of Petroleum Geologists Bulletin*, **67**, 684–91.

Singer, A. and Müller, G. (1983) Diagenesis of argillaceous sediments, in *Developments in Sedimentology*, vol. 25B, *Diagenesis of Sediments and Sedimentary Rocks, 2*, Chapter 3 (eds G. Larsen and G. V. Chilingarian) Elsevier, Amsterdam, pp. 115–212.

Skempton, A. (1954) The pore pressure coefficients A and B. *Géotechnique*, **4**, 143–7.

Smith, J. E. (1971) The dynamics of shale compaction and evolution of pore fluid pressures. *Mathematical Geology*, **3**, 239–63.

Spencer, C. E. (1987) Hydrocarbon generation as a mechanism for overpressuring in Rocky Mountain Region. *American Association of Petroleum Geologists Bulletin*, **71**, 368–88.

Stuart, C. A. and Kozik, H. G. (1977) Geopressuring mechanism of Smackover Gas reservoirs, Jackson Dome area, Mississippi. *Journal of Petroleum Technology*, **29**, 579–85.

Surdam, R. C. and Crossey, L. J. (1985) Organic–inorganic reactions during progressive burial: key to porosity and permeability enhancement and preservation. *Philosophical Transactions of the Royal Society of London, Series A*, **315**, 135–56.

Surdam, R. C., Crossey, L. J., Hagen, E. S. and Heasler, E. P. (1989) Organic–inorganic interactions and sandstone diagenesis. *American Association of Petroleum Geologists Bulletin*, **73**, 1–23.

Swartzen-Allen, S. L. and Matijevic, E. (1974) Surface and colloid chemistry of clays. *Chemical Reviews*, **74**, 385–400.

Terzaghi, K. and Peck, R. B. (1967) *Soil Mechanics in Engineering Practice*, 2nd edn, John Wiley, New York.

Thomeer, J. H. M. A. and Bottema, J. A. (1961) Increasing occurrence of abnormally high reservoir pressures in boreholes, and drilling problems resulting therefrom. *American Association of Petroleum Geologists Bulletin*, **45**, 1721–30.

Thorne, J. A. and Watts, A. B. (1989) Quantitative analysis of North Sea subsidence. *American Association of Petroleum Geologists Bulletin*, **73**, 88–116.

Underwood, L. B. (1967) Classification and identification of shales. *American Society of Civil Engineers, Journal of the Soil Mechanics and Foundation Division*, **93**, 97–116.

Ungerer, P., Bessis, F., Chenet, P. Y., Durand, B., Nogaret, E., Chiarelli, A., Oudin, J. L. and Perrin, J. F. (1984) Geological and geochemical models in oil exploration: principles and practical examples. *Memoirs, American Association of Petroleum Geologists*, **35**, 53–78.

Veblen, D. R., Guthrie, G. D., Jr., Livi, K. J. T. and Reynolds, R. C., Jr. (1990) High-resolution transmission electron microscopy and electron diffraction of mixed-layer illite/smectite: experimental results. *Clays and Clay Minerals*, **38**, 1–13.

Vockroth, G. B. (1974) Abnormal subsurface pressure. *American Association of Petroleum Geologists Reprint Series*, **11**, 205 pp.

Von Engelhardt, W. and Gaida, K. H. (1963) Concentration changes of pore solutions during the compaction of clay sediments. *Journal of Sedimentary Petrology*, **33**, 919–30.

Van Olphen, H. (1977) *Introduction to Clay Colloid Chemistry*, 2nd edn, Wiley Interscience, New York.

Walder, J. S. and Nur, A. (1984) Porosity reduction and crustal pore pressure development. *Journal of Geophysical Research B*, **89**, 11,539–84.

Watts, N. L. (1983) Microfractures in chalks of Albuskjell Field, Norwegian Sector, North Sea: possible origin and distribution. *American Association of Petroleum Geologists Bulletin*, **67**, 201–34.

Weaver, C. E. (1989) *Clays, Muds and Shales, Developments in Sedimentology*, vol. 44, Elsevier, Amsterdam.

Weaver, C. E. and Beck, K. C. (1971) Clay water diagenesis during burial: how mud becomes gneiss. *Geological Society of America, Special Paper*, **134**, 96 pp.

Weller, J. M. (1959) Compaction of sediments. *American Association of Petroleum Geologists Bulletin*, **43**, 273–310.

Whitney, G. (1990) Role of water in the smectite-to-illite reaction. *Clays and Clay Minerals*, **38**, 343–50.

Whittaker, A. (1985) *Theory and Evaluation of Formation Pressures*, Reidel, Dordrecht.

Yau, Y.-C., Peacor, D. R. and McDowell, S. D. (1987) Smectite-to-illite reactions in Salton Sea shales: a transmission and analytical electron microscopy study. *Journal of Sedimentary Petrology*, **57**, 335–42.

Young, A. and Low, P. F. (1965) Osmosis in argillaceous rocks. *American Association of Petroleum Geologists Bulletin*, **49**, 1004–8.

Zhao, W. L., Davis, D. M., Dahlen, F. A. and Suppe, J. (1986) Origin of convex accretionary wedges: evidence from Barbados. *Journal of Geophysical Research B*, **91**, 10,246–58.

CHAPTER TWELVE

Clays as chemical and hydraulic barriers in waste disposal: evidence from pore waters

A. H. Bath

12.1 Introduction

A unique combination of physical, chemical and mechanical properties makes clays valuable as barriers against the leakage of pollutants from wastes. Being composed of very fine-grained layered minerals gives compacted clays low permeabilities to fluid flow. The clay layers compact together to leave molecular-scale interstitial water; moreover the layered structure has remarkable anisotropy so that permeability across layering may be orders of magnitude lower than that parallel to layering. The important chemical property of clays, that of sorbing solutes, derives from the high charged surface area of the clay minerals. This sorbing, or ion-exchanging, behaviour is of course well known in natural processes. The high density of exchange sites on the clay surfaces and the low flux of solution equilibrating with the surface are the key properties which govern the ability of clay layers, both natural *in situ* and artificially emplaced, to retard migrating pollutants. Mineralogical studies have provided abundant data on the ion-exchange behaviour of the individual clay minerals, with respect to selectivity and capacity. However, the difficulty of obtaining representative samples of the indigenous pore fluid, and an even greater difficulty in making any *in situ* fluid measurements, has inhibited a full understanding of the physics and chemistry of clay pore waters. Geochemical and physical models for clays containing natural solutes are only now beginning to be tested against experimental and natural system data, and the predictions of their response to pollutant species are still, at best, only qualitative.

This paper gives a brief overview of the use to which clays are put as barriers against the leakage of pollutants from sites where hazardous wastes are being deposited at surficial or subsurface disposal facilities. A special case arises in the disposal of radioactive wastes which, though of relatively small volume, has stimulated much new research. The sources of the limited data for pore-water

Geochemistry of Clay-Pore Fluid Interactions. Edited by D. A. C. Manning, P. L. and C. R. Hughes. Published in 1993 by Chapman & Hall, London. ISBN 0 412 48980 5

chemistries are reviewed. Reactions which might diminish the sorbing capacity of the clays, enhance their permeabilities, or otherwise change the nature of solution flow, are particularly detrimental to clay barrier performance. Porewater measurements in clay formations of various ages and with different post-depositional histories are particularly valuable in testing physical models for groundwater and solute movement through these media.

12.2 Clay barriers in the disposal of hazardous waste to landfill

Solid wastes have been 'landfilled' in a wide range of geological or hydrogeological environments. Landfills are commonly located in disused quarries, particularly limestone, sand or other aggregate workings, or workings of clay for the brick industry which, though distributed less frequently, offer many immediate advantages for waste disposal. Man-made surface depressions, such as disused railway cuttings or similar natural topographic features, may also be suitable sites for disposal. The bulk of disposed wastes are solid, although in some countries (e.g. the UK) co-disposal is practised, in which controlled industrial liquid effluents can be added to the solids (Williams, 1988). Liquid wastes have in the past also been placed in seepage lagoons (Williams *et al.*, 1984). In all cases the authorization requires that the site satisfies a set of environmental criteria, of which a principal factor is hydrogeology and concerns the location with respect to groundwater flow and the concomitant pollution risk to nearby aquifers or discharges into surface water courses. Whereas past waste disposal practice frequently resulted in a 'dilute and disperse' or 'attenuation' concept, the present philosophy is one of 'contain and treat'. In the former case, either the unsaturated zone above the water table or a site base comprising low permeability and/or sorbing strata were considered adequate means of attenuating pollutant transport to the surroundings. The 'attenuation' concept also emphasized the importance of degradation of organic pollutants, as well as sorption and dilution (Mather, 1989). In contrast, modern sites are frequently designed to retain leachate completely within natural or artificial barriers so that it is collected and treated. In both cases, it is particularly important to minimize infiltration of rainwater during operation and after closure. To this end, compacted clays are used to form a 'clay cap' over the completed landfill or over filled compartments of an operating site. The integrity of this clay cover is also now recognized as being important in controlling methane migration and optimizing the collection of the 'off gas' for heat or power generation. Modern landfill sites use natural and/or emplaced clay bases and may also have bentonite-containing cutoff walls around the periphery to prevent inflow of groundwater if the water table is intersected (Figure 12.1). Increasing scarcity of sites with suitable natural barriers has led to rapidly growing application of emplaced clay liners and/or synthetic geomembranes which are required to satisfy the containment licensing condition.

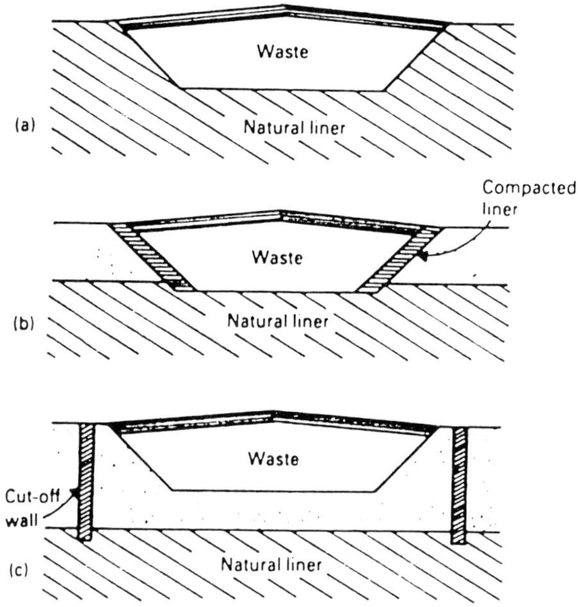

Figure 12.1 Schematic examples of natural and emplaced clay liners and cut-off walls (after Daniel and Shackelford, 1989).

Design specifications for the use of clays as barriers in UK landfills were initiated by the Department of the Environment's Waste Management Paper No. 4 (Department of the Environment, 1976). A model calculation using Darcy's Law of water flow (e.g. Freeze and Cherry, 1979) showed that a typical clay base 10 m thick with a hydraulic conductivity of $10^{-11}\,m.s^{-1}$ would transmit about $7 \times 10^{-5}\,m^3.y^{-1}.m^{-2}$ of leachate depending on the head of accumulated leachate. A revision of this guidance paper (Department of the Environment, 1988) indicates that license conditions should specify the thickness of clay at base and sides, from which representative samples should be laboratory tested to demonstrate that they meet a set permeability criterion. It also suggested that clay permeabilities should be measured as a function of moisture content. The Department of the Environment Waste Management Paper No. 26 (Department of the Environment, 1986) refers to the role of clay liners as being to impede flow and attenuate leachate, with a recommended maximum hydraulic conductivity of $10^{-9}\,m.s^{-1}$. The recommended capping thickness is 1 m of compacted clay also having hydraulic conductivity of no more than $10^{-9}\,m.s^{-1}$. The clays might be local clay, bentonite or colliery shale, although it is admitted that 'there is very little information on the behaviour of natural lining materials in contact with difficult wastes', i.e. pore solutions from some wastes might react deleteriously with the clay. For example, organic leachates from some industrial wastes could increase hydraulic conductivity due to dehydration and shrinkage of clays (Fernandez and

Quigley, 1985; Brown, 1988). Inorganic solutes are also likely to affect deleteriously the permeability of clay lining materials (Daniel and Shackelford, 1989). Waste Management Paper No. 26 states that bentonite-based liners should not be significantly affected by oils or organic solvents although deleterious effects might arise from either alkaline or acidic waste leachates, due to dissolution of the clays themselves or of other minor minerals such as carbonates (Figure 12.2). Reconstituted bentonite clay has the advantage of consistency and greater swelling capacity, and may therefore be used in a thin (100–150 mm) lining. In cases where leachates are likely to be >1000 ppm solutes, mixing the bentonite with polymer additives is recommended to prevent flocculation caused by invasion of the pores. The chemical compositions of leachates vary widely due to many factors: domestic wastes tend to produce leachates which are neutral to acidic, with high total organic carbon and ammoniacal nitrogen while salinity may be moderate, up to several thousand ppm chloride. Dissolved sulphate and iron may also be high. Redox conditions may be strongly reducing and capable of methanogenesis. Industrial wastes generate leachates closely related to their source, either alkali or acidic, possibly with high concentrations of volatile organics and/or metals for which complexation and precipitation are important in determining their mobility (Ross, 1980; Newman and Ross, 1985; Yanful *et al.*, 1988a; 1988b; Evans, 1989; Holm, 1990).

Regulatory conditions imposed in the USA are somewhat stricter than in the UK, requiring a composite clay liner 1–2 m thick, or thinner bentonite bed, compacted to give a hydraulic conductivity of less than 10^{-9} m.s^{-1}, often in

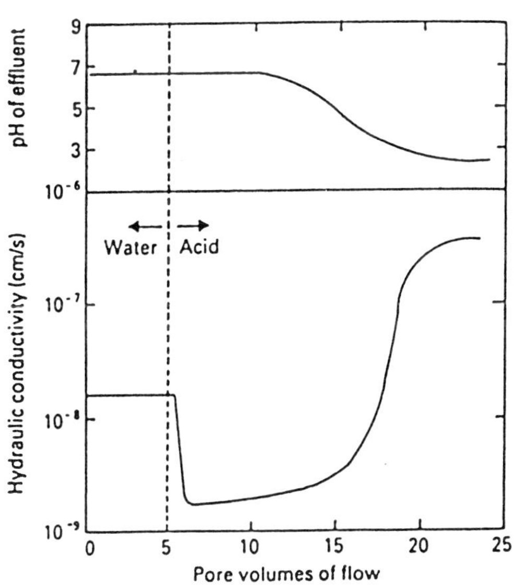

Figure 12.2 Variations in hydraulic conductivity for sample of compacted clay permeated by acid (after Daniel and Shackelford, 1989).

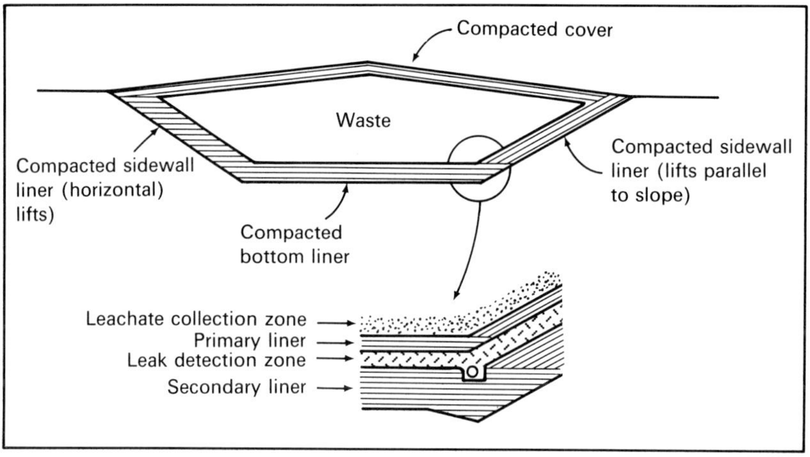

Figure 12.3 Schematic diagram of a double liner system with leak detection 'guard' zone; liners may be compacted clay and/or synthetic geomembranes (after Daniel and Shackelford, 1989).

combination with a greater than 1.5 mm geomembrane base to the landfill. Leachate is collected in a drainage layer having a hydraulic conductivity in excess of $10^{-5}\,m.s^{-1}$ whilst a further permeable 'guard' layer intercepts any leakage between the clay and geomembrane. Secondary 'failsafe' lining systems and Quality Assurance of materials and emplacement procedures are also being introduced to raise the integrity of the barrier system overall (Figure 12.3). Guidelines for the application and evaluation of clay liners and covers in the USA are covered in USEPA publications (Goldman *et al.*, 1988; USEPA, 1989).

12.3 Radioactive waste disposal into clay formations

Thick formations of low-permeability mudrocks, dominantly argillaceous sediments, are potential hosts for construction of repositories for long-term storage or disposal of radioactive waste. Such repositories may take the form of excavated trenches or surface 'vaults', as at the French site at Centre de la Manche and at Drigg in Cumbria in the UK. Alternatively, they may be deep mined tunnels as are proposed at the Belgian site at Mol or at a Swiss site. The facilities at Drigg take packaged bulk low-level radioactive waste into concrete-lined trenches excavated into glacial clays and sands, although earlier trenches were unlined except for localized bentonite sealing and relied on the natural barriers alone. Potential locations for mined repositories for radioactive wastes are being investigated in the Tertiary Boom Clay at Mol in Belgium (Figure 12.4), and in various mudrock or marl formations in Switzerland, Italy and France. Four English mudrock sites were being investigated by UK Nirex as potential hosts for construction of a repository for intermediate level waste until the investigations were terminated in 1987. The benefit of a mudrock host

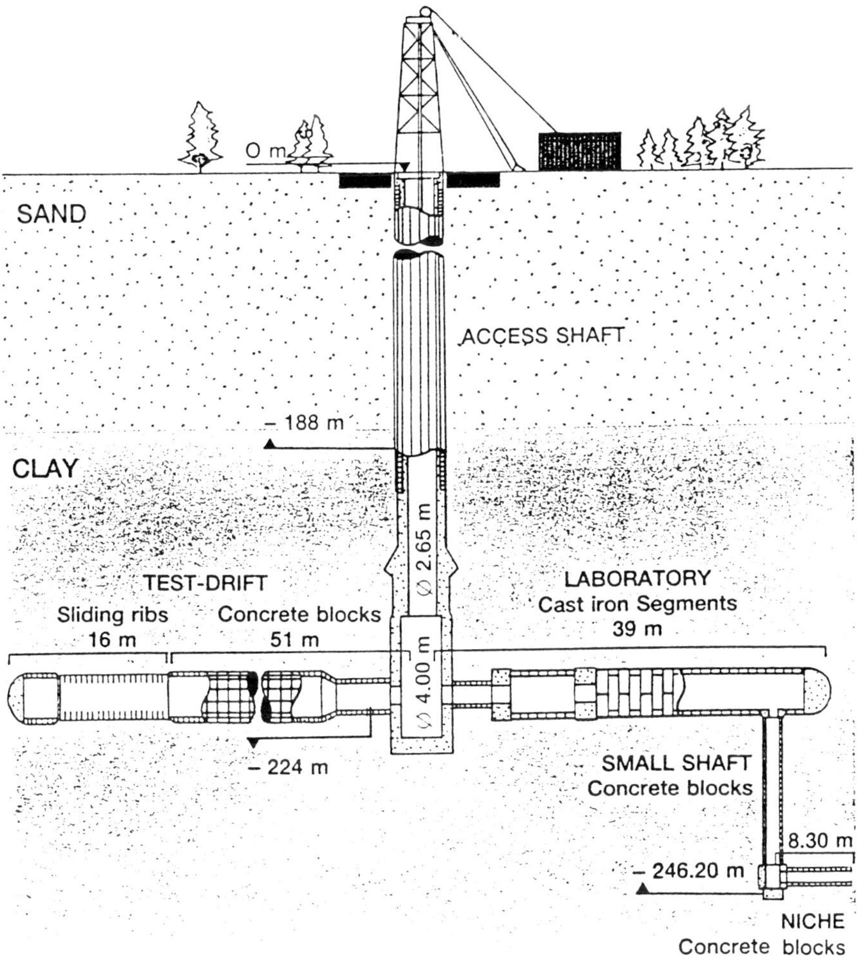

Figure 12.4 Underground experimental 'laboratory' tunnels in Tertiary Boom Clay at Mol, Belgium, operated by CEN/SCK Belgian Nuclear Energy Research Centre.

is, in addition to its low hydraulic conductivity and sorbing capacity, the likely homogeneity and structural simplicity of the sedimentary formation relative to crystalline basement rocks.

Licensing regulations for radioactive waste disposal are considerably more stringent than for hazardous waste disposal, reflecting both the longer time-scale of the hazard and also the perception of greater risk. The integrity of geological disposal relies on 'multibarrier' containment by which both engineered and natural systems, with some inherent redundancy, operate to contain the released radionuclides. The engineered barrier may comprise steel packaging and cement or bentonite grout and backfill, which provide both physical and chemical containment, whilst the natural barrier is provided by emplacement in rocks with minimal groundwater flow. Long-term safety is evaluated by predictive modelling of the performance of each of these systems and of interactions between them. Principal factors determining the release of radionuclides through adjacent rocks are their solubilities by dissolution from the waste matrix, and sorptive retardation of the radionuclides as they move through the repository and into the surrounding groundwater system. The chemistry of groundwater exerts a strong influence on these properties: aqueous chemistry within the repository or 'near field' will be a result of reaction between resaturating groundwater and the repository and waste materials which might include cements, clays, steel and organic substances.

As in disposal strategies for other hazardous wastes, the high sorptive capacity of clays leads to the use of bentonite as an engineering material for backfilling and sealing as well as to clay lithologies being suitable repository hosts. Knowledge of pore-water chemistry is important for engineering purposes and also as a baseline for the aqueous environment in which radionuclides will be transported. A further important aspect of pore waters in mudrocks is that they record chemical evidence of their hydrogeological history, for example in their salinities and natural isotopic compositions. The integrated hydrochemical evolution is complementary to the physical measurement of present-day hydrogeology, and offers some reassurance concerning changes to the system over long timescales.

12.4 Concepts of solute transport in clay pore waters

The very low permeability of compacted clays raises questions about the mechanisms that govern solute transport (Brightman *et al.*, 1987). Gravity-driven advection is the normal process by which groundwater moves through rocks; this is described by Darcy's Law (Freeze and Cherry, 1979), which relates the flux through a porous medium to the applied pressure or hydraulic gradient. Solute transport in this groundwater flux depends on the retardation of dissolved species relative to water flow. Retardation of solutes in intergranular flow is primarily determined by sorption onto mineral surfaces, which is high for reactive solutes in clay media. However, it may also be effectively enhanced by diffusive exchange of solutes with stagnant pore water in less permeable pore spaces. Clay rock masses, or artificially emplaced bulk clay, may have such low hydraulic conductivity that diffusion becomes the dominant transport mechanism for water and solutes under all normal hydraulic gradients. This may be so for clays with bulk hydraulic conductivity in the order of 10^{-9} m.s^{-1} or lower, in which the effective diffusion coefficient will be 10^{-9} to 10^{-11} m^2.s^{-1}. The net diffusive movement of solutes occurs along the concentration gradient rather than the hydraulic gradient; this places a lower limit on migration of solutes including pollutants and also indigenous solutes which have a heterogeneous distribution within the clay pore waters. Fick's 1st and

2nd Laws (e.g. Freeze and Cherry, 1979) describe the flux and concentration changes over time caused by the diffusive mechanism. Solute transport mechanisms and the likely effects of leachate on the hydraulic properties of clay liners for landfills have been reviewed by Daniel and Shackelford (1989).

Hydrochemical data, principally chloride and stable oxygen and hydrogen isotope compositions, record physical migration processes undisturbed by chemical reactions and thus allow solute transport mechanisms to be tested. Diffusion of reactive or sorbing solutes will be retarded in the same way that advective transport is retarded. However, it has been suggested that enhanced diffusion may occur as a consequence of electrical interactions within the charged layer close to the mineral surface. The operation of such a process (and several other potential solute transport phenomena in clays arising in extreme cases from the coupling of electrical, thermal and chemical forces) has yet to be fully tested and quantified (Charles *et al.*, 1986; Cook, 1988).

Because of the low fluxes across clay formations, it is expected that clays which were deposited in marine conditions would retain vestiges of the original salinity. This is confirmed by pore water profiles across various marine mudrocks in English sedimentary basin sequences. The extent of dilution from original seawater salinity gives a qualitative indication of the post-depositional flushing of connate waters by the influx of meteoric water after basin uplift, with solutes having been redistributed among the different lithologies in proportion to cross-formational fluxes. The end result of these processes is well demonstrated by profiles of salinity through the Cretaceous–Jurassic mudrock and limestone sequence on the western rim of the London Basin (Brightman *et al.*, 1985) and by similar profiles through other Jurassic mudrock sequences in southern England (Figure 12.5; Bath *et al.*, 1989b; Ross *et al.*, 1989a; Metcalfe *et al.*, 1990). Pore waters in the Tertiary London Clay at a coastal site have chloride concentrations and stable O and H isotope compositions which have been interpreted as reflecting fluctuating seawater intrusion (Figure 12.6; Bath *et al.*, 1989a). These profiles have been fitted to the palaeoclimatic record of Holocene sea-level change using diffusion calculations (Bath *et al.*, 1988; Falck and Bath, 1989a; 1989b; Falck *et al.*, 1990). A similar history of diffusive flushing of saline pore waters has been found in glacial clays near Montreal (Figure 12.7; Desaulniers and Cherry, 1989).

Overall, the evidence from a growing number of studies of pore waters in clay and mudrock profiles suggests that diffusion has governed solute transport over substantial length scales and timescales in many, but not all, cases. Data from pore water profiles through, or adjacent to, zones of faulting in mudrocks suggest that localized advection may be occurring in fault zones with diffusion into the clay matrix (Figure 12.8; Metcalfe *et al.*, 1990). Thus, vertical solute concentration gradients are found to be less steep where this form of localized advection is suspected. Heterogeneous and poorly sorted lithologies are also likely to have advective solute transport dominating over diffusive movement. This has been demonstrated particularly in glacial clays/sands such as those

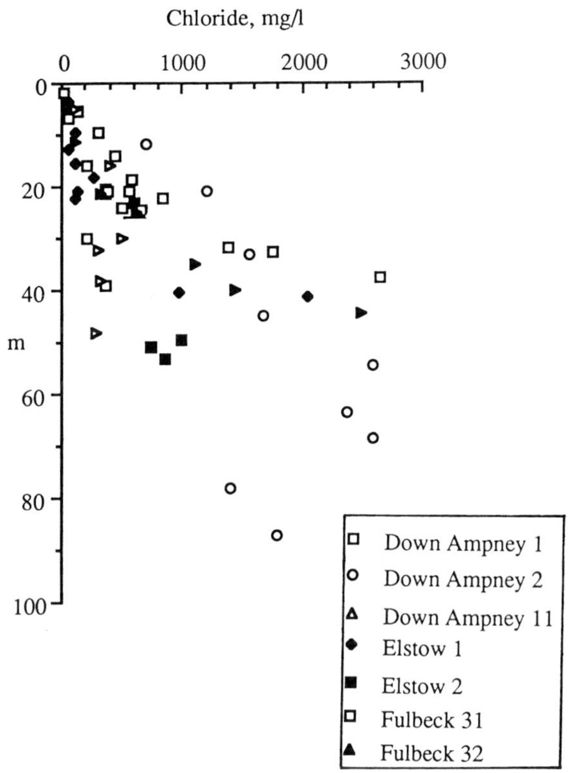

Figure 12.5 Chloride profiles through Jurassic mudrock sequences in southern England (data from Bath *et al.*, 1989b; Ross *et al.*, 1989a; Metcalfe *et al.*, 1990).

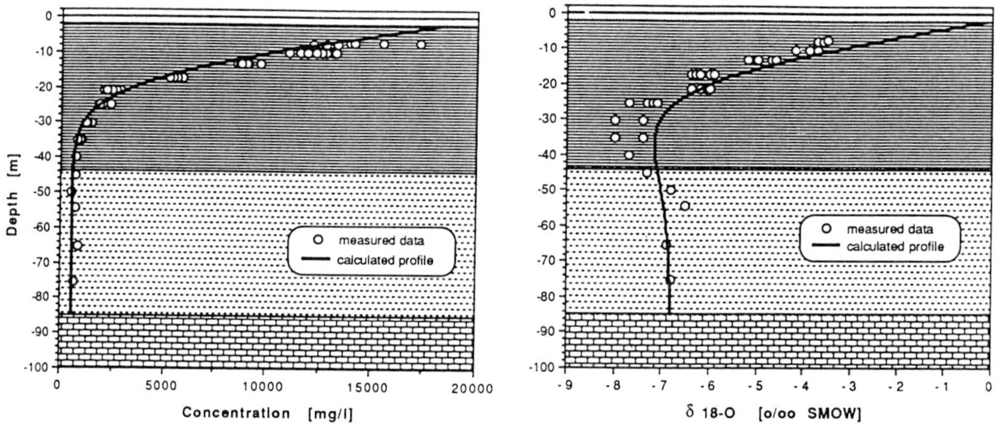

Figure 12.6 Chloride and stable O isotope profiles in coastal Tertiary mudrocks in eastern England, showing seawater intrusion; the calculated curve has used a diffusion model with boundaries set by palaeoclimatic conditions (Falck *et al.*, 1990).

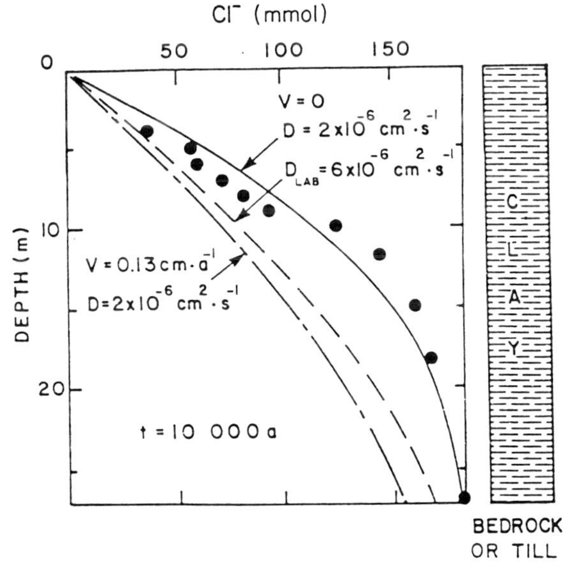

Figure 12.7 Simulated and observed profiles for chloride in pore waters in glacial clays at Varennes, near Montreal, showing that the $v=0$ diffusion curve gives a better fit to data than the advection curve $v>0$ (Desaulniers and Cherry, 1989).

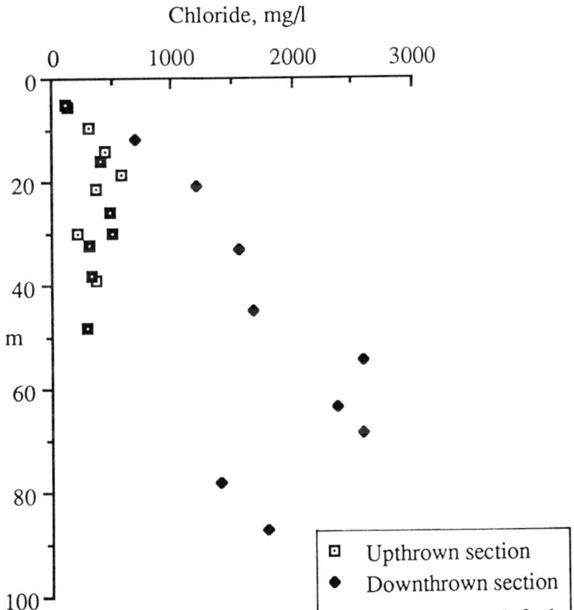

Figure 12.8 Comparison between chloride in faulted and unfaulted sections of Jurassic Oxford Clay at a site in central England, suggesting that advection associated with faulting causes a reduction in concentration gradient (data from Metcalfe *et al.*, 1990).

overlying the Chalk in eastern England where both solute and thermonuclear tritium profiles suggest vertical flow rates comparable with those from Darcy's Law calculations (Ross *et al.*, 1989b). However, hydrochemical and isotopic studies of pore waters in Quaternary glaciolacustrine clays at various sites in Ontario show that diffusion tends to dominate over advection at the scale of investigation (Desaulniers *et al.*, 1981; 1986).

The results of studies such as these carried out in undisturbed mudrocks are significant for inferring the likely behaviour, and limits of 'performance', of both natural and emplaced clays as barriers to migration of pollutants from waste disposal sites. A few studies of solute migration have been carried out by direct sampling within clay barriers below landfills or by experiments using synthetic leachate. Effective diffusion coefficients for some anion and cation species in leachate infiltrating compacted clay were determined by Shackelford *et al.* (1989). Studies of clay linings and bases to landfills in Ontario, Canada have shown that chloride migrates by diffusion (Quigley *et al.*, 1987; 1990; Johnson *et al.*, 1989). Analyses of extracted pore waters from a natural clay base have also showed that concentration profiles of various organic leachate compounds were consistent with diffusive migration retarded relative to chloride due to sorption (Figure 12.9; Johnson *et al.*, 1989).

12.5 Summary

This paper has given a brief overview of the ways in which clays are used as natural or artificial barriers in waste disposal strategies. Hazardous wastes are placed in landfills which have natural or emplaced clay bases and covers for which regulations impose permeability requirements; measurements of pore water chemistry could provide an *in situ* test of performance. The feasibility of geological disposal of radioactive wastes into thick clay or mudrock formations is being investigated in various countries. Long-term safety requires compliance with a stricter measure of performance based on risk, and this requires data for pore waters in the 'geosphere barrier' which demonstrate both solute migration mechanisms and also the hydrochemical influences on radionuclide solubility and retardation.

The reviewed work has focused on the importance of pore waters for investigating the operation of the 'barrier' processes and particularly for demonstrating the physical mechanism of solute migration in the groundwater system. Diffusion is clearly the limiting mechanism and dominates in many cases where the clay or mudrock is relatively homogeneous and unfractured. However, fractures or other discontinuities introduce heterogeneity in which significant advection may occur, bypassing the interstitial pores although this matrix will be influenced by diffusive exchange. Solutes, including potential pollutants, are retarded by sorption or irreversible reaction with the clays; the exchange reactions are characterized by the compositions of coexisting clay surfaces and pore waters.

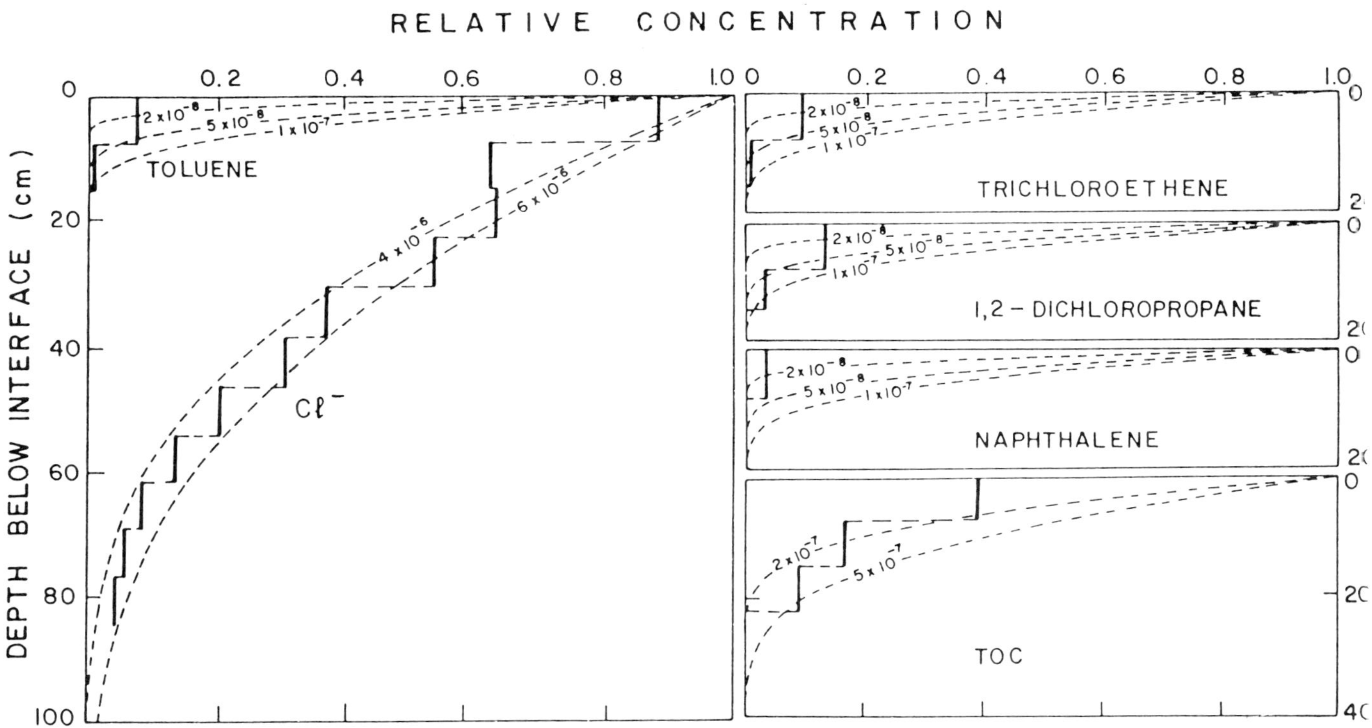

Figure 12.9 Measured profiles for chloride and various organic contaminants in pore waters from the natural clay base to a landfill in Ontario; calculated diffusion curves for $t = 1800$ days are also shown with various degrees of retardation (i.e. decreased effective diffusion coefficients, shown as $cm^2.s^{-1}$) for organics (Johnson *et al.*, 1989).

There is a paucity of data for clay pore waters, particularly where these have been influenced by the chemical or mechanical disturbances of excavation and emplacement of wastes. There are practical difficulties in recovering representative pore-water samples for analysis and in making *in situ* measurements, which normally hinder the collection of reliable data (Entwisle *et al.*, 1989; Entwisle and Reeder, 1993). However, such measurements will be essential if the physical and chemical processes within pore waters are to be modelled so that clay barrier behaviour can be predicted. The coupling of these processes due to the extreme proximity of pore waters and solid matrix introduces further complexity and a challenge to future studies.

Acknowledgements

The assistance of various colleagues in the Fluid Processes Group at BGS is gratefully acknowledged. In particular the help and advice of Geoff Williams and Rob Ward was invaluable in accessing the literature on the use of clays as landfill liners. Much of the BGS work on the extraction and analyses of pore waters from mudrocks was due to the efforts of Christine Ross, Dave Entwisle, Mark Cave and other colleagues. This paper is published with approval of the Director, British Geological Survey (NERC).

References

Bath, A. H., Entwisle, D., Ross, C. A. M., Cave, M. R., Falck, W. E., Fry, M., Reeder, S., Green, K. A., McEwen, T. J. and Darling, W. G. (1988) Geochemistry of porewaters in mudrock sequences: Evidence for groundwater and solute movements, in *Proceedings of the International Symposium on Hydrogeology and Safety of Radioactive and Industrial Hazardous Waste Disposal*, Orleans, France, Bureau de Recherches Géologiques et Minières, Paris, Documents du B.R.G.M. No. 160, pp. 87–97.

Bath, A. H., Ross, C. A. M., Entwisle, D., Cave, M. R., Fry, M., Green, K. A. and Reeder, S. (1989a) Hydrochemistry of Porewaters from Mudrocks at the Bradwell site, *Nirex Safety Studies Research Report*, NSS/R/170, UK Nirex Ltd, Harwell, Oxon.

Bath, A. H., Ross, C. A. M., Entwisle, D., Cave, M. R., Fry, M., Green, K. A. and Reeder, S. (1989b) Hydrochemistry of Porewaters from Lower Lias Silstones and Limestones at the Fulbeck site, Lincolnshire, *Nirex Safety Studies Research Report*, NSS/R171, UK Nirex Ltd, Harwell, Oxon.

Brightman, M. A., Alexander, J. and Gostelow, T. P. (1987) Groundwater Movement Through Mudrocks – Measurement and Interpretation, *Technical Report, British Geological Survey Fluid Processes Series*, FLPU 87-1, British Geological Survey, Keyworth, Nottingham.

Brightman, M. A., Bath, A. H., Cave, M. R. and Darling, W. G. (1985) Pore Fluids From the Argillaceous Rocks of the Harwell Region, *Technical Report, British Geological Survey Fluid Processes Series*, FLPU 85-6, British Geological Survey, Keyworth, Nottingham.

Brown, K. W. (1988) Review and Evaluation of the Influence of Chemicals on the Conductivity of Soil Clays, *U.S. Environmental Protection Agency*, EPA-600-2-88-016, USEPA, Hazardous Waste Engineering Research Laboratory, Washington, DC.

Charles, R. J., Cook, A. J. and Ross, C. A. M. (1986) Solute Transport Processes in a Saturated Clay, *Technical Report, British Geological Survey Fluid Processes Series*, FLPU 86-11, British Geological Survey, Keyworth, Nottingham.

Cook, A. J. (1988) A Desk Study of Surface Diffusion and Mass Transport in Clay, *Technical Report, British Geological Survey Fluid Processes Series*, WE/88/34, British Geological Survey, Keyworth, Nottingham.

Daniel, D. E. and Shackelford, C. D. (1989) Containment of landfill leachate with clay liners, in *Sanitary Landfilling: Process, Technology and Environmental Impact*, (eds T. H. Christensen, R. Cossu and R. Stegmann), Academic Press, New York, pp. 323–42.

Department of the Environment. (1976) The Licensing of Waste Disposal Sites, *Waste Management Paper*, No. 4, HMSO, London.

Department of the Environment. (1986) Landfilling Wastes – A Technical Memorandum, *Waste Management Paper*, No. 26, HMSO, London.

Department of the Environment. (1988) The Licensing of Waste Disposal Sites, *Waste Management Paper*, No. 4, 2nd edn, HMSO, London.

Desaulniers, D. E. and Cherry, J. A. (1989) Origin and movement of groundwater and major ions in a thick deposit of Champlain Sea clay near Montreal. *Canadian Geotechnical Journal*, **26**, 80–9.

Desaulniers, D., Cherry, J. A. and Fritz, P. (1981) Origin, age and movement of pore water in argillaceous Quaternary deposits at four sites in southwestern Ontario. *Journal of Hydrology*, **50**, 231–57.

Desaulniers, D., Kaufmann, R. S., Cherry, J. A. and Bentley, H. W. (1986) ^{37}Cl–^{35}Cl variations in a diffusion-controlled groundwater system. *Geochimica et Cosmochimica Acta*, **50**, 1757–64.

Entwisle, D. C. and Reeder, S. (1993) Extraction of pore fluids from mudrocks for geochemical analysis. This volume pp. 365–86.

Entwisle, D. C., Reeder, S., Bath, A. H. and Ross, C. A. M. (1989) Techniques for the Characterisation of Solutes in Drillcores from Mudrocks, *Nirex Safety Studies Research Report*, NSS/R173, UK Nirex Ltd, Harwell, Oxon.

Evans, L. J. (1989) Chemistry of metal retention by soils. *Environmental Science and Technology*, **23**, 1046–56.

Falck, W. E. and Bath, A. H. (1989a) Vertical Changes in Chloride and Stable Isotope Compositions of Porewaters in Coastal London Clay: Numerical Modelling and Data Evaluation, *Nirex Safety Studies Research Report*, NSS/R174, UK Nirex Ltd, Harwell, Oxon.

Falck, W. E. and Bath, A. H. (1989b) DISPERS — A Fortran Program to Solve the Advection–Dispersion Equation With Varying Boundary Conditions and Inhomogeneous Distribution of Parameters, *Technical Report, British Geological Survey Fluid Processes Series*, WE/89/11, British Geological Survey, Keyworth, Nottingham.

Falck, W. E., Bath, A. H. and Hooker, P. J. (1990) Long-term solute migration profiles in clay sequences. *Zeitschrift der Deutschen Geologischen Gesellschaft*, **141**, 415–26.

Fernandez, F. and Quigley, R. M. (1985) Hydraulic conductivity of natural clays permeated with simple liquid hydrocarbons. *Canadian Geotechnical Journal*, **22**, 205–14.

Freeze, R. A. and Cherry, J. A. (1979) *Groundwater*, Prentice Hall, Englewood Cliffs, New Jersey.

Goldman, L. J., Greenfield, L. I., Damle, A. S., Kingsbury, G. L. and Northeim, C. M. (1988) Design, Construction and Evaluation of Clay Liners for Waste Management Facilities, *U.S. Environmental Protection Agency*, EPA-530-SW-86-007F, USEPA, Office of Solid Waste and Emergency Response, Washington, DC.

Holm, T. R. (1990) Trace-metal speciation in landfill leachate. *Applied Geochemistry*, **5**, 367–9.

Johnson, R. L., Cherry, J. A. and Pankow, J. F. (1989) Diffusive contaminant transport in natural clay: a field example and implications for clay-lined waste disposal sites. *Environmental Science and Technology*, **23**, 340–9.

Mather, J. D. (1989) The attenuation of the organic component of landfill leachate in the unsaturated zone: a review. *Quarterly Journal of Engineering Geology*, **22**, 241–6.

Metcalfe, R., Ross, C. A. M., Cave, M. R., Green, K. A., Reeder, S. and Entwisle, D. C. (1990) Porewater and Groundwater Geochemistry at the Down Ampney Fault Research Site, *Technical Report, British Geological Survey Fluid Processes Series* WE/90/46, British Geological Survey, Keyworth, Nottingham.

Newman, J. R. and Ross, C. A. M. (1985) Mineralogical and Geochemical Controls on Heavy Metal Pollution in Monolith Lysimeters, *Technical Report, British Geological Survey Fluid Processes Series*, FLPU 85-5, British Geological Survey, Keyworth, Nottingham.

Quigley, R. M., Yanful, E. K. and Fernandez, F. (1987) Ion transfer by diffusion through clayey barriers, in *Geotechnical Special Publication* No. 13, *Geotechnical Practice for Waste Disposal '87, Proceedings of the Conference of the Geotechnical Engineering Division of the American Society of Civil Engineers*, Ann Arbor, Michigan, American Society of Civil Engineers, New York, pp. 137–58.

Quigley, R. M., Yanful, E. K. and Fernandez, F. (1990) Biological factors influencing laboratory and field diffusion, in *FEMS Symposium*, No. 59, *Microbiology in Civil Engineering*, (ed. P. Howsam), *Proceedings of the Federation of European Microbiological Societies Symposium*, Cranfield Institute of Technology, September 1990, E. & F. N. Spon, London, pp. 261–73.

Ross, C. A. M. (1980) Experimental assessment of pollutant migration in the unsaturated zone of the Lower Greensand. *Quarterly Journal of Engineering Geology*, **13**, 177–87.

Ross, C. A. M., Bath, A. H., Entwisle, D., Cave, M. R., Fry, M., Green, K. A. and Reeder, S. (1989a) Hydrochemistry of Porewaters From Jurassic Oxford Clay, Kellaways Beds, Upper Estuarine and Lower Lias Formations at the Elstow site, Bedfordshire, *Nirex Safety Studies Research Report*, NSS/R172, UK Nirex Ltd, Harwell, Oxon.

Ross, C. A. M., Bath, A. H., Entwisle, D., Cave, M. R., Fry, M., Green, K. A. and Reeder, S. (1989b) Hydrochemistry of Porewaters From Glacial Till and Chalk at the Killingholme Site, South Humberside, *Nirex Safety Studies Research Report*, NSSR/R169, UK Nirex Ltd, Harwell, Oxon.

Shackelford, C. D., Daniels, D. E. and Liljestrand, H. M. (1989) Diffusion of inorganic chemical species in compacted clay soil. *Journal of Contaminant Hydrology*, **4**, 241–73.

USEPA. (1989) Technical Guidance document: Final Covers on Hazardous Waste Landfills and Surface Impoundments, *U.S. Environmental Protection Agency*, EPA-530-SE-89-047, U.S. Environmental Protection Agency, Office of Solid Waste, Washington, DC.

Williams, G. M. (1988) Hazardous waste management and landfill research in the U.K., in *Proceedings of the Seminar on Hazardous Waste: Production, Italy, Control and Disposal*, Bergamo, 1986, Istituto Sperimentale Modellie Strutture (ISMES) S.p.A., Bergamo, pp. 277–89.

Williams, G. M., Ross, C. A. M., Stuart, A., Hitchman, S. P. and Alexander L. S. (1984) Controls on contaminant migration at the Villa Farm lagoons. *Quarterly Journal of Engineering Geology*, **17**, 39–55.

Yanful, E. K., Nesbitt, H. W. and Quigley, R. M. (1988a) Heavy metal migration at a landfill site, Sarnia, Ontario, Canada—1. Thermodynamic assessment and chemical interpretations. *Applied Geochemistry*, **3**, 523–33.

Yanful, E. K., Nesbitt, H. W. and Quigley, R. M. (1988b) Heavy metal migration at a landfill site, Sarnia, Ontario, Canada—2. Metal partitioning and geotechnical implications. *Applied Geochemistry*, **3**, 623–9.

CHAPTER THIRTEEN

Mass transfer mechanisms in compacted clays

*P. J. Bourke, N. L. Jefferies, D. A. Lever and T. R. Lineham**

13.1 Introduction

Compacted clay formations form low-permeability environments in which groundwater flow is slow, and which are described by hydrogeologists as forming aquicludes. Under such conditions, the quantity of mass of a solute transferred by convection in moving groundwater is likely to be small, whereas the movement of mass under concentration gradients by molecular diffusion may be more significant.

A low-permeability clay formation could be used as a host formation for the disposal of radioactive waste. The Oligocene Boom Clay in Belgium is such a formation and is undergoing investigation at the underground facility at CEN/SCK, Mol (Bonne and Neerdael, 1986). This programme comprises both field experiments (Monsecour *et al.*, 1991) and laboratory experiments (Put and Henrion, 1987) to measure mass transfer. In the UK, the disposal of low-level radioactive waste in vaults at shallow depth in clay was under consideration in 1986 and early 1987. The experimental data presented in this paper were obtained as part of the programme to quantify mass transfer properties of compacted clays with respect to radioactive waste disposal. Measurements of diffusion, convection, hydrodynamic dispersion and accessible porosity were made using tracers in the aqueous phase.

The material studied, London Clay from Bradwell in Essex, was cored during the site investigations of potential locations for the shallow disposal of low-level radioactive waste by UK Nirex Ltd. These site investigations were subsequently abandoned in May 1987.

The core was recovered using a triple-core barrel rotary coring system, with samples confined in 1 m plastic Mazier tubes which were sealed with paraffin wax to prevent sample degradation. Tubes were stored in controlled environmental conditions at 4°C until required for experimentation.

The London Clay is an Eocene marine deposit, which has been described by Anderton *et al.* (1979) as a typical North-Sea-type shale facies. The clay is stiff

*To whom correspondence should be addressed.

Geochemistry of Clay-Pore Fluid Interactions. Edited by D. A. C. Manning, P. L. Hall and C. R. Hughes. Published in 1993 by Chapman & Hall, London. ISBN 0 412 48980 5

and silty in appearance and has been subjected to over-consolidation during its geological history. It has a compactional fabric, which is displayed by splitting parallel to the bedding planes.

Mineralogical analyses on the material by Bourke *et al.* (1989) have shown that the coarser fractions are dominated by quartz, feldspar and dolomite, whereas the finer fractions are dominated by illite, smectite and kaolinite. Illite is the most abundant mineral and may constitute as much as 35% by volume of the clay. Table 13.1 lists some of these XRD mineralogical analyses.

Porosimetry studies have also been undertaken on the clay using high-pressure mercury intrusion techniques (Gilling *et al.*, 1987). Data from these studes have yielded pore size distributions which have indicated that half of the total porosity has dimensions less than 4 nm, assuming a cylindrical porosity model. It is considered that the majority of this microporosity is associated with the phyllosilicate component of the clay, where water is absorbed between clay platelets. Such an association would yield pores with a slit-like shape and aperture of approximately 0.7 nm. There is, therefore, the potential that electrostatic exclusion processes could operate to exclude charged solutes from the microporosity of the clay. Evidence to support anion exclusion is presented in this paper.

13.2 Experimental techniques and data analysis

The experimental techniques used in this study are commonly used to assess the mass transfer properties of rocks in the context of radioactive waste disposal, where mass transfer over long timescales and at slow rates through low-permeability rocks is important. The techniques are consequently specialized in application. For instance, the same type of diffusion experiments used in this study have been applied to biotite gneiss by Skagius and Neretnieks (1985) and similar convection experiments have been undertaken on Darley Dale Sandstone by Bradbury and Stephen (1985).

The main experimental techniques can be subdivided into those for addressing diffusive mass transport and those for addressing convective mass transport. Additional experiments have been undertaken to assess the permeability and diffusibility of the material as a function of applied stresses.

13.2.1 Diffusion experiments

Measurements of the diffusivity of the clay and the porosity accessible to conservative tracers have been undertaken, using two types of diffusion experiment. These experiments are known as 'through-diffusion' and 'in-diffusion', indicating diffusion of tracers through or into a sample.

'Through-diffusion' experiments were performed using a perspex cell (Figure 13.1), consisting of three parts. These included a central section to hold the sample, and end sections for the storage of measurement and reservoir

Table 13.1 Mineralogical analyses of the London Clay: X-ray diffraction data.

Mineral	Sample depth 32.70 m					Sample depth 33.23 m				
	$<2 \mu m$		$>2 \mu m$		Bulk composition	$<2 \mu m$		$>2 \mu m$		Bulk composition
	A	B	A	B		A	B	A	B	
Smectite/vermiculite	31	8.6	6	4	13	31	9.9	5	3	13
Illite	34	9.5	46	33	43	28	8.9	40	27	36
Chlorite	tr	tr	0	0	tr	12	3.8	0	0	4
Kaolinite	35	9.9	26	19	29	29	9.2	24	16	25
Plagioclase	0	0	3	2	2	0	0	8	5	5
Quartz	0	0	16	11	11	0	0	20	14	14
Pyrite	0	0	3	2	2	0	0	3	2	2

A = weight % relevant size fraction.
B = weight % bulk sample.
tr = trace ($<1\%$).

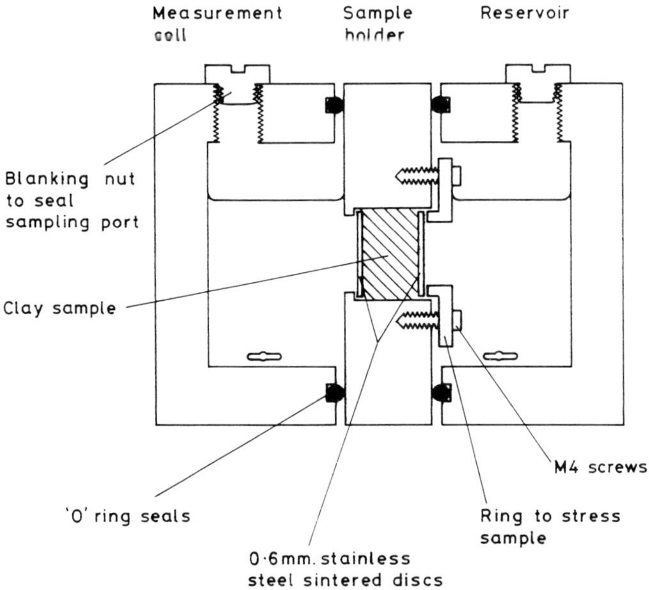

Figure 13.1 Perspex through-diffusion cell.

solutions. The sample was confined between stainless steel sinters and was sealed into the sample holder using a perspex back ring that applied stress directly onto one of the sinters. The sinter in contact with the opposite face of the sample was in contact with the retaining flange of the sample holder. The back ring was screwed to the holder and the seal between the holder and the sample was produced by the slight deformation of the sample. The formation of this seal was observed through the perspex and once this was complete, no further force was applied to the screws. The clay samples tested had dimensions up to 15 mm in length and 20 mm in diameter.

Diffusion experiments were undertaken by setting up a concentration gradient across the sample. Care was taken to ensure that no hydraulic gradient existed, so that mass transfer took place solely by a diffusive mechanism. For the experiments conducted in this study, the measurement side of the cell was filled with an artificial clay groundwater, and the reservoir side was filled with the same groundwater spiked with tracers. The composition of the artificial clay groundwater is shown in Table 13.2 and was based on data supplied by the British Geological Survey (Bath, 1987, personal communication) for extracted clay pore water. Three conservative, non-sorbing tracers were used in this study. The tracers were tritiated water (HTO), deuterated water (HDO), and iodide (I^-) as the sodium salt.

Periodically, samples were taken from the measurement side, with replacement of an equal volume of artificial clay groundwater. Samples were then analysed for tracer concentration using various analytical techniques. These

Table 13.2 Simulated groundwater composition.

Element	Concentration (ppm)	(molar)
Ca^{2+}	25.5	6.4×10^{-4}
Mg^{2+}	31.4	1.3×10^{-3}
Na^{+}	715	3.0×10^{-3}
K^{+}	13.0	3.3×10^{-2}
SO_4^{2-}	892	9.4×10^{-3}
HCO_3^{-}	293	3.9×10^{-3}
Cl^{-}	264	7.4×10^{-3}
NO_3^{-}	3.2	5.5×10^{-3}
pH	9.0	

Simulated clay pore water. Analysis based upon BGS data for extracted clay pore water (first fraction squeezed from sample). Bradwell site. Borehole 101, depth 29.5 m.

included liquid scintillation counting for tritium analysis, infrared spectrometry for deuterium and ion-chromatography for iodide (I^-). Experiments were continued until sufficient steady-state data were obtained. Frequent sampling of the solution in the measurement cell allowed the maintenance of the concentration of tracer in the measurement cell at a level less than 10% of that in the reservoir. If the tracer concentration in the reservoir fell below 90% of the initial value, an additional spike of tracer was added to the reservoir solution to restore the concentration to the initial value.

Data from these experiments were analysed by plotting graphs of tracer concentration in the measurement cell (corrected for sample removal) versus time (Figure 13.2). The graphs produced have two parts, a transient region and a steady-state region. The latter region was used to yield a diffusion coefficient for the tracer and a porosity accessible to the tracer. Assuming a 'single-porosity' model where all the porosity is equally accessible and access is rapid, the increase in tracer concentration in the measurement cell, C_m, has been expressed mathematically by Lever (1986) using the following equation:

$$C_{\mathrm{m}} = \frac{ALC_0}{V}\left[\frac{D_{\mathrm{i}}t}{L^2} - \frac{\alpha}{6} - \frac{2\alpha}{\pi^2}\sum_{1}^{\infty}\frac{(-1)^n}{\mathrm{n}^2}\exp\left(\frac{-D_{\mathrm{i}}n^2\pi^2 t}{L^2\alpha}\right)\right],$$

where A is the cross-sectional area of the sample, L is the sample length, V is the volume of the measurement cell, C_0 is the concentration in the reservoir, D_i is the diffusion coefficient and where α, the rock capacity factor, is equivalent to the accessible porosity (ϕ) for conservative, non-sorbing tracers.

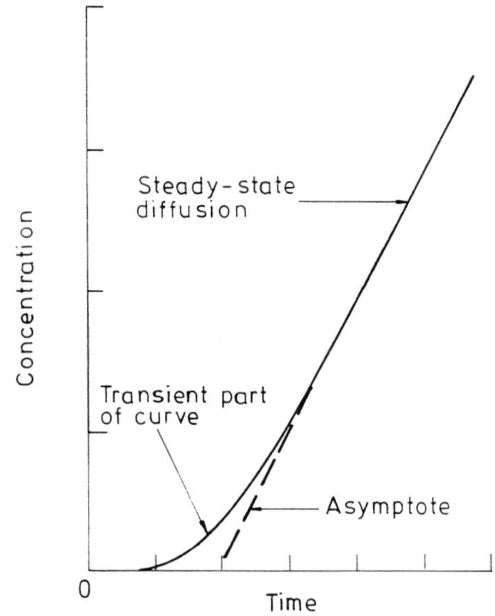

Figure 13.2 Schematic representation of the breakthrough into the measurement cell in the through-diffusion experiment.

This reduces at long times to

$$C_m = \frac{ALC_0}{V}\left(\frac{D_i t}{L^2} - \frac{\phi}{6}\right)$$

as the exponentials diminish to zero. This equates to the steady-state region and thus the diffusion coefficient can be calculated from the slope of the curve. The porosity accessible to the tracer can be calculated from the intercept of an asymptote extrapolated from the curve to the x-axis.

The diffusion coefficient measured is known as the intrinsic diffusion coefficient (D_i) and quantifies diffusive transport in a porous medium. It is defined as a steady-state flux in the pore water (F) per unit surface area of porous medium (A) per unit concentration gradient (C/L).

$$F = D_i AC/L$$

On completion of a 'through-diffusion' experiment, the clay sample was removed and placed into a known volume of ultra-pure water, to disaggregate the material and thus permit the 'out-diffusion' of the tracer from the clay. After a set period of time, typically 1 month, a sample of the solution was taken and filtered through a 0.45 μm filter to remove particulate material. The sample was then analysed to determine the tracer concentration. The porosity in the clay accessible to the tracer was then calculated from the mass balance. Thus, two measurements of accessible porosity were obtained for each experiment.

'In-diffusion' experiments were undertaken using the same perspex cell as used in the 'through-diffusion' experiments, but with the addition of solutions containing high tracer concentrations in both sides of the cell. Thus, diffusion of tracer from the solutions to the pore water took place simultaneously through both faces of the sample. The tracers used were deuterated water (HDO) and iodide (I^-).

These experiments required no continuous sampling and were left for set periods of time to allow tracer 'in-diffusion'. Experiments were stopped on a staggered basis, with some terminated after short times whereas others continued over timescales comparable with 'through-diffusion' experiments. Once the experiments were completed, the clay samples were removed and an 'out-diffusion' measurement was undertaken, as described above. The use of experiments over different timescales allowed the rate of access to the porosity of the clay by the tracer to be established. Modelling of the data using a 'single-porosity' porous medium model was then undertaken. This modelling was based on solving the following equation, quantifying the relationship between the amount of porosity (ϕ) accessed by the tracer and the time (t):

$$\phi \frac{\partial C}{\partial t} = D_{\mathrm{i}} \frac{\partial^2 C}{\partial x^2}$$

with $C = C_0$ at $x = 0$ and $x = L$ and, $\partial C/\partial x = 0$ at $x = L/2$, where L was the length of the clay sample.

13.2.2 *Convection experiments*

Convection experiments have been undertaken to quantify the hydrodynamic dispersion properties of the clay. These experiments can yield dispersivity coefficients, mixing lengths and accessible porosity data for the tracers used. Two different types of experiments have been performed using different boundary conditions. One type of experiment used mixed reservoirs and re-circulation systems to detect tracer breakthrough, whereas in the other type, samples of eluates were collected directly by a fraction collector. In both these experiments a solution containing a high concentration of tracer was driven through clay samples under applied pressure gradients. Experiments were continued until a few pore volumes of solution had been convected through the sample.

13.2.2.1 MIXED RESERVOIRS

Convection experiments using mixed reservoirs were employed when the clay sample tested was small in volume. The reason for this was that the quantity of pore water in the sample was negligible and thus a pore volume was limited in size. Small samples were used so that a high-pressure stainless steel cell could be employed for the experiments. This allowed the application of high hydraulic gradients across the sample (10^5), which were required to ensure that the ratio of convective to diffusive transport was high. The boundary conditions for this experiment were $C = C_0$ at the high-pressure face of the sample and $C \ll C_0$ at the low-pressure face of the sample.

The high-pressure stainless steel pressure vessel illustrated in Figure 13.3 was capable of accommodating clay samples up to 10 mm in length and 10 mm in diameter. The sample was confined between stainless steel sinters as in the diffusion experiments. The sample was sealed in the cell as a result of the deformation of the clay due to axial loading.

The experiment was performed using the equipment illustrated in Figure 13.4. This consists of two re-circulation systems, one low pressure and one high pressure, with associated measurement and reservoir cells. The re-circulation systems were connected to opposite sides of the stainless steel cell and thus opposite faces of the clay sample. The breakthrough of the tracer into the measurement cell was determined by periodic sampling with the replacement of an equal volume of artificial clay groundwater. On completion of the experiment, a graph of the flux of tracer into the measurement cell (corrected for sample removal and volume increase) versus time was

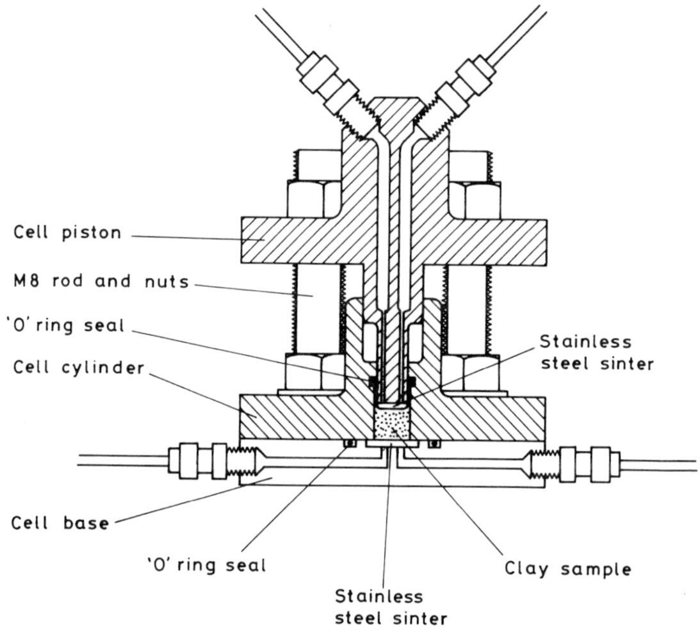

Figure 13.3 Stainless steel high-pressure cell for permeation experiments.

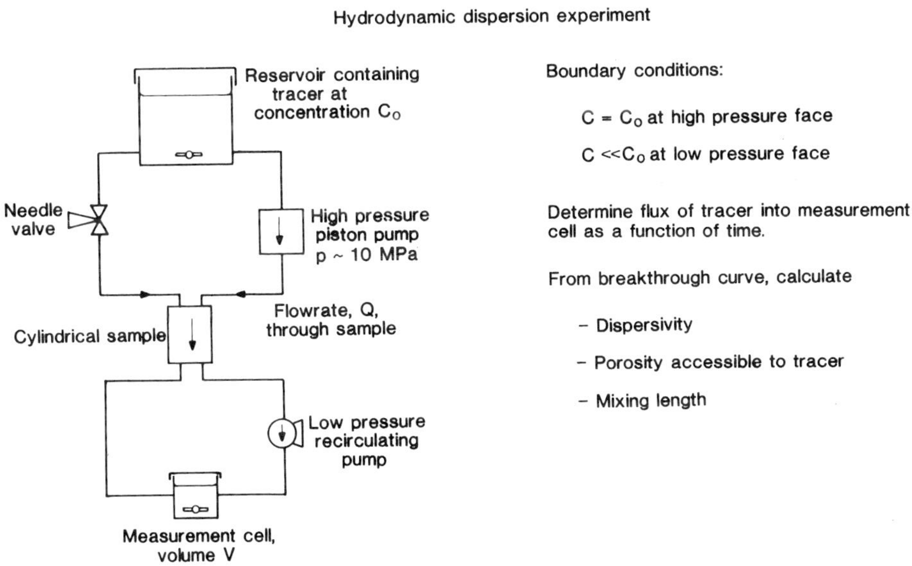

Figure 13.4 Apparatus for convection experiments utilizing stirred reservoirs and re-circulation systems.

produced. The data were then modelled by solving the advection–dispersion equation (ADE) for a porous medium by numerical techniques. The equation is shown below:

$$R\frac{\partial C}{\partial t}+u\frac{\partial C}{\partial x}=D\frac{\partial^2 C}{\partial x^2}$$

where $D=D_P+D_L$ and,

$$R=\frac{\alpha}{\phi},\ u=\frac{Q}{A\phi},\ D_P=\frac{D_i}{\phi},\ D_L=a_L u,$$

and Q is the volumetric flow rate. D_P represents molecular diffusion and D_L represents hydrodynamic dispersion. Hydrodynamic dispersion in the clay is modelled by a diffusion-like term with a dispersion length, a_L.

The techniques used to solve this equation included the application of Laplace transforms, and use of Talbot's method to allow the numerical inversion of the transforms as illustrated by Garratt (1987). For the modelling performed in this study values for the porosity accessible to the tracer (ϕ) and the Peclet Number (Pe), the ratio of convective to dispersive fluxes, were varied until the best fit to the experimental data was obtained.

13.2.2.2 DIRECT COLLECTION

Convection experiments utilizing the direct collection of eluates by a fraction collector were performed when larger clay samples were tested. The dimensions of these samples were typically 30 mm in length and 52 mm in diameter. These experiments made use of a diffusion/permeation cell (Figure 13.5), which was mounted on an oedometer rig. This cell has a lower pressure rating than the high-pressure cell detailed above and thus restricts the magnitude of hydraulic gradients to less than 4000. Mounting the cell on the oedometer rig allows the stress applied to the sample to be fixed at a set level. In order that the ratio of convective to molecular diffusive flux was as large as possible, samples of clay were obtained by subcoring parallel to the compactional fabric. This then allowed the experiment to utilize the known anisotropic permeability relationship displayed by the London Clay (Gilling *et al.*, 1987), where the permeability parallel to the fabric is approximately three times that perpendicular to the fabric.

In these experiments a solution of artificial clay groundwater spiked with tracer(s) was forced under pressures of less than 1 MPa through the sample. Eluates were collected automatically by a fraction collector and were then analysed for tracer concentration. Similar analyses of data to that performed

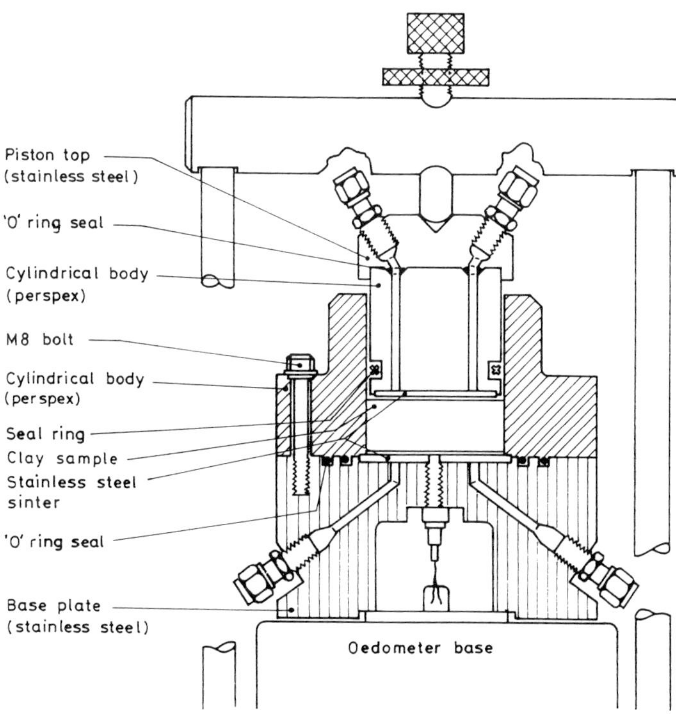

Figure 13.5 Diffusion/permeation cell.

for the other type of convection experiment were undertaken. These analyses took account of the different boundary conditions used, which were as follows:

$$QAC - DA\frac{\partial c}{\partial x} = QC_0 \text{ at } x = 0$$

$$\frac{\partial c}{\partial x} = 0 \text{ at } x = L$$

13.2.3 Additional experiments

Additional experiments have been undertaken to measure the hydraulic conductivity of the clay under a range of applied stresses. These experiments have used a commercially available soil mechanics consolidation cell and the stainless steel high-pressure cell described earlier.

The consolidation cell was used to perform uniaxial consolidation experiments on the clay. The consolidation data were analysed to yield values of hydraulic conductivity and porosity as a function of applied stress. The stainless steel high-pressure cell was used to directly measure the hydraulic conductivity of the clay under varying consolidation stresses.

In addition, one 'through-diffusion' experiment was undertaken to measure the vertical diffusion properties of the London Clay under different applied stresses. This experiment used a triaxial pressure vessel similar to those commercially available, but modified to allow solutions from reservoirs to be re-circulated against both sinters, which sandwiched the sample during the experiment. This cell has a maximum working pressure of 10 MPa and this was used during the experiment.

One 'through-diffusion' experiment was undertaken to measure the horizontal component of diffusion. This experiment made use of the permeation/diffusion cell used in some of the convection experiments. Again, re-circulation of solutions from reservoirs was utilized and as before the sample was sandwiched between stainless steel sinters.

13.3 Results

13.3.1 Diffusion experiments

Ten 'through-diffusion' experiments were undertaken in this study and all measured the diffusive properties of the clay perpendicular to the compactional fabric. Two of these experiments were performed under anaerobic conditions in a nitrogen-filled glove box. Table 13.3 summarizes the parameters determined from these experiments and Figure 13.6 shows an example of the data from one experiment with theoretical fits to the data using the 'single-porosity' model.

Table 13.3 Summary results from through-diffusion experiments.

Tracer	n	$D_i(m^2.s^{-1}) \pm 1\sigma$	Porosity, ϕ	
			From through-diffusion $\pm\sigma$	From out-diffusion $\pm\sigma$
Aerobic conditions				
HDO	8	$7.6\times10^{-11}\pm0.9\times10^{-11}$	0.61 ± 0.05	0.59 ± 0.05
HTO	4	$7.6\times10^{-11}\pm0.5\times10^{-11}$	0.56 ± 0.03	0.62 ± 0.02
I^-	3	$3.9\times10^{-11}\pm0.6\times10^{-11}$	0.21 ± 0.06	0.31 ± 0.06
Anaerobic conditions				
HDO	2	7.2×10^{-11}	0.63	0.55
I^-	2	–	–	0.15

n = number of experiments.

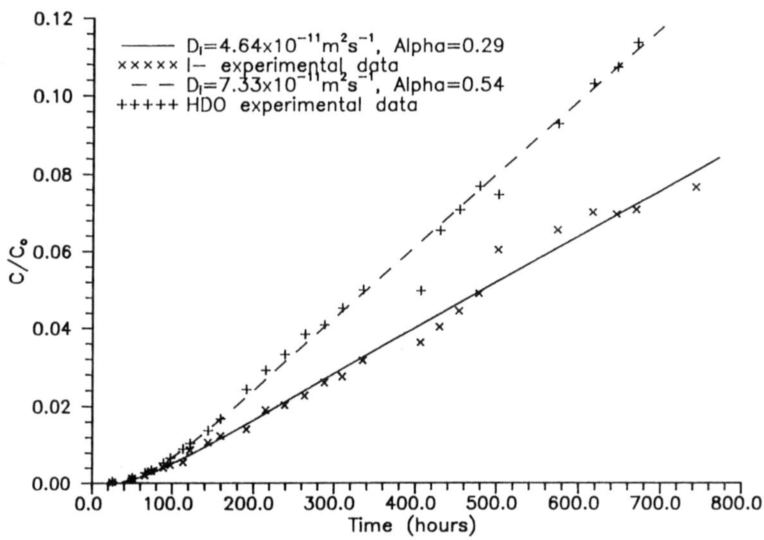

Figure 13.6 An example of data from a through-diffusion experiment with mass transfer perpendicular to the compactional fabric of the clay.

Comparing the limited data collected on samples held under anaerobic conditions with those samples held under aerobic conditions there appears to be no significant difference in the values of D_i and ϕ obtained.

Direct comparison of values of D_i for HDO and HTO with those for I^- is not possible as the intrinsic diffusion coefficient (D_i) is a property of the whole system being investigated. The system comprises the diffusing species, the aqueous phase in which it is present and the porous solid material through which it is diffusing. The single property which describes the diffusion potential of a porous medium is known as the diffusibility, Ψ (Lever, 1986). This property is related to the diffusion coefficient of the species in free water (D_0) as follows:

$$D_i = D_0 \Psi.$$

For the tracers used in the study the following values of D_0 were used: $D_0(HDO) = 2.3 \times 10^{-9}\ m^2.s^{-1}$ (Trappeniers *et al.*, 1965); $D_0(NaI) = 1.5 \times 10^{-9}\ m^2.s^{-1}$ (Li and Gregory, 1974). The diffusibility of the clay towards each of the tracers is similar, with reduction of the free water diffusive flux by a factor of approximately 30.

This correlation for values of D_i for different tracers is not repeated in the values of the porosity accessible to the tracers. A statistical difference is seen in the results presented in Table 13.3. Tritiated water and HDO accessed a porosity of approximately 0.6, equivalent to the total porosity determined by oven drying, whereas the porosity accessible to I^- was approximately 0.3.

The data from the 'in-diffusion' experiments are illustrated in Figure 13.7. These data yield accessible porosity values of approximately 0.5 for HTO and between 0.14 and 0.24 for I^- after 400 h elapsed time. These values are in good agreement with data from 'through-diffusion' tests.

The data have been fitted with theoretical transients based on a porous medium approach. Values of D_i obtained from the 'through-diffusion' experiments and accessible porosity from the 'in-diffusion' experiments were used to construct the transients. Scatter in I^- data does not allow the differences between the transients and the data to be adequately visualized. The HDO data are more stable and at early times the porosity accessed is consistently less than the predictions. The discrepancy between the model and the observed data would suggest that access by HDO to some portion of the total porosity takes place over timescales of approximately 100 h. This result is surprising considering the small distances over which the tracer has to diffuse. However, small-angle neutron scattering (SANS) studies of mass transfer in the London Clay (Allen *et al.*, 1989) have suggested that approximately 10% of the total porosity of the material remains inaccessible, or only slowly accessible, to deuterated water. This porosity could be associated with structured water in the clay. The SANS study, therefore, supports the observation from the in-diffusion experiments.

13.3.2 Convection experiments

The convection experiments also provided data that were consistent with the findings from the diffusion experiments. The experiments using stirred reservoirs and re-circulation systems attempted to quantify the convective fluxes of tracers perpendicular to the fabric. Unfortunately the quality of the data was not good enough to justify detailed modelling. Figure 13.8 illustrates an example of this, with the concentration of tracer in the measurement cell

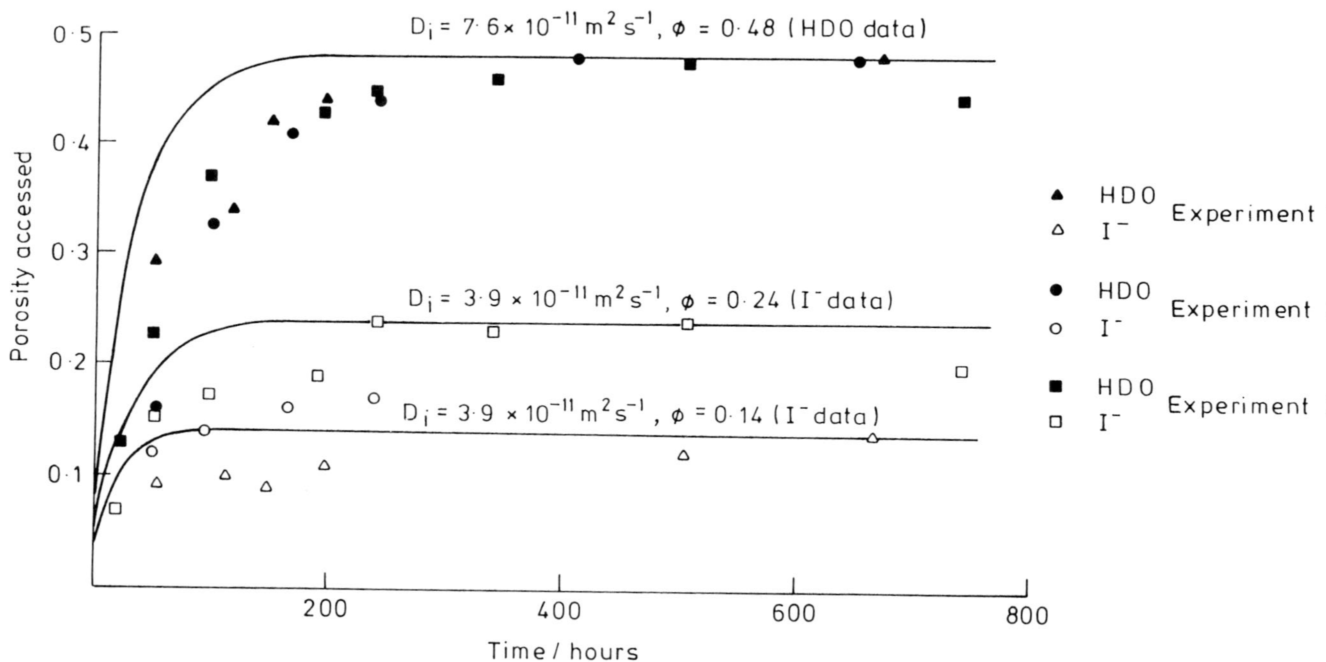

Figure 13.7 In-diffusion experiments. The theoretical transients are fitted for D_i values, obtained from the 'through-diffusion' experiments, and ϕ values determined from the 'in-diffusion' experiments.

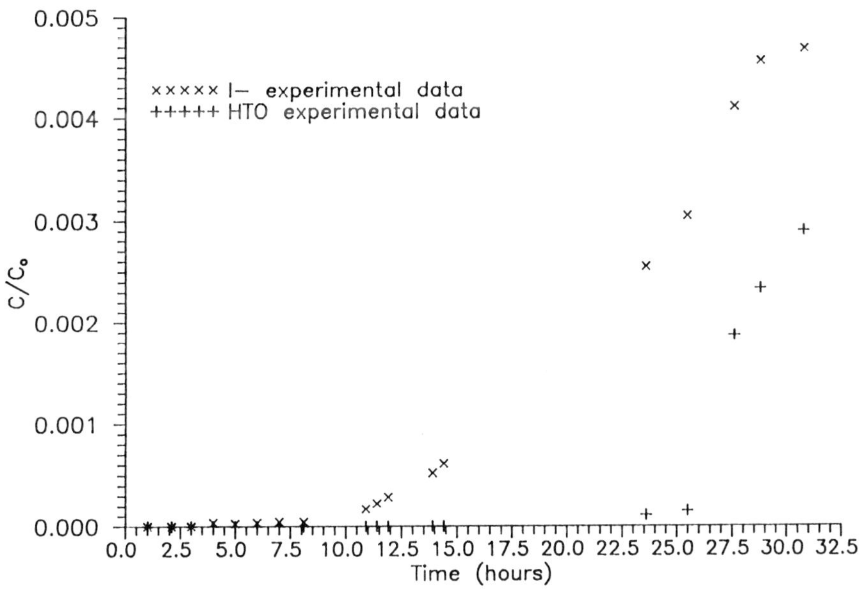

Figure 13.8 Example of data from a dispersion experiment using stirred reservoirs and re-circulation systems.

plotted against time. To justify modelling, the data shown in Figure 13.8 had to define a curve with distinct transient and steady-state regions, which would be similar to the graphs of 'through-diffusion' data. As these regions cannot be defined it was not possible to isolate the transient data, which was of primary importance for the modelling. Although the transient data are poor, the earlier breakthrough of I^- suggests that less porosity is accessible to I^- than to HTO.

Following the results of these tests it was decided to use the permeation/diffusion cell with samples of clay mounted to allow convection of tracers parallel to the fabric, with direct collection of eluates. Figure 13.9 shows the results from one of these experiments with theoretical fits to the data. Once again the I^- data are modelled more satisfactorily using a lower value of accessible porosity (0.32) than that accessible to HTO (0.50).

Figure 13.9 also illustrates that the experimental data can be modelled using a simple porous medium approach with a Peclet Number (Pe) of approximately 4. This indicates that the dispersive fluxes are less than the convective fluxes by a factor of 4. The dispersive fluxes comprise a diffusive component and a hydrodynamic dispersive component. It is possible to calculate the size of the hydrodynamic dispersive component by using the results from the earlier 'through-diffusion' experiments to provide an estimate of the diffusive component. This is achieved as follows:

$$\mathrm{Pe} = \frac{QL}{\phi AD}$$

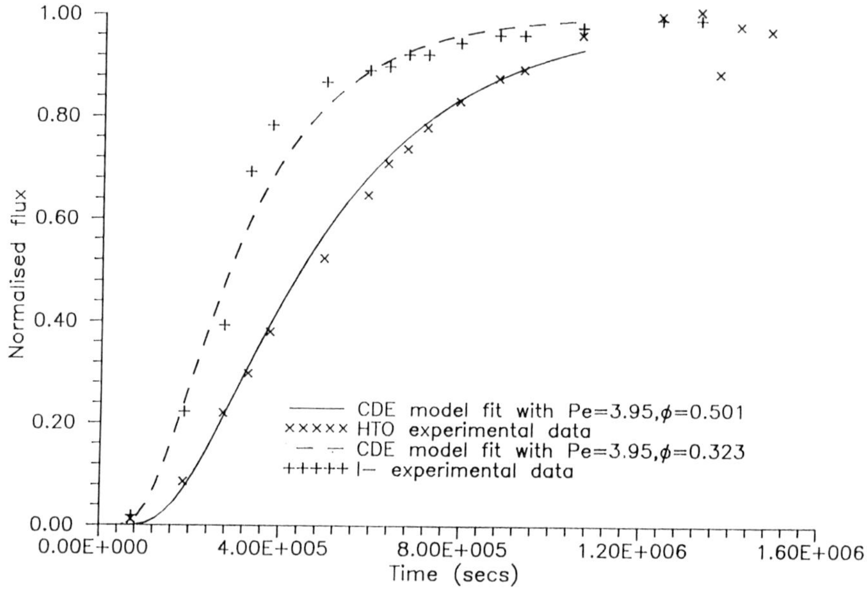

Figure 13.9 A dispersion experiment using direct collection of eluates with the direction of mass transfer parallel to the compactional fabric.

D can be calculated as Q, L, A and ϕ are known; then $D = D_p + D_L$, where

$$D_P = \frac{D_i}{\phi}.$$

Therefore, D_L can be calculated using an average value of D_i from the 'through-diffusion' experiments. In addition, as $D_L = ua_L$, a_L can be calculated and represents the dispersion length. For the HTO data shown in Figure 13.9, $D\,[\mathrm{HTO}] = 2.6 \times 10^{-10}\,\mathrm{m^2.s^{-1}}$, and therefore taking

$$D_P[\mathrm{HTO}] = \frac{8 \times 10^{-11}}{0.5},$$

gives $D_L[\mathrm{HTO}] = 1.0 \times 10^{-10}\,\mathrm{m^2.s^{-1}}$. Finally a_L can be calculated and for the data from Figure 13.9, which gives $a_L[\mathrm{HTO}] = 0.1 \times L$. This result compares well with the results obtained by Neretnieks *et al.* (1987) of dispersion lengths in fissured crystalline rock. In that study, dispersion lengths were found to be approximately 10% of the travel distance, and to increase as the migration distance increased.

Ideally, 'through-diffusion' experiments should have been performed on the samples in these tests, prior to undertaking convection experiments. This would have allowed a determination of D_i to be made on each sample used for a convection experiment. This cannot be achieved in practice as it requires the use of the HTO or I^- tracer on the same sample in two experiments. Taking this limitation into account, the results for the HTO tracer from Figure 13.9, indicate that the diffusive component was of the same magnitude as the dispersive component (D_P[HTO]$=1.6\times10^{-10}$ m^2.s^{-1}, D_L[HTO]$=1.0\times10^{-10}$ m^2.s^{-1}), where a hydraulic gradient of approximately 4000 had been used.

13.3.3 Supplementary experiments

The additional experiments to quantify the relationship of hydraulic conductivity and porosity to applied stress yielded the expected reduction in permeability and porosity as stress increased. Consolidation tests carried out over a 0.095–4.180 MPa stress range produced data which quantified a porosity reduction from 0.52 to 0.36. Experiments to determine the hydraulic conductivity of the clay perpendicular to the fabric were performed over a range of 0.4–30 MPa and the results are illustrated in Figure 13.10. Once a stress of 1 MPa was exceeded only small reductions in the hydraulic conductivity were observed.

A 'through-diffusion' experiment was performed under a confining stress of between 0.1 and 10 MPa, using HTO and I^- in artificial clay groundwater as tracers. The intrinsic diffusion coefficient for HTO was reduced from 1.1×10^{-10} to 5.4×10^{-11} m^2.s^{-1} over this stress range, whereas that for I^- decreased from 7.9×10^{-11} to 3.9×10^{-12} m^2.s^{-1}. The values obtained from this experiment at low stress (0.1 MPa) were slightly higher than from

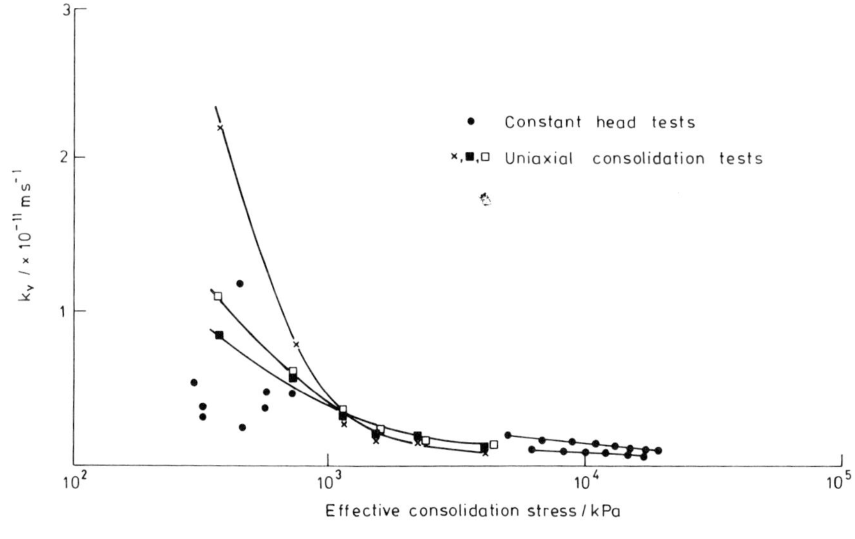

Figure 13.10 The relationship between hydraulic conductivity of the London Clay and confining stress.

'through-diffusion' experiments undertaken in the perspex cells. This could indicate that stresses in excess of 0.1 MPa were applied to the samples loaded into the perspex cells. The diffusibility of clay at 0.1 MPa was approximately 0.05 for both tracers, whereas at 10 MPa it was approximately 0.02 for HTO and 0.003 for I^-. This result may indicate that at higher stresses not only was I^- excluded from some part of the microporosity in the clay, but that the porous structure of the clay was acting as a semi-permeable membrane.

The 'through-diffusion' experiment performed to quantify D_i for HTO parallel to the clay fabric yielded a value of $2.1 \times 10^{-10}\,m^2.s^{-1}$. This value is a factor of 2–3 higher than values deduced from any other 'through-diffusion' experiment, where mass transfer was measured perpendicular to the fabric, and therefore illustrates the anisotropic nature of the clay.

13.4 Conclusions

Samples of London Clay have been tested to quantify mass transfer properties of the rock, using non-sorbing tracers in pore water. Different experimental techniques using different mass transfer processes have been employed and have yielded an internally consistent set of results.

'Through-diffusion' experiments under low stress conditions have yielded D_i values of approximately $8 \times 10^{-11}\,m^2.s^{-1}$ for HTO and $4 \times 10^{-11}\,m^2.s^{-1}$ for I^-. Values of the diffusivity of the material for both of the tracers have yielded reductions in the free water diffusive flux by a factor of about 30.

Data from 'through-diffusion', 'in-diffusion' and convection experiments have quantified different amounts of porosity accessible to non-charged and anionic tracers. Deuterated water or HTO access a porosity of approximately 0.5 which was equivalent to the total porosity determined by oven drying, whereas the porosity accessible to I^- was approximately 0.2–0.3. Data from 'in-diffusion' experiments suggest that a fraction of the porosity accessed by HDO or HTO may be accessed over longer timescales.

Both the convection experiments and the 'through-diffusion' experiments can be modelled using a simple porous medium approach. In contrast, the 'in-diffusion' experiments, which are sensitive to very short timescales, show a discrepancy between model prediction and experimental result. This would suggest that local access to some pore space in the clay may take up to 100 h. However, such short timescales are not relevant to long-term transport processes within the clay.

Measurements of the porosity, hydraulic conductivity and solute diffusion coefficients of the clay under different stress conditions show reduction in values as stress increases. Diffusibility values for both I^- and HTO were comparable at low stress but differed under high stress conditions. Under high applied stress the diffusibility of the clay to HTO was reduced slightly, whereas that for I^- was reduced by an order of magnitude. This indicates possible ion exclusion effects.

The results presented in this paper suggest that, for mass transfer considerations, the porosity of the London Clay can be divided into two components. The first component is the 'transport porosity' through which the steady-state flux of solute is effected, by either convective or diffusive mechanisms. The second component has been referred to as 'dead-end' porosity by previous authors (Bradbury and Green, 1985). It is accessed only by diffusion and acts as a 'sink' for solutes during transient conditions. For the clay investigated in this work, we consider the 'dead-end' porosity to consist largely of the micropores within the individual clay mineral particles. Most of the surface area in the system is associated with the internal surfaces of the clay minerals. Therefore, access to the 'dead-end' porosity will control important parameters of the rock, such as its sorption capacity for radionuclides.

At low consolidation stresses the diffusibility of the clay towards I^- and HTO was similar. This demonstrates that, although I^- cannot access the total porosity of the clay, it is not excluded from the 'transport' porosity. Therefore, the clay will not exhibit any osmotic or reverse-osmotic effects. However, at higher consolidation stresses (up to 10 MPa) the diffusibility of the clay towards I^- is less than 10% that of the diffusibility towards HTO. The implication is that I^- is now excluded from a large part of the transport porosity, presumably because of a reduction in pore size under the higher stress. Under these conditions the clay has the potential to act as a semi-permeable membrane and to exhibit osmotic behaviour. This conclusion is supported by a study of convective mass transfer through compacted cakes of smectite clay (Benzel and Graf, 1984). They showed that convecting a moderately concentrated brine through thin smectite cakes resulted in the ultrafiltration of Na^+ and Ca^{2+}, indicated that the clay acts as a semi-permeable membrane.

The data obtained in this study indicate that in low-permeability clay formations, where natural hydraulic gradients are commonly fractions of unity, aqueous phase mass transfer will be dominated by diffusion. If the clay formation is under high confining stress, the material may act as a semi-permeable membrane.

Acknowledgement

This work was funded by UK Nirex Ltd.

References

Allen, A. J., Baston, A. H., Bourke, P. J. and Jefferies, N. L. (1989) The application of small-angle neutron scattering to the study of mass transfer in clays, in *Proceedings of the 6th International Symposium on Water–Rock Interaction*, Rotterdam (ed. D. L. Miles) Balkema, Rotterdam, pp. 9–13.

Anderton, R., Bridges, P. H., Leeder, M. R. and Sellwood, B. W. (1979) *A Dynamic Stratigraphy of the British Isles*, George Allen & Unwin, London.

Benzel, M. W. and Graf, D. L. (1984) Studies of smectite membrane behaviour: Importance of layer thickness and fabric in experiments at 20°C. *Geochimica et Cosmochimica Acta*, **48**, 1769–78.

Bonne, A. and Neerdael, B. (1986) Disposal of radwaste in clay: the experience gained in Belgium, in *Transactions of the European Nuclear Conference* (ENC '86), Geneva, Switzerland, 1–6 June 1986.

Bourke, P. J., Jefferies, N. L., Lever, D. A. and Lineham, T. R. (1989) Mass transfer through clay by diffusion and advection. *Nirex Safety Studies Report*, NSS-R159, *Description of Intraval Test Case 1A*, UK Nirex Ltd, Harwell, Oxon.

Bradbury, M. H. and Green, A. (1985) Measurement of important parameters determining aqueous phase diffusion rates through crystalline rock matrices. *Journal of Hydrology*, **82**, 39–55.

Bradbury, M. H. and Stephen, I. G. (1985) Diffusion and permeability based sorption measurements in intact rock samples. *Material Research Society Symposium Proceedings*, **50**, 81–90.

Garratt, T. J. (1987) Talbot's Method for the Numerical Inversion of Laplace Transforms: An Implementation for Personal Computers, *Nirex Safety Studies Report*, NSS/R153, UK Nirex Ltd, Harwell, Oxon.

Gilling, D., Jefferies, N. L. and Lineham, T. R. (1987) An Experimental Study of Solute Transport in Mudstones, *Nirex Safety Studies Report*, NSS-R109, UK Nirex Ltd, Harwell, Oxon.

Lever, D. A. (1986) Some Notes on Experiments Measuring Diffusion of Sorbed Nuclides Through Porous Media, *UKAEA Harwell Report*, AERE-R12321, UKAEA, Harwell, Oxon.

Li, Y. H. and Gregory, S. (1974) Diffusion of ions in sea water and in deep-sea sediments. *Geochimica et Cosmochimica Acta*, **38**, 703–14.

Monsecour, M., Put, M. and Fonteyne, A. (1991) Migration experiments in the underground facility at Mol to validate assessment model, in *Proceedings of the Geoval-90 Symposium on Validation of Geosphere Flow and Transport Models, OECD/NEA*, Stockholm, 1990. Office for Economic Cooperation and Development, Paris.

Neretnieks, I., Abelin, H., Birgersson, L., Moreno, L., Rosmuson, A. and Skagius, K. (1987) Chemical transport in fractured rocks, in *Advances in Transport Phenomena in Porous Media*, (eds J. Bear and M. Y. Corapcioglu) Martinus Nijhoff, pp. 473–550.

Put, M. J. and Henrion, P. N. (1988) An improved method to evaluate radionuclide migration model parameters from flow-through diffusion tests in reconsolidated clay plugs. *Radiochimica Acta*, **44/45** (Part II), 343–7.

Skagius, K. and Neretnieks, I. (1985) Diffusion Measurements of Cesium and Strontium in Biotite Gneiss, *Report*, SKB-TR-85-15, Svensk Körnbrönslehantering AB (Swedish Nuclear Fuel & Waste Management Co.), Stockholm.

Trappeniers, N. J., Gerritsma, C. J. and Oosting, P. H. (1965) The self-diffusion coefficient of water, at 25°C, by means of spin-echo technique. *Physics Letters*, **18** (3), 256–7.

CHAPTER FOURTEEN

Monitoring the evolution of the clay environment beneath a concrete vault at the Drigg Low Level Waste Disposal Site

D. P. Trivedi and R. G. G. Holmes*

14.1 Introduction

The Drigg Low Level Waste Disposal Site has operated since 1959. The disposal facility accepts low level waste from the nuclear and radiochemical industries, hospitals, etc. and is operated by British Nuclear Fuels PLC. Historically the waste has been disposed of in trenches on a base of a low-permeability boulder clay designated the G5 clay horizon. Since 1988 containerized waste has been emplaced in a concrete-lined facility, termed Vault-8.

A comprehensive technical programme is underway in support of present and future developments at Drigg (Grimwood, 1989). One aspect of the technical programme associated with the vault is the long term-monitoring of the environment beneath the vault. Monitoring will continue both during the operational and post-closure phases of the vault, until such time as the monitoring apparatus ceases to function. This paper describes the installed equipment, together with some preliminary results obtained during the first year of operation of the facility.

14.2 Description of the vault

The vault is constructed predominantly of reinforced concrete and has a usable volume of 180 000 m^3. Directly beneath the vault is a gravel drainage blanket of approximately 0.2–0.5 m depth. Groundwater flowing around the vault and any water that may infiltrate the structure of the vault are intercepted by the drainage blanket and directed to the site drainage system (Holmes, 1989).

*To whom correspondence should be addressed.

Geochemistry of Clay-Pore Fluid Interactions. Edited by D. A. C. Manning, P. L. and C. R. Hughes. Published in 1993 by Chapman & Hall, London. ISBN 0 412 48980 5

Underlying the drainage blanket is the low-permeability natural G5 clay stratum that forms part of the glacial deposits covering the underlying St. Bees Sandstone. In places, the natural clay has been augmented by an imported or 'engineered' clay consisting of a graded limestone aggregate with sodium bentonite.

Currently the vault is being filled with 'ISO-freight' containers loaded with low level waste. When the vault is full it will be covered with a low-permeability cap to limit the ingress of water into the vault.

14.3 The monitoring facility

An array of instruments was installed beneath the vault at the time of emplacement (Holmes, 1989). The purpose of the apparatus installed was to:

1. Monitor the evolving chemical environment beneath the concrete structure in the underlying gravel drainage blanket and in the upper 3 m of the engineered or G5 clay.
2. Monitor any changes in the permeability of the underlying clay or infer modifications to the ability of the clay to sorb radionuclides, as a result of changes in the chemical environment. This applies in particular to interactions with cementitious leachates from the concrete vault.
3. Detect the advent of radionuclides in groundwater within the clay and in the drainage blanket.
4. Provide information to support future developments at Drigg.

The instrumentation was installed in boreholes at three locations beneath the vault (Figure 14.1). These locations, termed 'windows', are:

Window 1—Beneath a non-porous drain.
Window 2—Beneath unjointed concrete.
Window 3—Beneath a concrete expansion joint.

It is envisaged that situating monitoring equipment at these locations will cover the range of physical and chemical environments occurring beneath the vault.

The instrumentation consists of:

1. Temperature probes.
2. Electrical conductivity probes.
3. Pneumatic piezometers to measure the hydrostatic head beneath the vault.
4. Gas-sampling apparatus.
5. Pressure vacuum lysimeters (PVLs) to take water samples.
6. Pumping apparatus to remove large-volume samples from sumps, which will enable sampling in undersaturated conditions.
7. *In situ* clay-sampling apparatus which operates by means of a falling head test.
8. Clay-sampling apparatus which enables clay core samples to be taken.

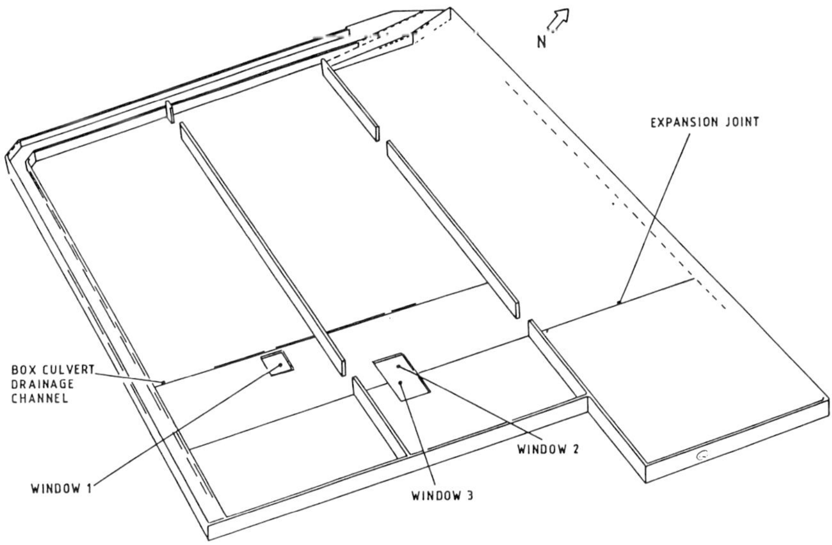

Figure 14.1 Drigg storage depot Vault 8.

The temperature, conductivity probes, gas sampling and PVLs are installed in the gravel drainage blanket and at depth intervals down to 3 m from the top of the G5 or engineered clay. The location of these instruments is shown schematically in Figure 14.2.

At each sampling depth the equipment is emplaced in coarse sand. Sets of apparatus at each depth are separated by a layer of bentonite. Sample pipes and cables from the equipment are routed through a series of ducts to an instrumentation facility located close to the side of the vault. Provision has been made to allow samples of clay from beneath the vault to be removed using an augered coring tool.

The sampling schedule is indicated in Table 14.1. The PVLs operate by entrapping groundwater in an evacuated chamber, after passing through a low-porosity ceramic filter cup. Samples are taken by pressurizing the chamber with nitrogen gas, forcing water through pipes to the instrumentation facility. Water samples are analysed for major anions and cations and gross radioactivity content (alpha, beta and gamma).

Gas samples are analysed for bulk gases and any trace volatiles which may be present.

As one of the priorities of this stage of the programme is to determine the background levels of radioactivity in pore water as a reference for long-term monitoring, the pH values of water samples and then the gross radioactivity levels are measured. This is followed by dissolved solute analysis. In cases where less than 200 ml of sample are recovered, the whole of the sample is analysed for radionuclide content. In this case the chemical compositions are not determined and the results are labelled ‘insufficient sample’.

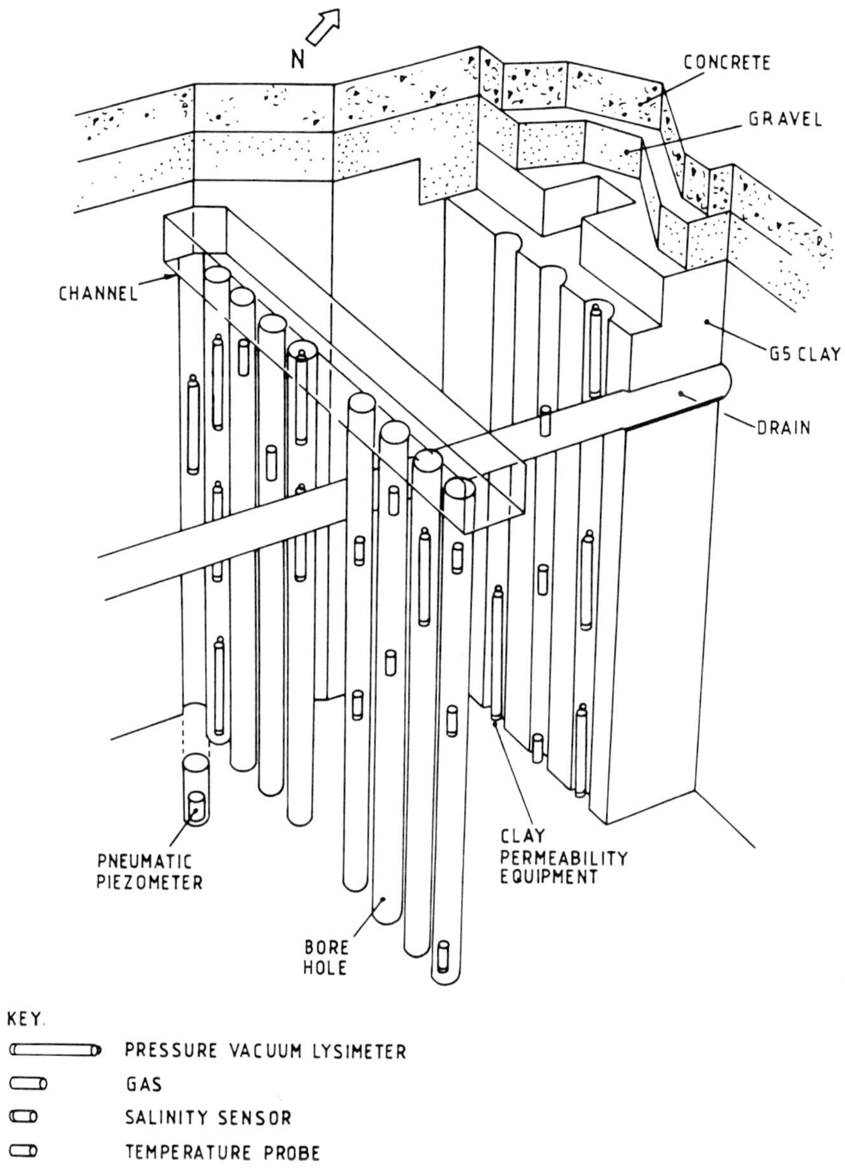

Figure 14.2 Location of instrumentation in window 1.

The analytical techniqes employed for groundwater and gas-sample analysis are given in Table 14.2.

Most apparatus has been installed in duplicate and it is hoped that the monitoring programme will extend for a significant period into the post-closure phase of the vault.

Table 14.1 Vault-8 monitoring programme sampling schedule.

Apparatus	Monitoring frequency	Type of determination
Temperatures	Weekly	Sub-vault temperatures
Electrical conductivity probes	Weekly	Conductivity/salinity
Pneumatic piezometers	Weekly	Head of water
Pressure vacuum lysimeters	3 monthly	Water samples for major and trace elements
Gas samplers	3 monthly	Bulk and trace gases
Clay samplers	Occasional	Sub-vault clay samples
Well pumps	Occasional	Large-volume samples from sumps in the clay for use on PVL sampling failure

Table 14.2 Analytical methods employed for the sub-vault water and gas phase analysis

Element	Analytical method
Na, K, Mg, Ca	ICP–AES
Fe(total)	Polarography
Cl, SO_4	Ion-chromatography
Total inorganic carbon	Total carbonate precipitation as $BaCO_3$ followed by carbonate assay by potentiometric titration
pH	Calibrated glass electrode used immediately on taking a sample
Gases	Mass spectrometry

14.4 Results of the monitoring programme

14.4.1 Temperature

The temperature variations over the first year of monitoring for window 3 are illustrated in Figure 14.3. There is little significant variation between windows. The results are typical of a clay stratum which is a poor thermal conductor. The surface of the clay responds rapidly to changes in the ambient air temperature, while the deeper clay locations are well insulated and exhibit only a seasonal temperature variation.

14.4.2 Conductivity

Variations in conductivity are again illustrated with reference to window 3 (Figure 14.4). There is little temporal and no systematic variation in the majority of samples. This was expected as the low hydraulic conductivity of the

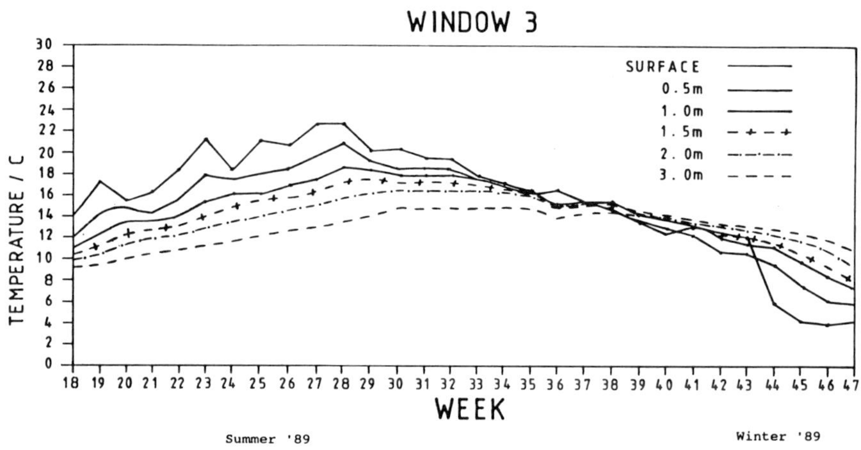

Figure 14.3 Temperature variations in the sub-vault environment over the first year of monitoring.

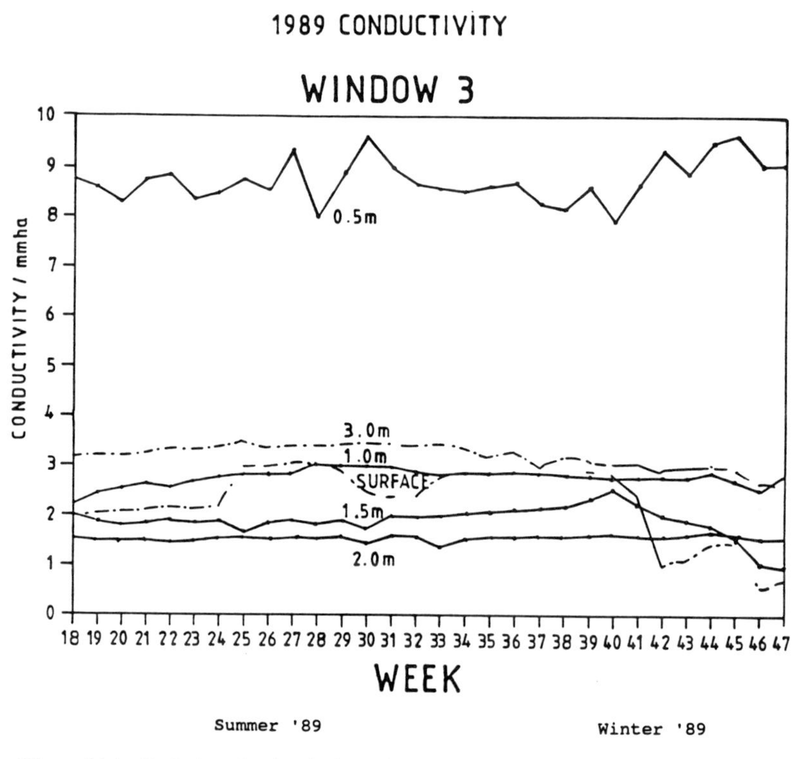

Figure 14.4 Variations in electrical conductivity during the first year of monitoring.

G5 clay (Howarth *et al.*, 1992) indicates that the groundwater flow through the clay is likely to be very slow. Changes in the pore-water chemistry as a result of interaction with the vault will be equally slow.

14.4.3 Water table measurements

The pneumatic piezometers measure the head of water at the monitoring locations. The piezometer in window 1 is located at a depth of 5.8 m from the clay surface, and has consistently indicated that the ground is unsaturated at this location.

The pneumatic piezometers in windows 2 and 3 are adjacent to each other and located at a depth of 12.5 m below the clay surface. A relatively constant head of 2 m is observed at these locations, indicating that the water table is located at a depth of 10.5 m below the clay surface. These observations are consistent with the conditions prevailing at the time of piezometer installation when it was observed that piezometer 1 was located in unsaturated conditions, and those in windows 2 and 3 were located in saturated conditions. These observations are also consistent with the volumes of samples collected from the PVLs; most PVLs are located in unsaturated ground. Only PVL 38, located at a depth of 12.5 m specifically to intercept groundwater consistently gives large-volume samples.

14.4.4 Gas sampling

Gas samples have been taken on an occasional basis at present. The results shown in Table 14.3 are typical of the results so far obtained. They indicate that the gas phase at present is typical of the compositional range of atmospheric air. This is probably due to air entrapped in the borehole sand at the time of equipment installation. The presence of volatile gases such as

Table 14.3 Gas compositions beneath the vault.

Sample location	CO (%vol/vol)	N_2 (%vol/vol)	O_2 (%vol/vol)	Ar (%vol/vol)	CO_2 (%vol/vol)
Window 1					
Port 1 (gravel)	0	85.87	12.87	1.0	0.26
Port 2 (0.5 m)	0	80.57	18.20	0.95	0.28
Window 2					
Port 1 (gravel)	—	—	—	—	—
Port 2 (0.5 m)	0.05	79.25	19.68	0.94	0.08
Port 3 (1.0 m)	0.01	78.88	20.37	0.92	0.04
Window 3					
Port 1 (gravel)	0.01	78.73	20.23	0.93	0.10

methane is also being monitored to indicate the onset of reducing conditions, but these gases are currently below analytical detection limits.

14.4.5 Groundwater sampling

The results of groundwater sampling during May 1989 are presented in Table 14.4. The results are representative of the type of groundwater compositions obtained during the first 18 months of monitoring. Sample volumes typically vary between 10 and 1500 ml, the low sample volumes occurring for the reasons outlined previously.

The pH measurements show that pH does not systematically vary between windows. High pH values are observed both in some of the clay and drainage

Table 14.4 Composition of groundwater samples taken in May 1989 as part of the monitoring programme.

PVL No.	Location	Sample (ml)	pH	Ca^{2+} (ppm)	Fe (ppm)	Mg^{2+} (ppm)	Na^{+} (ppm)	Cl^{-} (ppm)	SO_4^{2-} (ppm)	TIC (ppm)
Window 1										
1	Surface	20	7.77							
2	0.5	150	7.32	180	1.7	16	180	59	175	680
3	1	300	7.38	170	2.2	21	140	69	230	480
4	1.5	100	7.62	155						
5	2	500	7.25	360	1.8	54	140	76	970	440
6	3	300	8.99	86	0.9	1.7	280	40	790	85
12	Sump	1000	7.52	130	1.3	12	170	41	580	130
Window 2										
13	Surface	1400	9.77	21	2.1	0.7	800	130	1180	280
14	0.5	10								
15	1	70	8.55							
16	1.5	110	8.53							
17	2	150	7.46							
18	3	1000	10.5	120		3.3	130	39	370	120
24	Sump	1150	8.7	130		3.3	130	39	370	120
Window 3										
25	Surface	10	7.95							
26	0.5	10	8.3							
27	1	300	9.14	48	1.2	2.8	1700	200	710	2500
28	1.5	900	7.74	47		11	590	90	300	1000
29	2	700	7.6	48		17	590	124	510	750
30	3	500	9.82	45		0.5	21	30	370	220
36	Sump	1400	9.12	100		1.4	130	22	390	80

Where sample volumes were less than 200 ml, pH was determined and the samples were analysed for gross radioactive contents only.

All Fe results are for total iron. The results left blank (subject to note 1) are all at detection limits of 0.5–1 ppm.

TIC refers to total inorganic carbon.

blanket PVL locations, which should be a chemically distinct environment from the clay pore waters.

The chemical composition of the waters are shown on a Piper diagram (Figure 14.5). This illustrates that the waters range in composition from Ca–SO_4–CO_3 to Na–CO_3 dominated. There has been a limited number of published clay pore-water compositions, but those available from the Belgium Boom Clay (Beatsle *et al.*, 1986) and from the Oxford Clay (Bonne *et al.*, 1985) are also plotted on the Piper diagram. These show that the majority of compositions found in this study are not dissimilar to those of other clay pore waters.

To date, the levels of gross alpha, beta and gamma in sub-vault waters have always been measured at instrumental background levels. This was expected due to the integrity of the waste containers, the reinforced concrete floor and the efficiency of the sub-vault drainage system.

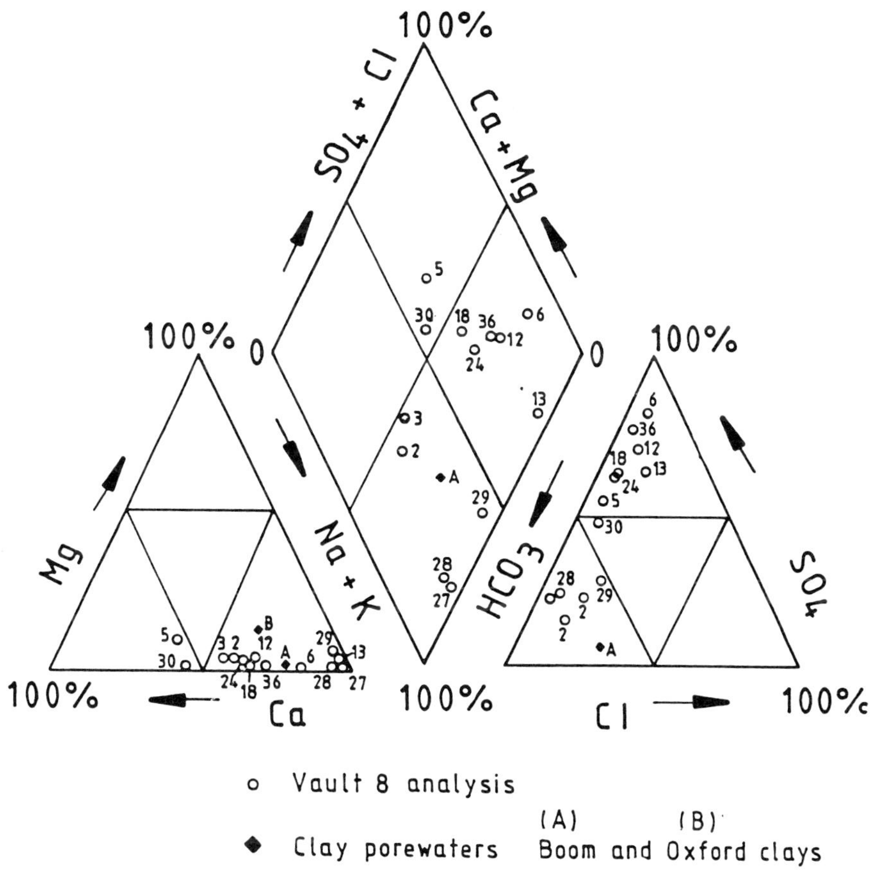

Figure 14.5 Geochemical classification of pore waters (based on May 1989 PVL samples).

14.5 Discussion

Some relations between dissolved solutes measured over the first year of monitoring are represented in Figures 14.6–14.8. As shown by Figure 14.6(a), there is poor correlation between Na^+ and Cl^-, with a far higher Na/Cl ratio than that which would be produced by halite dissolution. Although Na^+ and Cl^- are considered to be some of the most conserved of groundwater solutes, the lack of a constant Na/Cl ratio suggests that one or both of these ions is not conserved in these waters.

The relationship of calcium and magnesium with total inorganic carbon (TIC) is illustrated by Figure 14.7. This suggests that calcium and magnesium concentrations are controlled by equilibrium with carbonate minerals. This was shown to be the case by groundwater speciation calculations performed using the PHREEQE geochemical speciation code linked to the CHEMVAL database (Parkhurst *et al.*, 1980). The resulting mineral saturation indices (SI) are illustrated in Figure 14.8. The results show that initially both calcite and dolomite rise from undersaturation to supersaturation with increasing dissolved TIC. After returning back to values close to carbonate mineral saturation, the SI values again begin to rise with increasing TIC. The groundwaters finally become Na–CO_3 dominated as shown by Figure 14.7.

The relationship between pH and TIC also indicates that the groundwater chemistry is dominated by carbonate mineral equilibria (Figure 14.8). At the

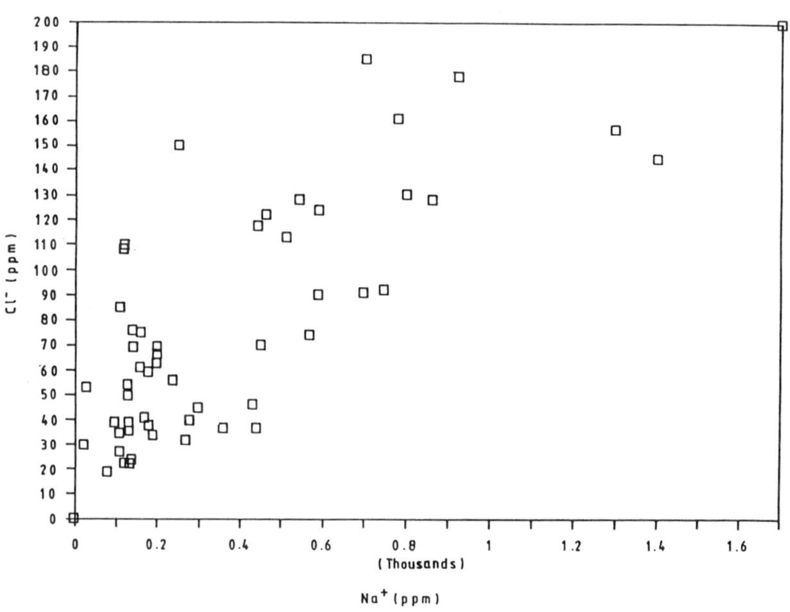

Figure 14.6 The relationship between Cl^- and Na^+ in sub-vault groundwater samples.

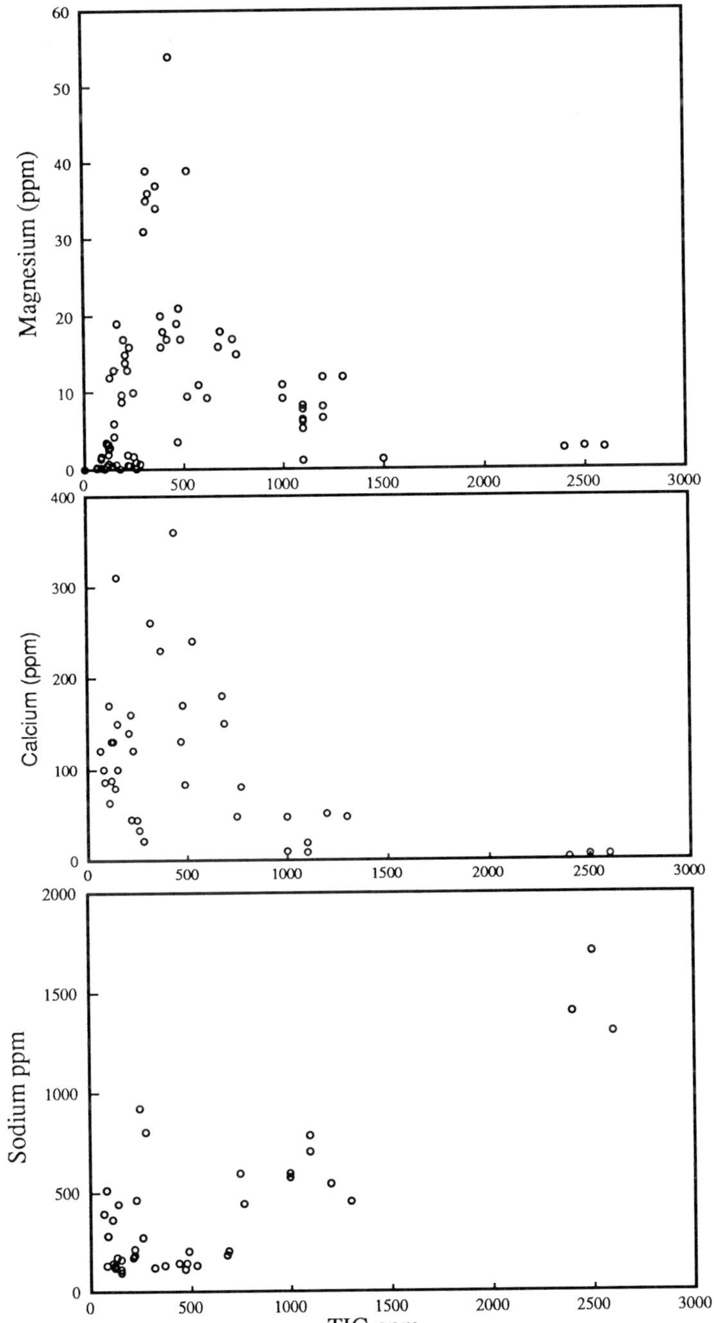

Figure 14.7 Variation of sodium, magnesium and calcium with TIC (samples taken between May and November 1989).

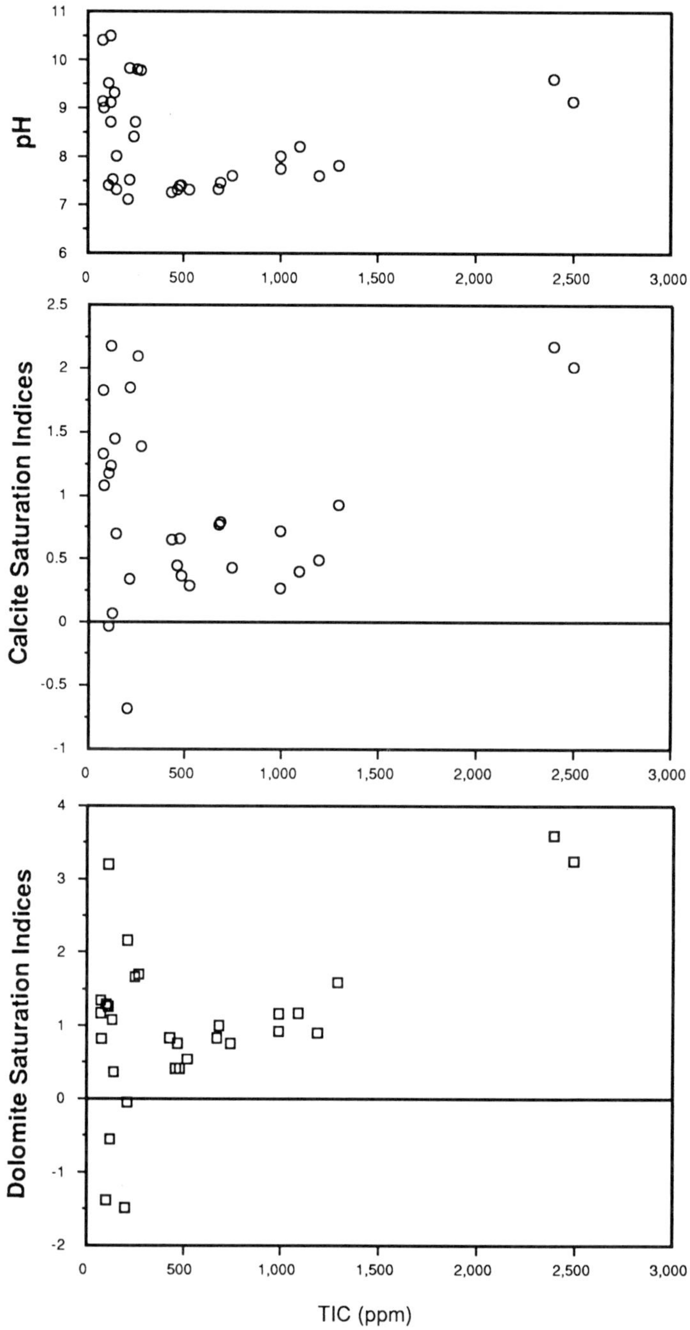

Figure 14.8 Variation of pH, calcite and dolomite saturation indices with TIC (samples taken between May and November 1989).

onset of carbonate mineral saturation as shown by Figure 14.8 the pH is shown to decrease. As with the relationship between sodium and TIC, the pH is observed to increase again after 500 ppm TIC, at which stage the waters are $Na-CO_3$ dominated.

A study of the interaction between a $Ca(OH)_2$ saturated solution and the G5 and engineered clays as found beneath Vault-8 has shown that the resulting solutions from engineering clay interactions were equivalent to dilute Na_2CO_3 (Howarth *et al.*, 1992). It was found that the ions were supplied into solution as a result of dissolution of Na_2CO_3 which was used as a stabilizer for the sodium bentonite. Should this be the case then the sodium carbonate levels in the waters should rise to a maximum with time as a result of complete dissolution of the stabilizer.

At least a 0.5 m thick layer of bentonite was emplaced over the area containing all three windows, with up to 1 m thickness over parts of windows 2 and 3. This means that the gravel drainage blanket, the 0.5 m and some of the 1 m depth piezometers are located in or close to engineered clay. This is confirmed by water sample compositions from these locations which have some of the highest pH values and highest sodium and TIC concentrations.

14.6 Conclusions

The results from the first year of monitoring the sub-vault environment show that the environment is still dominated by the disturbances caused during the construction of the vault and the installation of the monitoring apparatus. The evidence for this is the similarity between the sub-vault gas samples and atmospheric air and the interaction between the sub-vault engineered clay and the clay pore waters.

The groundwater chemistry is dominated by carbonate mineral equilibria as demonstrated by the relationship between Ca^{2+}, Mg^{2+} and pH with TIC, and also by the saturation of these waters with calcite and dolomite.

To date, no radioactivity levels have been detected above natural background levels, as would be expected due to the integrity of the waste containers and the efficiency of the sub-vault drainage system.

Acknowledgements

This work is being funded as part of an extensive programme of work to assess the performance of the Drigg LLW Disposal Site. The work is being funded by Waste Management Unit, BNFL.

References

Beatsle, L. H., Henrion, P., Put, M. and Cremers, A. (1986) Study of the Boom Clay layer as a geochemical barrier for long lived radionuclides, in *High Level Disposal*, (ed. H. C. Buckholder), Battelle Press, Columbia, USA.

Bonne, A., Black, J., Gera, F., Gonze, P., Tassoni, E. and Thimus, J. F. (1985) Characterisation and behaviour of argillaceous rocks, in *Radioactive Waste Management, Proceedings of the Second European Communities Conference* (ed. R. Simon), pp. 487–505.

Grimwood, P. (1989) Modelling and assessment in support of the Drigg Low Level Radioactive Waste Disposal site, in *Proceedings of the 2nd International Conference on Radioactive Waste Measurement*, Vol. 2, British Nuclear Engineering Society, London, pp. 136–7.

Holmes, R. G. G. (1989) Monitoring the evolving chemical and radiochemical environment beneath a low level waste repository, in *Chemical and Migration Behaviour of Actinides and Fission Products in the Geosphere, Proceedings of the Second International Conference*, Monterey, California, USA.

Howarth, A., Jefferies, N. L., Lineham, T. R., Smith, A. C. and Holmes, R. G. G. (1992) An experimental study of the interaction of alkaline solutions and clays from the Drigg Low Level Waste Disposal site, Cumbria. *Radiochemica Acta* (in press).

Parkhurst, D. L., Thorstenson, D. C. and Plummer, L. N. (1980) *PHREEQE—A Computer Program for Geochemical Calculations*, U.S. Geological Survey, Richmond, USA, USGS/WRI-80-96.

CHAPTER FIFTEEN

New apparatus for pore fluid extraction from mudrocks for geochemical analysis

D. C. Entwisle and S. Reeder

15.1 Introduction

The understanding of water movement through rock can be improved by the chemical characterization of water solutes. Sampling of pore water from specific depths of aquifer rocks (in Britain mainly sandstones and chalk) is regularly carried out by centrifugation (Kinniburgh and Miles, 1983). However, this technique is not successful for most clays and mudrocks mainly due to the strong adsorption of the aqueous phase and the low hydraulic conductivity ($<10^{-11}$ m. s^{-1}) of these materials.

The solute distribution in pore water from mudrocks reflects the hydrochemical evolution in changing regimes over long periods of time. The solutes evolve by diffusion and/or advection as well as reaction with the host rock. Similar processes will influence migrating pollutants, such as pesticides and solvents, and material from waste disposal sites in general. The interactions between the bound water, the pore water and the minerals are important controls on the mechanical behaviour and transport properties of these sediments.

The extraction of pore water for geochemical analysis is reviewed together with work on argillaceous rocks from the succession at the Harwell Research Site, carried out by the British Geological Survey (Brightman *et al.*, 1985). The squeezing technique has been used to obtain pore water samples mainly from unconsolidated marine silts and clays. Manheim (1966) developed a heavy duty squeezer capable of applying a stress of up to 150 MPa based on the earlier designs of a number of Soviet workers, most notably Kriukov (1971). A similar design to that of Manheim was used by Morgenstern and Balasubramonian (1980) to extract pore water from Bearpaw Shale from Saskatchewan and Morden shale from Manitoba; both these rock types are marine deposits of the Upper Cretaceous. The natural moisture contents of the two shales were 30.3–31.8% and 18.8–21.0%, respectively (defined as (initial

Geochemistry of Clay-Pore Fluid Interactions. Edited by D. A. C. Manning, P. L. Hall and C. R. Hughes. Published in 1993 by Chapman & Hall, London. ISBN 0 412 48980 5

weight – oven-dried weight)/(oven-dried weight and expressed as a percentage). The maximum stress used was 52 MPa (7500 p.s.i.). Brightman *et al.* (1985) and Morgenstern and Balasubramonian (1980) both found that there was a decrease in sodium or potassium and an increase in calcium and magnesium concentrations with increased pore-water extraction and increased stress. Morgenstern and Balasubramonian also found that there was a decrease in the total salinity at higher stresses. They suggested that these changes were due to dilution of pore water with water from the double layer and also a membrane effect exhibited by highly consolidated clayey sediments. They also suggested that in cases where the total salinity remained the same the dilution by the double layer was not significant. Krahn and Fredlund (1972) found the squeezing technique to be a satisfactory way of obtaining pore fluid for the determination of osmotic suction.

This paper describes the extraction of pore water from drill cores of argillaceous rocks and associated strata by squeezing using a new apparatus designed at the British Geological Survey. The pore waters have been analysed for a wide range of cations and anions, stable isotopes of oxygen and hydrogen and the isotopes of uranium.

15.2 Mechanism of pore water extraction by squeezing

The extraction of pore water by squeezing is analogous to the natural process of consolidation, due to the deposition of material during geological time, but at a greatly accelerated rate. The squeezing process involves the expulsion of pore water from the sediments being compressed. The sediments consist of solid particles, or the mineral phase, and the spaces, or voids, which in a saturated rock are filled with water. When the squeezing stress is applied to a water-saturated clay or mudrock its volume decreases by three main mechanisms:

1. compression of the solid grains;
2. compression of the pore water between the grains;
3. escape of water from the voids.

In most circumstances the compression of the solid phase and water phase is negligible. Most of the changes in volume are caused by the escape of pore water as illustrated by the spring analogy described below.

A hydromechanical analogy for load changing and squeezing is shown in Figure 15.1 (Lambe and Whitman, 1979). The resistance of the internal phase in compression is represented by a spring and the rate at which the pore water flows is dependent on the size of the valve aperture. In (a) the valve is closed and in equilibrium. When a load or stress is added (b) the piston load is apportioned by the water and the spring in relation to the stiffness of each. There is little movement in the piston because the water is relatively incompressible. Most of the load is carried by the water and this increases the water pressure.

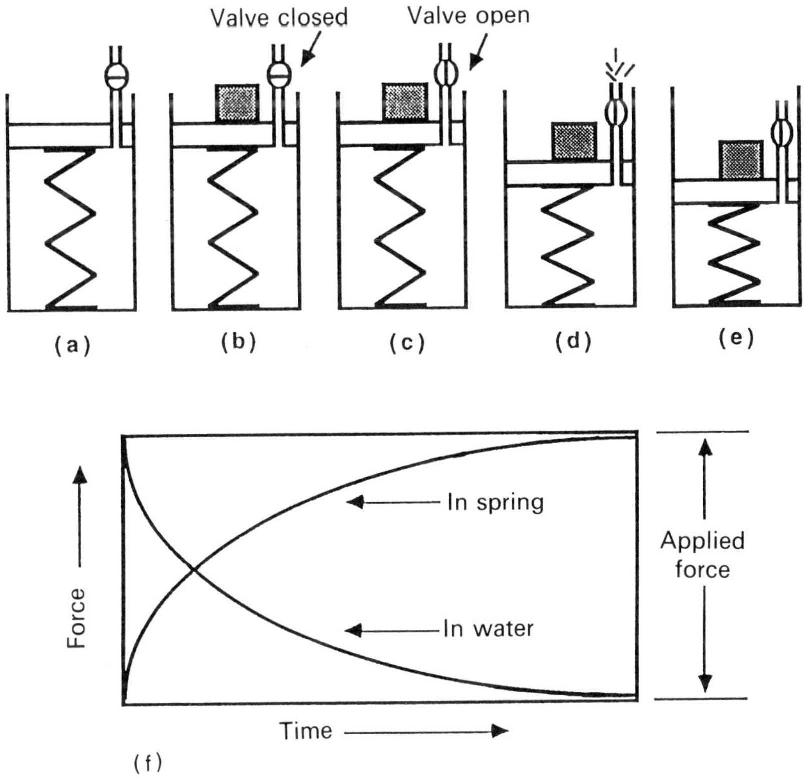

Figure 15.1 A hydromechanical analogy for load changes during squeezing.

If the valve is now opened (c) the excess pore pressure will dissipate by water escaping through the valve (d). The piston drops and the volume of the chamber decreases until there is a new equilibrium when the applied load is carried by the spring and the water pressure has returned to the original hydrostatic condition (e); (f) shows the gradual transfer of load from the water to the spring. The dissipation of the pore-water pressure is called primary consolidation.

The volume change which occurs is related to the stress applied and the difference between the stress and the pore pressure (Yong and Warkentin, 1975). The difference is the effective stress:

$$\sigma' = \sigma - u, \tag{1}$$

where σ' is the effective stress, σ is the total stress and u is the pore pressure.

The rate at which the pore fluid is expelled is related to the height of the sample and the pore size (in unfissured material). A graph of the rate of settlement for each stress measurement is given in Figure 15.2 and shows both

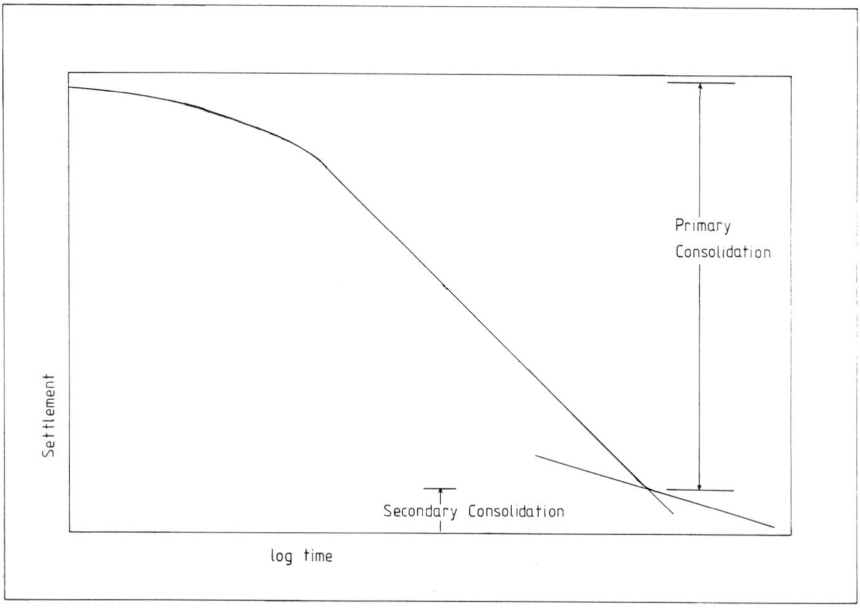

Figure 15.2 The rate of settlement for increasing stress.

primary (dissipation of pore pressure) and secondary consolidation (movement of particles adjusting to the increase in effective stress). The pore water extracted during squeezing is mainly due to primary consolidation. The increasing stress, which is normally added incrementally in an experiment, results in a reduction of pore size thus reducing the rate of pore-water extraction.

The water extracted under each load increment, or in total, can be related to the change in voids ratio, *e*, the ratio of the volume of the voids to the volume of the grains (cf. Rieke and Chilingarian, 1974).

$$e = \frac{W_w}{W_s} G, \tag{2}$$

where W_s is the weight of solids, W_w is the weight of water and G is the specific gravity of the solid. Also

$$e = \phi/(1-\phi), \tag{3}$$

where ϕ is the porosity.

The change in voids ratio increases with the log of the stress in a clay above the pre-consolidation stress. The gradient of the line is the compression index, *Cc*. Figure 15.3 shows the theoretical behaviour and Figure 15.4 is an example of a one-dimensional consolidation test. The compression index, *Cc*, has been

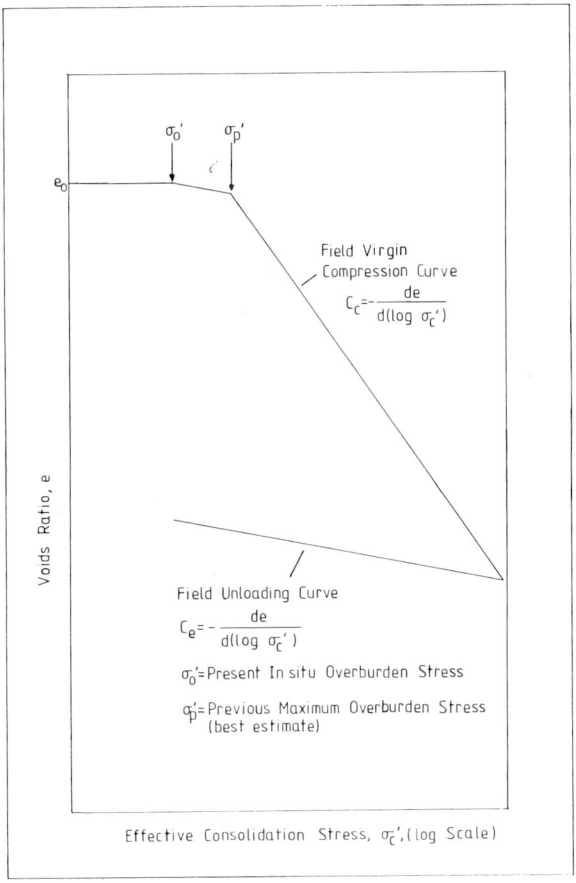

Figure 15.3 An idealized graph of the voids ratio and stress of a clay or mudrock.

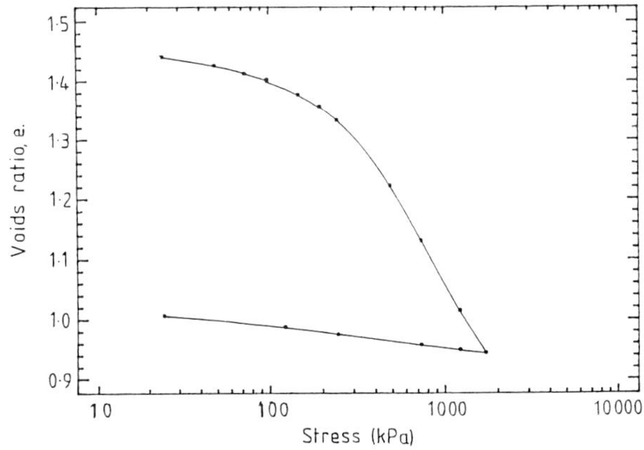

Figure 15.4 A plot of voids ratio and log effective stress for a clay.

related to a commonly measured geotechnical parameter, the liquid limit of the sample by Skempton (1944):

$$Cc = 0.009\ (P_L - 10), \tag{4}$$

where P_L is the liquid limit.

This empirical relationship can be used to estimate the maximum water output during clay squeezing. The volume will be reduced in practice if the clay is cemented or partly saturated.

15.3 Apparatus

The new squeezing apparatus utilizes a commercially available hydraulic pump, which has a maximum output stress of 70 MPa, and hydraulic hoses and fittings. A diagram of the hydraulic system is given in Figure 15.5. The hydraulic pump selected was the 70 MPa autocontrol hydraulic pump from Wykeham Farrance Engineering Ltd of Slough, which is a balanced system

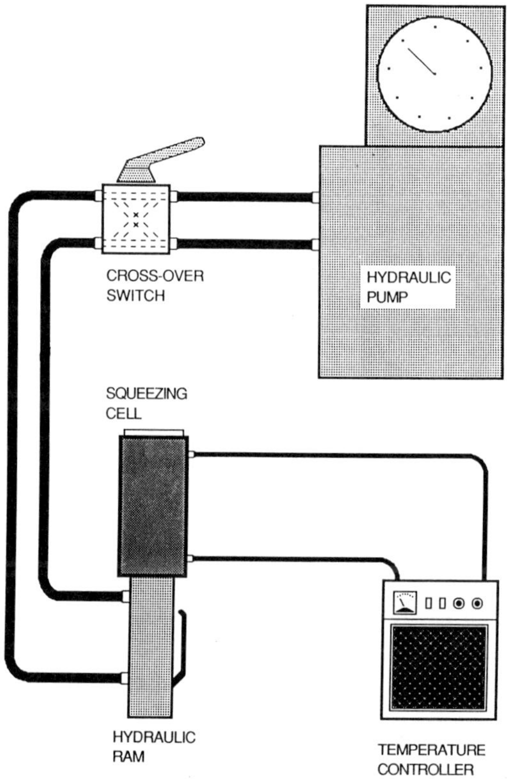

Figure 15.5 A schematic diagram of the hydraulic system of the squeezing apparatus.

and can run for long periods, generally about 2 weeks. The hydraulic ram selected was the R.S.R. 50–20 R hydraulic return cylinder from Tangye Ltd of Birmingham, which has a maximum normal working pressure of 70 MPa. The hydraulic ram has a long stroke which allows the test sample to be pushed out of the cell after test completion. A 3/8″–18 NPT quick fitting, flexible hose and a high-pressure direction valve connect the pump to the hydraulic ram. The valve is used to change the direction of flow of the hydraulic oil to raise or lower the ram.

The main body of the cell and other metal parts in contact with the test sample or pore water are manufactured from Type 316 stainless steel which was selected because of its resistance to corrosion and high tensile strength. However, this metal is difficult to machine and threads can stick if they are too tight. The cell body sample chamber (Figure 15.6) is 75 mm in diameter and 100 mm high. The outside of the cell has a spiral trough through which temperature control fluid circulates inside a plastic insulation jacket. Temperature control of the cell is by a heater/chiller which is capable of temperature control between $-10°$ and 50°C.

The bottom platen screws into the top of the hydraulic ram and is sealed with a 12V Pack Seal which is retained with a nylon packer. The top platen screws into the top of the cell and is sealed with an O-ring. The pore water collection pipe screws into the top platen. Pore fluid is usually collected directly into disposable polypropylene syringes. Two syringe taps with a 0.45 μm Acrodisc filter in between are pushed onto the pore water pipe allowing flexibility of pore water collection. The metal filter, which has a diameter of 90 mm, is also made of Type 316 stainless steel.

15.4 Method

The cell and all its components are washed and dried prior to starting a test. Disposable plastic gloves are worn during the cleaning process and changed regularly to reduce contamination. Copper grease is smeared on to the ram thread and the cell screwed on to the ram, reducing the likelihood of sticking. The lower platen and scraper seal are assembled and screwed into the top of the hydraulic ram. The cooler/heater system pipes are pushed into the fitting on the insulation jacket and the temperature control switched on. The chiller is not switched on until the sample is in the cell and a low stress (1 MPa) is applied; this reduces the chance of condensation. The cell is cleaned with a paper wipe. The sample can now be prepared.

The wide range of sample types and preservation methods has resulted in three sample preparation techniques. The simplest method used is for soft or disaggregated materials (peats or sand, respectively) to be poured or placed into the cell from their container. Stiffer materials, such as London Clay or Oxford Clay, are prepared by cutting a right-cylinder with a clean knife on a

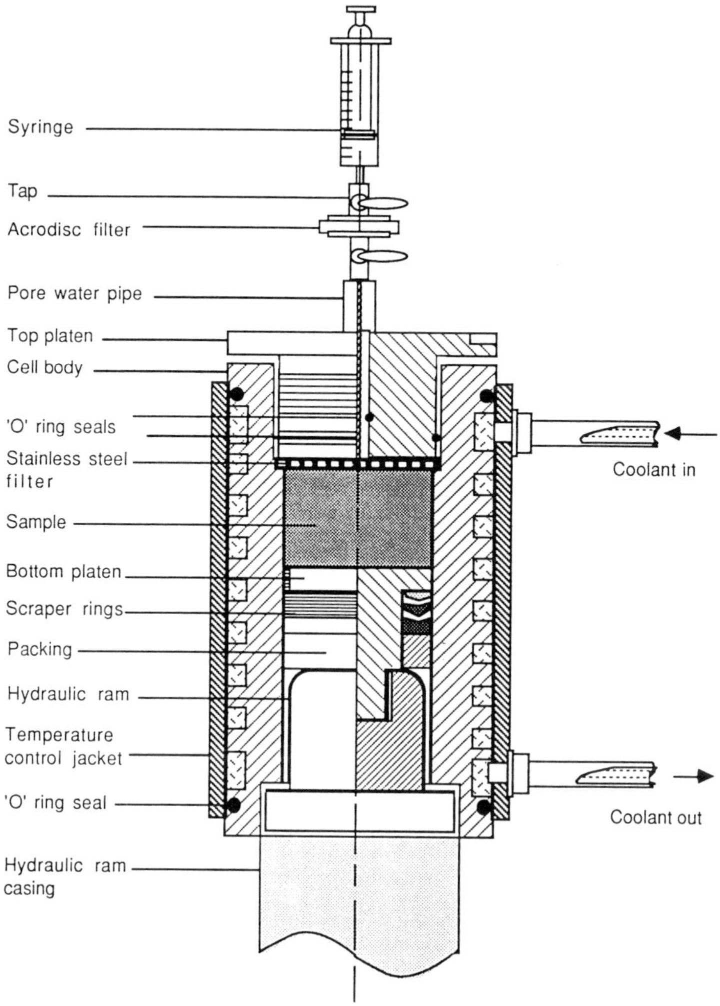

Figure 15.6 A schematic diagram of the cell of the squeezing apparatus.

soil lathe which produces 70 mm diameter cylinders. Harder material is broken up with a large knife on a clean glass plate. The pieces are placed into a beaker, weighed and packed into the cell, ensuring that the cell body threads are not contaminated. In all cases the weight of material used and, if possible, the volume is noted. Clean plastic disposable gloves are used at all times during sample preparation.

After the sample is placed into the cell a 90 mm diameter Whatman filter paper is usually placed on to the shoulder of the sample chamber and a clean steel filter placed on top of the filter paper. The thread of the upper platen is lightly smeared with copper grease. The top platen is screwed tightly into the cell to contact the metal filter. This stops the sample extruding above the shoulder of the cell chamber. The temperature control unit cooler is then switched on, if required.

A small nominal stress is applied to remove most of the air (reducing oxidation) from the cell and to allow the sample and components to bed in. When the selected cell temperaturc (+ or -2°C) is attained, the stress can be increased and water collected. The stress is added incrementally, the rate of increase being dependent upon the physical properties of the specimen and the type of test undertaken. The test may take from 1 h to in excess of 2 weeks.

Most of the pore water collected is analysed for the major elements (listed in Table 15.1), total alkalinity, pH and stable isotopes (if sufficient pore water is recovered). The cations are analysed by an inductively coupled plasma-optical emission spectrometer (ICP-OES). Bicarbonate and pH are measured on 1 ml of unpreserved sample using a Radiometer system incorporating PHM 84 Research pH meter, TTT 80 Titrator, ABU 80 Autoburette and REC 80 Servograph. Fluoride is determined by ion selective electrode using 1 ml of unpreserved sample. All other anions are determined by an ion chromatography and autosampler. Total organic carbon (TOC) contents are determined instrumentally. All analyses are subject to stringent quality control checks which include the analysis of tap and saline water with each run and periodic checks with standards, made in house, and outside, i.e. AquaCheck (Cave, personal communication) within the formal quality assurance system of the British Geological Survey. This complete suite of tests can be obtained using between 16–20 g of water; in some cases, insufficient pore water is available restricting the scope of analysis done and reducing the accuracy of the test. Other tests require a greater mass of water: 50 g for minor element analysis (listed in Table 15.1) and over 100 g each for uranium/thorium isotopes and carbon-14 determinations. A number of adjacent samples may have to be squeezed to provide sufficient water.

After the test is completed the stress is unloaded, the ram is slightly lowered and the top platen unscrewed and removed. The squeezed sample is removed from the cell and the apparatus dismantled and cleaned. The test specimen is measured to calculate its volume, and weighed and oven dried at 105°C for density and moisture content determinations.

A number of repeat tests were carried out using two adjacent samples of approximately the same mass and with similar stress regimes.

The effects of oxidation during sample preparation and pore-water extraction were investigated in a small number of tests carried out in a specially designed 1 ppm O_2 anaerobic chamber. The chamber, constructed by Millar Howe of Watlington, lies beneath the main body which contains the ram and squeezing cell. The comparison of the 'aerobic' and anaerobic tests is not complete and is not discussed in this chapter.

Table 15.1 Chemical species analysed

Major analysis	Minor analysis	Isotopes
Calcium	Major analysis +	Hydrogen-2
Magnesium	Ammonium	Oxygen-18
Sodium	Sulphide	Carbon-14
Potassium	Reduced iron	Uranium
Total alkalinity	Reduced sulphur	Thorium
Sulphate	Lithium	
Chloride	Boron	
Nitrate	Copper	
Nitrite	Zinc	
Fluoride	Chromium	
Sulphur	Molybdenum	
Iron		
Manganese	Lead	
Silicon	Yttrium	
Aluminium	Zirconium	
Cobalt	Tin	
Barium	Caesium	
Nickel	Lanthanum	
Strontium	Cerium	
Bromide	Neodymium	
Phosphate	Samarium	
Total organic carbon	Erbium	
Total inorganic carbon	Gadolinium	
	Dysprosium	
	Europium	
	Praseodymium	
	Ytterbium	
	Holmium	
	Uranium	
	Thorium	

15.5 Results

About 170 squeezing tests have been completed in the last 3 years, using five sets of apparatus, producing over 500 pore-water samples for geochemical analysis. During this period only three tests were unsuccessful in producing any pore water. These were one test each of Mercia Mudstone, Upper Greensand and Opalinus Clay. A wide variety of materials was tested and the results are listed in Table 15.2 with the number of tests undertaken, the number of unsuccessful tests, the range of initial moisture contents, the maximum stress, the time taken to complete the test, the weight of water extracted and the total efficiency. The initial weight of sample generally varied between 400 and 800 g. The percentage pore water extract efficiency was calculated as:

$$\text{extract efficiency} = 100\% \times M_p/(M_0 - M_d), \tag{5}$$

where M_p is the weight of pore water extracted (g), M_0 is the weight of pre-squeezed sample (g) and M_d is the weight of oven-dried sample (g).

Moisture content was calculated as:

$$\text{moisture content} = 100\% \times M_w / M_d, \tag{6}$$

where $M_w = M_0 - M_d$.

The initial moisture content of the samples varied between 7.4% for a sample of Mercia Mudstone and 471% for a remoulded sample of sodium montmorillonite. However, most samples were in the range of 10–50%. The weight of extracted pore water varied between 0 and 314 g. Most tests produced between 15 and 100 g of water. Some tests were undertaken to extract as much water as possible to produce a single bulk sample or a number of 'sequential' water samples, which were analysed separately to study the change in composition with progressive pore water extraction. In other tests only one pore water sample of weight about 16 g was required (Bath *et al.*, 1988; 1989a; 1989b; Ross 1989a; 1989b).

Figure 15.7 shows typical clay squeezing results in which a number of pore-water samples were taken from a sample of London Clay. There are three graphs: (a) shows the stresses and the time elapsed for each pore water sample taken and the stress history of the test; (b) shows the weight of pore water sampled and the efficiency of pore water extraction with time; and (c) shows the weight of pore water extracted and the efficiency of pore water extracted with load.

Figure 15.8 shows the weight of water collected, in grams, plotted against the initial moisture content. It shows that there is an increase in pore water extraction with increasing moisture content. There is also an apparent cutoff of about 7–8% below which no water is collected. The graph indicates the weight of pore water which may be extracted for a specific moisture content. This could be carried out for each lithology. Most of the calcareous, cemented samples (see Table 15.2) are in general less productive and glacial tills more productive. Many tests are grouped around 14–20 g as this was the amount of water required for the geochemical analysis. This figure also gives an upper envelope above which no more water could easily be collected using this apparatus.

The extract efficiency (equation (5)) was devised to account for the differences in the size of each test specimen. Extract efficiency varied between 0 and 71.5% but was mainly in the range 15–60%. A plot of the total extract efficiency against the initial moisture content is given in Figure 15.9. Samples with an initial moisture content of less than 10% also had a low extract efficiency, whereas samples with an initial moisture content of more than 15% tended to have a higher total extract efficiency.

The time taken to complete a test varied from 1 to 1555 h but most tests were completed in 24–250 h. Figure 15.10 is a plot of time elapsed against maximum

Table 15.2 Summary results of squeezing tests

Lithostratigraphy or sample type	Lithology/ symbol	No. of tests	Failed tests	Initial M.C. (%)	Maximum stress (MPa)	Time elapsed (h)	Water extracted (g)	Total efficiency (%)
Sodium montmorillonite	Clay, c	1	—	471	70	1555	314.4	71.5
Spoil tip sand	Sand, s	2	—	8.9–9.1	35–60	20.4–27.5	9.7	17.9–24.5
Peat	Peat, p	6	—	78.2–465	1–20	1.9–4.6	48.3–80.2	18.0–23.5
Glacial till	Till, t	9	—	14.3–42.4	5–65	65.8–167	55.7–69.9	40.9–55.6
Orte Clay, Italy	Clay, c	8	—	14.9–17.5	20–30	17.2–28.0	20.8–31.0	19.0–27.5
London Clay	Clay, c	54	—	20.0–52.8	10–70	24.0–52.8	16.1–99.2	15.3–63.5
Opalinus Clay, Switzerland	Clay, c	8	1	8.1–12.2	45–70	96.7–533	0–21.2	0–32.2
Oldhaven Beds	Clay, c	1	—	34.7	70	140.8	81.4	60.0
Woolwich and Reading Beds	Clay, c	5	—	17.9–31.9	50–70	28.8–190	4.9–53.7	4.8–39.6
Thanet Beds	Sand, s	1	—	29.8	40	24.9	68.8	45.4
	Silt, m	1	—	30.9	70	116.8	79.0	50.8
	Clay, c	1	—	25.1	50	24.3	56.6	40.2
Chalk Marl	Chalk, l	3	—	10.2–10.9	40–50	91.8–96.0	1.4–21.4	2.0–20.2
Chalk	Chalk, l	2	—	19.6–32.5	10–65	1.0–235	22.2–23.3	16.5–21.2
Upper Greensand	Sand, s	7	1	8.3–15.1	40–55	24.4–138	0–26.8	0–13.0
Gault Clay	Clay, c	7	—	10.0–18.9	20–55	21.8–88.5	9.7–24.5	10.0–25.2
Oxford Clay	Clay, c	25	—	16.4–39.7	10–65	18.7–215	18.1–117	14.2–64.1
Kellaways Beds	Sand, s	5	—	11.4–28.6	30–65	47.7–142	18.1–61.3	17.8–44.1
	Clay, c	3	—	13.3–16.6	20–55	49.3–141	18.8–60.6	21.3–61.9
Forest Marble and Cornbrash	Clay, c	2	—	12.5–15.6	35–60	72.6–138	16.8–18.1	16.1–23.1
Esturine Clay	Clay, c	2	—	20.4–23.9	60–65	139–163	61.3–75.7	48.9–51.6
Upper Lias	Clay, c	3	—	20.7–24.5	55–70	93.6–261	51.0–65.6	41.9–46.5
Lower Lias	Clay, c	16	—	10.3–16.9	22–70	142–310	7.5–50.4	11.9–46.5
Mercia Mudstone	Clay, c	1	1	7.4	52	128	0	0

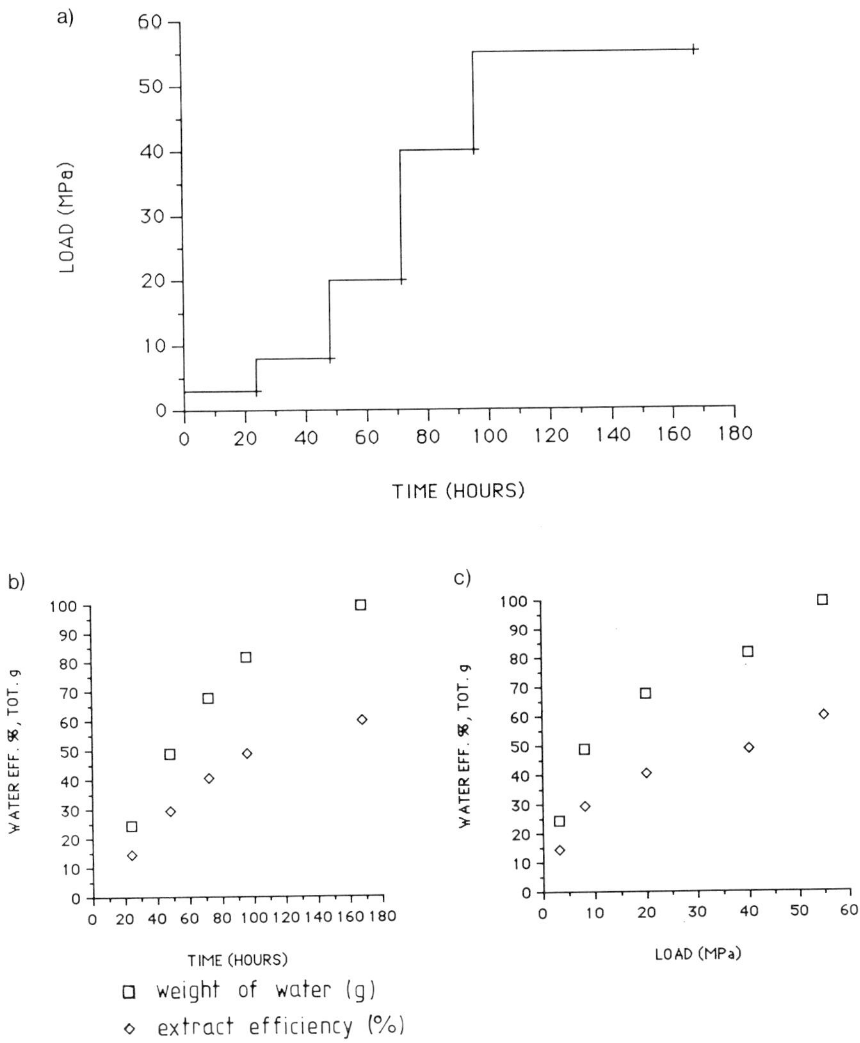

Figure 15.7 Squeezing test results: (a) stress and pore water sampling results; (b) extracted pore water weight and extract efficiency versus time; (c) extracted pore water weight and extract efficiency versus load.

stress, and shows the test conditions used to extract pore water. Some clay, till and calcareous samples required 50 MPa or more for over 150 h to produce the required weight of water. Peat and softer clays and some sands were tested very rapidly, at low maximum stress. Stiffer materials of low permeability require longer periods of squeezing.

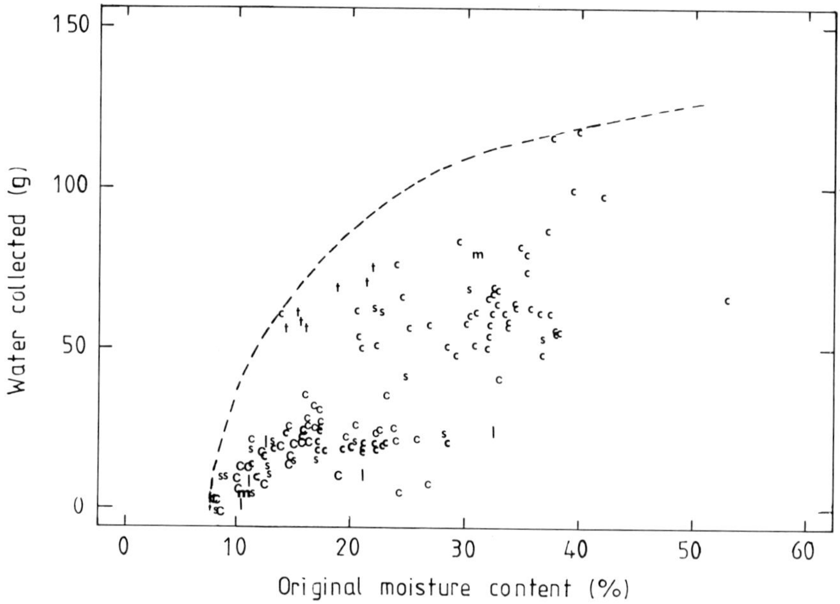

Figure 15.8 A plot of the mass of water collected and the original moisture content where c = clay or mudrock, m = silt, s = sand, t = glacial till, 1 = chalk.

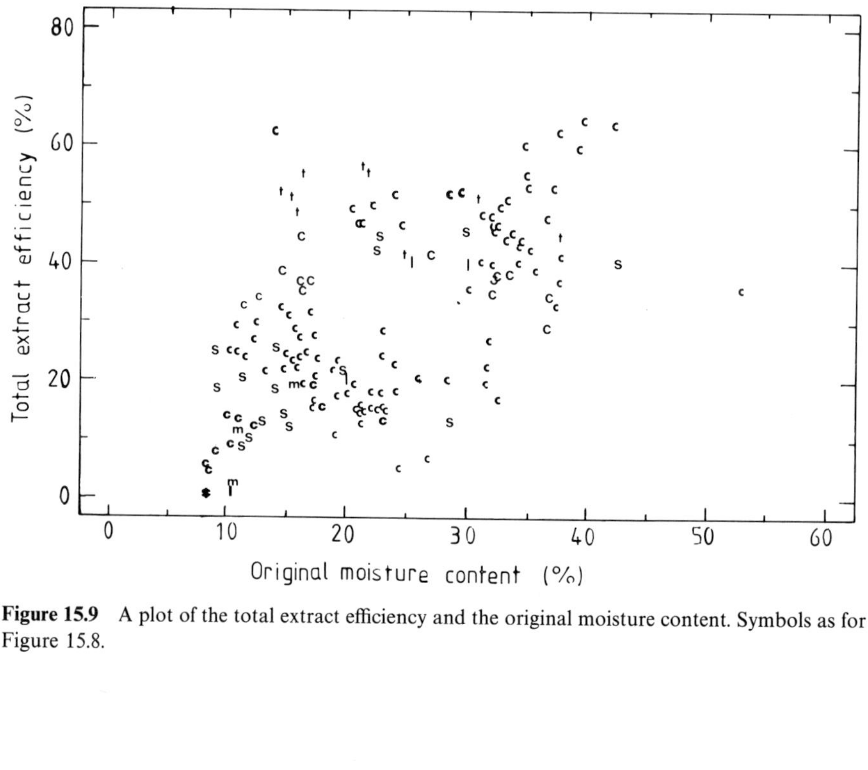

Figure 15.9 A plot of the total extract efficiency and the original moisture content. Symbols as for Figure 15.8.

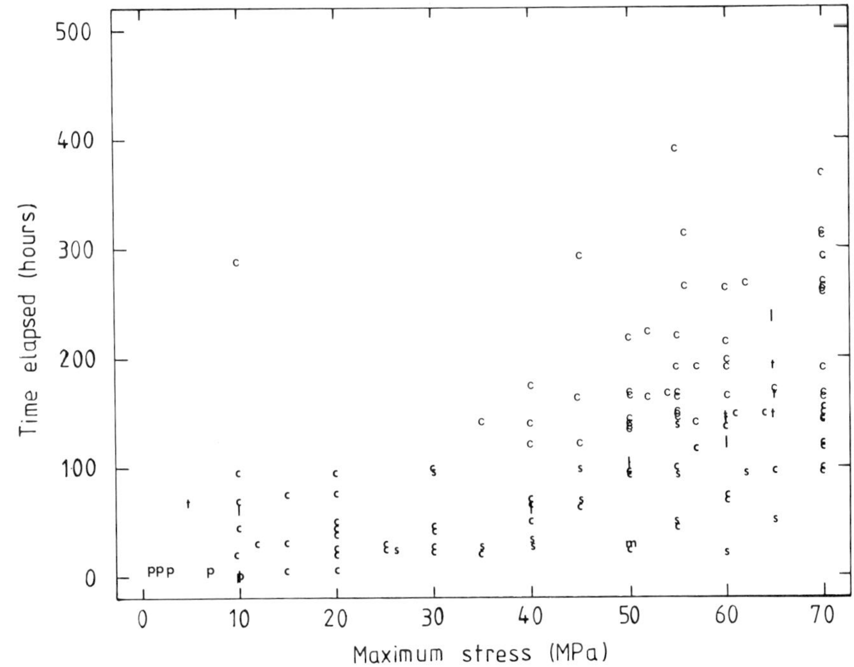

Figure 15.10 A plot of the test conditions—time elapsed and maximum stress. Symbols as for Figure 15.8.

15.6 Discussion

15.6.1 Apparatus

The apparatus developed for pore water extraction initially was devised for low-permeability materials, mainly over-consolidated clays. However, it has been used successfully to extract water from a much wider range of materials including sands and chalk.

A number of design problems have been encountered during development from the prototype to the present version. Most of the mechanical problems have concerned the hydraulic pumps which have not performed as hoped, either not providing the maximum stress required or failing during operation.

The cells, even after many tests at high pressure over long periods of time, show little wear except for the bottom platens, which are slightly pitted due to corrosion. These platens are fairly simply manufactured and so can be replaced.

Operational temperatures of 10 ± 1°C have been consistently maintained even when there was large variation in room temperature.

15.6.2 Pore water extraction

The rate of pore water extraction is partly controlled by the stiffness and permeability of the material tested. Figure 15.10 and Table 15.2 indicate that it takes less time to complete a test in soft material, such as peats and high-moisture-content clays, and granular material, such as sand, than in stiff low-moisture clays. It was noted that when the stress was added too rapidly (a test carried out on three occasions) the pore pressure generated exceeded the capacity of the Acrodisc filter and the filter failed, producing water containing particulate matter.

The main constraint on the amount of water collected is the stiffness of the sample to be squeezed. The constrained modulus, D, of a saturated sample in primary consolidation can be used to give a maximum stiffness above which water extraction is unlikely. (D is the inverse of the coefficient of volume compressibility, m_v, which is commonly used in civil engineering and is derived from consolidation tests; Yong and Warkentin, 1975). The constrained modulus is the resultant deformation in the axial direction only of a confined sample giving an axial strain, ε, of

$$\varepsilon = \Delta H / H_0, \tag{7}$$

where ΔH is the change in height, and H_0 is the initial sample height.

The constrained modulus, D, is defined as

$$D = \Delta\sigma' / \Delta\varepsilon, \tag{8}$$

where $\Delta\sigma'$ is the change in effective stress and $\Delta\varepsilon$ is the change in axial strain.

A saturated sample of 70 mm height and 75 mm diameter, when squeezed to the shape of the chamber with a change in volume of about 310 000 mm^3, leads to a change in height of about 5 mm. This would provide about 22 g of water if the deformation occurs only as primary consolidation. If the change in effective stress is 50 MPa, then the constrained modulus is 750 $MN.m^{-2}$, or a coefficient of volume compressibility of 0.0013 $m^2.MN^{-1}$. This value probably overestimates the stiffness at which pore water could be extracted using this apparatus. A more realistic maximum constrained modulus is between 100 and 250 $MN.m^{-2}$.

The minimum moisture content from which pore water has been extracted using this apparatus is 8.2%. To collect enough pore water for a restricted geochemical analysis required repeat tests. Test samples with moisture contents greater than about 12.5% generally produce enough water for major geochemical analysis with one squeezed sample. Higher moisture content samples, above 15%, may allow more than one water sample to be taken. However, if a number of squeezing tests are carried out on the same bulk sample and the extracts combined then specimens of moisture content as low as 8% may be tested.

An example of pore water chemistry changes with depth in a borehole is illustrated in Figure 15.11, which shows a reduction in sodium concentration through the London Clay from saline to fresh water (Bath *et al.*, 1988).

The extraction of a number of sequential pore water samples from a single test has shown that there is a change in the concentration as the extraction proceeds. This is manifested as an increase in calcium, magnesium and silicon, and a decrease in sodium and potassium, resulting in an overall decrease in total salinity. The association of chemical changes and reduction in moisture content led us to investigate the relationship between pore size and the percentage fractionation of a number of chemical species. Pore size, d, is calculated thus (Yong and Warkentin, 1975):

$$d=(100\times m)/S, \tag{9}$$

where d is the mean interparticle spacing, or pore size, (10^{-10} m), m is the moisture content (%; defined in equation (6)) and S is the surface area ($m^2 . g^{-1}$) using 2-ethoxyethanol (Carter *et al.*, 1965).

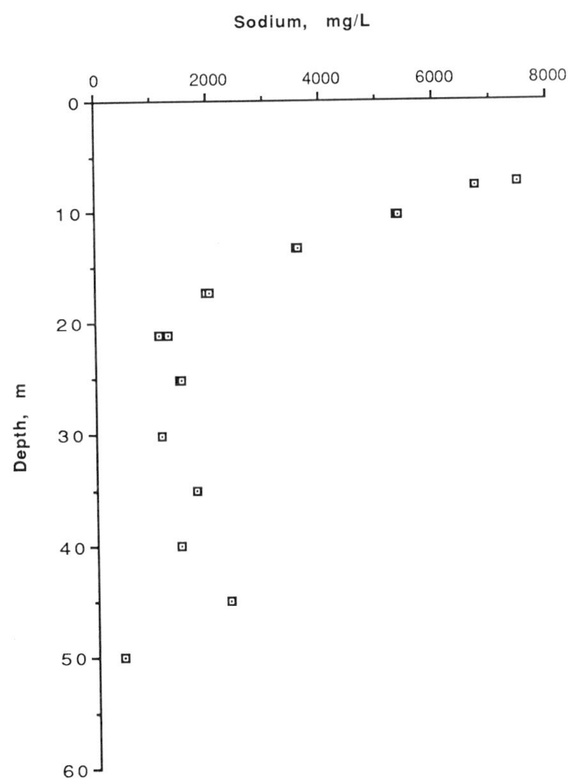

Figure 15.11 The change in sodium concentration in London Clay from saline to fresh water.

% fractionation is the % deviation from the weighted average and is calculated as:

$$\% \text{ fractionation} = (a_v - w_v) \times 100/w_v, \tag{10}$$

where a_v is the actual value for the individual fraction, w_v is the weighted average values of all the fractions from the single test calculated as:

$$w_v = [\Sigma(V_e \times \text{concentration})]/(V_t), \tag{11}$$

where V_e is the volume extracted for the individual fraction and V_t is the total volume extracted for the whole test.

A number of examples of fractionation are presented here from two boreholes in London Clay, one fresh water and the other saline at the surface becoming fresh with depth (Figure 15.11; Bath *et al.*, 1988). Figures 15.12 and 15.13 show the % fractionation of sodium, calcium and silicon from the freshwater and the saline to freshwater borehole, respectively. Figure 15.14 shows the % fractionation of the total cation content from the saline to freshwater borehole. In general, it was found that there is an increase in calcium, magnesium and silicon and a decrease in sodium and potassium concentration with decreasing pore size (progressive pore-water extracts). The variation in cation and anion concentrations complicates the interpretation of the term 'free' water. Changes in concentration of ions with the reduction of pore water content, and pore size, of the test specimen may be due to the interaction of 'free' water (which is probably the first to be extracted) and double-layer water as suggested by Morgenstern and Balasubramonian (1980).

Two sets of tests were carried out on adjacent samples in different cells under the same test conditions to investigate the repeatability of the chemical analysis of this method of pore-water extraction. Entwisle *et al.* (1989) demonstrated that the major element analysis of a number of separate pore-water samples from the same core showed good agreement between equivalent samples for calcium, magnesium, sodium and chloride (variation less than 5%) and potassium and sulphate (less than 10%) and an example is given in Table 15.3.

The geochemical results can be used to investigate contamination of the pore water from the cell. Possible contamination sources include the metal of the cell and the seals or the filters, plastic syringes and syringe taps. There was no indication of contamination of the pore water by corrosion of the metal cell. The values for nickel, an important constituent of the metal used, are, in most cases, below the analytical limit of detection, $0.5\,\text{mg.l}^{-1}$. The values for chromium, another important constituent, from minor element analysis were also below detection level, $0.002\,\text{mg.l}^{-1}$. Iron is a cation commonly found in pore water and therefore it is more difficult to assess possible contamination. However, a number of tests for iron were below detection level, indicating that contamination from the cell was not a problem.

Table 15.3 Parallel test results of a London Clay sample

Ions	Sample A ($mg.l^{-1}$)				Sample B ($mg.l^{-1}$)			
	Stress (MPa)				Stress (MPa)			
	10	35	55	55	13	35	35	55
Ca	468	462	500	488	491	495	532	519
Mg	603	707	930	903	651	745	956	925
Na	2426	2100	1596	1631	2545	2171	1658	1686
K	90	43	13	9	93	38	10	7
HCO_3	280	270	259	261	290	289	273	271
SO_4	1422	1390	1510	1498	1420	1443	1521	1570
Ci	5887	5354	5219	5254	5894	5622	5454	5200
NO_3	B/D	B/D	B/D	B/D	B/D	B/D	B/D	B/D
NO_2	B/D	B/D	B/D	B/D	B/D	B/D	B/D	B/D
NH_4	7.9	3.5	1.0	B/D	8.1	3.2	1.0	B/D
F	0.62	0.53	0.56	0.54	0.52	0.57	0.60	0.58
HPO_4	B/D	B/D	B/D	B/D	B/D	B/D	B/D	B/D
Total P	B/D	B/D	B/D	B/D	B/D	B/D	B/D	B/D
TIC	50	45	32	41	48	44	43	43
TOCua	23	12	8.4	7.0	35	14	7.0	19
Fe	0.17	1.19	0.64	0.26	0.97	1.71	1.04	1.08
Mn	0.12	0.12	0.15	0.15	0.13	0.14	0.17	0.16
Si	4.1	4.4	4.8	4.5	4.5	4.7	5.1	4.5
Al	B/D	B/D	B/D	B/D	B/D	B/D	B/D	B/D
Co	B/D	B/D	B/D	B/D	B/D	B/D	B/D	B/D
Ba	0.050	0.025	0.020	0.020	0.045	0.025	0.015	0.015
Ni	B/D	B/D	B/D	B/D	B/D	B/D	B/D	B/D
Sr	29.2	29.8	34.8	34.5	31.2	31.5	36.2	35.8
Br	B/D	B/D	23.7	22.9	19.1	23.5	24.1	23.8
Total cations ($mEq.1^{-}1$)	180.8	173.6	171.2	169.8	191.1	181.4	177.5	175.5
Total anions ($mEq.1^{-1}$)	199.9	184.1	182.6	183.4	200.2	193.1	189.7	183.6

Total P is the total phosphorous by ICP.
TIC is the total inorganic carbon.
TOCua is the organic carbon, unacidified subsample.
Total cations are the total major cations (Ca, Mg, Na, K).
Total anions are the total major anions (HCO_3, Cl, SO_4, NO_3).
B/D is below detection level.

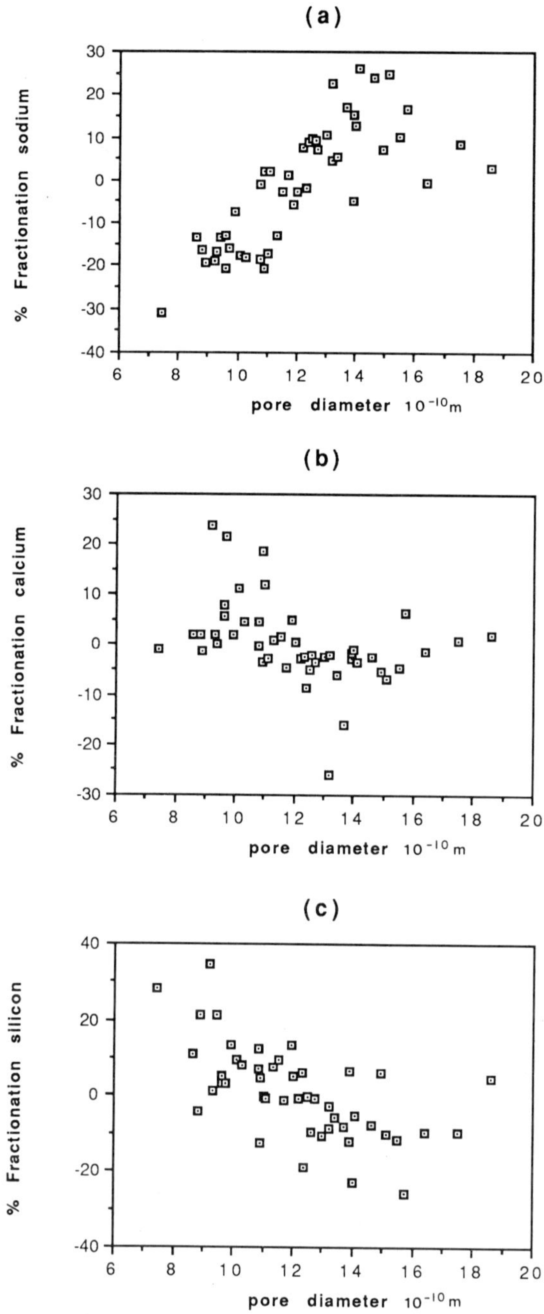

Figure 15.12 % fraction of (a) sodium, (b) calcium and (c) silicon from a freshwater borehole in London Clay.

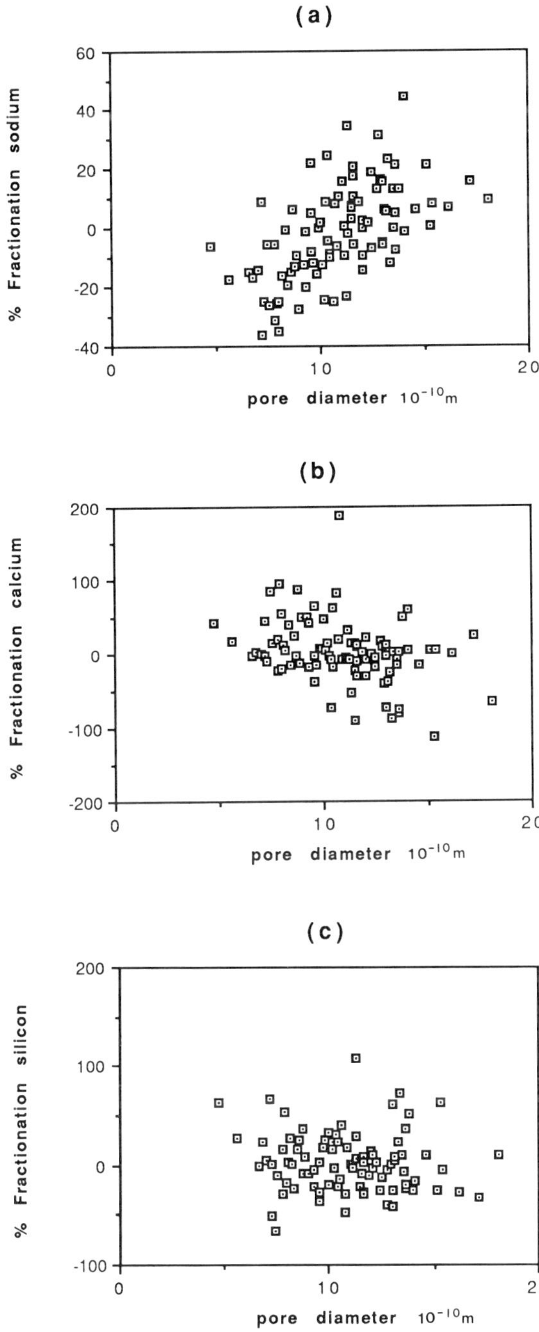

Figure 15.13 % fraction of (a) sodium, (b) calcium and (c) silicon from a saline to freshwater borehole in London Clay.

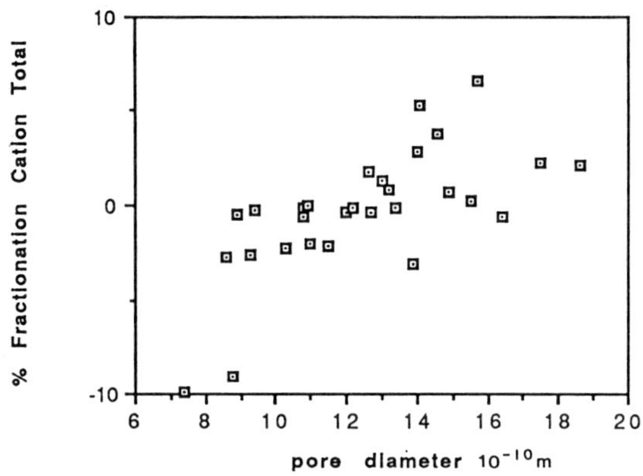

Figure 15.14 % fractionation of the total cation from a saline to freshwater borehole in London Clay.

It is considered that squeezing is a reliable method for extraction of pore waters from a variety of materials. For low-permeability material it is probably the only method that can be used to extract pore waters for a wide range of geochemical analyses. Permeable materials, especially those which are cemented or very stiff, should be centrifuged, which is a much more rapid method. The merits of different techniques of pore-water extraction, e.g. centrifugation, squeezing and residual solute leaching, are discussed in Entwisle *et al.* (1989).

15.7 Conclusions

This chapter presents a new clay squeezing apparatus designed by the British Geological Survey for extracting pore fluids for geochemical analysis. A wide variety of materials has been tested over the last 3 years and only three of over 170 samples have failed to produce any pore water. There is an increase in water collected with an increase in moisture content of the sample. A lower boundary on water collection appears to be between 7 and 8% moisture content. The main restriction on pore water extraction by squeezing of a saturated sample is its stiffness, which can be determined by using the constrained modulus. The maximum constrained modulus for pore water extraction is suggested to be between 100 and 250 $MN.m^{-2}$.

The geochemical analyses have shown that, in general, there is a decrease in total ion concentration with increasing pore-water extraction; sodium and potassium decrease whereas calcium and magnesium increase but to a lesser extent.

The geochemical data suggest that there is little direct contamination of the pore water from the cell.

This is a valid method of extracting pore water for geochemical analysis, especially from clay and mudrocks.

Acknowledgements

The authors would like to thank M. G. Culshaw, C. A. M. Ross, S. T. Horseman and A. H. Bath for their suggestions. This paper is published with the permission of the Director, British Geological Survey (NERC).

References

Bath, A. H., Entwisle, D. C., Ross, C. A. M., Cave, M. R., Falck, W. E., Fry, M. B., Reeder, S., Green, K. A., McEwen, T. J. and Darling, W. G. (1988) Geochemistry of pore-waters in mudrocks sequences—evidence for groundwater and solute movements, in *International Association of Hydrogeologists Symposium on Hydrogeology and Safety of Radioactive and Industrial Hazardous Waste Disposal*, Orléans, France, 7–10 June 1988, Vol. 1, pp. 87–97.

Bath, A. H., Ross, C. A. M., Entwisle, D. C., Cave, M. R., Fry, M. B., Green, K. A. and Reeder, S. (1989a). Hydrochemistry of Porewaters From London Clay, Lower London Tertiaries and Chalk at the Bradwell site, *British Geological Survey Report*, WE/89/26, British Geological Survey, Keyworth, Nottingham.

Bath, A. H., Ross, C. A. M., Entwisle, D. C., Cave, M. R., Fry, M. B., Green, K. A. and Reeder, S. (1989b). Hydrochemistry of Porewaters from Lower Lias Siltstones and Limestones at the Fulbeck Site, Lincolnshire, *Report of the British Geological Survey*, WE/89/27C, British Geological Survey, Keyworth, Nottingham.

Brightman, M. A., Bath, A. H., Cave, M. R. and Darling, W. G. (1985) Pore Fluids From the Argillaceous Rocks of the Harwell Region, *Report of the British Geological Survey*, FLPU 85–6. British Geological Survey, Keyworth, Nottingham.

Carter, D. L., Heilman, M. D. and Gonzalez, F. L. (1965) Ethylene glycol monoethyl ether for determining surface area for silicate minerals. *Soil Science*, **100**, 356–60.

Entwisle, D. C., Reeder, S., Bath, A. H. and Ross, C. A. M. (1989) Techniques for the Characterisation of Solutes in Drillcore From Mudrocks, *British Geological Survey Technical Report*, WE/89/29C, British Geological Survey, Keyworth, Nottingham.

Kinniburgh, D. G. and Miles, D. L. (1983) Extraction and chemical analysis of interstitial water from soils and rocks. *Environmental Science and Technology*, **17**(6), 362–8.

Krahn, J. and Fredlund, D. G. (1972) On total, matric and osmotic suction. *Soil Science*, **114** (5) 339–48.

Krinkov, P. A. (1971) *Gornye, pochverrye i Ilovye Rastvory* (*Interstitial waters of soil, rocks and sediments*). Izdatel, stvo Nauka, Moscow.

Lambe, T. W. and Whitman, R. V. (1979) *Soil Mechanics, SI Version*, John Wiley and Sons, New York.

Manheim, F. T. (1966) A hydraulic squeezer for obtaining interstitial waters from consolidated and unconsolidated sediments. *United States Geological Survey, Professional Papers* 550–556, USGS, Washington, pp. 256–61.

Morgenstern, N. R. and Balasubramonian, B. I. (1980) Effects of pore fluid on the swelling of clay shale, in *Proceedings of the Fourth International Conference on Expansive Soils*, Denver, vol. 1, pp. 190–205.

Rieke, H. H., III and Chilingarian, G. V. (1974) *Developments in Sedimentology*, vol. 16, *Compaction of Argillaceous Sediments*, Elsevier, Amsterdam.

Ross, C. A. M., Bath, A. H., Entwisle, D. C., Cave, M. R., Fry, M. B., Green, K. A. and Reeder, S. (1989a) Hydrochemistry of Porewaters from Jurassic Oxford Clay, Kellaways Beds, Upper

Estuarine and Upper Lias Formations at the Elstow Site, Bedfordshire, *British Geological Survey Report*, WE/89/28, BGS, Keyworth, Nottingham.

Ross, C. A. M., Bath, A. H., Entwisle, D. C., Cave, M, R., Fry, M. B., Green, K. A. and Reeder, S. (1989b) Hydrochemistry of Porewaters from Glacial Till and Chalk at the Killingholme Site, South Humberside, *British Geological Survey* WE/89/25, BGS, Keyworth, Nottingham.

Skempton, A. W. (1944) Notes on the compressibility of clays. *Quarterly Journal of the Geological Society*, **100**, 119–35.

Yong, R. N. and Warkentin, B. P. (1975) Soil properties and behaviour, in *Developments in Geotechnical Engineering*, Vol. 5, Elsevier, Amsterdam.

CHAPTER SIXTEEN

Chemical composition of pore waters in Italian Plio-Pleistocene clayey formations

A. Fontanive, R. Gragnani, C. Mignuzzi and G. Spat

16.1 Introduction

Migration of dissolved species through geological media depends on the geochemical characteristics of the species considered and the geological, mineralogical, hydrogeological and physico-chemical conditions of the geological formation concerned.

In this context, the geochemical characteristics of both clay formations and their pore waters are fundamental with regard to the migration of toxic elements in the geosphere, because they can condition the oxidation state and the speciation of the elements. Chemical composition of pore waters can additionally give information on diagenetic processes and water–rock interactions.

Clay formations do not generally contain aquifers because of their low permeability. However, water may be found in the interstitial space between the various particles or in some cases in interbedded silty-sandy levels, where isolated or open aquifers may exist, depending on the characteristics and dimensions of the interbedded strata.

After deposition of sediments it is usual for many reactions to occur between solid and liquid phases. For example, processes related to bacterial sulphate-reduction activity may begin very soon and the production of H_2S changes Eh conditions, thus triggering other important reactions. Some reactions are very slow but can cause major changes in mineral associations of sediments and the chemical composition of pore waters. Hence pore water composition is the result of geochemical equilibrium due to interaction between water trapped during sedimentation and clay minerals. In some cases, therefore, the study of pore waters can yield information both on chemical fluid composition during clay deposition and on diagenetic processes.

Geochemistry of Clay-Pore Fluid Interactions. Edited by D. A. C. Manning, P. L. Hall and C. R. Hughes. Published in 1993 by Chapman & Hall, London. ISBN 0 412 48980 5

16.2 Experimental procedures

The samples were collected in various Italian geotectonic basins (Figure 16.1) where clay formations vary widely in granulometric, mineralogical and geochemical composition (Brondi *et al.*, 1984). For the geological and palaeogeographical setting the reader is referred to Crescenti (1975), Ruggieri and Sprovieri (1977), Crescenti *et al.* (1980), Casnedi *et al.* (1981) and Colalongo *et al.* (1982).

The determinations of the grain-size characteristics of the samples were preceded by dispersion with sodium hexametaphosphate (0.025%); then the sample was wet-sieved at 0.063 mm and weighed. The fraction smaller than 0.063 mm was analysed by sedigraph.

Mineral identification was performed by X-ray diffraction (XRD) both on whole rock samples and the clay fraction ($\phi < 2\,\mu m$). Because with XRD analysis there is a considerable uncertainty as to the abundance of minerals present in the samples the results presented here can only be considered as semi-quantitative.

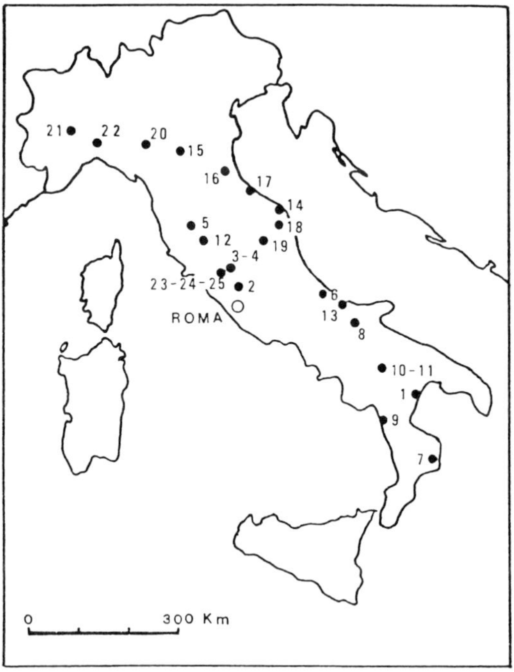

Figure 16.1 Sampling site; 1=Trisaia; 2=Monterotondo; 3=Narni 1; 4=Narni 2; 5=Poggibonsi; 6=Casalbordino; 7=Strongoli; 8=Lucera; 9=Scalea; 10=Tolve 1; 11=Tolve 2; 12=S. Quirico; 13=Vasto Marina; 14=Palombina; 15=Fiorano; 16=Castrocaro; 17=Orciano; 18=Macerata; 19=Ginesina; 20=Collecchio; 21=Chiusano; 22=Novi Ligure; 23=Orte 1; 24=Orte 2; 25=Orte 3.

The choice of method for the extraction of pore water depends on the type of sediments concerned. In recent sediments, generally with a high water content, satisfactory results can be obtained by centrifugation or by squeezing with inert gas (Edmunds and Bath, 1976). These methods are not normally suitable, however, for clayey samples with a low water content. In such cases leaching, with increasing water/sample ratios, has been used to acquire raw information on porewater composition, but more precise results can be obtained by means of a mechanical squeezer (Kryukov, 1960; Kryukov and Komarova, 1954; Manheim, 1966).

In the work reported here, pore waters were extracted by employing an oedometer press. The 50 mm diameter, 45 mm high specimens were first loaded gradually up to 1500 kg in this type of press and then up to 30 000 kg in a hydraulic press. The technical characteristics of the devices and the procedures used are reported in Fontanive *et al.* (1985).

A washing technique was also tested. The samples were dispersed in demineralized water and increasing clay/water ratios were adopted.

After squeezing or clay-washing the waters were analysed for pH, Eh and electric conductance and were then filtered through a 0.1 μm Millipore filter ($\phi = 25$ mm). The methods used for analysis were as follows:

pH—Orion mod. SA250 with needle electrode model MI 410 (Microelectrodes inc).
Electric conductance—Philips model PW 9504 with microcell Philips model PW 9513.
Eh—Orion model SA 250 with combined platinum electrode Orion 97.78.00.
HCO_3—Titration with 0.05 N HCl using a microburette and methyl orange as indicator (1 ml sample).
Cl, F, SiO_2, B and NO_3—Autoanalyser.
SO_4—Turbidimetry.
Ca, Mg, Na, K, and Sr—AAS.
Li—Flame emission spectrometry.
Fe and Mn—GFAAS.

16.3 Results and discussion

16.3.1 Grain-size and mineralogical composition

Moisture content, grain-size composition, geological age and point of origin of the samples are reported in Table 16.1.

As far as the grain-size is concerned, according to Pettijohn's diagram (1975), the samples are classified as follows (Figure 16.2):

silty clay: 3, 12 and 20;
clayey silt: 1, 2, 4, 5, 6, 7, 8, 9, 10, 11, 13, 14, 15, 16, 17, 18 and 22;
sand silt clay: 19;
sandy silt: 21.

Table 16.1 Grain-size composition, age of formation and geographic coordinates of the samples

No.	Moisture content (%)	Sand (%)	Silt (%)	Clay (%)	Age	Coordinates
1	19	1.7	53.1	45.2	2	400948163821E
2	26.2	0.7	58.6	40.7	2	420239123440E
3	34.9	1.3	45.4	53.3	2	423324123201E
4	21.1	3.6	48.2	48.2	2	423324123201E
5	22.6	2.7	53.5	43.8	3	432808134445W
6	19.1	10.6	50.2	39.3	1	420841143522E
7	18.5	7.2	56.6	36.2	4	391531170455E
8	19	1.5	50.2	48.2	1	413023151913E
9	26.5	2.5	51.7	45.8	1	394808154909E
10	11.7	1	50.5	48.5	3	404121155949E
11	14.8	7.4	51.8	40.7	3	404121195949E
12	15.8	1.8	48.1	50.1	5	430319131717W
13	16.1	0.6	53.8	45.7	1	420549144247E
14	20.5	4.5	59.2	36.3	4	433636122515E
15	10.9	0.4	62.7	36.8	4	443132140533W
16	20.6	2.1	58.7	39.2	5	440916125940E
17	18.4	4.5	47.8	47.8	4	434044125715E
18	17.7	16.7	51.1	32.6	3	431800132513E
19	17.5	24.5	45.3	30.2	1	430652132207E
20	24.3	3.6	46.3	50.1	1	444511144107W
21	23.3	38.1	52.6	9.3	4	445703164833W
22	20	12.3	70.2	17.5	1	444457160654W
23	—	—	—	—	2	422650123059W

Age: 1 = Pleistocene; 2 = Plio-pleistocene; 3 = Upper Pliocene; 4 = Middle Pliocene; 5 = Lower Pliocene.

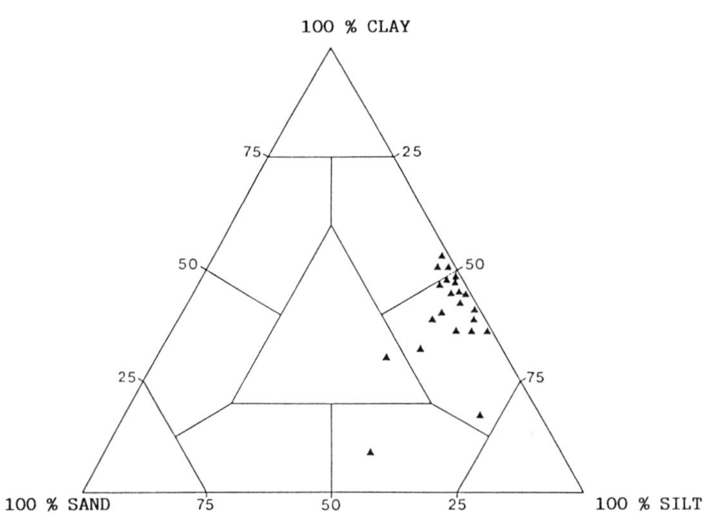

Figure 16.2 Grain-size composition of the samples.

Regarding mineralogical composition (Table 16.2) quartz ranges from 8 to 30%, sample 5 having the highest content and samples 2 and 3 the lowest. The feldspar content is rather low, with only sample 8 reaching 20–30%. The maximum calcite content is about 20–25%. The percentage of clay minerals in the whole rock is fairly uniform. Kaolinite and illite are the most abundant types, with sample 9 exhibiting the highest contents.

The following considerations can be drawn from analyses of the smaller than 2 μm fraction:

1. Smectite is most abundant in samples 3, 4, 5, 6, 9, 10, 11 and 12 and much less so in samples 16 and 17.
2. Chlorite does not vary to any great extent, though it is more abundant in sample 3 from Narni 2 than in sample 4 from Narni 1.
   The distribution of illite and kaolinite is uniform except for sample 15 and to a lesser extent samples 21 and 22 where these minerals are more abundant.

16.3.2 Chemical composition of pore waters

16.3.2.1 SQUEEZED WATERS

Tables 16.3, 16.4, 16.5 and 16.6 report the results of geotechnical data and squeezed water percentage in various types of clay. The quantity of extracted water ranges from 52 to 80% of the total water content of the samples. Variability was found both in different samples from the same formations and in different specimens of the same sample.

McKelvey and Milne (1962) and Chilingarian and Rieke (1968) obtained experimental confirmation that selective ionic migration can occur when clay is subjected to a pressure gradient. To check on this, chemical analyses were carried out on every 3 ml or so of water during squeezing. The results obtained are reported in Table 16.7.

Appreciable differences in all analysed ions were observed during squeezing. Variability exists in different samples from the same formation and in different specimens of the same sample. Chlorine, sodium and SO_4, and in some cases calcium, magnesium and potassium are generally found to have limited variations, whereas SiO_2, boron and manganese exhibit marked differences. In many cases the variability in chemical composition between the various specimens is greater than that observed during the squeezing of the same specimen.

Average chemical composition of pore waters is given in Table 16.8, while the mean value and standard deviation of the relevant parameters are reported in Table 16.9.

Pore-water salinity ranges from 0.45 $g.l^{-1}$ (sample 19) to 24.9 $g.l^{-1}$ (sample 15) and is lower than observed in present seawater. The decrease in the salt

Table 16.2 Mineralogical composition (in %) of whole rock

No.	Quartz	Na–Ca Feldspar	K-Feld-spar	Calcite	Dolomite	Smectite	Chlorite	Interstratified minerals	Illite	Kaolinite
1	12	<5	<5	15	P					
2	8	<5	—	15	P					
3	8	<5	—	9	P					
3	8	P	—	<5	—					
4	15	5	5	8	—					
4	15	5	5	<5	P					
5	25–30	10	<5	15–20	P	<5	<5	5	10–15	10–15
6	15–20	5	5	15–20	<5	<5	5	<5	20–25	10–15
7	20	5	5	10–15	<5	<5	5	5	20–25	25–30
8	20	10–15	10–15	20–25	<5	5	5	5	10–15	10–15
9	15–20	5	5	<5	5	—	10–15	<5	25–30	40–45
10	20	5	5	15–20	<5	5	5	5	15–20	15–20
11	20–25	5–10	<5	15–20	<5	5	5	5	15–20	20–25
12	20–25	<5	—	15–20	—	—	5	5	10–15	20–25
13	20	5	5	15–20	<5	5	5	5	20–25	15–20
15	20–25	5–10	5	15–20	P	<5	5	5	20–25	15–20

P = present.

Table 16.3 Results of geotechnical tests on sample 1

Test no.	Load (kg)	Time load (h)	Compression strain Part. (mm)	Compression strain Tot. (mm)	Extracted H_2O Part. (ml)	Extracted H_2O Tot. (ml)	Extract. H_2O (%)	Moisture content (%)
1	1500	22	1.61	1.61	0.75	0.75		
	3000	51	3.94	5.55	4.25	5.00		
	6000	40	3.15	8.70	3.50	8.50		
	12000	54	2.00	10.70	3.00	11.50		
	15000	72	0.91	11.61	0.50	12.00	64.36	19.10
2	1500	24	2.10	2.10	0.75	0.75		
	3000	25	5.96	8.06	5.25	6.00		
	6000	46	4.13	12.19	3.50	9.50		
	12000	66	1.52	13.71	0.05	9.55		
	15000	49	0.59	14.30	0.25	9.80	52.23	19.70
3	1500	24	3.50	3.50	2.00	2.00		
	3000	24	2.35	5.85	3.70	5.70		
	6000	32	1.15	7.00	2.80	8.50		
	12000	27	2.45	9.45	2.10	10.60		
	15000	28	1.61	11.06	1.30	11.90	62.92	19.14
4	1500	21	1.58	1.58	0.50	0.50		
	3000	24	2.22	3.80	2.50	3.00		
	6000	24	2.81	6.61	4.00	7.00		
	12000	27	2.69	9.30	3.00	10.00		
	15000	51	0.30	9.60	0.30	10.30	58.69	18.30

content of pore waters of clays deposited in the sea has been explained by Rieke *et al.* (1964) and by Chilingarian and Rieke (1968) who observed that during the first stage of sediment compaction ions can migrate outside the sedimentary formation. As compaction proceeds the migration of ions may cease and in some cases there is a new increase in pore-water salinity. Laboratory experiments carried out by Engelhardt and Gaida (1963) showed that the processes may depend on the mineralogical composition of the clay minerals concerned. For instance, no change in pore water composition was observed with kaolinite whereas variation was evident with montmorillonite.

Most of the samples studied here are from formations deposited in the marginal part of marine basins. Hence during geological evolution the pressure load did not vary greatly. This could produce a decrease in salt consistent with the processes referred to above. When salinity is much lower than that of seawater, there may be fresh groundwater recharge. In the case of a continental lacustrine basin (samples 3 and 4) pore-water salinity could be higher than that of the lacustrine water because of the interaction between pore water and minerals.

Table 16.10 shows correlation coefficients for the various parameters studied. The highest significant values are found for Elect. Cond. – Na, Elect. Cond. – Cl and Na–Cl, confirming that NaCl is one of the most important

Table 16.4 Results of geotechnical tests on sample 2

Test no.	Load (kg)	Time load (h)	Compression strain		Extracted H_2O		Extract. H_2O (%)	Moisture content (%)
			Part. (mm)	Tot. (mm)	Part. (ml)	Tot. (ml)		
1	1500	7	2.73	2.73	2.50	2.50		
	3000	55	5.85	8.58	5.00	7.50		
	6000	97	2.87	11.45	6.50	14.00		
	12000	25	1.00	12.45	1.48	15.48		
	15000	20	0.42	12.87	0.02	15.50	70.41	21.26
2	1500	40	3.87	3.87	3.60	3.60		
	3000	46	4.03	7.90	4.90	8.50		
	6000	94	2.75	10.65	3.50	12.00		
	12000	72	2.07	12.72	2.30	14.30		
	15000	72	1.09	13.81	0.40	14.70	73.15	25.12
3	1500	29	3.83	3.83	2.00	2.00		
	3000	22	5.00	8.83	5.80	7.80		
	6000	44	1.95	10.78	2.50	10.30		
	12000	51	1.54	12.32	1.70	12.00		
	15000	112	1.16	13.48	1.20	13.20	69.56	24.70
4	1500	29	3.64	3.64	3.00	3.00		
	3000	7	2.38	6.02	3.00	6.00		
	6000	42	2.37	8.39	4.00	10.00		
	12000	72	2.06	10.45	2.30	12.30		
	15000	112	0.86	11.31	1.40	13.70	73.11	24.63

components of pore waters. The correlation between sodium and chlorine is reported in Figure 16.3. This shows that in some samples the Na/Cl ratio is similar to that of the seawater, while in others the Na/Cl ratio is higher than that of seawater, possibly due to an exchange with clay minerals. The same holds good for the Mg–Cl correlation but in this case there is greater scatter in samples with an Mg/Cl ratio higher than that of seawater (Figure 16.4).

The Ca–SO_4 correlation (Figure 16.5) shows that a lot of samples are oversaturated with respect to $CaSO_4$ and the linear correlation coefficient ($r=0.25$) cannot be significant.

The pore waters have been classified according to their chemical composition as shown in Figure 16.6. They are characterized by a sodium-chloride and calcium-magnesium-sulphate chemistry, only samples 19 and 20 having a high percentage of HCO_3. The Figure 16.6 diagrams indicate continued transition between waters with a high percentage of alkaline elements and those with a high percentage of alkaline earth elements. Samples 19, 20, 21 and 25 have the highest percentage of calcium. It is evident from the anionic composition that HCO_3 percentages are generally low, because of the poor $CaCO_3$ and $MgCO_3$ solubility due to the low CO_2 partial pressure. In addition chlorine and SO_4 percentages exhibit marked scatter. The HCO_3 percentage is relatively high in

Table 16.5 Results of geotechnical tests on sample 3

Test no.	Load (kg)	Time load (h)	Compression strain Part. (mm)	Compression strain Tot. (mm)	Extracted H_2O Part. (ml)	Extracted H_2O Tot. (ml)	Extracted H_2O (%)	Moisture content (%)
1	1500	108	10.05	10.05	10.90	10.90		
	3000	33	1.79	11.84	1.60	12.50		
	6000	12	1.52	13.36	1.46	13.96		
	12000	24	0.96	14.28	2.00	15.96		
2	15000	33	0.72	15.00	1.67	17.63	76.98	30.32
	1500	69	21.20	21.20	17.19	17.19		
	3000	33	6.02	27.17	2.00	19.19		
	6000	96	2.81	29.98	1.50	20.69		
	12000	75	4.89	34.87	0.22	20.91		
3	15000	48	1.87	36.74	0.00	20.91	63.10	30.10
	1500	120	8.10	8.10	4.70	4.70		
	3000	48	3.48	11.58	4.35	9.05		
	6000	48	4.67	16.25	4.51	13.56		
	12000	48	4.43	20.68	1.44	15.00		
4	15000	36	0.82	21.50	2.28	17.28	67.96	22.80
	1500	96	19.98	19.98	11.58	11.58		
	3000	56	1.52	21.40	5.37	16.95		
	6000	62	5.55	26.95	5.21	22.16		
	12000	144	3.30	30.25	1.00	23.16		
	15000	72	3.43	33.68	0.00	23.16	80.05	29.60

Table 16.6 Results of geotechnical tests on sample 5

Test no.	Load (kg)	Time load (h)	Compression strain Part. (mm)	Compression strain Tot. (mm)	Extracted H_2O Part. (ml)	Extracted H_2O Tot. (ml)	Extracted H_2O (%)	Moisture content (%)
1	1500	53	5.90	5.90	2.00	2.00	53.85	22.63
	3000	62	1.75	7.65	7.20	9.20		
	6000	65	2.25	9.90	3.78	12.98		
	12000	32	1.30	11.20	2.52	15.50		
	15000	78	0.45	11.65	0.51	16.01	67.82	22.59
2	1500	87	6.36	6.36	5.00	5.00		
	3000	98	1.37	7.73	2.20	7.50		
	6000	42	1.27	9.00	3.62	11.12		
	12000	125	1.88	10.88	3.00	14.12		
	15000	168	0.86	11.74	0.47	14.59	64.90	21.40
3	1500	152	5.33	5.33	6.00	6.00		
	3000	24	1.72	7.05	3.20	9.20		
	6000	72	1.77	8.82	3.30	12.50		
	12000	96	1.80	10.62	2.50	15.00		
	15000	230	1.09	11.71	2.00	17.00		
	30000	560	4.27	15.98	2.00	19.00		

Table 16.7 Variation of pore-water chemical composition during squeezing of different specimens of the same samples

No.	pH	Eh (mV)	Cond. (mS)	Ca	Mg	Na	K	HCO_3	SO_3	Cl	SiO_2	F	B	NO_3	Sr	Li	Fe	Mn
				(mEq.l^{-1})							(mg.l^{-1})				(mg.l^{-1})			
1.1.1	7.9	220	14	1.9	7.4	130	3.7	14	30	110	20	n.d.	5	n.d.	1	83	<1	320
1.1.2	8	13	1.5	6.8	120	3.5	14	31	99	9.6	n.d.	5.2	n.d.	0.92	71	<1	130	—
1.1.3	8	13	0.9	5.1	140	3.5	17	30	120	20	n.d.	5.7	n.d.	0.79	59	<1	68	—
1.1.4	8.6	n.d.	n.d.	0.9	5.1	170	3.7	15	32	120	n.d.	n.d.	6.5	n.d.	n.d.	59	<1	59
1.2.1	8.3	340	14	1.5	11	150	4.8	19	28	110	5.4	n.d.	6.2	n.d.	0.81	70	<1	100
1.2.2	8.5	n.d.	14	1.3	10	150	4.1	19	31	120	7.2	n.d.	6.9	n.d.	0.81	86	<1	60
1.2.3	8.7	n.d.	14	1.1	10	150	3.8	21	31	120	11	n.d.	8.7	n.d.	0.61	87	16	54
1.2.4	8.1	n.d.	14	1.3	5.5	160	3.8	18	33	140	11	n.d.	n.d.	n.d.	0.81	98	n.d.	n.d.
1.3.1	7.7	270	15	1.2	8	160	6.2	20	51	110	3.6	n.d.	4.1	n.d.	0.81	72	39	140
1.3.2	7.6	n.d.	15	0.99	7.7	160	6	20	49	120	3.6	n.d.	4.1	n.d.	0.81	70	15	130
1.3.3	8	14	0.73	7.3	160	5.5	20	49	110	4.3	n.d.	4.5	n.d.	0.81	75	16	88	
1.3.4	8.1	n.d.	14	0.78	7.8	170	5.5	21	44	130	4.3	n.d.	3.7	n.d.	0.61	88	n.d.	200
1.4.1	8.1	n.d.	15	1.5	6.7	170	7.3	19	61	110	5.4	n.d.	5.2	n.d.	1	100	52	150
1.4.2	8.1	n.d.	15	1.3	6.7	170	6.8	20	62	110	7.2	n.d.	7.1	n.d.	0.81	100	90	160
1.4.3	8.2	n.d.	16	1.6	6.5	160	6.2	19	61	110	4.5	n.d.	10	n.d.	0.81	100	33	92
1.4.4	8.4	n.d.	16	1.8	8	170	6.5	20	62	120	7.2	n.d.	7.6	n.d.	0.81	110	n.d.	280
2.1.1	6.9	n.d.	6.7	22	26	25	6.6	4.7	71	8.3	12	n.d.	3.7	240	7.2	500	<1	600
2.1.2	7.1	n.d.	6.4	20	29	24	5.7	4.7	70	7.8	9	n.d.	2.8	250	7	610	<1	690
2.1.3	7.2	270	6.7	15	35	26	5.7	5.7	72	7.3	18	n.d.	2.5	270	6.6	660	<1	720
2.1.4	7.2	n.d.	6.5	19	32	25	4.8	5.7	69	7	9	n.d.	2.5	260	6.3	660	<1	360
2.1.5	n.d.	6.3	18	36	24	3.6	5.5	69	7.7	21	n.d.	3.7	250	6.3	680	<1	600	
2.2.1	6.8	n.d.	6.4	16	34	27	3.7	3.7	55	14	7.8	n.d.	4.4	270	5.5	390	<1	450
2.2.2	7.1	320	6.8	15	34	27	3.6	3.4	57	16	11	n.d.	3.3	270	5.5	440	<1	480
2.2.3	7	6.5	16	34	27	3.9	4	57	14	11	n.d.	5.3	230	5.5	390	<1	320	
2.2.4	7.2	n.d.	6.4	14	33	27	3.4	3.3	55	14	10	n.d.	3.8	230	5.2	390	<1	320
2.2.5	7.4	n.d.	6.5	16	37	27	2.7	3.6	60	14	11	n.d.	5.7	220	5.5	320	<1	320
3.1.1	7.8	n.d.	4.6	28	28	10	0.61	4.9	58	5.6	23	n.d.	1.6	n.d.	5	87	<1	700
3.1.2	7.9	340	4.3	29	28	7.9	0.48	4.9	55	4.7	23	n.d.	1.1	n.d.	5.5	94	<1	1400
3.1.3	7.7	n.d.	4.4	27	28	6.9	0.3	3.2	58	5.9	n.d.	n.d.	1.1	n.d.	5.4	100	<1	950
6.2.2	8.7	n.d.	28	3	35	290	7.6	11	130	220	4.8	2.5	n.d.	n.d.	n.d.	n.d.	n.d.	n.d.

Table 16.7 (*contd*)

No.	pH	Eh (mV)	Cond. (mS)	Ca	Mg	Na	K	HCO_3	SO_3	Cl	SiO_2	F	B	NO_3	Sr	Li	Fe	Mn
				($mEq.l^{-1}$)							($mg.l^{-1}$)				($mg.l^{-1}$)			
6.3.1	8.6	n.d.	28	5	37	280	7.6	8.5	100	220	5.4	0.55	7	n.d.	n.d.	n.d.	<1	100
6.3.2	8.5	n.d.	27	4	32	280	6	9	100	220	3.8	0.55	7.1	n.d.	n.d.	n.d.	<1	86
6.3.3	8.5	n.d.	n.d.	3	38	320	6	10	100	220	11	n.d.	n.d.	n.d.	n.d.	n.d.	160	n.d.
6.4.1	8.1	n.d.	27	6	35	270	6.8	6	100	210	5.4	0.25	6.3	n.d.	n.d.	n.d.	n.d.	n.d.
6.4.2	8.4	n.d.	27	4	36	300	7.6	6.7	98	210	1.8	0.95	5.4	n.d.	n.d.	n.d.	n.d.	n.d.
7.1.1	8	270	34	26	180	450	9	5	350	330	11	0.34	2.4	n.d.	n.d.	n.d.	n.d.	n.d.
8.1.1	8	320	8.8	38	35	52	4	2.5	73	56	20	0.8	0.87	n.d.	n.d.	n.d.	n.d.	n.d.
9.1.1	7.9	340	7.6	37	37	37	0.72	3	84	17	11	1.4	4.2	n.d.	n.d.	n.d.	n.d.	n.d.
10.1.1	7.5	350	29	74	120	250	3.6	2	140	250	7	n.d.	6.4	n.d.	n.d.	n.d.	n.d.	n.d.
11.1.1	7.9	350	25	31	76	190	5.5	4	190	130	14	0.78	6.9	n.d.	n.d.	n.d.	n.d.	n.d.
12.1.1	8.4	340	24	22	32	220	3.9	5	190	84	4	0.46	13	n.d.	n.d.	n.d.	n.d.	n.d.
13.1.1	8.1	320	30	17	100	260	11	5	230	170	82	0.46	1.4	n.d.	n.d.	n.d.	n.d.	n.d.
14.1.1	7.5	370	2.8	12	9.5	14	0.74	2.5	10	28	14	0.65	2.5	n.d.	n.d.	n.d.	n.d.	n.d.
14.1.2	7.7	390	2.7	12	7.7	10	0.54	2.5	7	21	18	0.65	2.5	n.d.	n.d.	n.d.	n.d.	n.d.
14.1.3	7.3	n.d.	2.7	10	7.7	11	0.51	2.7	6	19	n.d.	n.d.	n.d.	n.d.	n.d.	n.d.	n.d.	n.d.
15.1.1	8.8	290	40	4.5	68	320	6.6	10	47	380	12	0.3	12	n.d.	n.d.	n.d.	n.d.	n.d.
16.1.1	8.3	310	19	13	20	190	5.4	5.5	100	140	16	n.d.	8	n.d.	n.d.	n.d.	n.d.	n.d.
16.1.2	7.8	n.d.	19	20	21	210	2.3	5.3	100	120	13	0.76	11	n.d.	n.d.	n.d.	n.d.	n.d.
17.1.1	8.1	380	30	23	52	320	5.7	4.5	200	190	10	0.59	13	n.d.	n.d.	n.d.	n.d.	n.d.
17.1.2	7.5	390	30	22	55	320	5.9	5.5	170	220	9	n.d.	17	n.d.	n.d.	n.d.	n.d.	n.d.
18.1.1	8	390	20	72	100	85	2.3	3.5	96	150	11	0.44	6.5	n.d.	n.d.	n.d.	n.d.	n.d.
18.1.2	8	390	24	73	91	77	2.1	3.5	100	160	15	n.d.	6.1	n.d.	n.d.	n.d.	n.d.	n.d.
19.1.1	7.8	475	0.48	3.9	1.6	1.4	0.75	2.7	2.3	1.2	14	n.d.	n.d.	n.d.	n.d.	n.d.	n.d.	n.d.
20.1.1	7.4	390	0.60	4	1.4	0.7	0.28	3.3	2	1.2	n.d.	n.d.	n.d.	n.d.	n.d.	n.d.	n.d.	n.d.
21.1.1	7.3	440	4.5	40	26	1.7	0.7	3.6	60	4.5	n.d.	n.d.	n.d.	n.d.	n.d.	n.d.	n.d.	n.d.
22.1.1	8	380	6.5	32	59	14	5.8	2.6	93	11	n.d.	n.d.	n.d.	n.d.	n.d.	n.d.	n.d.	n.d.
23.1.1	7.6	445	4.4	23	22	3.1	0.75	3	46	1.4	n.d.	n.d.	n.d.	n.d.	n.d.	n.d.	n.d.	n.d.
24.1.1	7.9	440	2.1	8.4	8.4	3.9	0.75	4	15	1.4	n.d.	n.d.	n.d.	n.d.	n.d.	n.d.	n.d.	n.d.
25.1.1	7.6	355	2.5	20	12	2.5	0.55	2.7	31	1.5	n.d.	n.d.	n.d.	n.d.	n.d.	n.d.	n.d.	n.d.
3.1.4	7.9	n.d.	4.2	23	25	7.2	0.46	3	55	4.5	n.d.	n.d.	0.69	n.d.	5.4	120	<1	550
3.2.1	7.2	n.d.	4.4	29	26	8.7	0.49	3.9	55	5.6	n.d.	1.3	1.2	n.d.	5.2	92	<1	800
3.2.2	7.5	n.d.	4.1	28	26	9	0.49	3.5	55	5.9	12	n.d.	1	n.d.	5	82	<1	550
3.2.3	7.4	n.d.	4.1	27	26	9.3	0.36	3.4	53	5.5	20	n.d.	1.3	n.d.	5.1	97	<1	550
3.2.4	7.4	n.d.	4.1	27	27	9.4	0.57	2.7	56	5.6	20	n.d.	1.3	n.d.	5.4	110	<1	700
3.3.1	6.8	n.d.	3.7	28	18	4.7	0.42	2.6	48	3.5	13	n.d.	0.52	n.d.	4.2	59	<1	1000

Table 16.7 *(contd)*

No.	pH	Eh (mV)	Cond. (mS)	Ca	Mg	Na	K	HCO_3	SO_3	Cl	SiO_2	F	B	NO_3	Sr	Li	Fe	Mn
				($mEq.l^{-1}$)							($mg.l^{-1}$)				($mg.l^{-1}$)			
3.3.2	6.9	470	3.4	30	20	4.9	0.34	2.6	48	3.7	n.d.	n.d.	0.52	n.d.	4.3	68	<1	450
3.3.3	6.7	n.d.	3.5	29	20	4.9	0.27	2.4	50	3.7	12	n.d.	0.69	n.d.	4.4	73	<1	1700
3.3.4	6.9	n.d.	3.7	29	21	4.6	0.25	2.1	47	3.5	14	0.29	0.69	n.d.	4.1	82	<1	700
3.4.1	7.6	n.d.	3.6	26	22	5.3	0.27	3.5	49	3.7	15	0.48	0.54	n.d.	4.4	45	<1	170
3.4.2	7.7	n.d.	3.6	26	22	5.1	0.21	2	46	4.8	21	n.d.	0.39	n.d.	4.2	50	<1	84
3.5.1	7.6	n.d.	3.7	26	23	5.3	0.22	3.4	47	3.7	22	n.d.	1	n.d.	4.6	55	<1	200
3.5.2	7.4	375	3.6	26	23	5.7	0.34	3.1	45	3.9	13	n.d.	0.65	n.d.	4.6	45	<1	140
3.5.3	7.5	n.d.	3.5	24	22	5	0.17	2.6	51	3.1	15	0.42	0.56	n.d.	4.2	40	<1	130
3.5.4	7.5	n.d.	4	26	23	5.5	0.29	1.8	45	3.7	18	n.d.	0.74	n.d.	4.2	50	<1	94
4.1.1	7.1	n.d.	5.2	35	36	8	0.37	3.2	71	4.8	8.4	0.57	0.91	n.d.	7	130	<1	1600
4.1.2	7.3	n.d.	5	35	35	8.3	0.37	3.1	66	4.2	7.2	n.d.	0.91	n.d.	6.8	140	<1	1700
4.1.3	8	385	5	33	36	7.9	0.32	3	73	4.5	7.2	n.d.	0.91	n.d.	6.6	140	<1	1300
4.1.4	7.4	n.d.	4.9	34	36	10	0.5	2.6	75	5.3	9.6	n.d.	1	n.d.	6.7	150	<1	900
5.1.1	7.8	n.d.	n.d.	33	35	15	1.5	3.9	70	12	9.6	0.15	2	n.d.	n.d.	n.d.	n.d.	n.d.
5.1.2	6.8	n.d.	n.d.	35	37	17	1.8	3.5	74	9.6	9.6	0.13	2.4	n.d.	n.d.	n.d.	n.d.	n.d.
5.1.3	7.5	n.d.	n.d.	34	37	17	1.7	3.7	72	3.7	6	0.13	1.6	n.d.	n.d.	n.d.	n.d.	n.d.
5.1.4	7.2	n.d.	n.d.	35	37	17	1.7	2.2	78	1.9	13	0.08	1.7	n.d.	n.d.	n.d.	n.d.	n.d.
5.2.1	5.7	n.d.	n.d.	32	39	19	3.7	0.5	84	2.8	4.8	0.3	1.7	n.d.	n.d.	n.d.	n.d.	n.d.
5.2.2	7.4	n.d.	n.d.	29	40	18	3.2	2	83	2.1	6	0.11	1.7	n.d.	n.d.	n.d.	n.d.	n.d.
5.2.3	7.4	n.d.	n.d.	31	44	18	2.8	1.6	75	1.5	6	0.11	1.7	n.d.	n.d.	n.d.	n.d.	n.d.
5.3.1	7.9	n.d.	6.2	32	32	18	1.1	2.5	78	4.2	14	0.29	1.3	n.d.	n.d.	n.d.	<1	82
5.3.2	7.7	n.d.	4.7	25	25	13	1	2.2	62	1.7	8.4	n.d.	0.35	n.d.	n.d.	n.d.	<1	72
5.3.3	7.8	390	5.7	27	32	17	1	4	76	1.8	11	0.63	0.43	n.d.	n.d.	n.d.	<1	80
6.1.1	8.3	n.d.	29	9	43	310	12	8.5	150	240	3	0.84	13	n.d.	n.d.	n.d.	n.d.	n.d.
6.1.2	8.6	270	26	6	42	310	11	9.2	140	240	3.6	0.61	12	n.d.	n.d.	n.d.	n.d.	n.d.
6.1.3	8.7	n.d.	30	4	41	300	8	11	150	230	5.4	0.68	n.d.	n.d.	n.d.	n.d.	n.d.	n.d.
6.2.1	8.3	n.d.	28	3	35	290	11	6.7	130	220	4.8	1.4	7.8	n.d.	n.d.	n.d.	n.d.	n.d.

Under the heading No. the first figure indicates the sample site, the second the specimen number and the third the squeezed water fraction: n.d. means not detected.

Table 16.8 Average chemical composition of pore waters at the various sampling sites

No.	pH	Eh (mV)	Cond. (mS)	Ca	Mg	Na	K	HCO_3	SO_4	Cl	SiO_2	F	B	NO_3	Sr	Li	Fe	Mn
				($mEq.l^{-1}$)							($mg.l^{-1}$)				($\mu g.l^{-1}$)			
11	8.12	220	13.3	1.3	6.1	140	3.6	15	31	112.2	16.53	n.d.	5.6	n.d.	0.9	68	<1	144.2
12	8.4	340	14	1.3	9.1	152.5	4.12	19.2	31	122.5	8.65	n.d.	7.27	n.d.	0.76	85.2	6	71.3
13	7.85	270	14.5	0.93	7.7	162.5	5.8	20.2	48	117.5	3.95	n.d.	4.1	n.d.	0.76	76.2	23.3	139.5
14	8.2	n.d.	15.5	1.55	7	167.5	6.7	19.5	62	112.5	6.07	n.d.	n.d.	n.d.	0.86	102.5	58.3	170.5
21	7.1	270	6.5	18.8	31.6	24.8	5.28	5.3	70	7.6	13.8	n.d.	n.d.	254	6.68	622	<1	594
22	7.1	320	6.5	15.4	34.4	27	3.46	3.6	57	14.4	10.16	n.d.	n.d.	244	5.44	386	<1	378
31	7.83	340	4.4	26.75	27.2	8	0.46	4	56	5.2	23	n.d.	n.d.	n.d.	5.33	100.2	<1	900
32	7.38	n.d.	4.2	27.75	26.2	9.1	0.48	3.4	55	5.7	17.33	1.3	n.d.	n.d.	5.17	95.2	<1	650
33	6.83	470	3.6	29	19.8	4.8	0.32	2.4	48	3.6	13	0.29	n.d.	n.d.	4.25	70.5	<1	962.5
34	7.65	n.d.	3.6	26	22	5.2	0.24	2.7	48	4.2	18	0.48	n.d.	n.d.	4.3	47.5	<1	127
35	7.5	375	3.7	25.5	22.8	5.4	0.25	2.7	47	3.6	17	0.42	n.d.	n.d.	4.4	47.5	<1	141
41	7.45	385	5	34.25	35.8	8.6	0.39	3	71	4.7	8.1	0.57	n.d.	n.d.	6.78	140	<1	1375
51	7.33	n.d.	n.d.	34.25	36.5	16.5	1.67	3.3	74	6.8	9.55	0.12	n.d.	n.d.	n.d.	n.d.	n.d.	n.d.
52	6.83	n.d.	n.d.	30.67	41	18.3	3.23	1.4	81	2.1	5.6	0.17	n.d.	n.d.	n.d.	n.d.	n.d.	n.d.
53	7.8	390	5.5	28	29.7	16	1.03	2.9	72	2.6	11.13	0.46	n.d.	n.d.	n.d.	n.d.	<1	78
61	8.53	270	28.3	6.33	42	306.7	10.33	9.6	147	236.7	4	0.71	n.d.	n.d.	n.d.	n.d.	n.d.	n.d.
62	8.5	n.d.	28	3	35	290	9.3	8.8	130	220	4.8	1.95	n.d.	n.d.	n.d.	n.d.	n.d.	n.d.
63	8.53	n.d.	27.5	4	35.7	293.3	6.53	9.2	100	220	6.73	0.55	n.d.	n.d.	n.d.	n.d.	54	93
64	8.25	n.d.	27	5	35.5	285	7.2	6.3	99	210	3.6	0.6	n.d.	n.d.	n.d.	n.d.	n.d.	n.d.
71	8	270	34	26	180	450	9	5	350	330	11	0.34	n.d.	n.d.	n.d.	n.d.	n,d.	n.d.
81	8	320	8.8	38	35	52	4	2.5	73	56	20	0.8	n.d.	n.d.	n.d.	n.d.	n.d.	n.d.
91	7.9	340	7.6	37	37	37	0.72	3	84	17	11	1.4	n.d.	n.d.	n.d.	n.d.	n.d.	n.d.
101	7.5	350	29	74	120	250	3.6	2	140	250	7	n.d.	n.d.	n.d.	n.d.	n.d.	n.d.	n.d.
111	7.9	350	25	31	76	190	5.5	4	190	130	14	0.78	n.d.	n.d.	n.d.	n.d.	n.d.	n.d.
121	8.4	340	24	22	32	220	3.9	5	190	84	4	0.46	n.d.	n.d.	n.d.	n.d.	n.d.	n.d.
131	8.1	320	30	17	100	260	11	5	230	170	82	0.46	n.d.	n.d.	n.d.	n.d.	n.d.	n.d.
141	7.5	380	2.7	11.33	8.3	11.7	0.6	2.6	8	22.7	16	0.65	n.d.	n.d.	n.d.	n.d.	n.d.	n.d.
151	8.8	290	40	4.5	68	320	6.6	10	47	380	12	0.3	n.d.	n.d.	n.d.	n.d.	n.d.	n.d.
161	8.05	310	19	16.5	20.5	200	3.85	5.4	100	130	14.5	0.76	n.d.	n.d.	n.d.	n.d.	n.d.	n.d.
171	7.8	385	30	22.5	53.5	320	5.8	5	185	205	9.5	0.59	n.d.	n.d.	n.d.	n.d.	n.d.	n.d.
181	8	390	22	72.5	95.5	81	2.2	3.5	98	155	13	0.44	n.d.	n.d.	n.d.	n.d.	n.d.	n.d.
191	7.8	475	0.48	3.9	1.6	1.4	0.075	2.7	2	1.2	14	n.d.	n.d.	n.d.	n.d.	n.d.	n.d.	n.d.
201	7.4	390	0.6	4	1.4	0.7	0.28	3.3	2	1.2	n.d.	n.d.	n.d.	n.d.	n.d.	n.d.	n.d.	n.d.
211	7.3	440	4.5	40	26	1.7	0.7	3.6	60	4.5	n.d.	n.d.	n.d.	n.d.	n.d.	n.d.	n.d.	n.d.
221	8	380	6.5	32	59	14	5.8	2.6	93	11	n.d.	n.d.	n.d.	n.d.	n.d.	n.d.	n.d.	n.d.
231	7.6	445	4.4	23	22	3.1	0.75	3	46	1.4	n.d.	n.d.	n.d.	n.d.	n.d.	n.d.	n.d.	n.d.
241	7.9	440	2.1	8.4	8.4	3.9	0.75	4	15	1.4	n.d.	n.d.	n.d.	n.d.	n.d.	n.d.	n.d.	n.d.
251	7.6	355	2.5	20	12	2.5	0.55	2.7	31	1.5	n.d.	n.d.	n.d.	n.d.	n.d.	n.d.	n.d.	n.d.

Under the heading No. the last figure indicates the specimen number: n.d. means not detected.

Table 16.9 Mean value, standard deviation and coefficient of variation percentage of the determined parameters

Variables	Number of determinations	Mean value	Standard deviation	Coeff. of variation (%)
pH	25	7.8	0.41	5
Eh	25	360	56	16
Elec. Cond.	25	13	12	93
Ca	25	25	19	76
Mg	25	45	42	94
Na	25	120	140	116
K	25	3.4	3.1	90
HCO_3	25	4.6	3.4	73
SO_4	25	95	82	86
Cl	25	92	110	122
SiO_2	19	15	17	111
F	15	0.62	0.29	47
B	19	5.8	3.9	67
Sr	17	10	100	95
Li	4	200	200	101
Fe	6	13	22	164
Mn	6	470	510	109

samples 19 and 20 because of the low total dissolved salts (TDS) of the pore water.

When the pore waters are grouped in four conductivity classes (Figure 16.6) it is apparent that the chemical composition of the samples with smaller than 12 mS values differs from that of the greater than 12 mS class. In fact the NaCl content prevails in water with higher salinity whereas calcium and magnesium sulphate are dominant in those with lower salinity. This is attributable either to the geochemical conditions that occurred in the sedimentation basin and/or to the leaching of the more soluble components during the geological evolution of the formations. For instance, the pore waters of Pleistocene fluvio-lacustrine clays (samples 19 and 20) exhibit the lowest salinity of the samples analysed. These two samples have the highest HCO_3 percentage. Interstitial waters of lacustrine clay (samples 3 and 4) are distinguished by calcium-magnesium-sulphate chemistry. In a shallow-basin clay (sample 5) that was alternately sedimented in marine and brackish waters, the pore waters also have calcium-magnesium-sulphate chemistry, while the presence of NaCl becomes significant. Pore water of marine clays generally has a higher NaCl content.

Nevertheless in some cases the chemical composition of the pore water differs from that which might be expected on the basis of the palaeogeographic and/or geological evolution of the basins concerned (e.g. samples 23, 24 and 25). This may sometimes be due to the leaching of more soluble components.

The mineral equilibria have been studied in some samples by the 'Solminec' program. Several saturation indexes are reported in Table 16.11. Index values

Table 16.10 Correlation coefficients for the studied parameters

	pH	Eh	Cond.	Ca	Mg	Na	K	HCO_3	SO_4	Cl	SiO_2	F	B	Sr	Li	Fe
Mn	−.44	0.46	−.50	0.61	0.25	−.53	−.63	−.45	−.20	−.51	0.20	0.02	−.59	0.78	−.08	−.48
	6	6	6	6	6	6	6	6	6	6	6	4	6	4	4	6
Fe	0.94	−.70	1.00	−.76	0.00	0.99	0.88	0.52	0.71	0.99	−.65	0.78	0.91	−.95	−.38	
	6	6	6	6	6	6	6	6	6	6	6	4	6	4	4	
Li	−.67	−.47	−.20	−.05	0.50	−.27	0.43	−.30	0.48	−.33	0.02	****	0.19	0.46		
	4	4	4	4	4	4	4	4	4	4	4	2	4	4		
Sr	0.12	−0.2	0.47	0.59	0.44	0.39	0.18	−.26	0.52	0.28	−.11	−.32	0.45			
	17	17	17	17	17	17	17	17	17	17	17	13	17			
B	0.66	−.45	0.84	−.11	0.29	0.80	0.57	0.36	0.50	0.70	−.17	−.11				
	19	19	19	19	19	19	19	19	19	19	19	15				
F	0.02	−.11	−.22	0.02	−.37	−.26	−.23	−.14	−.21	−.33	−.12					
	15	15	15	15	15	15	15	15	15	15	15					
SiO_2	0.07	−.05	−.34	−.10	0.21	0.09	0.45	−.08	0.25	0.04						
	19	19	19	19	19	19	19	19	19	19						
Cl	0.63	−.61	0.90	0.07	0.72	0.93	0.74	0.43	0.61							
	25	25	25	25	25	25	25	25	25							
SO_4	0.31	−.47	0.59	0.25	0.83	0.79	0.73	0.03								
	25	25	25	25	25	25	25	25								
HCO_3	0.52	−.55	0.37	−.46	−.06	0.42	0.43									
	25	25	25	25	25	25	25									
K	0.59	−.71	0.59	−.12	0.63	0.82										
	25	25	25	25	25	25										
Na	0.61	−.65	0.87	−.04	0.69											
	25	25	25	25	25											
Mg	0.24	−.39	0.62	0.50												
	25	25	25	25												
Ca	−.24	0.14	0.17													
	25	25	25													
Cond.	0.61	−.56														
	25	25														
Eh	−.47															
	25															

****=No value.

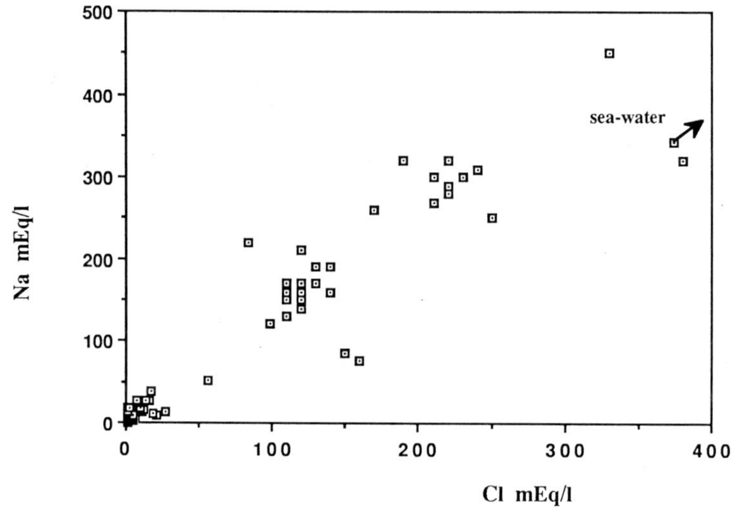

Figure 16.3 Cl–Na correlation.

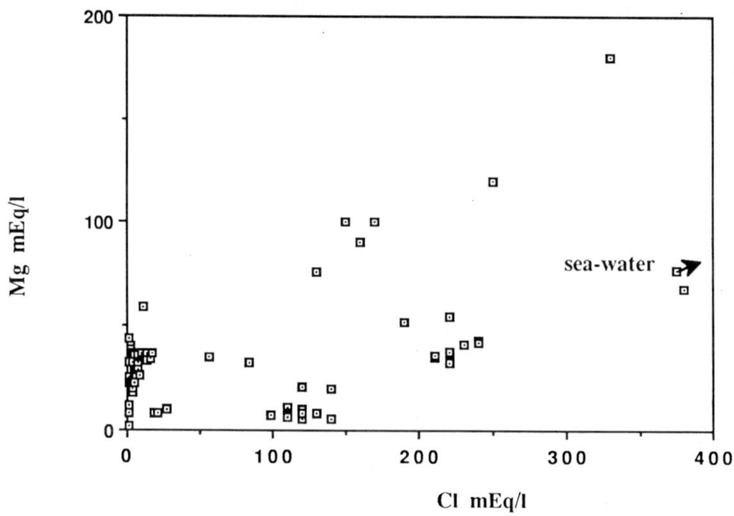

Figure 16.4 Cl–Mg correlation.

of >0 and <0 denote saturation and undersaturation, respectively. Note, for example, the saturation and undersaturation indices for gypsum.

16.3.2.2 WASH TECHNIQUES

With a view to checking on the possibility of using washing methods, a laboratory experiment was run on a clay from Trisaia (no. 1). This involved dispersing samples with an increasing quantity of distilled water. The water/clay ratio ranged from 1.46 to 49. After a few minutes of dispersion, the

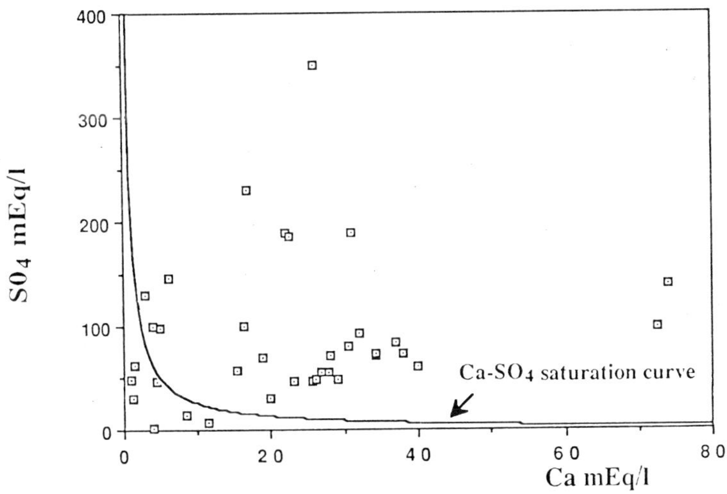

Figure 16.5 Ca–SO_4 correlation.

solution was filtered and analysed. The results are listed in Table 16.12 and plotted in Figure 16.7.

Taking into account the water/clay ratio of natural samples according to Devine *et al.* (1973), extrapolation of our results should give the composition of the outer solution around the clay minerals. What actually happens is that the calcium, sodium, potassium, chlorine and SO_4 contents are similar to those obtained by squeezing, whereas the HCO_3 contents are lower (see point N in Table 16.12). It is worth noting that calcium, magnesium, potassium and HCO_3 exhibit non-linear behaviour with increasing water/clay ratio. This could be due to exchange processes between water and clay.

This method could be improved to make an in-depth study of the pore fluid composition outside and inside the clay micelles. However, the method is time-consuming and is not reliable because of the dissolution of easily soluble minerals frequently present in clayey samples.

16.4 Conclusions

The TDS in the pore waters of the clays studied ranges from 0.45 to 24.9 $g.l^{-1}$. This variability is often consistent with the environmental conditions occurring in the sedimentation basin. Clays sedimented in marginal or continental basins and fluvio-lacustrine environments exhibit a lower pore-water salinity than do marine clays.

The fact that the Mg/Cl and subordinately Na/Cl ratios in the pore waters are higher than in sea water denotes exchange processes between interstitial fluids and minerals.

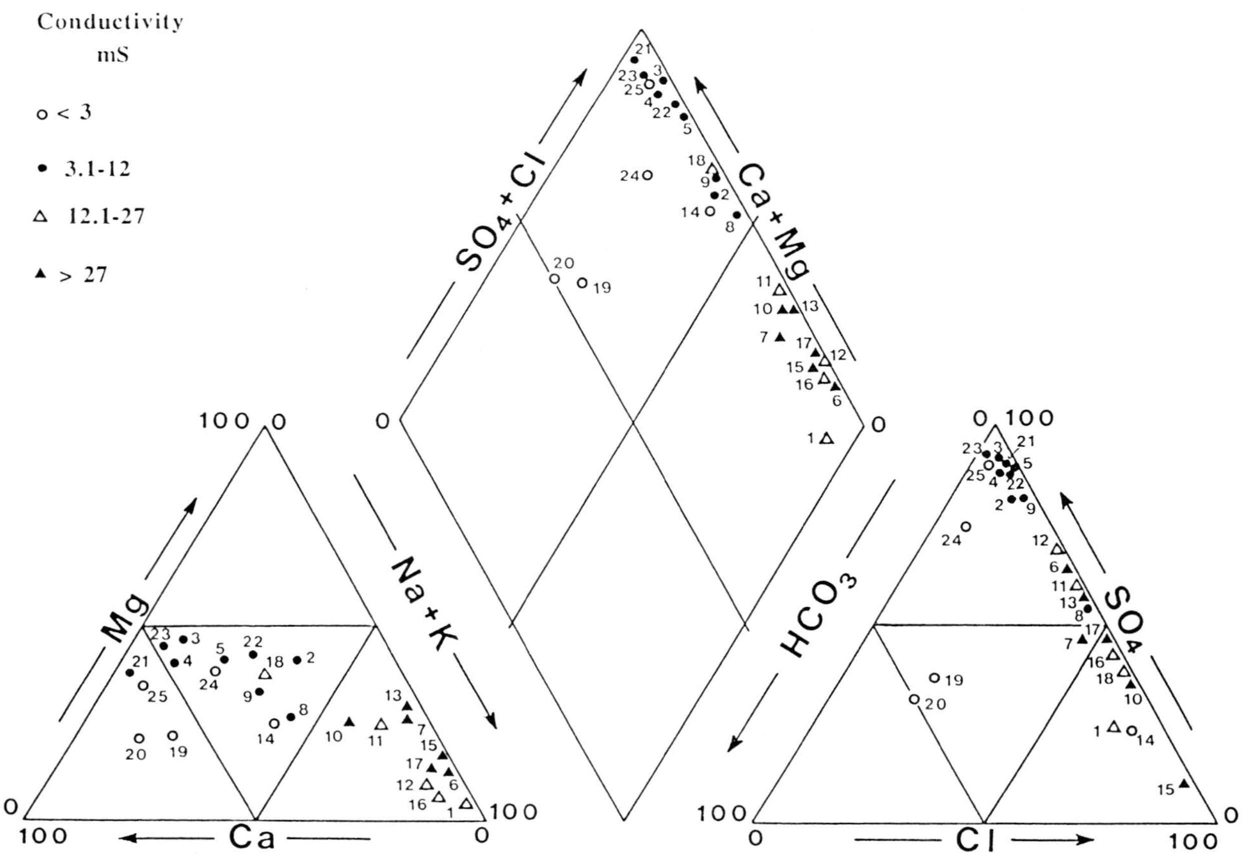

Figure 16.6 Chemical composition of the pore waters.

Table 16.11 Gibbs free energy difference between the actual and equilibrium states of a mineral (see text)

Mineral	Sample									
	1 $\Delta G(\text{kJ.mol}^{-1})$	4	5	6	7	8	9	10	15	16
Anhydrite	−10.989	−1.492	−1.547	−6.997	−2.048	−1.634	−1.250	−0.761	−9.518	−3.716
Aragonite	1.772	1.350	0.033	4.013	3.285	4.138	3.908	1.496	5.906	3.925
Calcite	2.596	2.174	0.852	4.836	4.109	4.962	4.732	2.316	6.730	4.748
Celestite	−8.928	−2.107		−1.020						
Chalcedony	0.242	0.192	0.401	−1.020	1.062	2.345	0.915	−0.067	1.032	1.605
Dolomite	15.273	10.086	7.641	20.499	18.672	15.437	15.107	11.725	26.046	15.796
Fluorite						−2.236	0.368		−13.510	−4.602
Goethite	35.785	28.582	28.450	37.917						
Gypsum	−8.945	0.568	0.514	−4.974	−0.058	0.418	0.811	1.258	−7.511	−1.680
Haematite	76.448	62.031	3.107	80.724						
Magnesite	3.862	−0.903	−2.027	6.847	5.748	3.173	1.559	0.598	10.500	2.232
Magnetite	56.518	28.274	28.106	61.785						
Quartz	1.074	1.020	1.233	−0.192	1.889	1.664	1.743	0.761	1.864	2.433

Table 16.12 Leaching waters chemical composition with various water/clay ratios (sample no. 1)

H_2O/Clay	pH	Eh (mV)	Cond. (mS)	Ca	Mg	Na	K	HCO_3	SO_4	Cl	SiO_2	Li	Fe	Mn	Al
				(mEq.l^{-1})							(mg.l^{-1})	(μg.l^{-1})			
1.46				0.32	0.88	26	1.1	5.0	6.1	17	1.8	9.5	13	16	
6.4	8.3	+325	0.96	0.10	0.14	7.6	0.42	3.7	1.6	4.1	2.2	4.0	71	24	105
12.46	8.4	+345	0.55	0.064	0.068	4.6	0.30	2.7	0.81	2.0	2.0	2.5	28	3.5	45
23.50	8.3	+345	0.35	0.047	0.039	2.9	0.20	1.9	0.35	1.0	1.9	1.8	13	1.0	41
36.84	9.6	+325	0.26	0.031	0.026	1.9	0.19	1.5	0.19	0.70	1.8	1.4	7.5	1.0	45
49.12	9.4	+325	0.21	0.035	0.033	1.5	0.17	1.2	0.12	0.49	1.4	1.1	8.5	1.0	64
N				1.9	17	110	6.5	6.5	34	110					

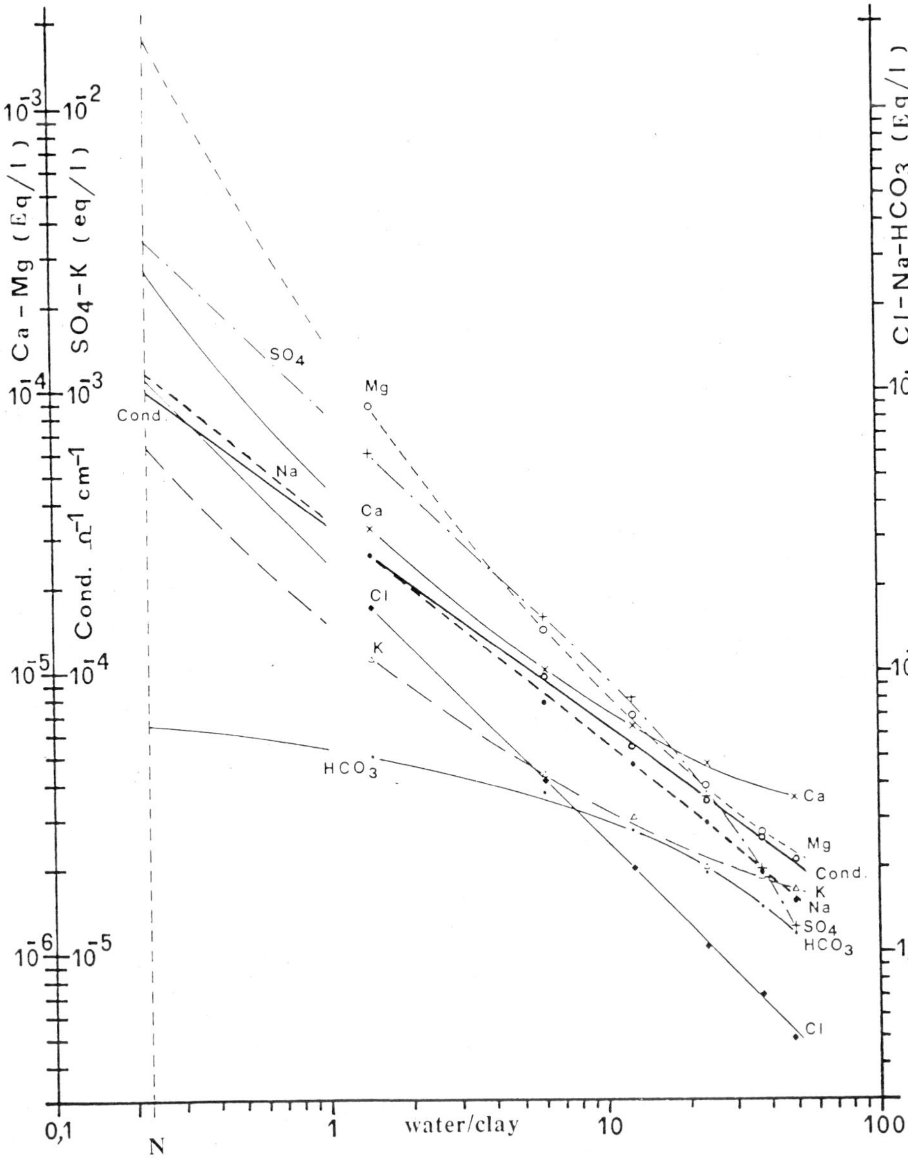

Figure 16.7 Washing water composition in relation to water/clay ratios. N = extrapolate pore-water concentration.

Pore waters with electrical conductivity lower than 12 μS exhibit a calcium-magnesium-sulphate chemistry, whereas those with higher electrical conductivity have a sodium-chloride chemistry.

In marine clay the progressive leaching of the more soluble salts (e.g. NaCl), that are not supported by minerals present in the clay, could cause a change in pore-water composition towards calcium-magnesium-sulphate chemistry.

Regarding the relationship between pore-water composition and clay mineral content, Long and Neglia (1968) observed that as potassium increases in the pore water, montmorillonite is able to form random or ordered interstratified minerals and, when the potassium content reaches high values illite becomes the stable mineral. In the present study, although the potassium content of the samples for which the mineralogical composition was determined ranges between 0.72 and 12 mEq.l^{-1}, there was no significant variation in the illite content. However, as the pore-water analysis used here was different from that adopted by Long and Neglia (1968) it is impossible to verify whether the range of the potassium content found in the present study is similar to that reported by these researchers.

In the clay deposited in the same palaeogeographic conditions, there is no significant variation in pore-water composition even though the mineralogical composition may be variable. However, definitive conclusions cannot be drawn on possible relations between the mineralogical composition of clay and pore-water chemistry, because of the relatively small number of samples studied and the semi-quantitative nature of the mineralogical analyses.

References

Brondi, A., Anselmi, B., Antonioli, F., Benveniu, F., Ferretti, O., Gerini, V., Giannotti, G. and Lenzi, G. (1984) Deduzione delle caratteristiche dei bacini argillosi sepolti attraverso indagini superficiali di basso costo. *Report of the CEE, Scienze e Techniche Nucleari*, EUR 9361 IT, Commissione delle Comunità, Luxembourg.

Casnedi, R., Crescenti, U., D'Amato, C., Mostardini, F. and Rossi, U. (1981) Il Plio-pleistocene del sottosuolo molisano. *Geologica Romana*, **20**, 1–42.

Chilingarian, G. V. and Rieke, H. H., III (1968) Data on consolidation of fine-grained sediments. *Journal of Sedimentary Petrology*, **38**(3), 811–16.

Colalongo, M. L., Pasini, G., Pelosio, G., Raffi, S., Rio, D., Ruggieri, G., Sartoni, S., Selli, R. and Sprovieri, R. (1982) The Neogene/Quaternary boundary definition: a review and proposal. *Geografia Fisica Dinamica Quaternaria*, **5**, 59–68.

Crescenti, U. (1975) Sul substrato pre-pliocenico dell'avanfossa appenninica dalle Marche allo Ionio. *Bollettino Società Geologica Italiana*, **94**, 583–684.

Crescenti, U., D'Amato, C., Balduzzi, A. and Tonna, M. (1980) Il plio-pleistocene del sottosuolo abruzzese-marchigiano tra Ascoli Piceno e Pescara. *Geologica Romana*, **19**, 63–84.

Devine, S. B., Ferrell, R. E. and Billings, G. K. (1973) The significance of ion exchange to interstitial solutions in clayey sediments. *Chemical Geology*, **12**, 219–28.

Edmunds, W. M. and Bath, A. N. (1976) Centrifuge extraction and chemical analysis of interstitial waters. *Environmental Science and Technology*, **10**(5), 467–72.

Engelhardt, W. V. and Gaida, K. H. (1963) Concentration changes of pore solutions during the compaction of clay sediments. *Journal of Sedimentary Petrology*, **33**, 919–30.

Fontanive, A., Gragnani, R., Mignuzzi, C. and Spat, G. (1985) Determinazione delle caratteristiche geochimiche delle acque interstiziali delle argille pliopleistoceniche italiane. *Report of the CEE, Scienze e tecniche nucleari*, EUR 10008 IT, Commissione delle Comunità, Luxembourg.

Kryukov, P. A. (1960) Towards procedures of squeezing out of solutions from sedimentary rocks. *Gidrokhimicheskie Materialy*, **33**, 191–7.

Kryukov, P. A. and Komarova, N. A. (1954) Ob otzhimanii Vody iz glin pri sverkhvysokikh davleniyakh (Concerning squeezing out of water from clays at very high pressures). *Akademii Nauk SSSR Doklady*, **99**(4), 617–19.

Long, G. and Neglia, S. (1968) Composition de l'eau interstitielle des argiles et diagenèse des minèraux argileux. *Revue de l'Institut Français du Pétrole et Annales des Combustibles Liquides*, **23-1**, 53–69.

Manheim, F. T. (1966) A hydraulic squeezer for obtaining interstitial water from consolidated and unconsolidated sediments. *U.S., Geological Survey, Professional Papers*, **550-C**. 256–61.

McKelvey, J. G. and Milne, I. H. (1962) The flow of salt solutions through compacted clay, in *Proceedings of 9th National Conference on Clay and Clay Minerals, Clays and Clay Minerals*, **9**, 248–59.

Pettijohn, F. J. (1975) *Sedimentary Rocks*, Harper and Row, New York.

Rieke, H. H., III, Chilingarian, G. V. and Robertson, J. O. (1964) High-pressure (up to 500,000 psi) compaction studies on various clays, in *22nd International Geological Congress*, New Delhi, Part XV, R. K. Sunderan, Calcutta, pp. 22–38.

Ruggieri, G. and Sprovieri, R. (1977) A revision of Italian Pleistocene stratigraphy. *Geologica Romana*, **16**, 131–9.

CHAPTER SEVENTEEN

Pore water–groundwater relationships in Jurassic shales and limestones of northern Switzerland

A. Gautschi, C. Ross and A. Scholtis

17.1 Introduction

As part of the current Swiss nuclear waste disposal concept, Nagra (National Cooperative for the Disposal of Radioactive Waste) is investigating the potential of sedimentary formations to host a repository for high-level radioactive waste.

The main option under consideration at present is the Opalinus Clay, a shale formation of Lower Dogger age. Studies to determine civil engineering constraints on repository construction, extensive hydrological modelling, and preliminary safety assessment analyses are currently in progress (Nagra, 1989).

An opportunity for a case study in the Opalinus Clay was offered by the excavation of the Mt. Terri motorway tunnel in north-western Switzerland. The work was organized as a joint project between Nagra and the Swiss National Hydrological and Geological Survey.

17.2 Geology

The Mt. Terri motorway tunnel penetrates two tectonic units, the Folded Jura and the Tabular Jura (Figure 17.1). The Folded Jura is characterized by folds and thrust planes, mainly formed during the late Miocene and Pliocene, with the main detachment in the evaporitic part of the Middle Triassic. The Folded Jura is thrusted on to the Tabular Jura, a sequence of slightly faulted Mesozoic–Tertiary sedimentary rocks.

The stratigraphy of the investigated section is summarized in Figure 17.2. The Opalinus Clay is a well-consolidated Aalenian (Lower Dogger) clay formation (or shale) with 70–80% clay minerals and micas, 10–15% quartz, 5–10% calcite, 1–5% siderite, 1% pyrite and 0.7% organic carbon. The total

Geochemistry of Clay-Pore Fluid Interactions. Edited by D. A. C. Manning, P. L. Hall and C. R. Hughes. Published in 1993 by Chapman & Hall, London. ISBN 0 412 48980 5

porosities range from 3 to 12%. The total thickness of the formation is 160 m.

The overlying formation (Blaukalke) consists of marly, sandy limestones, while the underlying Lias comprises shales (e.g. Posidonienschiefer), marls and a basal limestone (Gryphitenkalk).

17.3 Hydrogeology

In the tunnel, fractured, partly karstic limestone and dolomite aquifers are separated by low-permeability argillaceous and evaporitic formations (Figures 17.1 and 17.2). The main groundwater recharge results from precipitation in the mountain range above the tunnel at 650–940 m above sea-level and the discharge occurs to the nearby river Doubs at 440 m above sea-level. The hydraulic heads measured in different formations in exploration boreholes and in the tunnel are generally 100–200 m above the level of the tunnel. However, fissures encountered in the tunnel within the Triassic limestones and dolomites are dry, indicating that hydraulic heads in these formations are at a lower level (Tripet *et al.*, 1990).

17.4 Field investigations

The section investigated is situated in the southern limb of the Mt. Terri anticline and has an overburden of roughly 300 m (Figure 17.1). Detailed hydrogeological mapping of the lower part of the Blaukalke, the Opalinus Clay and the underlying Lias has been carried out in the reconnaissance tunnel (Tripet *et al.*, 1990).

The 240 m section of the Opalinus Clay was completely dry. One ‘wet zone’ reported by Tripet *et al.* (1990) was most probably caused by contamination with drilling water, because the minipiezometers installed in this zone remained dry. However, pore waters were recovered from the most argillaceous horizons of the Opalinus Clay by squeezing in a compression rig (cf. Section 17.5).

Groundwater samples could be obtained from the formations above and below the Opalinus Clay in the reconnaissance tunnel and in the pilot tunnel (Figure 17.2).

The hydrochemical and isotopic analyses of the pore waters and groundwaters are presented in Table 17.1. A description of the sampling and analytical procedures is given below.

17.5 Sampling and analytical procedures

17.5.1 Groundwaters

The groundwater samples were taken from the tunnel wall under atmospheric conditions. The analytical work was performed by ‘Versuchsstationschweizerischer Brauereien (VBS), Zürich’ (hydrochemistry) and ‘Forschungs-

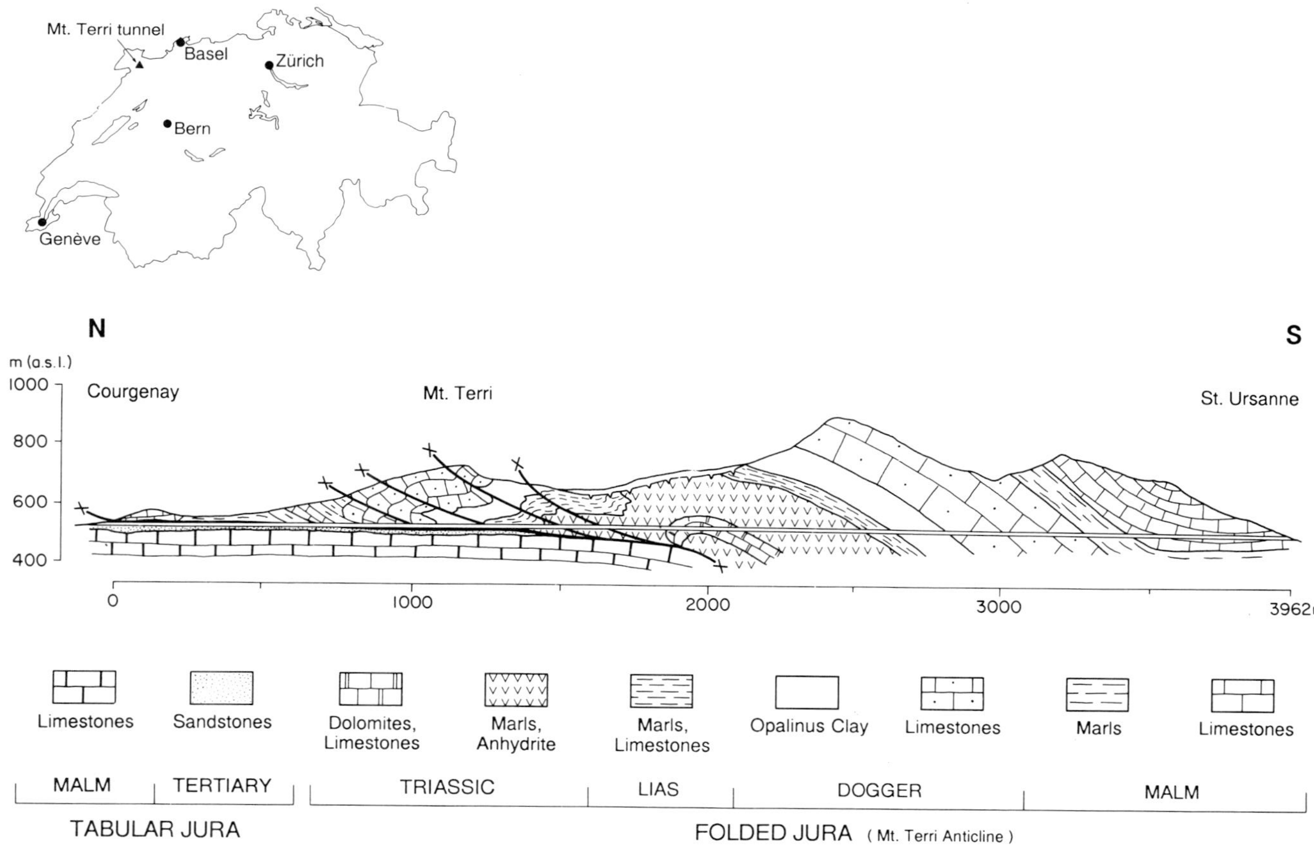

Figure 7.1 Location and simplified geological cross-section of Mt. Terri tunnel (provisional, after Shaeren and Norbert, 1989).

Table 17.1 Hydrochemical analysis and isotopic composition of Mt. Terri groundwaters and pore-waters.

Sample no.		A	B	1a	1b	1c	2	3a	4
Description		Squeezed pore-water	Squeezed pore-water	Tunnel source	Tunnel source	Tunnel source	Tunnel source	Tunnel source	Tunnel source
Date		1989	1989	25.07.89	22.01.90	13.03.90	09.08.89	13.03.90	13.03.90
Formation		Opalinus Clay	Opalinus Clay	Posidonien-schiefer	Posidonien-schiefer	Posidonien-schiefer	Gryphitenkalk	Gryphitenkalk	Blaukalke
Tunnel		R	R	R	R	R	R	Pilot	R
(m)		913	913	1038	1038	1038	1124	580	752
pH		6.78	7.51	7.4	7.6	7.62	7.5	7.86	7.47
Temperature (°C)		—	—	—	—	17.6	18.3	18.1	12.8
Moisture cont. (% dry wt)		11.3	12.2	—	—	—	—	—	—
Na^+	$(mg.l^{-1})$	5000	5370	4900	1220	1530	44	1210	293
K^+	$(mg.l^{-1})$	94.3	104	47	19.6	14	4.1	24	17
Ca^{2+}	$(mg.l^{-1})$	790	759	398	44	59	76	80	38
Mg^{2+}	$(mg.l^{-1})$	720	753	258	28	38	15	35	25
Cl^-	$(mg.l^{-1})$	7000	7250	8470	1520	2000	10	1270	97
F^-	$(mg.l^{-1})$	0.32	0.34	—	1.2	—	—	0.84	—
Br^-	$(mg.l^{-1})$	22.5	22.4	—	4.7	6.6	—	4.1	0.33
SO_4^{2-}	$(mg.l^{-1})$	6430	7150	480	240	364	48	900	180
HCO_3^-	$(mg.l^{-1})$	113	179	390	650	554	350	368	650
NO_3^-	$(mg.l^{-1})$	40.8*	40.2*	<1	<0.2	<0.1	<1	<0.1	4.7
H_2SiO_3	$(mg.l^{-1})$	6.9	7	14	15	14	10	13.5	14
TOC	$(mg.l^{-1})$	29.8*	40.6*	—	1.8	—	—	0.9	—
TDS	$(mg.l^{-1})$	20270	21686	14960	3766	4583	557	3940	1319
$\delta^{18}O$	‰(SMOW)	−8.2	−7.7	−8.98	−9.22	−9.48	−9.58	−9.76	−10.07
$\delta^{2}H$	‰(SMOW)	−55.0	−57.0	−60.0	−62.8	−65.5	−66.8	−67.5	−72.5
Tritium	TU	—	—	<0.9	4.1±0.7	3.3±0.7	27.1±2.3	7.6±4.4	6.8±0.8
Water type		Na–Cl–SO_4	Na–Cl–SO_4	Na–Cl	Na–Cl	Na–Cl	Ca–Na–HCO_3	Na–Cl–SO_4	Na–HCO_3–SO_4

*Possibly contaminated during tunnel blasting.
R = Reconnaissance tunnel.
Sample numbers refer to locations in Figure 17.2.

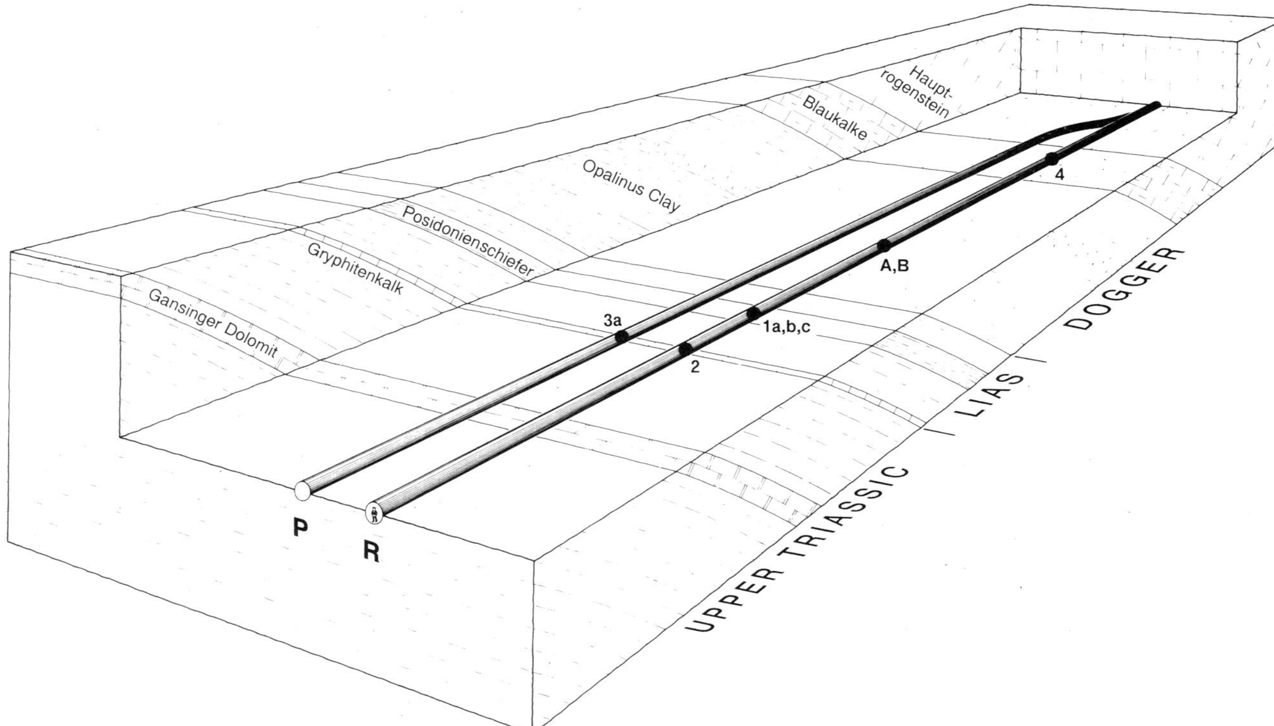

Figure 17.2 Sample locations in the reconnaissance tunnel (R) and in the pilot tunnel (P).

zentrum für Umwelt und Gesundheit (GSF), München–Neuherberg' (stable oxygen and hydrogen isotopes, tritium).

17.5.2 Pore waters

To prevent significant oxidation or moisture loss, the Opalinus Clay samples for pore water extraction were sealed in paraffin wax, about 1 h after excavation by blasting. Contamination by drilling water is unlikely to be important, since water was only used for drilling shotholes.

The pore waters were recovered from two samples of Opalinus Clay by squeezing in a specially constructed compression rig at the British Geological Survey (BGS), Keyworth, Nottingham. The pore water extraction method is described in detail in Entwisle *et al.* (1989), and in Chapter 15.

For the squeezing experiments large rock fragments with a total weight of 650 g were chipped from each sample and squeezed in two cells for 10 days, operating at loads up to 70 MPa and a temperature of 9.5°C. The two samples had moisture contents (drying at 105°C) of 11.3 and 12.2% dry weight, respectively, yielding total fluid weights of 21.2 and 18.0 g, which roughly correspond to 25% of the moisture content. A less argillaceous sample, with a moisture content of only about 8%, yielded no water even after squeezing for 23 days.

Each pore water sample was split into separate fractions, which were preserved for major, minor and trace species analyses, as well as for stable isotopic measurements. Total alkalinity and pH were measured using untreated sub-samples as soon as possible after collection.

The pore waters were analysed at BGS laboratories, for major cations and trace elements* by inductively coupled plasma optical emission spectrometry (ICP-OES) using a Perkin–Elmer Plasma II Emission Spectrometer. The acidified sub-samples were internally standardized, by the addition of a known concentration of scandium, prior to analysis. Chloride, sulphate, nitrate, nitrite*, phosphate* and bromide were analysed by ion chromatography using a Dionex 2000i Ion Chromatograph. Fluoride was quantified by direct analysis using an Orion 94-09 fluoride ion selective electrode with a Philips PW9415 ion selective meter. Total organic and inorganic carbon (TOC, TIC*) were determined using a Shimadzu TOC 500 analyser. Carbonate and bicarbonate alkalinity were determined by titration against 0.01 M sulphuric acid using an automatic Radiometer RTS 822 recording titration system comprising a high-precision pH meter and electrode assembly (PHM84), autoburette (ABU80), titration assembly (TTT80) and chart recorder (Servograph REC80). This system was also used to determine the sample pH.

*Data not reported in Table 17.1.

17.6 Results of groundwater and pore water investigations

The squeezed pore waters from the Opalinus Clay (samples A,B) are highly mineralized Na–Cl–SO_4-waters (TDS 21 g.l^{-1}) (Table 17.1). The high nitrate content is most probably a contaminant from tunnel blasting.

The chemistry of the groundwaters from the overlying and underlying limestones and shales is quite heterogeneous. It ranges from a tritium-poor, highly saline Na–Cl groundwater (sample 1a), which is similar to the squeezed pore-waters, to a tritium-rich, low-mineralized Ca–Na–HCO_3 groundwater (sample 2).

The stable isotopes of all waters are close to the meteoric water line (Figure 17.3). Generally, the waters with high total dissolved solids have low tritium contents and heavier stable isotopes than the low-mineralized groundwaters (Figure 17.4). The saline groundwaters from the Posidonienschiefer (1a,b,c) are intermediate in isotopic composition between the overlying Opalinus Clay pore-waters (A, B) and the underlying, lower mineralized, tritium-rich groundwaters from the Gryphitenkalk (2).

The Posidonienschiefer sample 1a, collected during tunnel excavation, is more highly mineralized and isotopically heavier than the samples 1b,c taken half a year later. This dilution by younger, lower mineralized waters is probably created by high hydraulic gradients in the tunnel.

Comparison with the local $\delta^{18}O$-altitude relationship (Figure 17.5) indicates that the Opalinus Clay pore-waters are isotopically much heavier than recent recharge and originate, therefore, from a warmer climatic period than today (at least before the last glaciation or >80 000 years).

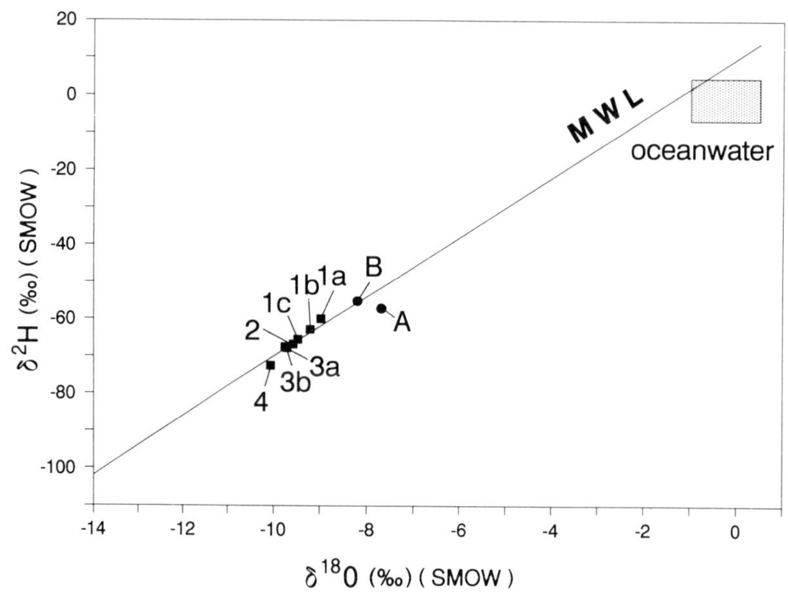

Figure 17.3 δ^2H–$\delta^{18}O$ diagram of the Mt. Terri tunnel waters (MWL = meteoric water line).

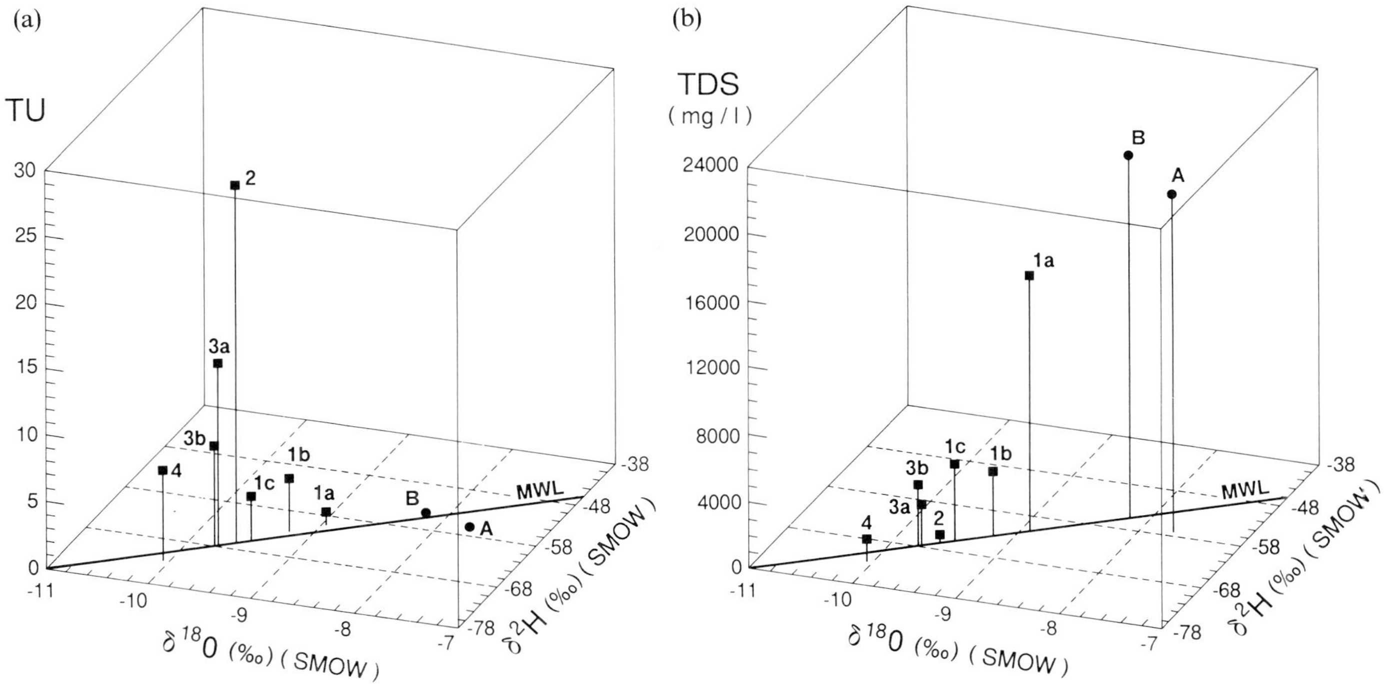

Figure 17.4 Stable isotopes versus tritium content (a) and total dissolved solids (b) from Mt. Terri tunnel waters. Tritium content of pore-waters A and B assumed to be < 1 TU.

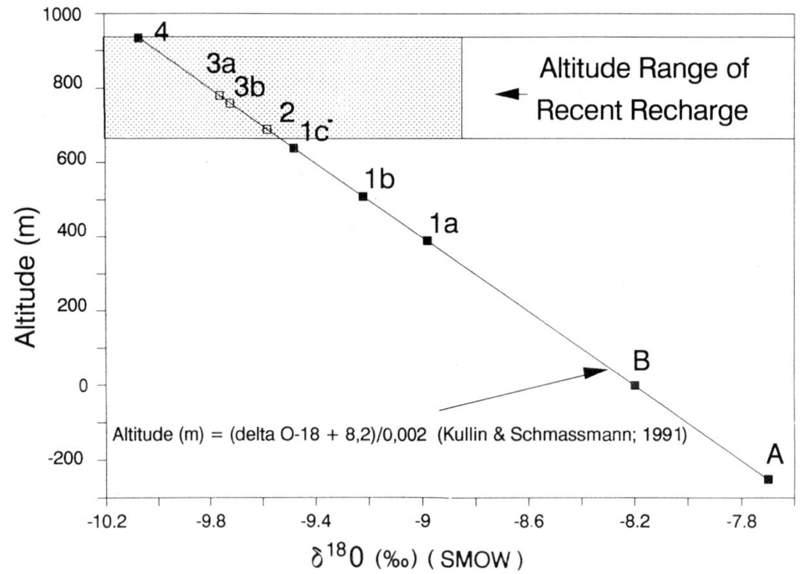

Figure 17.5 Recharge altitude of the Mt. Terri tunnel waters based on the local δ^{18}O-altitude relationship (Kullin and Schmassmann, 1991).

The Opalinus Clay pore waters seem to be derived from seawater/connate water. The mixing lines for the stable isotopes and, for example, the fairly constant Br/Cl ratio (close to seawater) in the pore waters and in all groundwaters, suggest that the suite of samples represents a continuum produced by progressive dilution – and some water–rock reaction – of essentially the same parent connate water by recharge of various ages.

17.7 Future plans

More information on the hydrogeological regime in the Opalinus Clay could be obtained from depth profiles of pore water composition as demonstrated, for example, by Brightman *et al.* (1985). Variations in pore water composition with depth have also been used by Bath *et al.* (1988) and Falck and Bath (1989) to evaluate solute transport mechanisms in mudrocks by comparison of observed data with modelled chloride and stable isotope profiles. In highly compacted or calcareous shales with low porosities, where pore water extraction by squeezing is not feasible, leaching techniques in combination with CEC measurements might be an alternative (see e.g. Entwisle *et al.*, 1989; Jones *et al.*, 1989), but would not yield stable isotope data.

17.8 Conclusions

Hydrogeological and geochemical investigations in the Mt. Terri exploration tunnel, analysing pore waters from the Opalinus Clay and groundwaters from the adjacent formations, have demonstrated that the water circulation in the Opalinus Clay is very limited despite the high hydraulic gradients likely to exist in the local flow systems.

Pore waters have been extracted successfully from two samples of Opalinus Clay by squeezing in a specially constructed compression cell. The shales were of low moisture content (11–12 wt%), but about 25% of the moisture content was recovered after prolonged compaction. The pore waters were much more saline than groundwaters from adjacent Dogger and Lias limestones. However, systematic trends in the isotopic and chemical compositions of the pore waters and groundwaters suggest that they have been derived from the same parent connate water by progressive dilution with younger meteoric waters up to modern recharge. None-the-less, the high salinity of pore waters from the Opalinus Clay indicates that very little flushing of this formation has occurred.

Acknowledgements

This paper is published by permission of Nagra, the Director of the Swiss National Hydrological and Geological Survey (SNHGS) and the Director of the British Geological Survey (BGS/NERC). We would like to thank J.-P. Tripet (SNHGS) for the administrative management of the field investigations, and Y. Brechbühler and T. Haarpaintner (Bureaux Géologiques Associés Norbert/Schindler) for the sample preparation in the field. We would also like to thank D. C. Entwisle (BGS, Keyworth) who performed the pore-water extraction experiments, G. Darling (BGS, Wallingford) for the pore-water isotopic analyses, M. R. Cave, K. Green, and S. Reeder (BGS, Keyworth) for the pore-water chemical analyses, H. Senften (VSB, Zürich) for the groundwater chemical analyses, and W. Rauert and P. Trimborn (GSF, München–Neuherberg) for the groundwater isotopic analyses.

The authors are grateful to the 'Office Fédéral des Routes' and the 'Service des Ponts et Chaussées (République et Canton du Jura)' for permission to carry out these investigations in the Mt. Terri tunnel. The logistic support of the 'Consortium d'étude IJA/BG' is also gratefully acknowledged.

References

Bath, A. H., Entwisle, D., Ross, C. A. M., Cave, M. R., Falck, W. E., Fry, M., Reeder, S., Green, K. A., McEwen, T. J. and Darling W. G. (1988) Geochemistry of pore-waters in mudrock sequences—evidence for groundwater and solute movements, in *International Association of Hydrogeologists Symposium on Hydrogeology and Safety of Radioactive and Industrial Hazardous Waste Disposal*, Orléans, France, 7–10 June 1988, vol. 1, pp. 87–97.

Brightman, M. A., Bath, A. H., Cave, M. R., and Darling, W. G. (1985) *British Geological Survey Report*, FLPU 85-6, Pore Fluids From the Argillaceous Rocks of the Harwell Region, Keyworth, Nottingham.

Entwisle, D. C., Reeder, S., Bath, A. H. and Ross, C. A. M. (1989) Techniques for the Characterisation of Solutes in Drillcore from Mudrocks, *British Geological Survey Technical Report*, WE/89/29C and *NIREX Safety Studies Research Report*, NSS/R173, British Geological Survey, Keyworth, Nottingham and UK Nirex Ltd, Harwell, Oxon.

Falck, W. E., and Bath, A. H. (1989) Numerical Modelling and Data Evaluation, *British Geological Survey Report*, WE/89/10C, and Vertical Changes in Chloride and Stable Isotope Compositions of Porewaters in Coastal London Clay *NIREX Safety Studies Research Report*, NSS/R174, British Geological Survey, Keyworth, Nottingham and UK Nirex Ltd, Harwell, Oxon.

Jones, T. G. J., Hughes, T. L. and Tomkins, P. (1989) The ion content and mineralogy of a North Sea Cretaceous Shale Formation. *Clay Minerals*, **24**, 394–410.

Kullin, M. and Schmassmann, H. (1991) Isotopic composition of modern recharge, in *Applied Isotope Hydrogeology–A Case Study in Northern Switzerland*, (eds F. J. Pearson *et al.*), Elsevier, Amsterdam, and *Nagra Technical Report*, NTB 88-01, Nagra, Wettingen, Switzerland.

Nagra. (1989) Sediment Study, Intermediate Report, Disposal Options for Long Lived Radioactive Waste in Swiss Sedimentary Formations (*Executive Summary*), *Nagra Technical Report*, NTB 88-25E, Wettingen, Switzerland.

Schaeren, G. and Norbert, J. (1989) *Société suisse des ingénieurs et des architectes, Documentation*, SIA D 037, *Tunnel du Mt. Terri et du Mt. Russelin–La Traversée des 'Roches à Risques': Marnes et Marnes à Anhydrite*, Zürich, Switzerland, pp 19–24.

Tripet, J. P., Brechbühler, Y. A., Haarpaintner, R. T. and Schindler, B. (1990) Hydrogéologie des milieux à faible perméabilité: étude des Marnes Aaléniennes dans la galerie de reconnaissance du Mont Terri (Canton du Jura). *Bulletin de la Société Neuchâteloise des Sciences Naturelles*, **113**, 179–89.

Index